AF344454

DNA Repair in Cancer Therapy

CANCER DRUG DISCOVERY AND DEVELOPMENT

Beverly A. Teicher, Series Editor

DNA Repair in Cancer Therapy

Edited by

Lawrence C. Panasci, MD

and

Moulay A. Alaoui-Jamali, DVM, PhD

Lady Davis Institute for Medical Research,
Sir Mortimer B. Davis-Jewish General Hospital,
McGill University, Montreal, Canada

Humana Press
Totowa, New Jersey

PREFACE

The field of DNA repair has been the subject of increasing interest at both the genetic and biochemical levels, leading to impressive progress in this area. DNA repair and its associated regulatory mechanisms lie at the heart of almost every fundamental aspect of cell biology, including transcription, cell cycle, apoptosis, and development. Thanks to the fascinating investigations of the inherent gene defects of specific components of DNA repair pathways found in rare human syndromes (e.g., xeroderma pigmentosum), we have been provided with the framework for subsequent studies on the translational aspects of DNA repair. Several genes have been cloned, and the crystal structures of some proteins are now reported. Polymorphisms in certain of the DNA repair genes are being identified in human populations. Furthermore, increased research efforts highlight the involvement of DNA repair mechanisms in the maintenance of genomic stability, mutagenesis and carcinogenesis, and resistance to endogenous and exogenous genotoxic stress.

In preparing *DNA Repair in Cancer Therapy*, we have been concerned with those practicing oncologists who are dealing on a daily basis with the hallmark "relapse" or "drug resistance" phenomena. Among the multifactorial mechanisms described so far, there is increasing evidence that impaired expression/ activity of at least some of the DNA repair proteins can account for tumor cell resistance to a particular therapeutic agent. Further interest has been stimulated by the demonstration that DNA repair is coupled to cell cycle checkpoint controls which, when impaired, could account for clinical drug resistance. Surprisingly, there have been relatively few comprehensive review articles and, as far as we know, no complete volume dedicated to the translational aspect of DNA repair in the clinic. This fostered the need to organize a set of timely, in-depth reviews covering the latest developments having potential for translational and clinical applications.

Chapter 1 by Dr. Leyland-Jones on the clinical implications of resistance to anticancer agents, including those whose primary mechanisms of cell death can be affected by DNA repair, introduces the important role that alterations in DNA repair play in limiting the therapeutic index of anticancer therapy. Experts in the field subsequently review the various mechanisms involved and their implications. Although the application of DNA repair pathways in therapeutics is still at the embryological stage, some inhibitors of DNA repair mechanisms (e.g., O^6-methylguanine-DNA methyltransferase [MGMT]) that would increase sensitivity/selectivity to kill tumor cells in a particular molecular context have reached the clinical stage, and the results will be discussed in the light of the clinical impact. Furthermore, inhibitors of other DNA repair enzymes, such as PARP and DNA-PK, are being developed, and clinical trials with such inhibitors alone or in combination with anticancer therapy (drugs and/or radiotherapy) should be completed in the foreseeable future. Thus, prospects are exciting, and the translation of bench research to the clinic is on the horizon. Some chapters deal with overlapping subjects, although from different experimental and personal perspectives; this reflects the complexity of a topic wherein there are sometimes conflicting data, but it also ensures that most of the current views are represented. We believe that *DNA Repair in Cancer Therapy* will prove to be valuable reading for a broad audience of clinicians, pharmacologists, medicinal chemists, and basic scientists.

We would like to thank the authors who have spent their valuable time in contributing to *DNA Repair in Cancer Therapy*. Their cooperation and expertise was crucial in obtaining this comprehensive, state-of-the-art synopsis of a complex area.

Lawrence C. Panasci, MD
Moulay A. Alaoui-Jamali, DVM, PhD

Contents

CONTRIBUTORS

MOULAY A. ALAOUI-JAMALI, DVM, PhD • *Departments of Medicine, Pharmacology, and Therapeutics, Lady Davis Institute of Medical Research, Sir Mortimer B. Davis-Jewish General Hospital, McGill University, Montreal, Canada*

RAQUEL ALOYZ, PhD • *Departments of Medicine, Pharmacology, and Therapeutics, Lady Davis Institute of Medical Research, Sir Mortimer B. Davis-Jewish General Hospital, McGill University, Montreal, Canada*

ADRIAN C. BEGG, PhD • *Division of Experimental Therapy, The Netherlands Cancer Institute, Amsterdam, The Netherlands*

KISHOR K. BHAKAT, PhD • *Department of Human Biological Chemistry and Genetics and Sealy Center for Molecular Science, University of Texas Medical Branch, Galveston, TX*

DORA BOCANGEL, PhD • *Baylor University, Houston, TX*

ISTVAN BOLDOGH, PhD • *Department of Microbiology and Immunology, University of Texas Medical Branch, Galveston, TX*

GOKUL C. DAS, PhD • *Department of Microbiology and Immunology, University of Texas Medical Branch, Galveston, TX*

M. EILEEN DOLAN, PhD • *Section of Hematology-Oncology, Department of Medicine, University of Chicago, Chicago, IL*

RAZQALLAH HAKEM, PhD • *Division of Cellular and Molecular Biology, Ontario Cancer Institute, University of Toronto, Toronto, Canada*

JOHN A. HARTLEY, PhD • *Cancer Research UK Drug–DNA Interactions Research Group, Department of Oncology, Royal Free and University College Medical School, University College London, London, UK*

CHAUNHUA HE, PhD • *Guy-Bernier Research Center, Maisonneuve-Rosemont Hospital, Montreal, Canada*

JENNIFER J. HU, PhD • *Departments of Cancer Biology and Public Health Sciences, Comprehensive Cancer Center, Wake Forest University School of Medicine, Winston-Salem, NC*

BERTRAND J. JEAN-CLAUDE, PhD • *Cancer Drug Research Laboratory, Division of Medical Oncology, Department of Medicine, Royal Victoria Hospital, McGill University Health Center, Montreal, Canada*

GARY M. KUPFER, MD • *Departments of Microbiology and Pediatrics, University of Virginia Health System, University of Virginia, Charlottesville, VA*

BRIAN LEYLAND-JONES, MD • *Department of Oncology, Faculty of Medicine, McGill University, Montreal, Canada*

MARTIN LOIGNON, PhD • *Departments of Medicine, Pharmacology, and Therapeutics, Lady Davis Institute of Medical Research, Sir Mortimer B. Davis-Jewish General Hospital, McGill University, Montreal, Canada*

PETER J. McHUGH, DPhil • *Cancer Research UK Laboratories, Weatherall Institute of Molecular Medicine, University of Oxford John Radcliffe Hospital, Oxford, UK*

SANKAR MITRA, PhD • *Department of Human Biological Chemistry and Genetics and Sealy Center for Molecular Science, University of Texas Medical Branch, Galveston, TX*

DESPINA MOSHOUS, MD, PhD • *Unite Developpement Normal et Pathologique du Systeme Immunitaire, INSERM U429, Hôpital Necker Enfants-Malades, Paris, France*

DAVID MURRAY, PhD • *Department of Oncology, Cross Cancer Institute, University of Alberta, Edmonton, Canada*

LAWRENCE C. PANASCI, MD • *Lady Davis Institute for Medical Research, Sir Mortimer B. Davis-Jewish General Hospital, Montreal, Canada*

ANTHONY E. PEGG, PhD • *Departments of Cellular and Molecular Physiology and of Pharmacology, The Milton S. Hershey Medical Center, Pennsylvania State University College of Medicine, Hershey, PA*

GUY G. POIRIER, PhD • *Faculty of Medicine, Laval University, Sainte-Foy, Quebec, Canada*

DINDIAL RAMOTAR, PhD • *Guy-Bernier Research Center, Maisonneuve-Rosemont Hospital, Montreal, Canada*

MICHÈLE ROULEAU, PhD • *Health and Environment Unit, Laval University Medical Research Center, CHUQ, Quebec, Canada*

GEORGE SHENOUDA, MBB Ch, PhD, FRCP • *Department of Radiation Oncology, McGill University, Montreal, Canada*

P. JAMES SCRIVENS, PhD • *Departments of Medicine, Pharmacology, and Therapeutics, Lady Davis Institute of Medical Research, Sir Mortimer B. Davis-Jewish General Hospital, McGill University, Montreal, Canada*

JEAN-PIERRE DE VILLARTAY, PhD • *Unite Developpement Normal et Pathologique du Systeme Immunitaire, INSERM U429, Hôpital Necker Enfants-Malades, Paris, France*

Huijie Wang, PhD • *Guy-Bernier Research Center, Maisonneuve-Rosemont Hospital, Montreal, Canada*

Zhi-Yuan Xu, MD • *Lady Davis Institute for Medical Research, Sir Mortimer B. Davis-Jewish General Hospital, Montreal, Canada*

1 Clinical Resistance to Alkylators
Status and Perspective

Brian Leyland-Jones, MD

Alkylating agents have been used in the treatment of cancer for over 50 yr, with the nitrogen mustard alkylating agent mechlorethamine being one of the first antitumor drug used in clinical practice half a century ago. These compounds can bind to a variety of cellular structures such as membranes, RNA, proteins, and DNA. It is, however, the ability to form DNA interstrand crosslinks that appears to be the most important event with regard to their antitumor activity. In addition to the nitrogen mustard agents (melphalan, chlorambucil, cyclophosphamide, and ifosfamide), the platinum drugs (cisplatin, carboplatin, and oxaliplatin) have become some of the most widely used cytotoxic anticancer agents.

Cisplatin (*cis*-diamminedicloroplatinum[II]) was the first of the platinum-containing drugs to be used in the therapy of cancer. More than 25 yr of experience with this agent has shown that it possesses clinically relevant cytotoxic effects in a wide range of solid tumors, including germ cell tumors, small-cell lung, head and neck, ovarian, and bladder cancers (for review see ref. *1*). Response to first-line treatment is generally high with overall response rates typically reported to be between 70% and 99% (*2*). The nitrogen mustard agents chlorambucil and cyclophosphamide have been the backbone of conventional treatment of chronic lymphocytic leukemia, with 60–80% of patients responding to initial therapy (*3*).

The early use of cisplatin, although possessing activity against a variety of tumors, was coupled with severe adverse effects such as nephrotoxicity, nausea/ vomiting, and peripheral neuropathy. The addition of hyperhydration to the treatment regimen and the use of serotonin 5HT$_3$ antagonists for nausea have contributed to the effective and tolerable use of cisplatin as a first-line therapy. To date, cisplatin remains the primary therapy for a wide variety of cancers, including ovarian, bladder, cervix, head and neck, esophageal, and small-cell lung cancer, among others (*1*). Despite the alleviation of these adverse events, there has been

From: *Cancer Drug Discovery and Development: DNA Repair in Cancer Therapy*
Edited by: L. C. Panasci and M. A. Alaoui-Jamali © Humana Press Inc., Totowa, NJ

a significant effort to develop platinum analogs that possess reduced toxicity. Carboplatin, the first of the cisplatin analogs to be approved worldwide for cancer therapy, was shown to have significantly reduced toxicity compared to its parent compound. Of particular importance, carboplatin treatment resulted in less nausea/vomiting, nephrotoxicity, and neurotoxicity *(4)* while still maintaining the same degree of response as its parent in the treatment of ovarian cancer *(5)*. Myelosuppression, although generally not severe with the use of cisplatin, is the dose-limiting toxicity of carboplatin *(6)*. Oxaliplatin, a relatively new analog with effectiveness in colorectal cancer and in platinum-pretreated advanced ovarian cancer *(7)*, has been shown to be well tolerated with no evidence of nephrotoxicity or ototoxicity, nausea/vomiting that is responsive to $5HT_3$ antagonists, and myelosuppression that is uncommon *(8)*.

Despite the effectiveness of alkylating agent treatment in a variety of cancers, some forms of cancer such as colorectal and non-small-cell lung cancer (NSCLC) have been shown to be inherently resistant. For example, the use of cisplatin in first-line therapy for NSCLC results in overall response rates of less than 20% *(9)*. Although the response rates for chemotherapy in this form of cancer is substantially lower than that seen with treatment of other cancers, platinum compounds continue to be the standard front-line therapy. Compared with supportive care alone, platinum-based chemotherapy can produce prolonged survival, symptom control, and improved quality of life *(10)*. A concerted effort has been made to improve chemotherapy response in NSCLC with treatment regimens that combine several new agents with platinum-based drugs. Although response rates in these trials have been moderately improved (up to 30%), no single regimen has demonstrated a significant superiority in the treatment of NSCLC over another form of therapy *(10)*. In addition, the effectiveness of alkylating agents as a second-line therapy in patients with recurrent disease can be problematic. A high proportion of patients relapse as a result of acquired resistance to cisplatin or its analogs, i.e., carboplatin *(11,12)*. As well, virtually all patients who respond to initial nitrogen mustard therapy experience incomplete remission and suffer a progression of their disease *(3)*. Typically, the further use of alkylating agents in many of these patients fails to substantially affect the recurrent disease, thus presenting a significant obstacle to the cure or control of their cancers.

In the case of platinum agents, the duration of the treatment-free period has been seen to be an important predictor of activity in second-line therapy *(13)*. Patients with recurrent disease after an interval greater than 12 mo have an increased response to platinum therapy compared to patients with recurrent disease within a 12-mo interval. For these latter patients, reinitiation of the primary chemotherapy regimen of platinum-based compounds has a significantly reduced effect. This is especially true for patients who relapse within 4 mo following termination of first-line therapy. Patients with small-cell lung cancer who relapse within 3 mo following first-line therapy with platinum drugs have less than a 10% response rate regardless of the chemotherapy regimen employed *(11)*. Similar response rates are seen in ovarian cancer patients, who are platinum refractory, those with a progression-free interval of less than 4 mo, or in patients who are platinum resistant *(14)*. The use of nonalkylating agents such as paclitaxel for patients who are platinum resistant during first-line therapy obtain only moderately improved response rates of 24–30% *(12)*. For such patients, the options for effective chemotherapy remain limited to investigational agents and second-line agents.

Given the serious impact that inherent or acquired chemoresistance has on the clinical outcome of cancer therapy, research efforts over the past decade have focused on overcoming this resistance. One such approach is characterized by the numerous attempts to discover new analogs of these agents not possessing cross-resistance with their parent compound. Approximately 28 different platinum complexes have been in various stages of clinical development as anticancer agents *(1)*. One such example is the third-generation cisplatin analog oxaliplatin (Eloxatin®). This compound, containing a 1,2-diaminocylohexane carrier ligand, has been shown to act as an alkylating agent on DNA, much like its parent cisplatin, but it has also been shown to lack cross-resistance to it (for a review, *see* ref. *7*). Preclinical evidence indicates that oxaliplatin showed non-cross-resistance to cisplatin in human tumor models of ovarian cancer *(15,16)*. Early clinical data have also shown that oxaliplatin treatment obtained objective responses in 5–17% of platinum-refractory patients *(17,18)*. However, this observation does not appear to be robust, as another study did not show a benefit of oxaliplatin therapy in patients with relapsed, clinically cisplatin-resistant ovarian cancer *(19)*.

JM-216 (bis-aceto-ammine-dichloro-cyclohexylamine platinum IV) is another platinum analog that has received some interest. JM-216, developed for oral administration, was observed to have a lack of cross-resistance with cisplatin when tested in vitro in several human cell lines; however, in vivo tests failed to confirm this finding *(20)*. Clinical trials with this compound are ongoing, and although early results suggest a favorable comparison with carboplatin, it remains to be seen whether JM-216 demonstrates improved response rates to those observed with more traditional platinum-based therapies *(1)*. Recently, a fourth platinum analog, a novel sterically hindered platinum complex, ZD0473 (cis-aminedichloro[2-methylpyridine] platinum [II], formerly known as JM473 and AMD473), was reported to show in vitro circumvention of acquired cisplatin resistance in human ovarian carcinoma cell lines *(21,22)*. Early phase I trials with this compound are also ongoing. Although cisplatin analogs have often shown less toxicity and/or activity against tumors of different origin than the parent compound, to date, the problem of clinical resistance to platinum therapy remains *(23)*.

Another approach that has been used to overcome the problem of inherent and/ or acquired resistance to alkylating agents has involved the substantial research effort to understand the cellular mechanisms of resistance. The discovery of molecular mechanisms that mediate this process as well as the existence of a drug-resistant phenotype would aid in the development of novel drugs and treatment regimens to circumvent chemoresistance, thereby having the promise of an enhancement in chemotherapeutic effectiveness. Knowledge of how tumor cells respond to exposure to alkylating agents will provide insight into how cells develop resistance and what chemotherapeutic options may be pursued to combat this problem. Several mechanisms of drug resistance in tumor are well recognized and include the multidrug-resistance gene MDR1 *(24)*, multidrug-resistance-associated protein *(24)*, and DNA repair processes *(25)*. In addition, changes in a diverse group of gene products that include cell cycle, transcription and cell death regulators, growth factor receptors, tumor suppressors, and oncogenes also affect cellular sensitivity to chemotherapeutic agents *(26)*. Given the evidence that a variety of cellular mechanisms can contribute to chemoresistance, it is conceivable that a combination of alterations in several sites within the tumor cell may play a substantial role in mediating the development of resistance. Increasingly, a wide variety of modern molecular techniques is being employed in the study of the cellular mechanisms of drug resistance. Techniques such as classical cytogenetics, differential display, fluorescent *in situ* hybridization (FISH), comparative genomic hybridization, spectral karyotyping, and cDNA microarray are providing valuable insights.

An example of this work is research from our laboratory examining resistant cell lines for gains and losses of DNA associated with the acquisition of resistance using comparative genomic hybridization. Our results, comparing seven cell lines and cisplatin and two analogs, showed that aberrations in specific genes associated with epidermal growth factor, high-mobility-group protein 2, cyclin B and C, DNA repair, programmed cell death, signal transduction, and glutathione S-transferase were observed *(27)*.

Complementary (DNA) (cDNA) microarray is another promising technique that is yielding important data on those genes related to acquired chemoresistance. In a study examining the gene expression profile changes of two 2-(4-aminophenyl) benzothiazole (CJM126)-resistant cell lines, MCF-7$^{10\,nM\ 126}$ and MCF-7$^{10\,\mu M\ 126}$, changes in the resistant MCF-7 cell lines were observed in genes involved in a variety of cell signaling pathways *(28)*. Gene expression changes common to MCF-7$^{10\,nM\ 126}$ and MCF-7$^{10\,\mu M\ 126}$ cells, compared to sensitive MCF-7^{wt} cells, were the shutdown of transcription factor Oct-2, the upregulation of the negative apoptosis regulator MCL-1, the G1-to-S-phase regulator ubiquitin carrier protein, and the GTP-binding protein GST1-HS. These findings indicate the association of a resistance phenotype with a profound gene transcription dysregulation, a decreased apoptosis activity, and an increased proliferation. Specific changes unique to each of the resistant cell lines were also observed. Genes involved in the DNA mismatch-repair pathway, such as MSH2, DNA repair protein RAD51, and damage-specific DNA-binding protein were downregulated in MCF-7[10 nM 126], whereas genes involved in the nucleotide-excision repair pathway, such as ERCC1, RFC, and PCNA were overexpressed in MCF-7[10 μM 126]. The differential changes in the DNA-repair pathways between MCF-7[10 nM 126] and MCF-7[10 μM 126] cell lines may be indicative of different processes employed to circumvent the growth inhibition produced by exposure to CJM126.

Complementary DNA microarrays are increasingly employed to examine gene expression of clinical biopsy materials, in order to facilitate both diagnosis and patient-treatment selection. Although a full review of the use of microarrays is beyond the scope of this chapter (for such a review, *see* ref. *29*), several studies illustrate its potential clinical utility. Golub et al. *(30)* demonstrated that DNA microarray analysis of bone marrow samples could differentiate acute myeloid from acute lymphocytic leukemia. Large-scale RNA profiling has also been used to predict the tissue origin of a variety of carcinomas, including prostate, breast, lung, ovary, colorectum, kidney, liver, pancreas, bladder/ureter, and gastroesophagus *(31)*. Microarrays have been used to classify breast tumors at the molecular level *(32,33)*; a number of different molecular phenotypes among breast tumors were observed, including ER+/luminallike, basallike, Erb-B2+, and normal breast *(32)*. The clinical importance of this finding was demonstrated in a study examining patient outcome correlated with gene expression patterns *(33)*. In this study, it was reported that survival analysis of uniformly treated patients with locally advanced breast cancer showed different clinical outcomes based on gene expression patterns, including a significant difference in outcome for two estrogen-receptor-positive groups.

Although these studies appear to demonstrate the utility of microarray analysis in the classification of cancers, the application of this technique is also yielding important information regarding treatment response. In a recent study *(34)*, gene expression profile changes were reported from the fine-needle aspiration of primary breast tumors before and after systemic chemotherapy (adriamycin and cyclophosphamide). That article reported that the number of genes that change after one cycle of chemotherapy was 10 times greater in the responding than in the nonresponding group. van't Veer et al. *(35)* used microarray analysis on primary breast tumors of 117 young patients; they identified an expression signature of 70 genes strongly predictive of a short interval to distant metastases. The MD Anderson team reported at the 12th International Congress on Anti-Cancer Treatment (Paris, 4–7 February 2002) a gene expression signature that identified patients who benefit most from Taxol/FAC adjuvant therapy.

In summary, gene expression profiles will be increasingly applied in vitro and in vivo studies to identify key genes or gene expression signatures associated with resistance to chemotherapeutic agents (or combinations thereof). At the clinical level, the studies cited here suggest that cDNA microarray technology has the potential of identifying patients most likely to benefit from specific chemotherapeutic regimens; moreover, they may also have the potential of providing an early indicator of responding versus nonresponding patients.

The foregoing represents examples of the nature of the information that is emerging on the molecular processes regulating tumor cell resistance to current chemotherapy regimens. The chapters that follow illustrate the breadth of approaches used in the study of these mechanisms. The data that they describe are providing significant insights not only into those molecular processes mediating resistance, but they are also providing the necessary clues for the future development of new anticancer compounds and/or adjuncts to the current stable of alkylating agents. The loss of survival, poor symptom control, and degraded quality of life associated with inherent and acquired resistant to alkylating agents is reflected in decreased initial response rates to therapy and nonresponsive recurrent disease. The development of new analogs and the possible pretreatment screening of resistant patients can only enhance the effectiveness of cancer therapy.

REFERENCES

1. Lebwohl D, Canetta R. Clinical development of platinum complexes in cancer therapy: an historical perspective and an update. *Eur J Cancer* 1998;34:1522–1534.
2. Giaccone G. Clinical perspective on platinum resistance. *Drugs* 2000;59(Suppl 4):9–17.
3. Foon KA, Rai KR, Gale RP. Chronic lymphocytic leukemia: new insights into biology and therapy. *Ann Intern Med* 1990;113:525–539.
4. Calvert AH, Newell DR, Gumbrell LA, et al. Carboplatin dosage: prospective evaluation of a simple formula based on renal function. *J Clin Oncol* 1989;7:1748–1756.
5. Alberts DS, Green S, Hannigan EV, et al. Improved therapeutic index of carboplatin plus cyclophosphamide vs cisplatin plus cyclophosphamide: final report by the Southwest Oncology Group of a phase III randomized trial in stages 3 and 4 ovarian cancer. *J Clin Oncol* 1992;10:706–717.
6. Evans BD, Raju KS, Calvert AH, et al. Phase II study of JM8, a new platinum analog, in ovarian cancer. *Cancer Treat Rep* 1983;67:997–1000.
7. Misset JL, Bleiberg H, Sutherland W, et al. Oxaliplatin clinical activity: a review. *Crit Rev Oncol/Hematol* 2000;35:75–93.
8. O'Dwyer PJ, Stevenson JP, Johnson SW. Clinical pharmacokinetics and administration of established platinum drugs. *Drugs* 2000;59:19–27.
9. Jassem J. Chemotherapy of advanced non-small cell lung cancer. *Ann Oncol* 1999;10:S77–S82.
10. Cortés-Funes H. New treatment approaches for lung cancer and impact on survival. *Semin Oncol* 29(Suppl 8):2002;26–29.
11. Huisman C, Postmus PE, Giaccone G, et al. Second-line chemotherapy and its evaluation in small cell lung cancer. *Cancer Treat Rev* 1999;25:199–206.
12. Thigpen T. *Second-line Therapy for Ovarian Carcinoma: General Concepts.* American Society of Clinical Oncology 1999 American Society of Clinical Oncology: Atlanta, GA, 1999.
13. Berek JS, Bertelsen K, du Bois A, et al. Advanced epithelial ovarian cancer : 1998 concensus statements. *Ann Oncol* 1999;10(Suppl 1):87–92.

14. McGuire WP, Ozols RF. Chemotherapy of advanced ovarian cancer. *Semin Oncol* 1998;25:340–348.

15. Pendyala L, Creaven P. In vitro cyctotoxcity, protein binding, red blood cell partitioning, and biotransformation of oxaliplatin. *Cancer Res* 1993;53:5970–5976.

16. Raymond E, Lawrence R, Izbicka E, et al. Activity of oxaliplatin against humon tumor colony-forming units. *Clin Cancer Res* 1998;4:1021–1029.

17. Chollet P, Bensmaine A, Brienza A, et al. Single agent activity of oxaliplatin in heavily pretreated advanced epithetial ovarian cancer. *Ann Oncol* 1996;7:1065–1070.

18. Bougnoux P, Dieras V, Petit T, et al. A multicenter phase II study of oxaliplatin (OXA) as a single agent in platinum (PT) and/or taxane (TX) pretreated advanced ovarian cancer (AOC) final results. *Proc Am Soc Clin Oncol* 1999;18:368a (abstract).

19. Piccart MJ, Green JA, Lacave AJ, et al., Oxaliplatin or paxlitaxel in patients with platinum-pretreated advanced ovarian cancer: a randomized phase II study of the European Organization for Research and Treatment of Cancer Gynecology Group. *J Clin Oncol* 2000;18:1193–1202.

20. Harrap KR. Initiatives with platinum and quinazoline-based antitumor molecules. Fourteenth Bruce F. Cain Memorial Award Lecture. *Cancer Res* 1995;55:2761–2768.

21. Holford J, Sharp SY, Murrer BA, et al. In vitro circumvention of cisplatin resistance by the novel sterically hindered platinum complex AMD473. *Br J Cancer* 1998;77:366–373.

22. Holford J, Beale PJ, Boxall FE, et al. Mechanisms of drug resistance to the platinum complex ZD0473 in ovarian cancer cell lines. *Eur J Cancer* 2000;36:1984–1990.

23. Guminski AD, Harnett PR, deFazio A. Scientists and clinicians test their metal—back to the future with platinum compounds. *Lancet Oncol* 2002;3:312–318.

24. Lautier D, Canitrot Y, Deeley RG, et al. Multidrug resistance mediated by the multidrug resistance protein (MRP) gene. *Biochem Pharmacol* 1996;52:967–977.

25. Godwin AK. Meister A, O'Dwyer PJ, et al. High resistance to cisplatin in human ovarian cancer cell lines is associated with marked glutathione synthesis. *Proc Natl Acad Sci USA* 1992;89:3030–3074.

26. Wafik S, El-Deiry MD. Role of oncogenes in resistance and killing by cancer therapeutic agents. *Curr Opin Oncol* 1997;9:79–87.

27. Leyland-Jones B, Kelland LR, Harrap KR, et al. Genomic imbalances associated with acquired resistance to platinum analogues. *Am J Pathol* 1999;155:77–84.

28. Yu Q, Hiorns LR, Bradshaw TD, et al. Profiling gene expression of 2-(4-aminophenyl)-benzothiazole-resistant MCF7 cells using cDNA microarray. Submitted.

29. MacGregor PF, Squire JA Application of microarrays to the analysis of gene expression in cancer. *Clin Chem* 2002;48:1170–1177.

30. Golub TR, Slonim DK, Tamayo P, Huard C, et al. Molecular classification of cancer: class discovery and class prediction by gene expression monitoring. *Science* 1999;286:531–537.

31. Su AI, Welsh JB, Sapinoso LM, et al. Molecular classification of human carcinomas by use of gene expression signatures. Cancer Res 2001;61:7388–7393.

32. Perou CM, Sørlie T, Eisen MB, et al. Molecular portraits of human breast tumours. *Nature* 2000;406:747–752.

33. Sørlie T, Perou CM, Tibshirani R, et al. Gene expression patterns of breast carcinomas distinguish tumor subclasses with clinical implications. *Proc Natl Acad Sci USA* 2001;98:10,869–10,874.

34. Sotiriou C, Powles TJ, Dowsett M, Jazaeri AA, et al. Gene expression profiles derived from fine needle aspiration correlate with response to systemic chemotherapy in breast cancer. *Breast Cancer Res* 2002;4:R3.

35. van't Veer, LJ, Dai H, van de Vijver MJ, et al. Gene expression profiling predicts clinical outcome of breast cancer. *Nature* 2002;415(6871):484–485.

2 Role of Nonhomologous End-Joining and Recombinational DNA Repair in Resistance to Nitrogen Mustard and DNA Crosslinking Agents

Lawrence C. Panasci, MD, Zhi-Yuan Xu, MD, and Raquel Aloyz, PhD

CONTENTS

From: *Cancer Drug Discovery and Development: DNA Repair in Cancer Therapy*
Edited by: L. C. Panasci and M. A. Alaoui-Jamali © Humana Press Inc., Totowa, NJ

1. RESISTANCE TO THE NITROGEN MUSTARDS

The nitrogen mustards are an important group of alkylating agents with activity against several human tumors *(1–4)*. Many nitrogen mustard analogs are transported by carrier-mediated systems into cells and alkylate DNA, RNA, and proteins *(5–7)*. Alkylation of DNA and, more specifically, the formation of DNA interstrand crosslinks have been considered to be responsible for their cytotoxicity *(8–10)*. Resistance to the nitrogen mustards in murine and human tumor cells has been reported to be secondary to (1) alterations in the transport of these agents *(11)*, (2) alterations in the kinetics of the DNA crosslinks formed by these agents *(9,10,12)*, (3) cytoplasmic metabolism of the chloroethyl alkylating moiety to the inactive hydroxyethyl derivative *(13)* via glutathione (GSH)/ glutathione-*S*-transferase (GST) *(14–16)*, (4) overexpression of metallothionein, which confers resistance to cis-platinum and cross-resistance to melphalan *(17)*, (5) changes in resistance to apoptosis *(18)*, and (6) altered DNA repair activity (*see* Fig. 1) *(19)*. There have been previous reports of alterations in the kinetics of DNA interstrand crosslink formation and removal associated with resistance to the nitrogen mustards *(9,10,12)*, although others have found no differences in the ability of sensitive or resistant cells to remove nitrogen mustard-induced crosslinks *(20,21)*. This review will concentrate on the involvement of DNA repair in nitrogen mustard drug resistance and cross-resistance to cisplatin. We will discuss results obtained in clinical samples and human cancer cell lines.

2. DNA CROSSLINKS VIS-À-VIS NITROGEN MUSTARD DRUG RESISTANCE

Nitrogen mustard-induced alkylation of DNA results predominantly in the development of purine–drug complexes *(22)*. The nitrogen mustards, including chlorambucil and melphalan, may also form intrastrand and/or interstrand crosslinks at N-7 guanines *(23)*. These interstrand crosslinks are considered to be important in the cytotoxicity of these drugs *(8–10)*. There are technical problems involved in quantitating nitrogen mustard-induced interstrand crosslinks. Nitrogen mustards produce thermolabile glycosylic bonds (N[7]-guanine adducts), which yield apurinic sites and which, in turn, can cause strand breaks and/or breaks of crosslinks (reviewed in ref. *24*). Strand breaks can interfere with molecular size-based assays. The ethidium bromide fluorescence assay has the advantage that strand breaks are less likely to influence the quantification of crosslinks *(25)*. A widely utilized assay to determine DNA crosslinks is the alkaline elution assay *(26–28)*. This technique involves molecular size differences. However, the strand breaks induced by nitrogen mustards may complicate interpretation of repair of interstrand crosslinks when utilizing this assay. More recently, the comet assay has been utilized to quantitate interstrand crosslinks, but there are similar problems with the alkaline assay, which may be less with a neutral assay *(29,30)*.

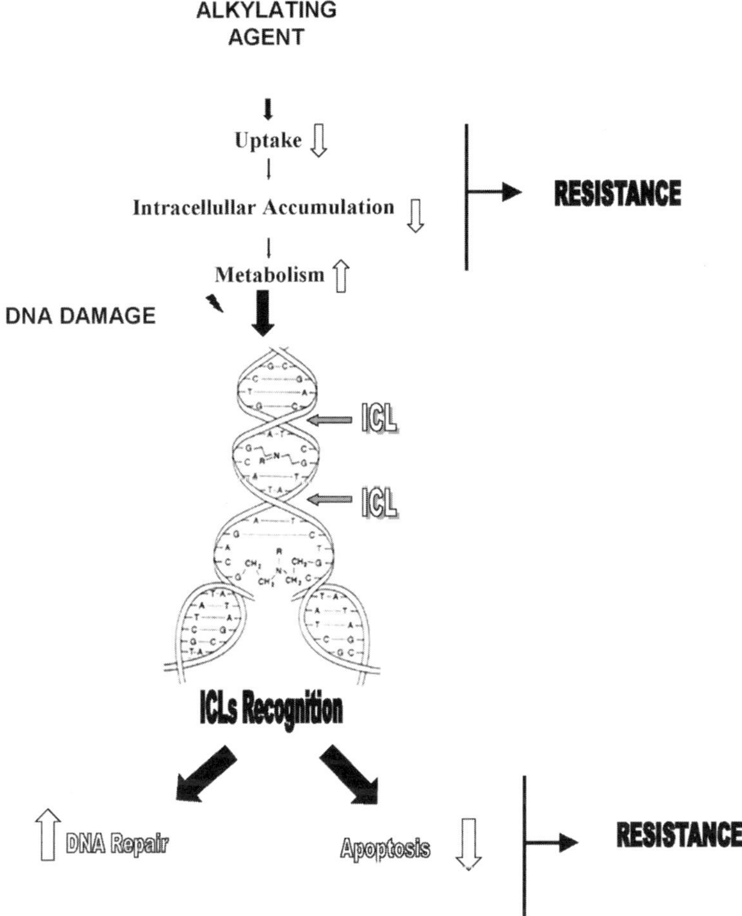

Fig. 1. Before DNA interstrand crosslink (ICL) are induced, decreased uptake (↓), decreased intracellular accumulation, or increased metabolism (↑) of the DNA-damaging agent may account for alkylating agent resistance. Downstream DNA damage recognition, alteration in the DNA repair, and/or apoptotic signaling pathways resulting in increased DNA repair (↑) and/or decreased apoptosis (↓) can mediate alkylating agent drug resistance.

3. NITROGEN MUSTARD DRUG RESISTANCE IN CHRONIC LYMPHOCYTIC LEUKEMIA

It is difficult to study clinical samples because of the heterogeneity of most tumor samples and the difficulty in obtaining serial samples from the same patient.

A model of drug resistance with direct relevance to clinical practice is a malignancy with easy access to a homogeneous population of malignant cells, which represents the clinical status of the patients. Chronic lymphocytic leukemia is characterized by the proliferation and accumulation of B-lymphocytes that appear to be mature but are biologically immature. In some patients, chronic lymphocytic leukemia has an indolent course and does not require treatment for many years. When treatment is necessary, single-agent chemotherapy with a nitrogen mustard, usually chlorambucil, is the standard initial therapy, although fludarabine, a new exciting agent, may be incorporated in front-line treatment. At least 60–80% of patients respond to nitrogen mustard therapy, often for years, but eventually all patients become resistant to these agents *(31)*. Furthermore, many patients with chronic lymphocytic leukemia respond well to low-dose chlorambucil treatment, indicating that this disease is initially often very sensitive (hypersensitive) to these anticancer agents, to a greater extent than virtually all epithelial malignancies. A homogeneous monocellular population of malignant B-lymphocytes is easily obtained from chronic lymphocytic leukemia patients, thus providing a relatively unique opportunity to study clinically derived cells. We and others have previously demonstrated that there is a strong correlation between in vitro cytotoxicity of chlorambucil (measured by the microtiter [MTT] assay) and in vivo response in chronic lymphocytic leukemia patients *(32–34)*. Therefore, chronic lymphocytic leukemia is an excellent malignancy for in vitro studies, which should have direct clinical applicability.

Our laboratory, utilizing the ethidium bromide fluorescence assay, originally reported that DNA interstrand crosslink formation at 4 h after melphalan incubation (a time-point believed to be associated with maximal crosslink formation) was decreased in malignant B-lymphocytes from resistant chronic lymphocytic leukemia patients *(35)*. However, when we examined crosslink formation and removal at 0, 4, and 24 h after a 35-min melphalan incubation, there was evidence of a greater amount of crosslinks at time 0 in malignant B-lymphocytes from resistant chronic lymphocytic leukemia patients as compared to those from untreated chronic lymphocytic leukemia patients. Moreover, the untreated patients' lymphocytes developed a greater amount of crosslinks at 4 h without evidence of removal at 24 h, whereas there was evidence of progressive removal of DNA crosslinks at 4 and 24 h in lymphocytes from resistant chronic lymphocytic leukemia patients. This suggests that enhanced DNA repair is implicated in this process *(36)*. In another study, a patient with advanced chronic lymphocytic leukemia was treated with iv cyclophosphamide and DNA interstrand crosslinks in the lymphocytes were measured by the alkaline elution technique. Maximal DNA interstrand crosslink formation occurred 12 h after injection. However, the level of crosslinks was just above the sensitivity of the assay at 12 and 24 h after drug administration *(37)*. Also, utilizing the alkaline elution technique, Johnston et al. examined DNA crosslink formation in chronic lymphocytic leukemia lymphocytes at 6 h after an in vitro incubation with chlorambucil. They found that the lymphocytes from two resistant chronic lymphocytic leukemia patients had as many DNA crosslinks as the lymphocytes from patients sensitive to chlorambucil *(38)*.

4. DNA CROSSLINKING AGENT DRUG RESISTANCE IN EPITHELIAL CELL LINES

As concerns epithelial cancer cell lines, DNA repair has been implicated in DNA crosslinking agent drug resistance (enhanced repair of DNA interstrand crosslinks in some investigations) *(39–44)*, whereas in other investigations, drug resistance appears to develop independent of altered DNA repair *(45–47)*. As previously stated, the assays utilized to quantitate interstrand DNA crosslinks have technical problems that may render difficult the interpretation of "repair" of DNA interstrand crosslinks *(25–28)*. Thus, some of the investigations in which DNA repair is not implicated may be the result of these technical problems or may represent alternative mechanisms of drug resistance as initially discussed.

5. DNA REPAIR OF NITROGEN MUSTARD DNA CROSSLINKS IN CANCER CELLS

The mechanism of removal of DNA interstrand crosslinks in mammalian cells is poorly understood. There are several different DNA repair systems that could be involved in the repair of nitrogen mustard-induced DNA interstrand crosslinks, including base excision repair, nucleotide excision repair, and recombinational repair (*see* Fig. 2). The mammalian base excision repair enzyme, alkyl-*N*-purine DNA glycosylase (3-methyladenine-DNA-glycosylase), can excise damaged guanine bases from DNA treated with chlorambucil *(48)*. We measured 3-methyladenine-DNA-glycosylase activity in chronic lymphocytic leukemia extracts and found a significantly higher activity (approx 1.7-fold) in lymphocytes from resistant chronic lymphocytic leukemia patients as compared to those from untreated chronic lymphocytic leukemia patients. Because this activity may vary with cell proliferation, it was corrected for differences in DNA synthesis utilizing (^{3}H)thymidine incorporation (there were differences in DNA synthesis between the two groups even though the vast majority of malignant B-lymphocytes are nonproliferative) and this resulted in no significant difference in enzyme activity between the two groups *(49)*. Moreover, overexpression of the human alkyl-*N*-purine DNA glycosylase in CHO cells did not result in nitrogen mustard resistance, suggesting that alkyl-*N*-purine DNA glycosylase was not a rate-limiting enzyme in nitrogen mustard drug resistance *(50)*. Furthermore, mouse embryonic stem cells bearing null mutations in this enzyme are not hypersensitive to the nitrogen mustards *(51)*.

Possible insights into mechanism(s) of interstrand crosslink repair are gained by examining nitrogen mustard hypersensitivity in DNA repair mutants.

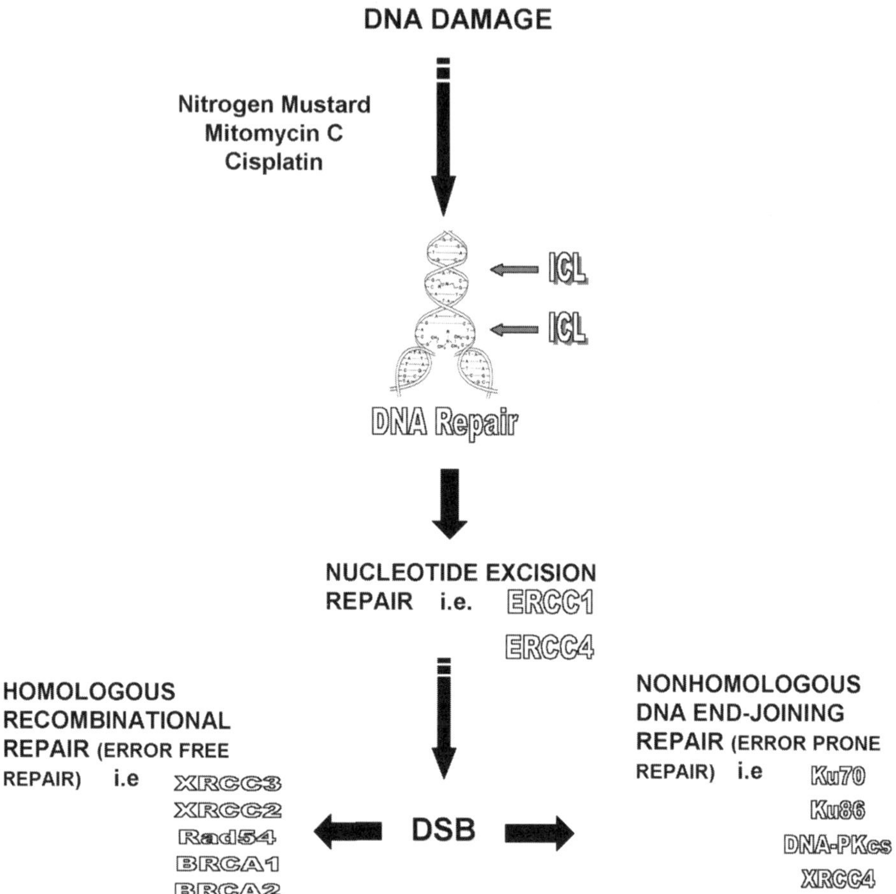

Fig. 2. The involvement of DNA repair proteins in DNA interstrand crosslinks (ICL) repair has been inferred by the fact that cells bearing mutations in such proteins are hypersensitive to ICL-inducing agents (nitrogen mustards, mitomycin c, and cisplatin). After DNA damage recognition, ICLs are unhooked by the nucleotide excision repair complex ERCC1/ERCC4, which could result in double-strand breaks (DSBs). The resulting DSBs would be repaired by homologous recombinational repair or nonhomologous DNA end-joining repair. The ICL-inducing agent's hypersensitive cell lines are defective in one of the following gene products: XRCC2, XRCC3, Rad54, BRCA1, BRCA2, Ku70, Ku86, DNA-PKcs, or XRCC4.

Significant DNA crosslinking agent hypersensitivity (varying from moderate to severe) is found in several DNA repair mutants, including ERCC-1, ERCC-4 (XPF), Xrcc-2, Xrcc-3, Rad54, Ku70, Ku86, and DNA-PK$_{cs}$ *(52–56)*.

This analysis, along with information gained from studying crosslink removal in both bacteria and *Saccharomyces cerevisiae* and the possibility that DNA double-strand breaks are repaired in a similar fashion to interstrand crosslinks, has resulted in the proposed model in mammalian cells that nucleotide excision repair, via the ERCC-1/ERCC-4 endonuclease, results in an incision 5' to the interstrand crosslink and that recombinational repair is involved in further processing of the lesion (reviewed in *57* and *58*). As concerns repair of double-strand breaks, nonhomologous DNA end-joining uses no, or very limited, sequence homology to rejoin ends directly, whereas homologous recombination requires extensive regions of DNA homology. Homologous recombinational repair would be necessary for error-free repair of interstrand crosslinks, whereas an illegitimate or nonhomologous DNA end-joining mechanism of repair could result in deletional repair of interstrand crosslinks. It is also conceivable that all three types of repair (nucleotide excision, homologous recombinational repair, and nonhomologous DNA end-joining) are implicated simultaneously or depending on the phase of the cell cycle, in the processing of interstrand crosslinks. The various genes implicated in nonhomologous DNA end-joining include the components of DNA-PK, Xrcc-4, and ATM, although ATM may be involved in homologous recombinational repair (*see* Fig. 3) *(58)*. Nonhomologous DNA end-joining is a major mechanism of double-strand breaks (DSB) repair in mammalian cells (reviewed in refs. *59* and *60*). Homologous recombinational repair in human cells implicates the HsRad51 family of proteins, including HsRad51, HsRad52, Rad51B, Rad51C, Rad51D, HsRad54, Xrcc-2, and Xrcc-3. Rad51 binding to DNA requires the precedent binding of Rad52. In addition, other Rad51 protein members are involved in the assembly of the Rad51 complex (*see* Fig. 3). Interactions of Rad51 with BRCA2, c-Abl kinase, and p53 have also been detected (reviewed in refs. *58* and *61*).

In order to gain insight into possible mechanisms of DNA crosslink removal in nitrogen mustard-resistant chronic lymphocytic leukemia, Bramson et al. analysed in vitro cross-resistance in chronic lymphocytic leukemia *(32)*. Chlorambucil-resistant chronic lymphocytic leukemia lymphocytes were completely cross-resistant to melphalan and mitomycin c, partially cross-resistant to cis-platinum, and not cross-resistant to ultraviolet (UV) light or methylmethane sulfonate. Because UV radiation damage is repaired by nucleotide excision repair and methylmethane sulfonate is repaired by base excision repair, it appears that these repair systems are not upregulated in nitrogen mustard drug resistance in chronic lymphocytic leukemia *(32)*. Also, ERCC-1 protein levels were not increased in nitrogen mustard drug-resistant chronic lymphocytic leukemia lymphocytes *(62)*. Nucleotide excision repair activity was very low in most chronic lymphocytic leukemia lymphocytes, including the majority of those obtained from previously treated chronic lymphocytic leukemia patients *(63)*.

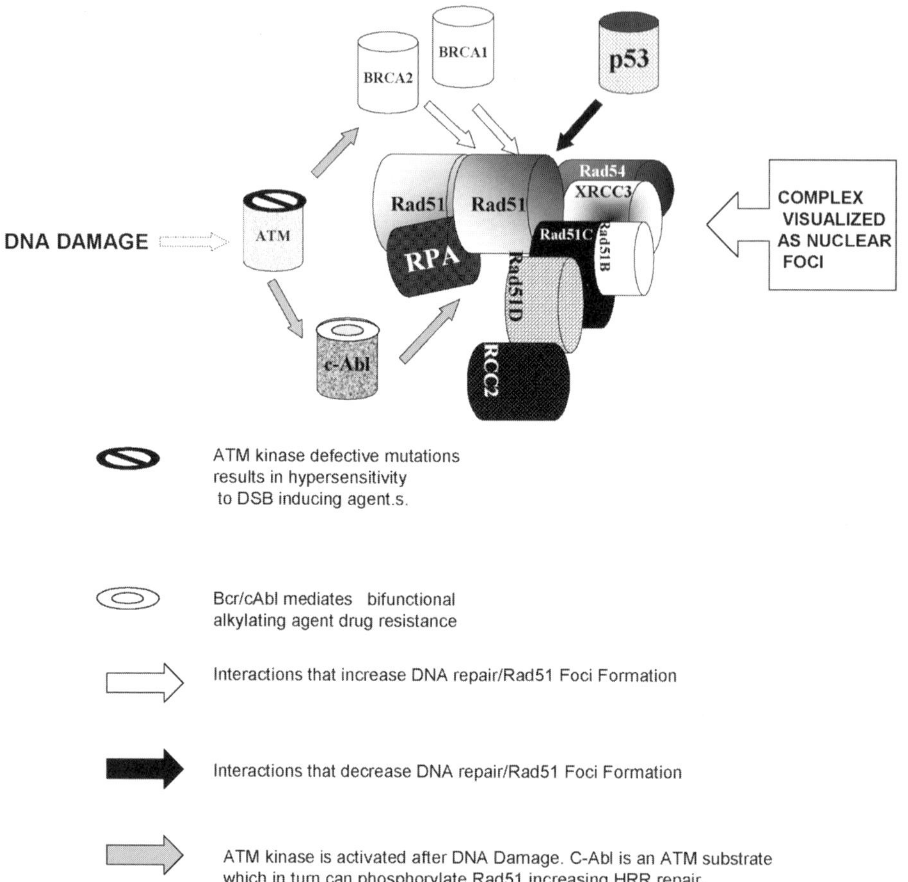

Fig. 3. The modulation of the homologous recombinational repair process can affect DNA damaging agent sensitivity. Homologous recombinational repair is mediated by multienzyme complexes, which includes the Rad51 paralogs (Rad51, Rad51B, Rad51C, Rad51 D, Xrcc2, Xrcc3 and Rad54). In response to DNA damage, the DNA repair complexes relocalize into the nucleus in discrete foci and can be visualized immunocytochemically using Rad51 antibodies. Other proteins known to interact and associate with these core complex include RPA, BRCA1, BRCA2, c-abl, and P53. Upon DNA damage, the serine–threonine kinase, ATM, is activated. Targets downstream of ATM include the BRCA1, BRCA2 and the c-abl tyrosine kinase proteins that are known to functionally interact with Rad51. BRAC 1/BRAC2—Rad51 interaction results in increased DNA repair and Rad51 foci induced by DNA damage. Constitutively active c-abl kinase (BCR/CABL) results in resistance to bifunctional alkylating agents resistance and induced Rad51 foci. The tumor suppresser p53 interacts with Rad51 decreasing homologous recombinational repair and Rad51 foci formation.

6. NONHOMOLOGOUS DNA END-JOINING

Because ionizing radiation results in DSBs that are largely repaired by nonhomologous DNA end-joining *(59)* and because DSBs are probably repaired in a similar fashion to interstrand crosslinks, cross-resistance studies between chlorambucil and ionizing radiation may be informative. Indeed, it appears that there is evidence of cross-resistance between ionizing radiation and chlorambucil in chronic lymphocytic leukemia lymphocytes *(64)*.

DNA-dependent protein kinase, a nuclear serine–threonine kinase, is a protein complex including a catalytic subunit of 460 kDa, DNA-dependent protein kinase$_{cs}$, and a DNA-binding subunit, the Ku autoantigen (a dimer of the Ku70 and Ku86 proteins). Ku binds to DSBs and other discontinuities in the DNA and recruits DNA-dependent protein kinase$_{cs}$ to the damaged site *(59,60)*. The active DNA-dependent protein kinase complex can then phosphorylate many DNA-bound proteins in the vicinity *(65)*. Because mutations in DNA-dependent protein kinase result in X-ray and alkylating agent sensitivity *(53,54)* and because X-ray resistance develops in parallel with chlorambucil resistance in chronic lymphocytic leukemia *(64)*, determination of DNA-dependent protein kinase activity in chronic lymphocytic leukemia should be informative. In a preliminary report with a small sample of chronic lymphocytic leukemia patients, an increase in DNA-dependent protein kinase activity was found in resistant samples *(66)*. In collaboration with Muller and Salles, our laboratory examined DNA-dependent protein kinase activity in a group of 34 patients (18 patients resistant to chlorambucil both in vitro and in vivo). There was an excellent linear correlation between DNA-dependent protein kinase activity and in vitro chlorambucil cytotoxicity $(r = 0.875, p = 0.0001)$ *(67)*. The increased DNA-dependent protein kinase activity was independent of other clinical and biological factors. The regulation of DNA-dependent protein kinase activity was associated with increased DNA-binding activity of its regulatory subunit, Ku, and increased Ku protein levels. Interestingly, most untreated chronic lymphocytic leukemia patients have very low levels of DNA-dependent protein kinase activity, suggesting that, initially, resistance in chronic lymphocytic leukemia may be simply a state in which tumor cells lose an abnormal sensitivity to alkylating agents. In approx 25% of the samples from untreated chronic lymphocytic leukemia patients, a variant (truncated) form of the Ku86 protein was associated with very low DNA-dependent protein kinase activity and hypersensitivity to chlorambucil *(67)*. Wortmannin, a nonspecific inhibitor of DNA-dependent protein kinase, which also inhibits other phosphatidylinositol 3-kinases, sensitized chronic lymphocytic leukemia lymphocytes to the effects of chlorambucil. Moreover, there was a significant correlation between the synergistic sensitization and fold decrease in DNA-dependent protein kinase activity, but because wortmannin also inhibits other phosphatidylinositol 3-kinases, these results must be interpreted with caution

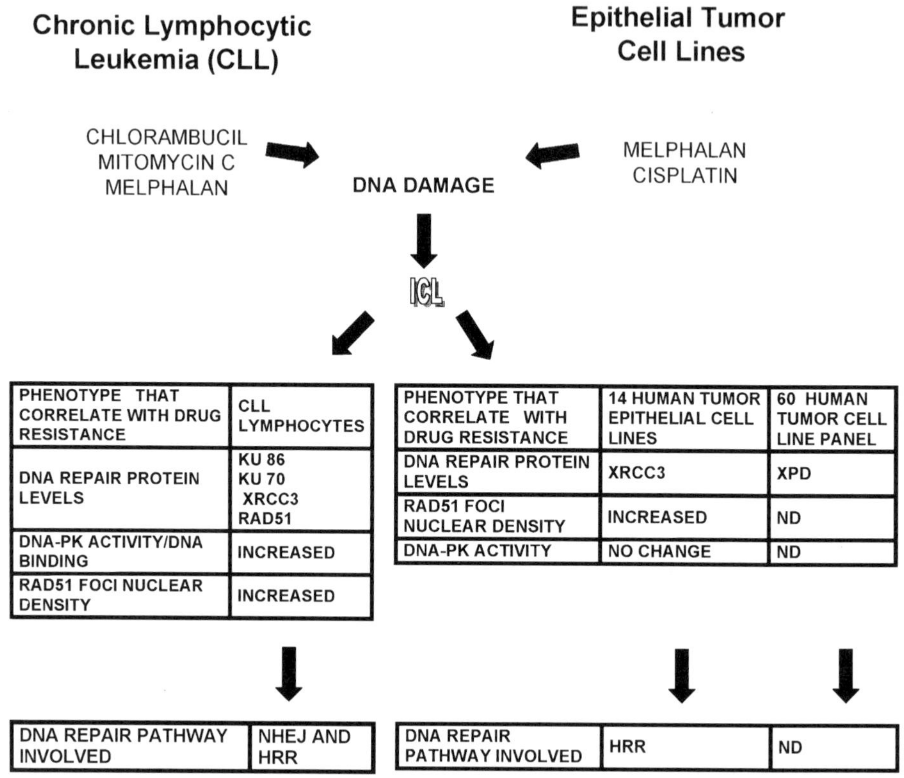

PHENOTYPE THAT CORRELATE WITH DRUG RESISTANCE	CLL LYMPHOCYTES
DNA REPAIR PROTEIN LEVELS	KU 86 KU 70 XRCC3 RAD51
DNA-PK ACTIVITY/DNA BINDING	INCREASED
RAD51 FOCI NUCLEAR DENSITY	INCREASED

PHENOTYPE THAT CORRELATE WITH DRUG RESISTANCE	14 HUMAN TUMOR EPITHELIAL CELL LINES	60 HUMAN TUMOR CELL LINE PANEL
DNA REPAIR PROTEIN LEVELS	XRCC3	XPD
RAD51 FOCI NUCLEAR DENSITY	INCREASED	ND
DNA-PK ACTIVITY	NO CHANGE	ND

DNA REPAIR PATHWAY INVOLVED	NHEJ AND HRR

DNA REPAIR PATHWAY INVOLVED	HRR	ND

Fig. 4. The expression levels of DNA repair proteins essential for ICL repair (summarized in Fig. 2) can affect drug sensitivity. XRCC3 protein levels and DNA-damage-induced Rad51 foci correlates with chlorambucil drug resistance in lymphocytes from chronic lymphocytic leukemia (CLL) patients and with melphalan and cisplatin resistance in epithelial tumor cell lines, indicating that increased homologous recombinational repair (HRR) can be involved in drug resistance. Moreover, in CLL lymphocytes but not in epithelial cell lines, drug resistance can be mediated by increased non-homologous DNA end-joining (NHEJ) because Ku70 and Ku86 protein levels correlated with drug resistance and DNA-PK activity. Interestingly, the levels of the nucleotide excision repair protein XPD correlated with drug resistance in a human tumor cell line panel.

(68). In contrast, neither Ku protein levels nor DNA-dependent protein kinase activity correlated with melphalan resistance in epithelial tumor cell lines, suggesting that DNA-dependent protein-kinase-related DNA repair is not a rate-limiting step in epithelial cancers (*see* Fig. 4) *(69)*.

The immunohistochemical expression of Ku autoantigen and DNA-dependent protein kinase$_{cs}$ was examined in various human tissues. There was a large variation in expression depending on the specific tissue type *(70)*. This supports our results that there is a variation in DNA-dependent protein kinase expression in human tissues.

Although it appears reasonable that increased DNA-dependent protein kinase activity is associated with increased repair of nitrogen mustard-induced interstrand crosslinks in chronic lymphocytic leukemia and thus, increased drug resistance, it is possible that other mechanisms are involved, including a role for DNA-dependent protein kinase with respect to apoptosis (71,72).

DNA-dependent protein kinase$_{cs}$ is a member of the phosphatidylinositol (PI) 3-kinase superfamily. Other members include the gene mutated in ataxia telangiectasia (ATM) and the cell cycle checkpoint protein ATR (60). Recently, loss of heterozygosity (LOH) or mutations of the ataxia telangiectasia gene and a decrease in ataxia telangiectasia protein levels have been found in approx 30–40% of B-chronic lymphocytic leukemia patients. These factors appear to be associated with a shorter survival, at least in younger patients (73–76). The association of ataxia telangiectasia with nitrogen mustard drug resistance in cancer has not been investigated to date.

7. HOMOLOGOUS RECOMBINATIONAL REPAIR

The involvement of nucleotide excision repair and homologous recombinational repair in the repair of interstrand crosslinks is inferred from the fact that the mutant cell lines with the greatest sensitivity (10- to 100-fold) to alkylating agents that produce interstrand crosslinks are those that are deficient in or lacking Xrcc-2, Xrcc-3, ERCC1, and ERCC4/XPF (52,56,58). The nucleotide excision repair complex (ERCC-1/ERCC-4) in mammalian cells makes dual incisions 22–28 bp apart, 5' to the interstrand crosslink on the same strand, (77). This would then be followed by homologous recombinational repair. Alternatively, it is possible that strand invasion mediated by the Rad51 repairasome, including Xrcc-2 and Xrcc-3, occurs prior to ERCC1/XPF endonuclease-induced incision (78).

Several human genes implicated in homologous recombinational repair have been characterized, including HsRad52, HsRad51, Rad51B, Rad51C, Rad51D, HsRad54, Xrcc-2, and Xrcc-3 (reviewed in refs. 58 and 61). A recent model of interaction in yeast proposes that Rad52 interacts with RPA, followed by Rad52 association with Rad51. This leads to the assembly of Rad51 and associated proteins onto single-stranded DNA (ssDNA), which then initiate recombinational DSB repair (see Fig. 3) (79). Xrcc-3 is necessary for the assembly of Rad51 foci and these proteins physically interact (56,80). In fact, if all of the interactions described occur in one complex, then HsRad51, Xrcc-3, Rad51C, Rad51B, Rad51D, and Xrcc-2 are complexed together (reviewed in ref. 58). Rad54 appears to be required after the association of the above-mentioned proteins, and Rad54 may assist Rad51 in interacting with damaged DNA (81).

In view of the critical role of the Rad51 protein in homologous recombinational repair and its probable involvement in repair of interstrand crosslinks, our laboratory investigated HsRad51 foci formation after in vitro chlorambucil treatment of chronic lymphocytic leukemia lymphocytes. In vitro chlorambucil treatment induced HsRad51 expression as measured by increased immunopositive staining in all chronic lymphocytic leukemia samples. In the chlorambucil-resistant chronic lymphocytic leukemia lymphocytes, there was a linear correlation between induction of HsRad51 foci at 5.4 μM chlorambucil and the in vitro LD_{50} concentration of chlorambucil *(82)*. Moreover, there was a significant correlation between Rad51 protein levels and, to a lesser extent, Xrcc-3 protein levels and chlorambucil cytotoxicity in chronic lymphocytic leukemia samples *(83)*. Thus, it appears that Rad51-directed homologous recombinational repair as evidenced by Rad51 foci, Rad51 protein levels, and Xrcc-3 protein levels is implicated in the development of nitrogen mustard drug resistance in chronic lymphocytic leukemia (*see* Fig. 4). Recent investigations of overexpression of fusion tyrosine kinases such as bcr/abl in myeloid cells results in DNA-crosslinking-agent drug resistance associated with increased homologous recombinational repair and Rad51 protein levels. Furthermore, overexpression of Rad51 results in DNA crosslinking agent drug resistance in these myeloid cells *(84,85)*.

In order to determine if our results in chronic lymphocytic leukemia were applicable to other malignancies, epithelial cell lines were investigated. We determined Rad51 foci formation in the epithelial cell lines. There was a good correlation between the density of Rad51 foci formation induced by 5.5 μM melphalan and melphalan drug resistance. Also, melphalan-induced Rad51 foci density correlated with cisplatin resistance *(69)*. Xrcc-3 may be a determining factor in Rad51-related recombinational repair and nitrogen mustard resistance in epithelial cell lines. There is a correlation between Xrcc-3 protein levels and melphalan cytotoxicity in the epithelial cell lines, suggesting that Xrcc-3 may be important in the induction of Rad51-mediated recombinational repair and drug resistance. Rad51 protein levels did not correlate with melphalan/cisplatin resistance in the 14 epithelial cell lines *(69)*. Additionally, overexpression of Rad51 in CHO cells produced minimal (1.5-fold to fold in synchronized cells in S-phase) resistance to cisplatin (*86*; M. DeFais, personal communication) suggesting that the role of Rad51 vis-à-vis DNA crosslinking agent drug resistance may be somewhat different in epithelial cells as compared to hematopoetic cells (myeloid and lymphocytic cells) (*see* Fig. 4). These data are consistent with the hypothesis that Rad51-mediated homologous recombinational repair is associated with DNA crosslinking agent drug resistance. These results suggest a novel mechanism of DNA crosslinking agent drug resistance with significant potential clinical implications.

8. OVEREXPRESSION OF Xrcc-3/XPD RESULTS IN DRUG RESISTANCE ASSOCIATED WITH ENHANCED Rad51-RELATED HOMOLOGOUS RECOMBINATIONAL REPAIR AND PROLONGED S-PHASE CHECKPOINT

Xrcc-3, a Rad51 paralog that binds to Rad51, was overexpressed in MCF-7 cells (a cell line with low Xrcc-3 protein levels and sensitive to cisplatin and melphalan). The Xrcc-3-transfected cells (Xrcc-3/MCF-7) have sixfold higher Xrcc-3 protein levels as compared to mock-transfected cells. The Xrcc-3/MCF-7 cells were twofold resistant to cisplatin/melphalan and fivefold resistant to mitomycin c utilizing the MTT assay. Initial results suggest that alkylating agent-treated Xrcc-3/MCF-7 cells demonstrate enhanced Rad51 foci density as compared to mock-transfected cells *(87)*.

Because nucleotide excision repair is also implicated in DNA crosslinking agent drug resistance, my laboratory in collaboration with the US National Cancer Institute determined the protein levels of XPA, XPD, XPB and ERCC-1 in their 60 cancer cell line panel utilized for drug screening and then correlated the protein levels with the cytotoxicity of 170 compounds screened in this panel. In this study, only XPD protein levels correlated significantly with alkylating agent drug resistance *(see* Fig. 4) *(88)*.

The XPD helicase is a component of the TFIIH transcription factor that participates in DNA unwinding to allow either gene transcription by RNA polymerase II and/or the removal of DNA lesions induced by a variety of genotoxic agents, including UV light and some anticancer drugs by nucleotide excision repair *(89)*. Our knowledge regarding the role of XPD in drug efficacy comes from correlations between loss of XPD function and changes in cell sensitivity to DNA damage. It has been reported that XPD mutations that impair nucleotide excision repair activity result in minimal or no DNA crosslinking agent hypersensitivity *(52,90)*.

In order to determine if XPD plays a role in DNA crosslinking agent drug resistance, my laboratory overexpressed XPD in the SK-MG-4 human glioma cell line. The XPD-overexpressing cell line (hereafter called XPD) was twofold to threefold resistant to cisplatin and melphalan but not to UV light as compared to mock-transfected cells (hereafter called PCD). As anticipated, there was no difference in nucleotide excision repair activity between XPD and PCD cells. Also, the basal doubling time and basal percentage of cells in the S-phase were similar. Following cisplatin treatment, XPD cells removed interstrand crosslinks faster than PCD cells. Consistent with these results, XPD overexpression increased homologous recombinational repair visualized as Rad51 foci density after DNA damage. Moreover, immunochemical and immunoprecipitation studies demonstrate that XPD and Rad51 interact constitutively and that this interaction is increased after cisplatin treatment. This is the first description of functional crosstalk between a nucleotide excision repair protein XPD and Rad51-homologous recombinational repair resulting in DNA crosslinking agent drug resistance and accelerated removal of interstrand crosslinks *(91)*.

Overexpression of XPD did not alter the doubling time or the percentage of cells in the S-phase in the basal state. However, cisplatin treatment significantly increased the percentage of cells in the S-phase in cells overexpressing XPD. This suggests that overexpression of XPD results in a prolongation of the S-phase checkpoint, allowing more time for Rad51-related DNA repair *(91)*. Similar studies are in progress with the Xrcc-3-overexpressing cells. Because Rad51-related homologous recombinational repair probably occurs mainly in the S-phase *(58)*, Rad51-related homologous recombinational repair and the S-phase checkpoint process are intimately interrelated. Thus, it is difficult to determine if enhanced Rad51-related HRR or prolonged S-phase arrest is primarily responsible for the DNA crosslinking agent drug resistant phenotype of the XPD-overexpressing cell lines (*see* Fig. 5).

9. SUMMARY OF RESULTS WITH RECOMBINATIONAL GENES

The regulation of DNA-dependent protein kinase activity appears to be tightly associated with the development of chlorambucil drug resistance in chronic lymphocytic leukemia. In particular, low DNA-dependent protein kinase activity is associated with hypersensitivity to chlorambucil. Furthermore, increased levels of DNA-dependent protein kinase activity are associated with chlorambucil resistance in chronic lymphocytic leukemia. Moreover, chlorambucil-induced HsRad51 foci, Rad51 protein levels, and Xrcc-3 protein levels correlate with chlorambucil drug resistance. The increased HsRad51 foci formation after chlorambucil treatment in resistant chronic lymphocytic leukemia samples may represent an active DNA repair process involving other Rad-51-related proteins. A plausible hypothesis to explain these results is that low DNA-dependent protein kinase activity defines a hypersensitive state, whereas high DNA-dependent protein kinase activity along with increased homologous recombination, as determined by HsRad51 foci formation, Rad51 protein levels, and Xrcc-3 protein levels contribute to the resistant state in chronic lymphocytic leukemia (*see* Fig. 4).

Regarding epithelial cancers (as represented by epithelial cancer cell lines), it appears that nonhomologous end-joining (i.e., DNA-dependent protein kinase) does not correlate with melphalan cytotoxicity but that HsRad51-related homologous recombinational repair is implicated in the development of nitrogen mustard and cisplatin drug resistance (*see* Fig. 4). Furthermore, Xrcc-3 protein levels are implicated in this process, as demonstrated by the correlation of Xrcc-3 protein with melphalan drug resistance and the fact that overexpression of Xrcc-3 results in DNA crosslinking agent drug resistance. The complexity of this situation is illustrated by the fact that overexpression of a nucleotide excision repair

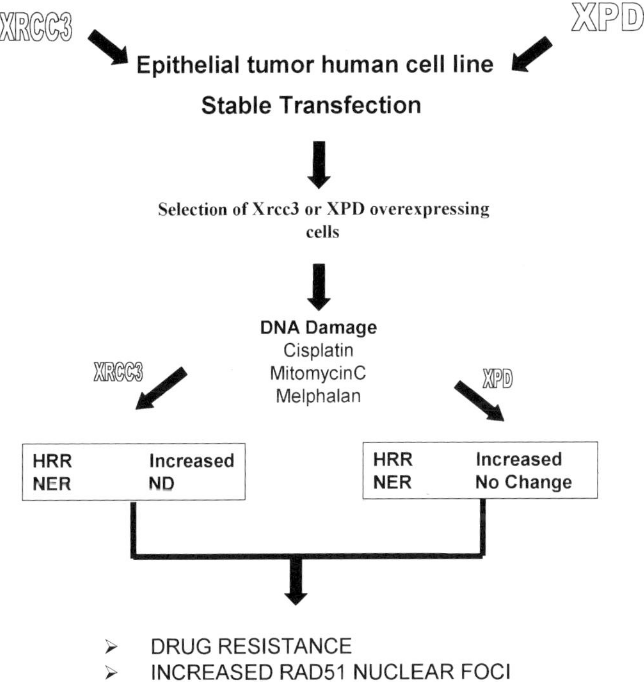

Fig. 5. To further assess the results summarized in Fig. 4, Xrcc3 or XPD open reading frames were stably transfected in human epithelial tumor cell lines. The cell lines were chosen from a cell line panel taking in account their relatively high sensitivity to alkylating agents and the low levels of the proteins to be overexpressed. Both cell lines XPD (XPD-overexpressing cells) and XRCC3 (Xrcc3-overexpressing cells) displayed similar phenotypes such as (1) resistance to alkylating agents (cisplatin, Mitomycin C, and melphalan), (2) increased HRR after DNA damage as assessed by Rad51 foci density, and (3) an increased percentage of cells in the S-phase after DNA damage.

protein, XPD, results in DNA crosslinking agent drug resistance via an interaction with Rad51-related homologous recombinational repair. The enhanced Rad51-related repair is also associated with cisplatin-induced S-phase arrest, suggesting that S-phase arrest may be a determining factor in enhanced Rad51-related repair. Because Rad51-related repair occurs largely in the S-phase, these two processes are intertwined and difficult to separate (*see* Fig. 5).

10. FUTURE PERSPECTIVES

The mechanism of low DNA-dependent protein kinase activity in sensitive chronic lymphocytic leukemia lymphocytes involves decreased Ku protein levels and a variant form of Ku86, as compared to resistant samples. The regulation of this process needs to be further investigated at both the translational and transcriptional levels. Inhibition of DNA-dependent protein kinase activity by small molecules in combination with nitrogen mustard chemotherapy may improve the therapeutic index of the latter compounds in chronic lymphocytic leukemia. As concerns homologous recombinational repair, the implication of other Rad51 paralogs and accessory proteins needs to be examined. Increased expression of bcr/abl in myeloid cells results in an associated increase in expression of Rad51 and Rad51 paralogs along with DNA crosslinking agent drug resistance *(84)*. Some of these proteins may be markers of drug resistance in clinical specimens. A model of repair of DSBs has been proposed in which either Ku or HsRad52 binds DSBs, thus directing entry into nonhomologous DNA end-joining or homologous recombinational repair, respectively *(92)*. The relationship between Ku and HsRad52 with respect to repair of DNA crosslinks should also be examined. Experiments designed to alter the expression of genes involved in nonhomologous DNA end-joining and/or homologous recombinational repair vis-à-vis DNA crosslinking agent drug resistance should help to clarify their respective roles in this process. Again, inhibition of homologous recombinational repair (e.g., Xrcc-3 or Rad51 inhibition) may result in sensitization of epithelial tumors and chronic lymphocytic leukemia lymphocytes that are resistant to cisplatin and the nitrogen mustards.

XPD and Xrcc-3 overexpression results in DNA crosslinking agent drug resistance associated with enhanced Rad51 foci density *(87,91)*. In addition, cisplatin-treated XPD-overexpressing cells are arrested in the S-phase *(91)*. Thus, it is not clear if DNA-crosslinking-agent drug resistance is a result of enhanced Rad51-related homologous recombinational repair or S-phase arrest or both. Because recombinational repair and S-phase arrest are intimately intertwined, the use of dominant-negative constructs and mutant cell lines should help to determine if S-phase arrest is essential for the development of DNA crosslinking agent drug resistance associated with XPD overexpression. There are a number of defective cell lines with mutated genes for ATM, CHK2, NBS, and MRE11. These genes are involved in the S-phase checkpoint as outlined in the chapter and these mutant cell lines are defective in S-phase arrest *(93–95)*. Furthermore, dominant-negative constructs of ATM, ATR, NBS, CHK1, and CHK2 can be utilized to determine the effect of loss of function on the XPD/Xrcc-3-DNA crosslinking agent phenotype *(96–99)*. These studies should help to clarify the role of S-phase arrest in Rad-51-related DNA crosslinking agent drug resistance.

REFERENCES

1. Ochoa M Jr. Alkylating agents in clinical chemotherapy. *Ann NY Acad Sci* 1969;163:921–930
2. Bergsagel DE, Griffith KM, Haut A, et al. The treatment of plasma cell myeloma. *Adv Cancer Res* 1967;10:311–359.

3. Fisher B, Carbone P, Economou SG, et al. L-Phenylalanine mustard (L-Pam) in the management of primary breast cancer. *N Engl J Med* 1975;292:117–122.

4. Young RC, Chabner BA, Hubbard SP, et al. Advanced ovarian adenocarcinoma. *N Engl J Med* 1978;299:1261–1266.

5. Goldenberg GJ, Land HB, Cormack DV. Mechanism of cyclophosphamide transport by L5178Y lymphoblasts *in vitro*. *Cancer Res* 1974;34:3274–3282.

6. Tew KD, Taylor DM. Studies with cyclophosphamide labelled with phosphorus-32:nucleic acid alkylation and its effect on DNA synthesis in rat tumor and normal tissues. *J Natl Cancer Inst* 1977;58:1413–1419.

7. Hoes I, Lemiere F, Van Dongen W, et al. Analysis of melphalan adducts of 2'-deoxynucleotides in calf thymus DNA hydrolysates by capillary high-pressure liquid chromatography-electrospray tandem mass spectroscopy. *J Chromatogr B Biochem Sci Appl* 1999;736:43–59.

8. Ross WE, Ewig RA, Kohn KW. Differences between melphalan and nitrogen mustard in the formation and removal of DNA crosslinks. *Cancer Res* 1978;38:1502–1506.

9. Zwelling L, Michaels S, Schwartz H, et al. DNA crosslinking as an indicator of sensitivity and resistance of L1210 leukemia cells to *cis*-diaminedichloroplatinum (II) and L-phenylalanine mustard. *Cancer Res* 1981;41:640–649.

10. Parsons PG, Carter FB, Morrison L, et al. Mechanism of melphalan resistance developed *in vitro* in human melanoma cells. *Cancer Res* 1981,41:1525 1534.

11. Moscow JA, Swanson CA, Cowan KH. Decreased melphalan accumulation in a human breast cancer cell line selected for resistance to melphalan. *Br J Cancer* 1993;68:32–37.

12. Parsons PC. Dependence on treatment time of melphalan resistance and DNA crosslinking in human melanoma cell lines. *Cancer Res* 1984;44:2773–2778.

13. Suzukake K, Vistica BP, Vistica DT. Dechlorination of L-phenylalanine mustard by sensitive and resistance tumor cells and its relationship to intracellular glutathione content. *Biochem Pharmacol* 1983;32:165–167.

14. Green JA, Vistica DT, Young RC, et al. Potentiation of melphalan cytotoxicity in human ovarian cancer cell lines by glutathione depletion. *Cancer Res* 1984;44:5427–5431.

15. Kramer RA, Greene K, Ahmad S, et al. Chemosensitization of L-phenylalanine mustard by the thiol-modulating agent buthionine sulfoximine. *Cancer Res* 1987;47:1593–1597.

16. Morgan AS, Ciaccio PJ, Tew KD, et al. Isozyme specific glutathione S-transferase inhibitors potentiate drug sensitivity in cultured human tumor cell lines. *Cancer Chemother Pharmacol* 1996;37:363–370.

17. Kelley SL, Basu A, Teicher BA, et al. Overexpression of metallothionein confers resistance to anticancer drugs. *Science* 1998;241:1813–1815.

18. Reed JC. Bcl-2 family proteins: regulators of apoptosis and chemoresistance in hematologic malignancies. *Semin Hematol* 1997;34:9–19.

19. Tan KB, Mattern MR, Boyce RA, et al. Elevated topoisomerase II activity and altered chromatin in nitrogen mustard-resistant human cells. *NCI Monogr* 1987;4:95–98.

20. Dean SW, Johnson AB, Tew KD. A comparative analysis of drug-induced DNA effects in a nitrogen mustard resistant cell line expressing sensitivity in nitrosoureas. *Biochem Pharmacol* 1986;35:1171–1176.

21. Robson CN, Lewis AD, Wolf CR, et al. Reduced levels of drug-induced DNA cross linking in nitrogen mustard-resistant Chinese hamster ovary cells expressing elevated glutathione-*S*-transferase activity. *Cancer Res* 1987;47:6022–6027.

22. Bank BB, Kanganis D, Liebes LF, et al. Chlorambucil pharmacokinetics and DNA binding in chronic lymphocytic leukemia lymphocytes. *Cancer Res* 1989;49:554–559.

23. Lawley PD, Brookes PJ. Interstrand crosslinking of DNA by difunctional alkylating agents. *J Mol Biol* 1967;25:143–160.

24. Ojwang JO, Grueneberg DA, Loechler EL. Synthesis of a duplex oligonucleotide containing a nitrogen mustard interstrand DNA-DNA crosslink. *Cancer Res* 1989;49:6529–6537.

25. de Jong S, Zijlstra JG, Timmer-Bosscha H, et al. Detection of DNA crosslinks in tumor cells with the ethidium bromide fluorescence assay. *Int J Cancer* 1986;37:557–561.

26. Kohn KW. DNA filter elution: a window on DNA damage in mammalian cells. *BioEssays* 1996;18:505–513.

27. Kohn KW. Principles and Practice of DNA filter elution. *Pharmacol Ther* 1991;49:55–77.

28. O'Connor PM, Kohn KW. Comparative pharmacokinetics of DNA lesion formation and removal following treatment of L1210 cells with nitrogen mustards. *Cancer Commun* 1990;2:387–394.

29. Merk O, Speit G. Detection of crosslinks with the comet assay in relationship to genotoxicity and cytotoxicity. *Environ Mol Mutagen* 1999;33:167–172.

30. Merk O, Reiser K, Speit G. Analysis of chromate-induced DNA-protein crosslinks with the comet assay. *Mutat Res* 2000;471:71–80.

31. Foon KA, Rai KR, Gale RP. Chronic lymphocytic leukemia: new insights into biology and therapy. *Ann Int Med* 1990;113:525–539.

32. Bramson J, McQuillan A, Aubin R, et al. Nitrogen mustard drug resistant B-cell chronic lymphocytic leukemia as an *in vivo* model for crosslinking agent resistance. *Mutat Res* 1995;336:269–278.

33. Hanson JA, Bentley DP, Bean EA, et al. *In vitro* chemosensitivity testing in chronic lymphocytic leukaemia patients. *Leuk Res* 1991;15:565–569.

34. Silber R, Degar B, Costin D, et al. Chemosensitivity of lymphocytes from patients with B-cell chronic lymphocytic leukemia to chlorambucil, fladarabine, and camptothecin analogs. *Blood* 1994;84:3440–3446.

35. Panasci L, Henderson D, Skalski V, et al. Transport, metabolism, and DNA interaction of melphalan in lymphocytes from patients with chronic lymphocytic leukemia. *Cancer Res* 1988;48:1972–1976.

36. Torres-Garcia SJ, Cousineau L, Caplan S, et al. Correlation of resistance to nitrogen mustards in chronic lymphocytic leukemia with enhanced removal of melphalan-induced DNA crosslinks. *Biochem Pharmacol* 1989;38:3122–3123.

37. DeNeve W, Valeriote F, Edelstein M, et al. *In vivo* DNA crosslinking by cyclophosphamide: comparison of human lymphatic leukemia cells with mouse L1210 leukemia and normal bone marrow cells. *Cancer Res* 1989;49:3452–3456.

38. Johnston JB, Israels LG, Goldenberg GJ, et al. Glutathione S-transferase activity, sulfhydryl group and glutathione levels and DNA crosslinking activity with chlorambucil in chronic lymphocytic leukemia. *J Natl Cancer Inst* 1990;82:776–779.

39. Batist G, Torres-Garcia S, Demuys JM, et al. Enhanced DNA crosslink removal: the apparent mechanism of resistance in a clinically relevant melphalan-resistant human breast cancer cell line. *Mol Pharmacol* 1989;36:224–230.

40. Bedford P, Fox BW. Repair of DNA interstrand crosslinks after busulphan. A possible mode of resistance. *Cancer Chemother Pharmacol* 1982;8:3–7.

41. Hill BT, Shellard SA, Hosking LK, et al. Enhanced DNA repair and tolerance of DNA damage associated with resistance to *cis*-diammine-dichloroplatinum (II) after *in vitro* exposure of a human teratoma cell line to fractionated X-irradiation. *Int J Radiat Oncol Biol Phys* 1990;19:756–83.

42. Johnson SW, Swiggard PA, Handel LM, et al. Relationship between platinum–DNA adduct formation and removal and cisplatin cytotoxicity in cisplatin-sensitive and -resistant human ovarian cancer cells. *Cancer Res* 1994;54:5911–5916.

43. Ali-Osman F, Rairkar A, Young P. Formation and repair of 1,3-bis-(2-chloroethyl)-1-nitrosourea and cisplatin induced total genomic DNA interstrand crosslinks in human glioma cells. *Cancer Biochem Biophys* 1995;14:231–241.

44. Dong Q, Bullock N, Ali-Osman F, et al. Repair analysis of 4 hydroperoxycyclophosphamide-induced DNA interstrand crosslinking in the c-myc gene in 4-hydroperoxy-

cyclophosphamide-sensitive and -resistant medulloblastoma cell lines. *Cancer Chemother Pharmacol* 1996;37:242–246.

45. Rawlings CJ, Roberts JJ. Walker rat carcinoma cells are exceptionally sensitive to *cis*-diamminedichloroplatinum(II) (cisplatin) and other difunctional agents but not defective in the removal of platinum-DNA adducts. *Mutat Res* 1986;166:157–168.

46. Petersen LN, Mamentaq EL, Stevnsner T, et al. Increased gene specific repair of cisplatin induced interstrand crosslinks in cisplatin resistant cell lines, and studies on carrier ligand specificity. *Carcinogenesis* 1996;17:2597–2602.

47. Roy G, Horon JK, Roy R, et al. Acquired alkylating drug resistance of a human ovarian carcinoma cell line is unaffected by altered levels of pro- and anti-apoptotic proteins. *Oncogene* 2000;19:141–150.

48. Mattes WB, Lee CS, Laval J, et al. Excision of DNA adducts of nitrogen mustards by bacterial and mammalian 3-methyladenine-DNA glycosylases. *Carcinogenesis* 1996;17:643–648.

49. Geleziunas R, McQuillan A, Malapetsa A, et al. Increased DNA synthesis and repair-enzyme expression in lymphocytes from patients with chronic lymphocytic leukemia resistant to nitrogen mustards. *J Natl Cancer Inst* 1991;83:557–564.

50. Bramson J, O'Connor T, Panasci LC. Effect of alkyl-*N*-purine DNA glycosylase overexpression on cellular resistance to bifunctional alkylating agents. *Biochem Pharmacol* 1995;50:39–44.

51. Allan JM, Engelward BP, Dreslin AJ, et al. Mammalian 3-methyladenine DNA glycosylase protects against the toxicity and clastogenicity of certain chemotherapeutic DNA crosslinking agents. *Cancer Res* 1998;58:3965–3973.

52. Hoy CA, Thompson LH, Mooney CL, et al. Defective DNA crosslink removal in Chinese hamster cell mutants hypersensitive to bifunctional alkylating agents. *Cancer Res* 1985;45:1737–1743.

53. Caldecott K, Jeggo P. Cross-sensitivity of gamma-ray-sensitive hamster mutants to crosslinking agents. *Mutat Res* 1991;255:111–121.

54. Tanaka T, Yamagami T, Oka Y, et al. The scid mutation in mice causes defects in the repair system for both double strand DNA breaks and DNA crosslinks. *Mutat Res* 1993;288:277–280.

55. Essers J, Hendriks RW, Swagemakers SMU, et al. Disruption of mouse RAD54 reduces ionizing radiation resistance and homologous recombination. *Cell* 1997;89:195–204.

56. Liu N, Lamerdin JE, Tebbs RS, et al. Xrcc-2 and Xrcc-3, new human Rad51-family members, promote chromosome stability and protect against DNA crosslinks and other damages. *Mol Cell* 1998;1:783–793.

57. Thompson LH. Evidence that mammalian cells possess homologous recombinational repair pathways. *Mutat Res* 1996;363:77–88.

58. Thompson LH, Schild D. Homologous recombinational repair of DNA ensures mammalian chromosome stability. *Mutat Res* 2001;477:131–153.

59. Anderson CW, Lees-Miller SP. The nuclear serine/threonine protein kinase DNA-dependent protein kinase. *Crit Rev Eukaryot Gene Expr* 1992;2:283–314.

60. Jeggo PA. DNA-dependent protein kinase at the cross-roads of biochemistry and genetics. *Mutat Res* 1997;384:1–14.

61. Shinohara A, Ogawa T. Rad51/RecA protein families and the associated proteins in eukaryotes. *Mutat Res* 1999;435:13–21.

62. Bramson J, McQuillan A, Panasci LC. DNA repair enzyme expression in chronic lymphocytic leukemia vis-à-vis nitrogen mustard drug resistance. *Cancer Lett* 1995;90:139–148.

63. Barret J-M, Calsou P, Laurent G, Salles B. DNA repair activity in protein extracts of fresh human malignant lymphoid cells. *Mol Pharmacol* 1996;49:766–771.

64. Bentley P, Salter R, Blackmore J, et al. The sensitivity of chronic lymphocytic leukaemia lymphocytes to irradiation *in vitro*. *Leuk Res* 1995;19:985–988.

65. Lees-Miller SP, Chen YR, Anderson CW. Human cells contain a DNA-activated protein-kinase that phosphorylates simian virus 40T antigen, mouse p53 and the human Ku autoantigen. *Mol Cell Biol* 1990;10:6472–6481.

66. Muller C, Salles B. Regulation of DNA dependent protein kinase activity in leukemic cells. *Oncogene* 1997;15:2343–2348.
67. Muller C, Christodoulopoulos G, Salles B, et al. DNA-dependent protein kinase activity correlates with clinical and *in vitro* sensitivity of chronic lymphocytic leukemia lymphocytes to nitrogen mustards. *Blood* 1998;92:2213–2219.
68. Christodoulopoulos G, Muller C, Salles B, et al. Potentiation of chlorambucil cytotoxicity in B-cell chronic lymphocytic leukemia by inhibition of DNA-dependent protein kinase activity using Wortmannin. *Cancer Res* 1998;58:1789–1792.
69. Wang ZM, Chen ZP, Xu ZY, et al. Xrcc-3 protein expression and induction of Rad51 foci correlate with melphalan resistance in human tumor cell lines. *J Natl Cancer Inst* 2001;93:1473–1478.
70. Moll U, Lau R, Sypes MA, et al. DNA-dependent protein kinase, the DNA-activated protein kinase is differentially expressed in normal and malignant human tissues. *Oncogene* 1999;18:3114–3126.
71. Jackson SP. DNA-dependent protein kinase. *Int J Biochem Cell Biol* 1997;29:935–938.
72. Song Q, Lees-Miller SP, Kumar S, et al. DNA-dependent protein kinase catalytic subunit: a target for an ICE-like protease in apoptosis. *EMBO* 1996;15:3238–3246.
73. Starostik P, Manshouri T, O'Brien S, et al. Deficiency of the ATM protein expression defines an aggressive subgroup of B-cell chronic lymphocytic leukemia. *Cancer Res* 1998;58:4552–4557.
74. Bullrich F, Rasio D, Kitada S, et al. ATM mutations in B-cell chronic lymphocytic leukemia. *Cancer Res* 1999;59:24–27.
75. Stankovic T, Weber P, Stewart G, et al. Inactivation of ataxia telangiectasia mutated gene in B-cell chronic lymphocytic leukaemia. *Lancet* 1999;353:26–29.
76. Schaffner C, Stilgenbauer S, Rappold GA, et al. Somatic ATM mutations indicate a pathogenic role of ATM in B-cell chronic lymphocytic leukemia. *Blood* 1999;94:748–753.
77. Bessho T, Mu D, Sancar A. Initiation of DNA interstrand crosslink repair in humans: the nucleotide excision repair system makes dual incisions 5' to the crosslinked base and removes a 22 to 28-nucleotide-long damage-free strand. *Mol Cell Biol* 1997;17:6822–6830.
78. De Silva IU, McHugh PJ, Clingen PH, et al. Defining the roles of nucleotide excision repair and recombination in the repair of DNA interstrand crosslinks in mammalian cells. *Mol Cell Biol* 2000;20:7980–7990.
79. Hays SL, Firmenich AA, Massey P, et al. Studies of the interaction between Rad52 protein and the yeast single-stranded DNA binding protein RPA. *Mol Cell Biol* 1998;18:4400–4406.
80. Bishop DK, Ear U, Bhattacharyya A, et al. Xrcc-3 is required for assembly of Rad51 complexes *in vivo*. *J Biol Chem* 1998;273:21482–21488.
81. Tan TL, Essers J, Citterio E, et al. Mouse Rad54 affects DNA conformation and DNA-damaged induced Rad51 foci formation. *Curr Biol* 1999;9:325–328.
82. Christodoulopoulos G, Malapetsa A, Schipper H, et al. Chlormabucil induction of HsRad51 in B-cell chronic lymphocytic leukemia. *Clin Cancer Res* 1999;5:2178–2184.
83. Bello VE, Aloyz RS, Christodoulopoulos G, et al. Homologous recombinational repair *vis-à-vis* chlorambucil resistance in chronic lymphocytic leukemia. *Biochem Pharmacol* 2002;63:1585–1588.
84. Slupianek A, Schmutte C, Tombine G, et al. BCR/ABL regulates mammalian RecA homologs, resulting in drug resistance. *Mol Cell* 2001;8:795–806.
85. Slupianek A, Hoser G, Majsterek I, et al. Fusion tyrosine kinases induce drug resistance by stimulation of homology-dependent recombination repair, prolongation of G(2)/M phase, and protection from apoptosis. *Mol Cell Biol* 2002;22:4189–4201.
86. Vispe S, Cazaux C, Lesca C, et al. Overexpression of Rad51 protein stimulates homologous recombination and increases resistance of mammalian cells to ionizing radiation. *Nucleic Acids Res.* 1998;26:2859–2864.
87. Xu R, Aloyz R, Panasci LC. Xrcc-3 overexpression results in melphalan/cisplatin drug resistance. *Proc Am Assoc Cancer Res* 2002;43:424.

88. Xu Z, Chen Z-P, Malapetsa A, et al. DNA repair protein levels vis-à-vis anticancer drug resistance in the human tumor cell lines of the National Cancer Institute drug screening program. *Anticancer Drugs* 2002;13:511–519.
89. Egly JM. The 14th Datta Lecture. TFIIH: from transcription to clinic. *FEBS Lett* 2001;498:124–128.
90. Damia G, Imperatori L, Stefanini M, et al. Sensitivity of CHO mutant cell lines with specific defects in nucleotide excision repair to different anti-cancer agents. *Intl J Cancer* 1996;66:779–783.
91. Aloyz R, Xu Z-Y, Bello V, et al. Regulation of cisplatin resistance and homologous recombination repair by the TFIIH subunit XPD. *Cancer Res.* 2002;62:5457–5462.
92. Van Dyck E, Stasiak AZ, Stasiak A, et al. Binding of double-strand breaks in DNA by human Rad52 protein. *Nature* 1999;398:728–731.
93. Walworth NC. Cell-cycle checkpoint kinases: checking in on the cell cycle. *Curr Opin Cell Biol* 2000;12:697–704.
94. Falck J, Mailand N, Syljuasen RG, et al. The ATM-Chk2-Cdc25A checkpoint pathway guards against radioresistant DNA synthesis. *Nature* 2001;410:842–847.
95. Abraham RT. Cell cycle checkpoint signaling through the ATM and ATR kinases. *Genes Dev* 2001;15:2177–2196.
96. Cliby WA, Roberts CJ, Cimprich KA, et al. Overexpression of a kinase-inactive ATR protein causes sensitivity to DNA-damaging agents and defects in cell cycle checkpoints. *EMBO J* 1998;17:159–169.
97. Falck J, Petrini JH, Williams BR, et al. The DNA damage-dependent intra-S phase checkpoint is regulated by parallel pathways. *Nature Genet* 2002;30:290–294.
98. Lim DS, Kim ST, Xu B, Maser RS, et al. ATM phorphorylates p95/nbs1 in an S-phase checkpoint pathway. *Nature* 2000;404:613–617.
99. Guo N, Faller DV, Vaziri C. Carcinogen-induced S-phase arrest is Chk1 mediated and caffeine sensitive. *Cell Growth Differ* 2002;13:77–86.

3 Repair of DNA Interstrand Crosslinks Produced by Cancer Chemotherapeutic Drugs

Peter J. McHugh, DPhil and
John A. Hartley, PhD

CONTENTS

1. DNA INTERSTRAND CROSSLINKING DRUGS

It has been clear for over 50 yr that bifunctional reactivity is an essential prerequisite for the potent cytotoxic and antitumor activity of agents such as the nitrogen mustards *(1)*. DNA was later identified as a target for these drugs *(2,3)*, and the covalent modification of DNA almost certainly accounts for the antitumor activity of these drugs *(1)*. The fact that a bifunctional covalent reaction with DNA (crosslinking) is essential for the toxicity of these agents is evident from studies employing monofunctional analogs; for drugs such as the nitrogen mustards, their monofunctional counterparts are many orders of magnitude less toxic *(4)*. Crosslinks can be formed on the same strand of DNA (intrastrand), between the two complementary strands of DNA (interstrand), or between a base on DNA and a reactive group on a protein (DNA–protein). For the bifunctional alkylating drugs, it now seems clear that the interstrand crosslink (ICL) is the critical lesion

From: *Cancer Drug Discovery and Development: DNA Repair in Cancer Therapy*
Edited by: L. C. Panasci and M. A. Alaoui-Jamali © Humana Press Inc., Totowa, NJ

(5). In contrast, for the platinum drugs, where the majority of crosslinks are intrastrand, interstrand adducts might not be the sole critical cytotoxic lesion *(6)*.

The repair of cisplatin intrastrand adducts is dealt with in detail elsewhere in this volume (*see* Chapter 4), and we will concentrate exclusively on the interstrand adducts (ICLs) and the problems they pose to the cellular DNA repair machinery. The emphasis will be on human and mammalian studies. ICL repair is well understood in *Escherichia coli* and to some extent characterized in yeast, but because these are less relevant to cancer therapy, readers are directed to another review for information on crosslink repair in these organisms *(7)*.

Different human tumor types differ in their inherent sensitivity to DNA crosslinking agents, and this appears to be the result, at least in part, of their differing abilities to repair drug-induced ICLs. Increased repair of ICLs is also emerging as a critical mechanism of clinical acquired resistance to agents such as the nitrogen mustards chlorambucil and melphalan. This was suggested some years ago in chronic lymphocytic leukaemia *(8–10)*, and has recently been established in myeloma *(11)*. In addition, the capacity to repair ICLs appears to decline with age in normal cells, which may be a factor in the poor tolerance of chemotherapy in the elderly *(12)*.

1.1. Crosslinking Agents Employed in Mechanistic Studies

Although there are a number of ICL anticancer drugs approved for clinical use, relatively few of these agents have been widely employed in the most detailed mechanistic studies of ICL repair. In fact, one of the crosslinking agents most commonly employed in such studies is not used in cancer treatment. Of the anticancer agents, the classic nitrogen mustard mechlorethamine (chemically, 2-chloro-*N*-[2-chloroethyl]-*N*-methylethanamine) is by far the best characterized. This agent crosslinks preferentially the opposed guanines in the sequence 5'-GNC-3'/3'-CNG-5' *(13)* and this crosslinking represents only a small fraction of the total DNA lesions that this drug produces (under 5%), the remainder being monofunctional alkylations at guanine N7 and adenine N3 *(1)*. The other anticancer agent commonly used in mechanistic studies is mitomycin C. This molecule requires metabolic reduction in order to generate the reactive species, which produces crosslinks in the minor groove through reaction with the N2 position of guanines, crosslinking the opposed guanines in the sequence 5'-GC-3'/5'-CG-3' (up to 13% of total adducts are ICLs) *(14)*. The non-anticancer compounds that have been studied in great detail are the psoralens *(15)*, particularly 8-methoxypsoralen. Following 405 nm visible radiation, the formation of DNA monoadducts is favored, whereas ultraviolet A (UVA) (365 nm) is required to convert these to abundant ICLs (up to 40% of the total adducts). The basis of the activity of psoralens is UVA-induced reactivity at 5'-AT-3'/5'-TA-3' basepairs to form ICLs. Because of the more complex, multiringed structure of the psoralens, they form asymmetric crosslinks that bear a furan-ringed side and pyrone-ringed side. This influences repair processes, as will be apparent later. From the researcher's point of view, nitrogen mustard and psoralens are also attractive because of the availability of monofunctional analogs that allow the crosslink-specific nature of the results to be clarified.

2. EXCISION REPAIR PATHWAYS AND CROSSLINKS

2.1. Nucleotide Excision Repair

Before attempting to describe putative ICL-specific incision and excision reactions, it will be useful to briefly review the process of nucleotide excision repair (NER). NER is the principal pathway used to eliminate bulky, helix-distorting DNA adducts *(16)* (e.g., the dipyrimidine lesions induced by UV light and intrastrand crosslinks resulting from the reaction of cisplatin with DNA).

Nucleotide excision repair is necessarily initiated through a lesion-recognition activity and this process is still not fully understood. Evidence points to the XPC–hHR23B complex as a damage sensor in humans because it demonstrates strong affinity for several types of damaged DNA *(17–20)*. It is also possible that other factors assist in recognition, notably XPA in conjunction with replication protein A (RPA) as well as UV-DDB (UV-DNA-damaged binding activity composed of two peptides of 127 kDa and 47 kDa) *(18–22)*.

Following damage recognition, a large complex that shares identity with the RNA polymerase II transcription factor TFIIH is recruited *(17,23–25)*. This multisubunit complex contains two ATP-dependent helicases, XPB and XPD, able to unwind the DNA in the vicinity of the adduct. XPB is a 3'- to-5' helicase, whereas XPD acts with the opposite polarity *(26,27)*. Clearly, this step might well be inhibited by an ICL, which may explain some of the unusual, experimen-tally determined, ICL incision reactions described later in this chapter. The "bubble" structure arising from unwinding the lesion bears the structure-specific landmarks, single-stranded to double-stranded DNA transitions, required for the repair nucleases to proceed. First, 3' incision located 2–9 phosphodiester bonds from the lesion (the precise location of the incisions apparently depends on the type of lesion) is made by the XPG protein *(24,28)* and, subsequently, a 5' inci-sion located 16–25 phosphodiester bonds from the lesion results from the action of the XPF–ERCC1 heterodimer *(29,30)*. XPF is thought to act as the nuclease *(30)*. A damage-containing oligonucleotide of between 24 and 32 nucleotides is displaced, probably by the repair factors themselves *(17)*. RPA protects the resulting gap from further nucleolytic resection. Resynthesis across the gap by DNA polymerase δ and/or ε plus proliferating cell antigen (PCNA) and replica-tion factor C (RF-C), followed by DNA ligase I sealing, completes the repair process *(17,31)*.

An additional feature of NER is that actively transcribed regions of the genome are repaired more rapidly, which can be attributed to preferential repair of the transcribed strand *(16,17)*. Current models for this process posit that elongating RNA polymerase II is obstructed by a lesion *(32)*. The presence of this stalled complex, which is associated with an already open "bubble" DNA structure around the lesion, permits recruitment of NER factors, bypassing the requirement for XPC–hHR23B in damage recognition *(17,33)*.

2.2. Interstrand Crosslink Incision Reactions

In both bacteria and yeast, there is evidence that a full, or near-normal, complement of the NER proteins act to incise ICLs, but it is beyond the scope of this chapter to review this material *(see* refs. *34–41)*. In contrast, an increasing body of evidence suggests that for mammalian cells, this is not always the case.

A recurrent observation concerning the mechanism of ICL repair in mammalian cells is the extreme sensitivity of many *XPF* and *ERCC1* defective cell lines to crosslinking agents, compared to cells bearing mutations in other components of the NER apparatus *(4,42,43)*. If the UV sensitivity of *XPF* or *ERCC1* rodent cells is compared to that of *XPG* cells, they are a similar order of magnitude *(16)*, which is expected because they are both required for the full excision of photoproducts. It is therefore very striking that for agents that induce ICLs (and these agents only), quite disparate sensitivity is often observed in such mutants. There are, however, notable exceptions *(44,45)*.

An extra dimension in the interpretation of this observation arises from the increasing body of evidence supporting a role for the XPF–ERCC1 heterodimer in repair processes other than NER. In yeast, it is well established that the homologous complex (the Rad1–Rad10 heterodimer) is required for certain types of mitotic homologous recombination—in particular single-strand annealing (SSA) *(46–48)*. This can occur when a chromosomal double-strand break (DSB) is flanked by homologous regions on the broken chromosome, and in mammalian genomes where there is a high proportion of repetitive DNA (e.g., *Alu* repeats), this may be a favorable recombination repair event (although there is no compelling evidence for this in human cells). SSA is initiated by resection of the DSB ends in the 5'-to-3' direction, allowing the complementary 3' ends to anneal in the regions of homology *(48)*. The overhanging 3' ends are removed and ligation completes repair *(46,48)*. Significantly, the structure-specific endonuclease required for the 5' incision in NER, the Rad1–Rad10 heterodimer, is the activity responsible for removing these 3' tails *(46)* (as in NER, it cleaves duplex DNA on the 5' side of substrates containing a single-stranded to double-stranded DNA junction *[49]*). There is now some evidence that the XPF–ERCC1 dimer plays a role during some mammalian intrachromosomal homologous recombination reactions and targeted gene replacement recombination *(50–53)*. Hence, it is possible that the ERCC1–XPF nuclease plays a role additional to damage incision during ICL repair. To account for this, it has been suggested that the primary recombination event during ICL repair is ERCC1–XPF dependent, perhaps SSA.

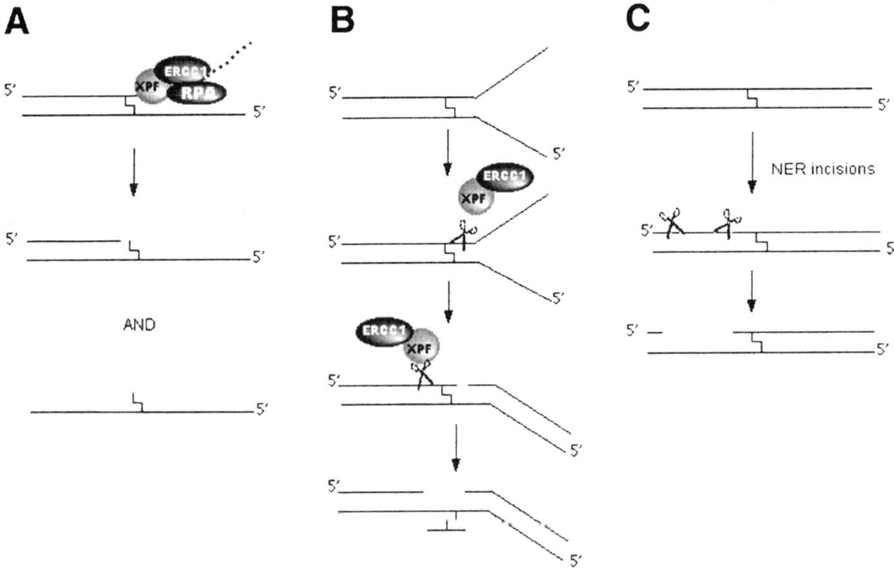

Fig. 1. Three reactions at psoralen-induced ICLs recently identified biochemically:
(**A**) Mu et al. have demonstrated that the XPF–ERCC1 heterodimer, in the presence of
RPA, degraded DNA bearing a site-specific psoralen crosslink in the 3'-to-5' direction.
Degradation may terminate just past the crosslink or continue beyond. (**B**) Kuraoka *et al*
showed that when a site-specific psoralen ICL was placed 4–6 bp from an unpaired 3' tail,
the XPF–ERCC1 nuclease was able to cleave on either side of the ICL, with the 3' side
cleaved first. (**C**) Bessho et al. demonstrated that the normal complement of NER factors
was able to incise DNA bearing a site-specific psoralen ICL on its 5' side on one strand only.

These possibilities have recently begun to be examined in a number of studies,
both genetic and biochemical. The accumulating evidence does favor a special
role for XPF–ERCC1 in ICL incision events under some circumstances and does
not exclude an important activity in recombinational repair of ICLs.

Studies of the ability of NER-competent and mutant mammalian cells to incise
and uncouple nitrogen mustard ICLs in vivo indicate that although *XPG, XPB,*
and *XPD* mutant cells all release the ICLs normally, this step is completely
eliminated in *XPF* and *ERCC1* cells *(4)*. Two recent biochemical studies also
demonstrated that these two activities alone are capable of processing psoralen
ICLs under some circumstances. In the first of these reports (*see* Fig. 1A) the
mechanism appears to be exonucleolytic, involving 3'-to-5' degradation of the
DNA toward the ICL by purified XPF–ERCC1 in conjunction with RPA *(54)*.
These workers found that the degradation was strongly attenuated by the
crosslink, but in a few cases, it was able to digest the entire DNA strand associated with the initial incision. There was no preference for the pyrone-adducted
strand over the furan side for these reactions, and the reaction was ICL-specific

because these products were not generated in a control substrate containing a psoralen monoadduct.

The second study employed purified XPF–ERCC1 and indicated that these two proteins alone are capable of the endonucleolytic incision of psoralen ICLs (*see* Fig. 1B) *(55)*. Importantly, this particular reaction could only occur efficiently when a specific DNA structure, an unpaired 3' tail region, was present adjacent to the ICL; hence, the DNA may have to be unwound in this manner prior to ERCC1–XPF incisions. The authors' suggest that such structures might arise frequently in growing cells because of DNA helicase activity or, perhaps more significantly, at stalled replication forks. Incisions were observed bracketing the furan side of the crosslink, leaving a small adducted oligonucleotide attached, produced by incision on the 3' side of the crosslink (at the third phosphodiester bond) and at the second (and some extent third) phosphodiester bond 5' to the crosslink on the same strand. It should be noted that the reaction described in this report was not ICL-specific because psoralen monoadducts could also be incised by XPF–ERCC1 in a similar manner in the 3'-unpaired-tail substrates used.

A single study has suggested one further possible crosslink-specific incision event (*see* Fig. 1C). Bessho et al. *(56)* employed Chinese hamster ovary cell extracts to examine incisions on defined psoralen crosslinked substrates. In contrast to monoadducted substrates, which, as expected, were subject to dual NER incisions bracketing the adduct, ICLs were subject to dual incisions (apparently involving the full complement of basal NER factors) 5' to the crosslink on one strand only with a strong preference for the pyrone side of the ICL. The XPG incision in this case was just one phosphodiester bond 5' to the ICL, whereas the XPF–ERCC1 incision was at the 27th phosphodiester bond 5' to the ICL. The authors speculate that this might be a signal for further processing of the crosslink, perhaps initiating recombination.

3. RECOMBINATION AND INTERSTRAND CROSSLINK REPAIR

Double-strand break formation has been reported in yeast cells treated with both nitrogen mustards and psoralen/UVA *(57–61)*. DSB formation is elevated in dividing cells *(59,61)*, suggesting that stalled replication forks might be the source of these breaks. Two reports published recently indicate that for both psoralen and nitrogen mustard ICLs, this process might be conserved in mammals. In the first report, pulsed field gel electrophoresis was used to monitor the induction and repair of DSBs in nitrogen mustard-treated Chinese hamster ovary (CHO) cells and high levels of DSBs were observed in dividing cells, but not in confluent cells *(4)*. It is striking that in both yeast and mammalian cells, these DSBs do not result from NER or ERCC1–XPF incisions because their frequency is not reduced in cells defective in these activities *(4,61)*. Another set of experiments demonstrated similar events in primary human fibroblasts (which have intact cell cycle checkpoints) bearing psoralen crosslinks, and in this case, the workers directly correlated the induction of DSBs with replication *(62)*. When cells were treated with sublethal levels of psoralen/UVA, no uncoupling of the induced ICLs was observed in cells treated in the G_1- or G_2-phase of the cell cycle. However, in the subsequent S-phase, the cells arrested and uncoupled the ICLs; this was associated with extensive chromosome breakage and apoptosis. Taken together, these studies strongly suggest that replication is important for ICL repair in mammalian cells, and replication-associated DSBs might be a trigger not only for recombination but also for incision reactions and cell death at higher doses. In this respect, the observations of Kuraoka et al. *(55)* are particularly interesting because the preferred XPF–ERCC1 incision substrate they identified might resemble a stalled replication-fork-associated DSB.

In *Escherichia coli*, recombination is known to follow ICL uncoupling, where RecA-mediated strand invasion into the gap created following UvrABC incisions provides the necessary genetic template information for the completion of ICL repair through the excision of the second strand adduct *(7,26,63)*. In contrast to the situation in eukaryotes, this does not involve any DSB intermediates *(34)*. However, DSBs are highly recombinogenic and, thus, it is possible that the DSBs induced in replicating eukaryotic cells initiate strand invasion reactions as the next stage in ICL repair following incision. A comparison of the DSB repair capabilities of *XRCC2* and *XRCC3* mutant cells (members of the mammalian *RAD51* group of homologous recombination genes) to that of *XRCC5* cells (Ku80 and, therefore, nonhomolgous end-joining [NHEJ] defective) following nitrogen mustard treatment indicated that there is a strong requirement for homologous recombination, but not NHEJ, during the repair of the ICL-associated DSBs *(4)*. Therefore, conserved from bacteria and yeast, to mammals, homology-driven events appear responsible for the processing of ICL-associated recombination intermediates in cells. Apart from a requirement for *XRCC2* and *XRCC3*, there is little information on which other components of the mammalian homology-driven recombination apparatus act at crosslinks and their repair intermediates. Because there has been an explosion in identifying such genes over the last few years *(64)*, information should be forthcoming in the near future. There are reports of cells defective for the vertebrate homologs of the yeast Rad51 *(65)* and Rad54 *(66)* and, of particular clinical relevance, BRCA1 defective cells *(67,68)*, being sensitive to crosslinking agents.

One explanation proposed for the hypersensitivity of *ERCC1* and *XPF* mutant cells is that ERCC1–XPF-dependent homologous recombination sub-pathways are strongly favored during ICL repair (*see* Subheading 2.2.). This is not proven in any functional studies and several lines of evidence suggest that we should be cautious in assuming that ICL-associated recombination events are highly

XPF–ERCC1 dependent. First, genetic studies in yeast indicate that recombination repair dependent on Rad1–Rad10 (the yeast homologs of XPF and ERRC1) is only one of several pathways into which DSBs are channelled and exists in competition with other recombination subpathways *(46,47)*. There is currently no reason to suppose that ICL-associated events would be absolutely dependent on any one subpathway. Second, recent studies have demonstrated that if a plasmid containing a single ICL is coincubated with an undamaged plasmid in the presence of mammalian cell extracts extensive DNA synthesis occurs *(69,70)*. This DNA synthesis is dependent on the homologous recombination proteins XRCC2 and XRCC3 and on XPF and ERCC1 *(70)* as well as PCNA and RPA *(71)*. This suggests that the ICL is strongly recombinogenic and that following release of the ICL extensive DNA synthesis occurs. Most crucially, this DNA synthesis appears to be largely independent of homology between the two plasmids. This mode of exchange is most reminiscent of break-induced replication (BIR) *(72)*. It would be premature to state that BIR is a mode of recombination induced by ICLs to complete repair, but this illustrates that the ERCC1–XPF-dependent recombination model alone may not explain the hypersensitivity of *XPF* and *ERCC1* cells. Perhaps the extreme sensitivity observed in *XPF/ERCC1* mutant cells stems from a combination of ICL incision defects and also reduced levels of some recombination events.

4. TRANSLESION POLYMERASES AND ICL REPAIR

In organisms that can propagate as haploids, such as yeast and *E. coli*, nondividing cells possess a single copy of each chromosome and are, therefore, unable to obtain the genetic information required to complete ICL repair by recombination. Evidence from the work of Berardini and colleagues *(73,74)*, and our own laboratory *(61)* illustrate how these two organisms might cope in this situation. In both cases, the data are consistent with a specialized DNA polymerase copying past the incised nitrogen mustard ICL repair intermediate, filling in the gap left following incision. This substitutes for recombination and supplies the template information, allowing the subsequent second-strand excision reaction. In the case of *E. coli,* this appears to involve DNA polymerase II *(74)* and, in yeast, the translesion synthesis polymerase ζ *(61)*. Many novel translesion synthesis polymerases have been identified in organisms from bacteria through to humans in recent years *(75)*, and genetic data from both *E. coli* and yeast suggest that we are yet to identify the full complement of activities that achieve translesion synthesis at ICL repair intermediates *(61,74)*. These enzymes might contribute to human ICL repair by providing genetic information following initial crosslink incision, in a manner analogous to the *E. coli* polymerase II and yeast polymerase ζ proteins. In this respect, there is a single, recent, report that indicates this may occur in mammalian cells *(76)*. Interestingly, the suggested pathway appears to be dependent on an intact NER apparatus (not just XPF and ERCC1), suggesting that an NER excision reaction might precede translesion synthesis, perhaps that suggested by Bessho and colleagues. This study also indicated that the translesion polymerase encoded by the *XPV* gene (polymerase η) *(77)* is not required for recombination-independent ICL repair/tolerance, suggesting that, like in yeast, polymerase ζ might be favored in mammals.

5. A MODEL FOR INTERSTRAND CROSSLINK REPAIR IN MAMMALIAN CELLS

Any model for ICL repair in mammalian cells must, currently be extremely tentative, but it is useful to present the various possibilities suggested by our present genetic and biochemical knowledge as the basis for further discussion. Although a good deal of work suggests that replication might be important in crosslink repair (*see* Subheading 4), it may not be a necessity. Based on what is known of crosslink-specific incision reactions, replication-induced recombination elicited by DSBs, and also the potential requirement for translesion polymerases in ICL repair/tolerance we present a model that has two routes (*see* Fig. 2). The first route is favored when recombination is possible, and it is most favorable in all eukaryotes during or following replication when a sister chromatid is present *(78)*. The alternative route is favored when such a preferred recombination substrate is not available (for instance, in the G_1-phase of the cell cycle) and the genetic information required to complete repair might be, at least partly, provided in a nonrecombinational manner. This is suggested to be through translesion synthesis.

In the recombinational branch, the approach of the DNA synthetic machinery toward a crosslink leads to disintegration of the replication fork, possibly associated with the formation of a DSB *(79)*. The resulting structure could be a substrate for an XPF–ERCC1 incision reaction, perhaps one of those detailed in Subheading 2.2. Resection of the incision gap, by unknown mechanisms, facilitates Rad51-driven strand exchange into the gapped incised ICL site, and an associated DSB might act to stimulate this recombination reaction. RAD51/XRCC2 complexes have recently been shown to be able to stimulate strand exchange into such gaps *(80)*, albeit not associated with incised ICLs. This recombination event places information opposite the adducted strand, which can then be excised and filled in.

The nonrecombinational alternative does not require the presence of a DSB, but probably does require that the ICL is incised in some way prior to translesion synthesis. It has been suggested that the full NER apparatus is required to initiate nonrecombinational crosslink repair, and it is possible that the incisions 5' to the ICL identified by Bessho et al. *(56)* initiates this pathway. As these authors proposed, the initial incised substrates could be further degraded to produce a gap that might then be filled in by a translesion polymerase, perhaps polymerase ζ.

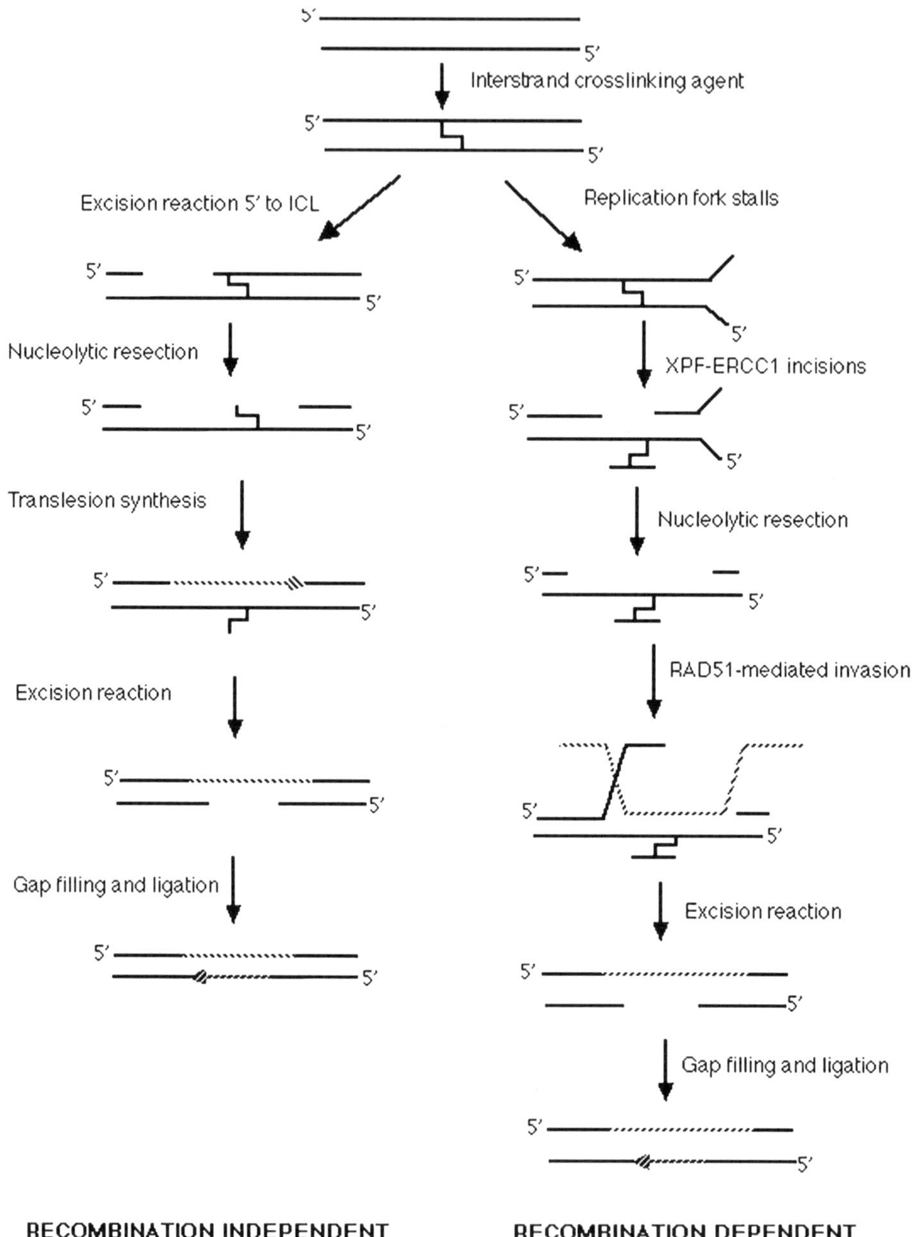

Fig. 2. Proposed model of ICL repair as applied to human cells. See text for detailed description of the processes shown.

It should be possible to test many aspects of this model. For instance, the action of purified DNA translesion polymerases on defined ICL/incised ICL substrates should determine whether the nonrecombinational pathway is feasible and if incision is an essential prerequisite to this. In addition the use of defined crosslinked substrates in conjunction with purified NER factors and recombination initiation factors such as RAD51 should shed light on the order of events in the recombination branch of repair.

6. ADDITIONAL FACTORS INVOLVED IN INTERSTRAND CROSSLINK REPAIR

6.1. The SNM1/PSO2 Family

The Saccharomyces cerevisiae SNM1 or PSO2 gene was simultaneously and independently identified by workers aiming to isolate novel genes required for the repair of DNA ICLs produced by nitrogen mustard and psoralen/UVA treatment (*SNM1* = sensitivity to nitrogen mustard, *PSO2* = psoralen sensitive) *(81–83)*. In contrast to several other mutants identified in the original screens, cells defective in this gene were uniquely sensitive to agents that induce ICLs (nitrogen and sulfur mustard, cisplatin, triaziquone, mitomycin C and crosslinking psoralens), but they were similarly sensitive to ionizing radiation and UVC as the parent strains from which they were derived *(81,82)*. Little is known about the role the product of this gene plays in the repair of ICLs, but the recent realization that there is a mouse homolog (*mSNM1*) and one further paralog (*mSNM1B*) *(84)*, a human homolog, and two further possible paralogs (*hSNM1A, hSNM1B* and *hSNM1C*) *(84,85)* emphasizes the importance of elucidating its role in DNA ICL repair. It has recently been shown that *SNM1* null mouse embryonic stem cells are sensitive to the crosslinking drug mitomycin C, as are the *mSNM1 /–* animals *(84)*. These mice are viable and fertile, and had no apparent developmental abnormalities. Importantly, the human *SNM1A* gene is able to functionally complement the mouse embryonic stem (ES) *SNM1–/–* cells for mitomycin C sensitivity *(84)*.

The yeast gene is located on chromosome XIII and encodes a 76-kDa protein bearing a single putative zinc-finger motif that is dispensable for its repair activity *(86)*. Antibodies have been raised to the Snm1 protein, and its nuclear localization confirmed. Constitutive transcription of *SNM1* leads to a steady state level of only 0.3 transcripts per cell *(87)*, however, the gene is induced about fourfold by a variety of DNA damaging agents including UV and monofunctional alkylating agents, as well as crosslinking agents *(88,89)*. *snm1* mutant yeast cells are able to incise crosslinks with normal efficiency *(90,91)*, but accumulate unrepaired DSBs in response to this damage *(91)*, leading to speculation that the Snm1 protein might play some role in coordinating intermediate steps between excision and recombination steps. Analysis of the predicted structure of

Snm1 has placed in it the metallo-β-lactamase superfamily of proteins that have diverse roles in cellular metabolism *(92)*. However, it is interesting to note that within the metallo-β-lactamase superfamily a protein involved in nucleolytic processing of the 3' termini of nascent mRNAs (CPSF-A) shares a similar domain with *SNM1/PSO2*, raising the possibility that these genes encode a novel nuclease activity *(92)*. Membership of the metallo-β-lactamase superfamily was the criterion used to ascribe two new human paralogs to the emerging *SNM1* group of genes. The first, *ELAC2 (SNM1C) (93)*, is a prostate cancer susceptibility gene located at chromosome 17p, whereas the second, *ARTEMIS (94)*, is mutated in a rare radiosensitive form of severe combined immune deficiency (RS-SCID). It is tempting to speculate that, together, these genes define a family of new DNA repair and/or genome stability factors.

6.2. Fanconi Anemia Genes

Eight distinct Fanconi anemia (FA) complementation groups have been identified (*FANCA, FANCB, FANC, FANCD1, FANCD2, FANCE, FANCF,* and *FANCG*), and of these six genes have been cloned (FANCB and *FANCD1* are the exceptions) *(95–97)*. FA is an autosomal recessive cancer susceptibility syndrome, and cells from affected individuals are highly sensitive to crosslinking agents *(95–97)*, but only slightly ionizing radiation sensitive *(98)*. Although there is evidence to support a role for FA proteins in cell cycle control, the apoptotic response, or oxygen radical detoxification *(96,97)*, there is increasing belief that they play a role in DNA repair, which is especially pronounced for crosslinking agents *(95)*. Several groups have reported that a multisubunit complex containing at least FANCA, FANCG, FANC, FANCE, and FANCF is presented to the nucleus *(99)* and becomes associated with chromatin following crosslinking DNA damage *(100)*. It has been shown that in irradiated cells, the FANC complex monoubiquitinates and activates the FANCD2 protein, resulting in the colocalization of FANCD2 with BRCA1 at ionizing radiation-induced nuclear foci and in meiotic synaptonemal complexes *(101)*. This suggests a scenario whereby the FANC proteins might influence the repair and genome stability functions of BRCA1. Given the association of BRCA1 with human recombination factors such as RAD51 *(64)*, it is possible that the FANC complex is involved in regulating the recombination repair arm of the ICL response, but no data are yet available to confirm this.

6.3. A Family of Novel Helicases

Over 10 yr, Harris et al. characterized a nitrogen mustard-sensitive *Drosophilla* mutant identified in a screen for strains defective in deoxyribonucleases *(102)*. Subsequent characterization of the mutant allele revealed that the gene affected encoded a novel combined DNA helicase–polymerase, *MUS308 (103)*. Sequence analysis suggested that the C-terminal of this protein encodes a polymerase, and the helicase domain is situated in the N-terminus. A putative polymerase that likely represents the protein product of the *MUS308* gene has been purified and shown to be associated with an ATPase and 3'-to-5' exonuclease activity *(104)*. As for the *SNM1/PSO2* gene, mutations in this locus conferred sensitivity to bifunctional alkylating agents and psoralen/UVA, but not monofunctional agents such as methyl methanesulfonate. Further experiments suggested that *mus308* mutants could eliminate UV induced pyrimidine dimers normally and that the helicase acts at a stage following the initial incision of ICLs *(104)*.

Recently, human and mouse homologs of the *MUS308* helicase domain (with no associated polymerase) *(HEL308) (105)* and a separate protein homologous to just the polymerase domain *(POLQ) (106)* have been identified. *HEL308* encodes a single-stranded DNA-dependent ATPase and helicase, which translocates along DNA with a 5'-to-3' polarity that can displace short duplex oligonucleotides *(105)*. The addition of RPA stimulates this activity, allowing longer substrates to unwind. In *E. coli,* it has been established that the 5' nuclease activity of DNA polymerase I cooperates with UvrD (also a 3'-to-5' helicase) to generate a gap at the site of the initial crosslink incisions, allowing RecA to initiate the subsequent strand-exchange reaction *(36)*. Marini and Wood *(105)* point out that the presence of a homologous helicase and polymerase *(HEL308* and *POLQ)* in separate subunits or in a single contiguous peptide *(MUS308)* could define a step analogous to the activities of polymerase I and UvrD in crosslink repair.

7. INTERSTRAND CROSSLINK REPAIR AND FUTURE ANTICANCER THERAPIES

As our understanding of the complex molecular mechanisms involved in the repair of ICLs in human cells and the critical determinants of cellular sensitivity to damage of this type increases, so does the potential to develop sensitive screens to predict clinical response. Key proteins involved in this specific process can also be considered as novel therapeutic targets, whose inhibition could increase sensitivity to crosslinking drugs in tumors normally inherently resistant or which have acquired resistance following initial therapy. Novel crosslinking agents continue to be developed in an attempt to produce more selective, less toxic drugs. Agents that produce crosslinks in the minor groove of DNA are of particular interest *(107)*. For example, the novel pyrrolobenzodiazepine dimer SJG-136 binds in the minor groove of DNA spanning six base pairs with a preference for binding 5'-purine-GATC-pyrimidine-3' sequences *(108)*. ICL occurs between the two guanine N2 positions and produces minimal distortion of the normal DNA structure. As a result, it appears to evade the recognition and repair mechanisms used for the processing of the distorting crosslinks produced in the major groove by conventional drugs *(109)*. This highly potent agent has significant antitumor activity in animal models and is currently in preclinical development. It remains to be established whether agents of this type will find a role in the clinic as a systemic therapy or the production of such potent and difficult to repair crosslinks will be more appropriate to a targeted approach such as antibody-directed or gene-directed prodrug therapy (ADEPT and GDEPT, which are currently using prodrugs based on conventional crosslinking agents) *(110,111)*.

REFERENCES

1. Hartley JA. Alkylating agents. In: Souhami RL, Tannock I, Hohenberger P, et al., eds. *Oxford Textbook of Oncology,* 2nd ed., Oxford University Press, Oxford, 2002:639–654.
2. Kohn KW, Spears CL, Doty P. Interstrand crosslinking of DNA by nitrogen mustard. *J Mol Biol* 1966;19:266–288.
3. Lawley PD, Brookes P. Interstrand crosslinking of DNA by difunctional alkylating agents. *J Mol Biol* 1967;25:143–160.
4. De Silva IU, McHugh PJ, Clingen PH, et al. Defining the roles of nucleotide excision repair and recombination in the repair of DNA interstrand crosslinks in mammalian cells. *Mol Cell Biol* 2000;20:7980–7990.
5. O'Connor PM, Kohn KW. Comparative pharmacokinetics of DNA lesion formation and removal following treatment of L1210 cells with nitrogen mustards. *Cancer Commun* 1990;2(12):387–394.
6. Trimmer EE, Essigmann JM. Cisplatin. *Essays Biochem* 1999;34:191–211.
7. Dronkert ML, Kanaar R. Repair of DNA interstrand crosslinks. *Mutat Res* 2001;486:217–247.
8. Torres-Garcia S.J., Cousineau L., Caplan S., et al. Correlation of resistance to nitrogen mustards in chronic lymphocytic leukemia with enhanced removal of melphalan-induced DNA crosslinks. *Biochem Pharmacol* 1989;38:3122–3123.
9. Panasci L, Paiement JP, Christodoulopoulos G, et al. Chlorambucil drug resistance in chronic lymphocytic leukemia: the emerging role of DNA repair. *Clin Cancer Res* 2001;7:454–461.
10. Wang ZM, Chen ZP, Xu ZY, et al. In vitro evidence for homologous recombinational repair in resistance to melphalan. *J Natl Cancer Inst* 2001;93:1473–1478.
11. Spanswick VJ, Craddock C, Sekhar M, et al. Repair of DNA interstrand crosslinks as a mechanism of clinical resistance to melphalan in multiple myeloma. *Blood* 2002;100:224–229.
12. Rudd GN, Hartley,JA, Souhami RL. Persistence of cisplatin-induced DNA interstrand crosslinking in peripheral blood mononuclear cells from elderly and young individuals. *Cancer Chemother Pharmacol* 1995;35:323–326.
13. Ojwang JO, Grueneberg DA, Loechler EL. Synthesis of a duplex oligonucleotide containing a nitrogen mustard interstrand DNA-DNA crosslink. *Cancer Res* 1989;49:6529–6537.
14. Tomasz M. Mitomycin C. Small, fast and deadly (but very selective). *Chem Biol* 1995;2:575–579.
15. Ben-Hur E, Song PS. The photochemistry and photobiology of furocoumarins (psoralens). *Adv Radiat Biol* 1984;11:131–171.
16. Friedberg, EC, Walker, GC, Siede, W. DNA Repair and Mutagenesis. American Society for Microbiology, Washington, D.C, 1995.
17. de Laat, W, Jaspers, NGJ, Hoeijmakers, JHJ. Molecular mechanism of nucleotide excision repair. *Genes Dev* 1999;13:768–785.
18. Wood, RD. DNA damage recognition during nucleotide excision repair in mammalian cells. *Biochimie* 1999;81:39–44.
19. Batty, D, Wood, RD. Damage recognition in nucleotide excision repair of DNA. *Gene* 2000,241:193–204.

20. Sugasawa, K, Ng, JM, Masutani, C, et al. Xeroderma pigmentosum group C protein complex is the initiator of global genome nucleotide excision repair. *Mol Cell* 1998;2:223–232.

21. Jones CJ, Wood RD. Preferential binding of the xeroderma pigmentosum group A complementing protein to damaged DNA. *Biochemistry* 1993;32:12,096–12,104.

22. Kusumoto, R., Masutani, C., Sugasawa, K., et al. Diversity of the damage recognition step in the global genomic nucleotide excision repair *in vitro*. *Mutat Res* 2001;485:219–227.

23. Wakasugi M, Sancar A. Order of assembly of human DNA repair excision nuclease. *J Biol Chem* 1999;274:18,759–18,768.

24. Evans E, Moggs JG, Hwang JR, et al. Mechanism of open complex and dual incision formation by human nucleotide excision repair factors. *EMBO J* 1997;16:6559–6573.

25. Wakasugi M, Sancar A. Assembly, subunit composition, and footprint of human DNA repair excision nuclease. *Proc Natl Acad Sci USA* 1998;95:6669–6674.

26. Sung P, Bailly V, Weber C, et al. Human xeroderma pigmentosum group D gene encodes a DNA helicase. *Nature* 1993;365:852–855.

27. Ma L, Siemssen ED, Noteborn HM, et al. The xeroderma pigmentosum group B protein ERCC3 produced in the baculovirus system exhibits DNA helicase activity. *Nucleic Acids Res* 1994;22:4095–5102.

28. O'Donovan A, Davies AA, Moggs JG, et al. XPG endonuclease makes the 3' incision in human DNA nucleotide excision repair. *Nature* 1994;371:432–435.

29. Sijbers AM, de Laat WL, Ariza RR, et al. Xeroderma pigmentosum group F caused by a defect in a structure-specific DNA repair endonuclease. *Cell* 1996;86:811–822.

30. McCutchen-Maloney SL, Giannecchini CA, Hwang MH, et al. Domain mapping of the DNA binding, endonuclease, and ERCC1 binding properties of the human DNA repair protein XPF. *Biochemistry* 1999;38:9417–9425.

31. Shivji MK, Podust VN, Hubscher U, et al. Nucleotide excision repair DNA synthesis by DNA polymerase epsilon in the presence of PCNA, RFC, and RPA. *Biochemistry* 1995;34:5011–5017.

32. Donahue BA, Yin S, Taylor JS, et al. Transcript cleavage by RNA polymerase II arrested by a cyclobutane pyrimidine dimer in the DNA template. *Proc Natl Acad Sci USA* 1994;91:8502–8506.

33. Mu D, Sancar A. Model for XPC-independent transcription-coupled repair of pyrimidine dimers in humans. *J Biol Chem* 1997;272:7570–7573.

34. Cole RS. Repair of DNA containing interstrand crosslinks in *Escherchia coli:* sequential excision and recombination. *Proc Natl Acad Sci USA* 1973;70:1064–1068.

35. Van Houten B, Gamper, H, Holbrook, SR, et al. Action mechanism of ABC excision nuclease on a DNA substrate containing a psoralen crosslink at a defined position. *Proc Natl Acad Sci USA* 1986;83:8077–8081.

36. Sladek FM, Munn MM, Rupp WD, et al. In vitro repair of psoralen-DNA crosslinks by RecA, UvrABC, and the 5'-exonuclease of DNA polymerase I. *J Biol Chem* 1989;264:6755–6765.

37. Cheng, S, Sancar, A, Hearst, J.E. RecA-dependent incision of psoralen-crosslinked DNA by (A)BC excinuclease. *Nucleic Acids Res* 1991;19:657–663.

38. Jachymczyk WJ, von Borstel RC, Mowat MR, et al. Repair of interstrand crosslinks in DNA of *Saccharomyces cerevisiae* requires two systems for DNA repair: the *RAD3* system and the *RAD51* system. *Mol Gen Genet* 1981;182:196–205.

39. Miller RD, Prakash L, Prakash S. Genetic Control of Excision of *Saccharomyces cerevisiae* interstrand crosslinks induced by psoralen plus near-UV light. *Mol Cell Biol* 1982;2:939–948.

40. Dalhardon M, Averbeck D. Pulsed field electrophoresis of the repair of psoralen plus UVA induced DNA photoproducts in *Saccharomyces cerevisiae. Mutat Res* 1995;336:49–60.

41. McHugh PJ, Gill RD, Waters R, et al. Excision repair of nitrogen mustard-DNA adducts in *Saccharomyces cerevisiae. Nucleic Acids Res* 1999;27:3259–3266.

42. Damia G, Imperatori L, Stefanini M, et al. Sensitivity of CHO mutant cell lines with specific defects in nucleotide excision repair to different anti-cancer agents. *Int J Cancer* 1996;66:779–783.

43. Hoy CA, Thompson LH, Mooney CL, et al. Defective DNA crosslink removal in Chinese hamster cell mutants hypersensitive to bifunctional alkylating agents. *Cancer Res* 1985;45:1737–1743.

44. Kaye J, Smith CA, Hanawalt PC. DNA repair in human cells containing photoadducts of 8-methoxypsoralen or angelicin. *Cancer Res* 1980;40;696–702.

45. Vuksanovic L, Cleaver JE. Unique crosslink and monoadduct repair characteristics of a xeroderma pigmentosum revertant cell line. *Mutat Res* 1987;184:255–263.

46. Paques F, Haber JE. Multiple pathways of recombination induced by double-strand breaks in *Saccharomyces cerevisiae*. *Microb Mol Biol Rev* 1999;63:349–404.

47. Haber JE. Lucky breaks: analysis of recombination in *Saccharomyces*. *Mutat Res* 2000;451:53–69.

48. Fishman-Lobell J, Haber JE. Removal of nonhomologous DNA ends in double-strand break recombination: the role of the yeast ultraviolet repair gene *RAD1*. *Science* 1992;258:480–484

49. de Laat WL, Appeldoorn E, Jaspers NG, et al. DNA structural elements required for ERCC1–XPF endonuclease activity. *J Biol Chem* 1998;273:7835–7842.

50. Sargent RG, Rolig RL, Kilburn AE, et al. Recombination-dependent deletion formation in mammalian cells deficient in the nucleotide excision repair gene *ERCC1*. *Proc Natl Acad Sci USA* 1997;94:13,122–13,127.

51. Adair GM, Rolig RL, Moore-Faver D, et al. Role of ERCC1 in removal of long non-homologous tails during targeted homologous recombination. *EMBO J* 2000;19:5552–5561.

52. Sargent RG, Meservy JL, Perkins BD, et al. Role of the nucleotide excision repair gene *ERCC1* in formation of recombination-dependent rearrangements in mammalian cells. *Nucleic Acids Res* 2000;28:3771–3778.

53. Niedernhofer LJ, Essers J, Weeda G, et al. The structure-specific endonuclease Ercc1-Xpf is required for targeted gene replacement in embryonic stem cells. *EMBO J* 2001;20:6540–6549.

54. Mu D, Bessho, T, Nechev, LV, et al. DNA interstrand crosslinks induce futile repair synthesis in mammalian cell extracts. *Mol Cell Biol* 2000;20:2446–2454.

55. Kuraoka, I, Kobertz, WR, Ariza, RR, et al. Repair of an interstrand DNA crosslink initiated by ERCC1–XPF repair/recombination nuclease. *J Biol Chem* 2000;275:26,632–26,636.

56. Bessho T, Mu D, Sancar A. Initiation of DNA interstrand crosslink repair in humans: the nucleotide excision repair system makes dual incisions 5' to the crosslinked base and removes a 22-to-28-nucelotide long damage-free strand. *Mol Cell Biol* 1997;17:6822–6830.

57. Jachymczyk W, Van Borstel RC, Mowat MRA, et al. Repair of interstrand crosslinks in DNA of *Saccharomyces cerevisiae* requires two systems for DNA repair: the *RAD3* system and the *RAD51* system. *Mol Gen Genet* 1981;182:196–205.

58. Magana-Schwencke N, Henriques J-AP, Chanet R, et al. The fate of 8-methoxypsoralen photoinduced crosslinks in nuclear and mitochondrial yeast DNA: comparison of wild-type and repair-deficient strains. *Proc Natl Acad Sci USA* 1982;79:1722–1726.

59. Dalhardon M, Averbeck D. Pulsed field electrophoresis of the repair of psoralen plus UVA induced DNA photoproducts in *Saccharomyces cerevisiae*. *Mutat Res* 1995;336:49–60.

60. Dardalhon M, de Massy MB, Nicolas A, et al. Mitotic recombination and localized DNA double-strand breaks are induced after 8-methoxypsoralen and UVA irradiation in *Saccharomyces cerevisiae*. *Curr Genet* 1998;43:30–42.

61. McHugh PJ, Sones WR, Hartley JA. Repair of intermediate structures produced at DNA interstrand crosslinks in *Saccharomyces cerevisiae*. *Mol Cell Biol* 2000;20:3425–3433.

62. Akkari YM, Bateman RL, Reifsteck CA, et al. DNA replication is required to elicit cellular responses to psoralen-induced DNA interstrand crosslinks. *Mol Cell Biol* 2000;20:8283–8289.

63. McHugh PJ, Spanswick VJ, Hartley JA. Repair of DNA interstrand crosslinks: molecular mechanisms and clinical relevance. *Lancet Oncol* 2001;483–490.

64. van Gent DC, Hoeijmakers JH, Kanaar R. Chromosomal stability and the DNA double-stranded break connection. *Nat Rev Genet* 2001;2:196–206.

65. Takata M, Sasaki MS, Sonoda E, et al. The Rad51 paralog Rad51B promotes homologous recombinational repair. *Mol Cell Biol* 2000;20:6476–6482.

66. Essers J, Hendriks RW, Swagemakers SM, et al. Disruption of mouse RAD54 reduces ionizing radiation resistance and homologous recombination. *Cell* 1997;89:195–204.

67. Husain A, He G, Venkatraman ES, et al. BRCA1 up-regulation is associated with repair-mediated resistance to cis-diamminedichloroplatinum(II). *Cancer Res* 1998;58:1120–1123.

68. Bhattacharyya A, Ear US, Koller BH, et al. The breast cancer susceptibility gene BRCA1 is required for subnuclear assembly of Rad51 and survival following treatment with the DNA crosslinking agent cisplatin. *J Biol Chem* 2000;275:23,899–23,903.

69. Zhang N, Zhang X, Peterson C, et al. Differential processing of UV mimetic and interstrand crosslink damage by XPF cell extracts. *Nucleic Acids Res* 2000;28:4800–4804.

70. Li L, Peterson CA, Lu X, et al. Interstrand crosslinks induce DNA synthesis in damaged and undamaged plasmids in mammalian cell extracts. *Mol Cell Biol* 1999;19:5619–5630.

71. Li L, Peterson CA, Zhang X, et al. Requirement for PCNA and RPA in interstrand crosslink-induced DNA synthesis. *Nucleic Acids Res* 2000;28:1424–1427.

72. Malkova A, Ivanov EL, Haber JE. Double-strand break repair in the absence of *RAD51* in yeast: a possible role for break-induced DNA replication. *Proc Natl Acad Sci USA* 1996;93:7131–7136.

73. Berardini M, Mackay W, Loechler, EL. A site-specific study of a plasmid containing a single nitrogen mustard crosslink: evidence for a second, recombination-independent pathway for the DNA repair of interstrand crosslinks. *Biochemistry* 1997;36:303–313.

74. Berardini M, Foster PL, Loechler EL. DNA polymerase II (*polB*) is involved in a new DNA repair pathway for DNA interstrand crosslinks in *Escherichia coli*. *J Bacteriol* 1999;181:2878–2882.

75. Livneh Z. DNA damage control by novel DNA polymerases: translesion replication and mutagenesis. *J Biol Chem* 2001;276:25,639–25,642.

76. Wang X, Peterson CA, Zheng H, et al. Involvement of nucleotide excision repair in a recombination-independent and error-prone pathway of DNA interstrand crosslink repair. *Mol Cell Biol* 2001;21:713–720.

77. Masutani C, Kusumoto R, Yamada A, et al. The XPV (xeroderma pigmentosum variant) gene encodes human DNA polymerase eta. *Nature* 1999;399:700–704.

78. Johnson RD, Jasin M. Sister chromatid gene conversion is a prominent double-strand break repair pathway in mammalian cells. *EMBO J* 2000;19:3398–3407.

79. Cox MM, Goodman MF, Kreuzer KN, et al. The importance of repairing stalled replication forks. *Nature* 2000;404:37–41.

80. Masson JY, Tarsounas MC, Stasiak AZ, et al. Identification and purification of two distinct complexes containing the five RAD51 paralogs. *Genes Dev* 2001;15:3296–3307.

81. Henriques JA, Moustacchi E. Isolation and characterization of *pso* mutants sensitive to photo-addition of psoralen derivatives in *Saccharomyces cerevisiae*. *Genetics* 1980;95:273–288.

82. Ruhland A, Kircher M, Wilborn F, et al. A yeast mutant specifically sensitive to bifunctional alkylation. *Mutat Res* 1981;91:457–462.

83. Cassier-Chauvat C, Moustacchi E. Allelism between *pso1-1* and *rev3–1* mutants and between *pso2–1* and *snm1* mutants in *Saccharomyces cerevisiae*. *Curr Genet* 1988;13:37–40.

84. Dronkert ML, de Wit J, Boeve M, et al. Disruption of mouse *SNM1* causes increased sensitivity to the DNA interstrand crosslinking agent mitomycin C. *Mol Cell Biol* 2000;20:4553–4561.

85. Demuth I, Digweed M. Genomic organization of a potential human DNA-crosslink repair gene, KIAA0086. *Mutat Res* 1998;409:11–16.

86. Haase E, Riehl D, Mack M, et al. Molecular cloning of *SNM1*, a yeast gene responsible for a specific step in the repair of crosslinked DNA. *Mol Gen Genet* 1989;218:64–71.

87. Richter D, Niegemann E, Brendel M. Molecular structure of the DNA crosslink repair gene *SNM1* (*PSO2*) of the yeast *Saccharomyces cerevisiae*. *Mol Gen Genet* 1992;231:194–200.

88. Angulo JF, Schwencke J, Fernandez I, et al. Induction of polypeptides in *Saccharomyces cerevisiae* after ultraviolet irradiation. *Biochem Biophys Res Commun* 1986;138(2):679–686.

89. Wolter R, Siede W, Brendel M. Regulation of *SNM1*, an inducible *Saccharomyces cerevisiae* gene required for repair of DNA crosslinks. *Mol Gen Genet* 1996;250:162–168.

90. Wilborn F, Brendel M. Formation and stability of interstrand crosslinks induced by *cis*- and *trans*-diamminedichloroplatinum (II) in the DNA of *Saccharomyces cerevisiae* strains differing in repair capacity. *Curr Genet* 1989;16:331–338.

91. Magana-Schwencke N, Henriques JA, Chanet R, et al. The fate of 8-methoxypsoralen photoinduced crosslinks in nuclear and mitochondrial yeast DNA: comparison of wild-type and repair-deficient strains. *Proc Natl Acad Sci USA* 1982;79:1722–1726.

92. Aravind L, Koonin EV. DNA-binding proteins and evolution of transcription regulation in the archaea. *Nucleic Acids Res* 1999;27:4658–4670.

93. Tavtigian SV, Simard J, Teng DH, et al. A candidate prostate cancer susceptibility gene at chromosome 17p. *Nature Genet* 2001;27:172–180.

94. Moshous D, Callebaut I, de Chasseval R, et al. Artemis, a novel DNA double-strand break repair/V(D)J recombination protein, is mutated in human severe combined immune deficiency. *Cell* 2001;105:177–186.

95. Grompe M, D'Andrea A. Fanconi anemia and DNA repair. *Hum Mol Genet* 2001;10:2253–2259.

96. D' Andrea AD, Grompe, M. Molecular biology of Fanconi anemia: implications for diagnosis and therapy. *Blood* 1997;90:1725–1736.

97. Joenje H, Patel KJ. The emerging genetic and molecular basis of Fanconi anaemia. *Nat Rev Genet* 2001;2:446–457.

98. Duckworth-Rysiecki G, Taylor AMR. Effects of ionising radiation on cells from Fanconi's anemia patients. *Cancer Res* 1985;45:416–420.

99. de Winter JP, van der Weel L, de Groot J, et al. The Fanconi anemia protein FANCF forms a nuclear complex with FANCA, FANCC and FANCG. *Hum Mol Genet* 2000;9:2665–2674.

100. Qiao F, Moss A, Kupfer GM. Fanconi anemia proteins localize to chromatin and the nuclear matrix in a DNA damage- and cell cycle-regulated manner. *J Biol Chem* 2001;276:23,391–23,396.

101. Garcia-Higuera I, Taniguchi T, Ganasan S, et al. Interaction of the Fanconi amemia proteins and BRCA1 in a common pathway. *Mol Cell* 2001;7:249–262.

102. Boyd JB, Sakaguchi K, Harris PV. mus308 mutants of *Drosophila* exhibit hypersensitivity to DNA crosslinking agents and are defective in a deoxyribonuclease. *Genetics* 1990;125:813–819.

103. Harris PV, Mazina OM, Leonhardt EA, et al. Molecular cloning of Drosophila mus308, a gene involved in DNA crosslink repair with homology to prokaryotic DNA polymerase I genes. *Mol Cell Biol* 1996;16:5764–5771.

104. Oshige M, Aoyagi N, Harris PV, et al. A new DNA polymerase species from *Drosophila melanogaster*: a probable mus308 gene product. *Mutat Res* 1999;433:183–192.

105. Marini F, Wood RD. A human DNA helicase homologous to the DNA crosslink sensitivity protein mus308. J Biol Chem 2001;277:8716–8723.

106. Harief FS, Vojta PJ, Ropp PA, et al. Cloning and chromosomal mapping of the human DNA polymerase theta (POLQ), the eighth human DNA polymerase. *Genomics* 1999;59:90–96.

107. Thurston DE. Nucleic acid targeting: therapeutic strategies for the 21st century. *Br J Cancer* 1999;80(Suppl 1):65–85.

108. Gregson SJ. Howard PW, Hartley JA, et al. Design, synthesis and evaluation of a novel pyrrolobenzodiazepine DNA-interactive agent with highly efficient crosslinking ability and potent cytotoxicity. *J Med Chem* 2001;44:737–748.

109. Hartley JA, Brooks N, McHugh PJ, et al. SJG-136 (NSC-D694501)—a novel DNA sequence specific minor groove crosslinking agent with significant antitumour activity. *Proc Am Assoc Cancer Res* 2000;41:Abs 2703.

110. Springer CJ, Niculescu-Duvaz I. Antibody-directed enzyme prodrug therapy (ADEPT): a review. *Adv Drug Deliv Rev* 1997;26:151–172.

111. Connors TA. The choice of prodrugs for gene directed enzyme prodrug therapy of cancer. *Gene Ther* 1995;2:702–709.

4

Chemosensitization to Platinum-Based Anticancer Drugs
Current Trends and Future Prospects

Bertrand J. Jean-Claude, PhD

CONTENTS

1. INTRODUCTION

Cis-diamminedichloroplatinum (cisplatin or *cis*-DDP) is one of the most potent agents used in the chemotherapy of many cancers, including testes, ovary, head, neck, and lung. Cisplatin shows considerable efficacy in the treatment of testicular cancers with cure rates of greater than 90% *(1)*. Despite its remarkable success in the treatment of cancer, its efficacy is limited by acquired or intrinsic resistance, and the mechanisms underlying chemoresistance are still under investigation. More importantly, novel strategies to reverse resistance and potentiate the antitumor action of cisplatin are actively being explored. Decreased cellular uptake and enhanced DNA repair activity are pointed to as the two major mechanisms of resistance to cisplatin *(2–6)*. In this chapter, we will focus on DNA repair-mediated resistance to platinum-based drug and on current strategies to increase their potencies in refractory tumors. We will cover the newly explored crosstalk between DNA repair mechanisms and cell signaling as a target for tumor cell sensitization to platinated adducts.

From: *Cancer Drug Discovery and Development: DNA Repair in Cancer Therapy*
Edited by: L. C. Panasci and M. A. Alaoui-Jamali © Humana Press Inc., Totowa, NJ

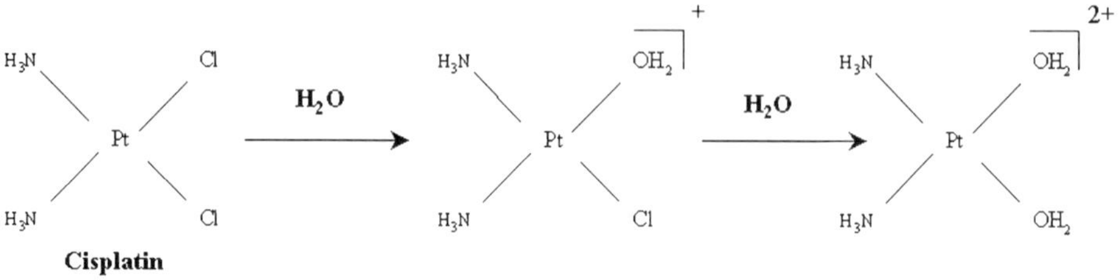

Cisplatin

Scheme 1

2. DNA ADDUCTS INDUCED BY CISPLATIN

Following cellular uptake, the chloride ligands of cisplatin are replaced by water molecules, leaving positively charged aquated species (*see* Scheme 1) that can react with biological nucleophiles (e.g., protein, RNA, and DNA) *(4,7,8)*. RNA and DNA adducts are believed to be primarily responsible for the cytocidal effects of platinum-based drugs *(9)*. The most abundant adducts detected in cells exposed to cisplatin are intrastrand crosslinks between two adjacent bases (1,2-d[GpG] and 1,2-d[ApG]) that represent approx 65% and 25%, respectively, of the total number of adducts *(4)*. Minor adducts such as monofunctional cisplatinated guanine bases, interstrand crosslinks between two guanines, and intrastrand crosslinks between two guanines separated by one or more bases are also considered to be important contributors to the cytotoxicity of cisplatin. Despite their significant contribution to the cytocidal effects of cisplatin, it is noteworthy that the interstrand DNA crosslinks account for only 3% of the total number of adducts *(9)*. Moreover, although the types of adduct most responsible for the cytotoxic effect of cisplatin are still a controversial issue, it is now commonly accepted that they all contribute to inhibition of RNA transcription, DNA replication, and chain elongation by DNA polymerization enzymes.

3. REPAIR MECHANISMS OF PLATINATED DNA ADDUCTS

The mechanisms underlying the repair of cisplatin-induced DNA lesions have been extensively studied. The two major mechanisms are nucleotide excision repair (NER) and recombinational repair. Correlations have also been established between loss of mismatch repair (MMR) and cell resistance to cisplatin *(10–14)* Intrastrand platinum DNA adducts are primarily repaired by the NER mechanisms, whereas double-strand breaks are repaired through recombinational repair. Mismatches are corrected by the complex MMR machinery, which is discussed in Subheading 3.3. Prior to analyzing the strategies designed to selectively sensitize cells to platinated adducts, a brief description of the basic steps involved in these various repair mechanisms is given in this section.

3.1. NER Mechanisms

Nucleotide excision repair requires over 20 polypeptides, including damage recognition factors such as XPA, XPC–hHR23B, replication protein A (RPA), and transcription factor TFIIH that comprises XPB and DNA helicases *(15–17)*. Two structure-specific endonucleases (ERCC1–XPF and XPG) are responsible for the incisions and DNA polymerase –/–, proliferating factor C, and RPA enzymes are needed for gap-filling DNA synthesis *(15,16)*.

Recent studies demonstrated that NER occurs in a stepwise mechanism. The proposed model suggests that XPC–hHR23B plays the role of an initiator of global genomic repair by recognizing and binding to the site of lesion *(15)*. This induces the recruitment of XPA, XPG, and the TFIIH complex. Further, RPA displaces the XPC–hHR23 complex, allowing the binding of XPF–ERCC1, which cleaves the damage 5' from the DNA lesion while XPG completes the double incision by cleaving 3' from the damage *(15)*.

Nucleotide excision repair mechanisms are primarily associated with the repair of intrastrand platinum DNA adducts. Expression of ERCC1, one of the components of NER, correlates with resistance to platinum-based therapy *(17,18)*. Recently, Selvakumaran et al. *(19)* showed that blockade of ERCC1 expression by antisense RNA strategies can sensitize human ovarian carcinoma cells OVCAR10 to cisplatin approx fourfold and immunocompromized mice bearing these cells exhibited prolonged survival when compared with mice bearing control cancer cells.

Nucleotide excision repair mechanisms can be categorized into two major types: transcription-coupled NER (TC-NER) and global genome NER (GG-NER). The TC-NER undergoes repairs of transcription-blocking region in transcribed DNA, whereas the GG-NER targets the nontranscribed strand of active genes. It has been demonstrated that cells deficient in TC-NER are hypersensitive to cisplatin, irrespective of the GG-NER status *(20)*. Recently, Wang et al. *(21)* demonstrated using site-specifically platinated DNA in mononucleosomes that NER in mammalian cell extracts is substantially diminished when compared with free DNA containing the same type of adducts, indicating that the histone core plays a significant role in the NER of platinated adducts.

3.2. Recombinational Repair

3.2.1. Homologous Recombination Mechanims

Interstrand platinum DNA adducts are handled by homologous recombination (HR). These DNA repair mechanisms start with partial degradation of the DNA next to the double-strand break, leaving single-strand ends. Next, exchanges occur with the sister chromatid in the following manner: the single-strand end of the damaged strand invades and binds to its complementary DNA sequence on the homologous duplex. This is associated with the displacement of the

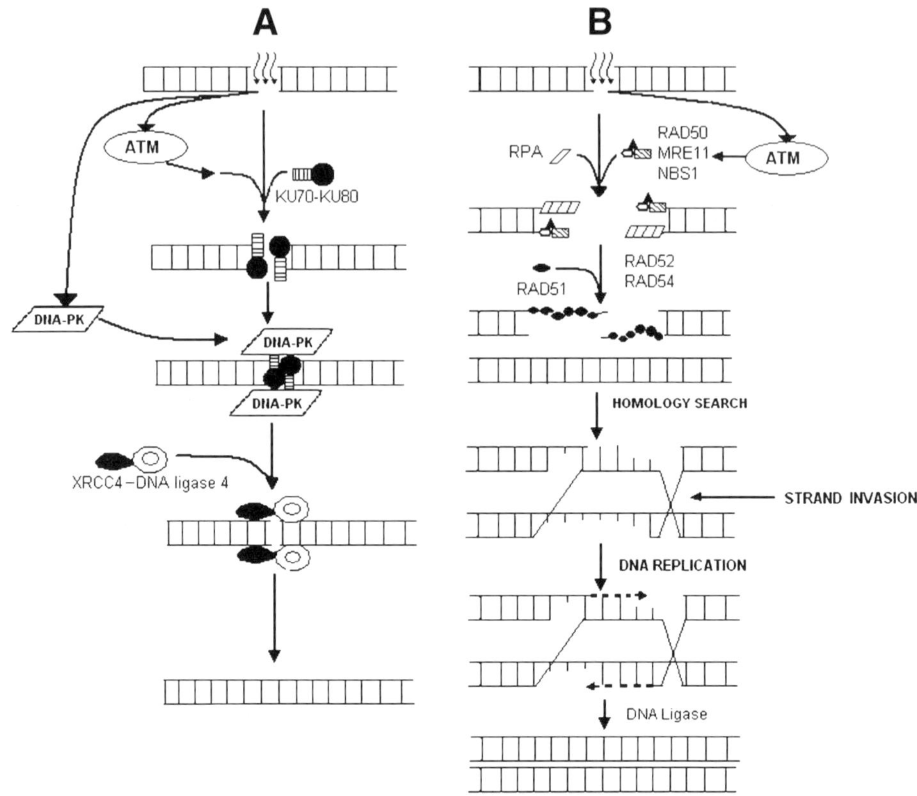

Fig. 1(A) Steps involved in the nonhomologous end-joining mechanism; **(B)** the homologous repair mechanism.

cross-complementary strand toward the gap site, where it serves as a template for gap-filling DNA synthesis (*see* Fig. 1A). The repair sequence is terminated by resolution of junctions termed "Holliday junctions" and religation of the repair patches.

This mechanism requires the presence of undamaged homologous DNA, and in many resistant cells, increased activity of this repair pathway has often been observed. HR requires Rad52 as a DNA-end-binding protein and Rad51, which forms filaments along the unwound DNA strand. The 3'-end of the damaged strand invades the homologous DNA duplex and is extended by DNA polymerase enzymes. The residual nicks are resealed by DNA ligase I and the Holliday junctions resolved, leading to crossover and noncrossover products. It is noteworthy that a recent study by Johnson and Jasin *(22)* suggests that this classical mechanism involving crossing over may not occur in mitotic cells. The authors suggested that the outcome of sister chromatid repair is primarily gene conversion *(22)*, not associated with reciprocal exchange. Moreover, a number of different mammalian proteins related to Rad51, Rad51B, Xrcc2, and Xrcc3 have been identified and shown to correlate with cisplatin chemoresistance in humans, indicating their significant involvement in the repair of cisplatin adducts *(15,16,23,24)*.

3.2.2. NONHOMOLOGOUS END-JOINING MECHANISM

Another repair mechanism of double-strand breaks is the nonhomologous end-joining mechanism (NHEJ), a fast process that requires DNA-dependent kinase (DNA-PK) and its regulatory subunits. DNA-PK is a nuclear serine–threonine kinase with a 460-kDa catalytic subunit (DNA-PKcs) coupled with a heterodimer formed by Ku70 and Ku80 *(5,17,25)*. The latter heterodimer plays the role of a regulatory subunit. DNA-PK is believed to be activated by double-strand breaks and to transmit DNA damage signals to other players that characterize the stress response pathway *(5)*. The Ku complex possesses high affinity for DNA ends. Its binding to DNA induces a conformational change in the subunit that enhances its affinity for the DNA-PKcs. The resulting DNA-PKcs–Ku80–Ku70 complex recruits XRCC4/DNA ligase that performs the ligation of the two strands (*see* Fig. 1B) *(5,26)*. DNA-PK inactivity induces double-strand-break repair deficiency and this has been shown to sensitize cells to cisplatin *(27,28)*.

3.3. The Mismatch Repair System

The MMR machinery is critical for ensuring replication fidelity and, like other DNA repair systems, it requires the tandem action of multiple proteins *(29)*. MMR corrects mismatches, insertion and deletion loops (IDLs), introduced by DNA polymerases. A dimeric protein MutS recognizes and binds to the mismatch or IDLs with an affinity 10- to 20-fold higher than that of binding to a nondamaged homoduplex. Recently, the crystal structures of procaryotic mutS have been reported by Obmolova et al. *(30)* and Lamers et al. *(31)*, who demonstrated that the general architecture consists of an induced-fit mechanism of recognition between four domains of a MutS dimer and a heteroduplex kinked at the mismatch.

The structure contains an ATPase active site composed of residues belonging to both subunits. In eucaryotes, the MutS homolog consists of MSH2/MSH6 (MutSα) that repairs mismatches and short IDLs and MSH2/MSH3 (MutSβ) that correct long IDLs. Recognition of a mismatch by MutS heterodimers is followed by the ATP-dependent binding of the ATPase MutLα complex, a MLH1/PMS heterodimer. It is believed that this may provide the energy for enhancing bidirectional DNA scanning until a strand discrimination (mismatch or IDL) signal is detected and another protein MutH binds to MutL and introduces nicks into the daughter strand. This is followed by binding of helicase II that unwinds the DNA and the segment of daughter strand is released by 3'–5' and 5'–3' exonucleases just

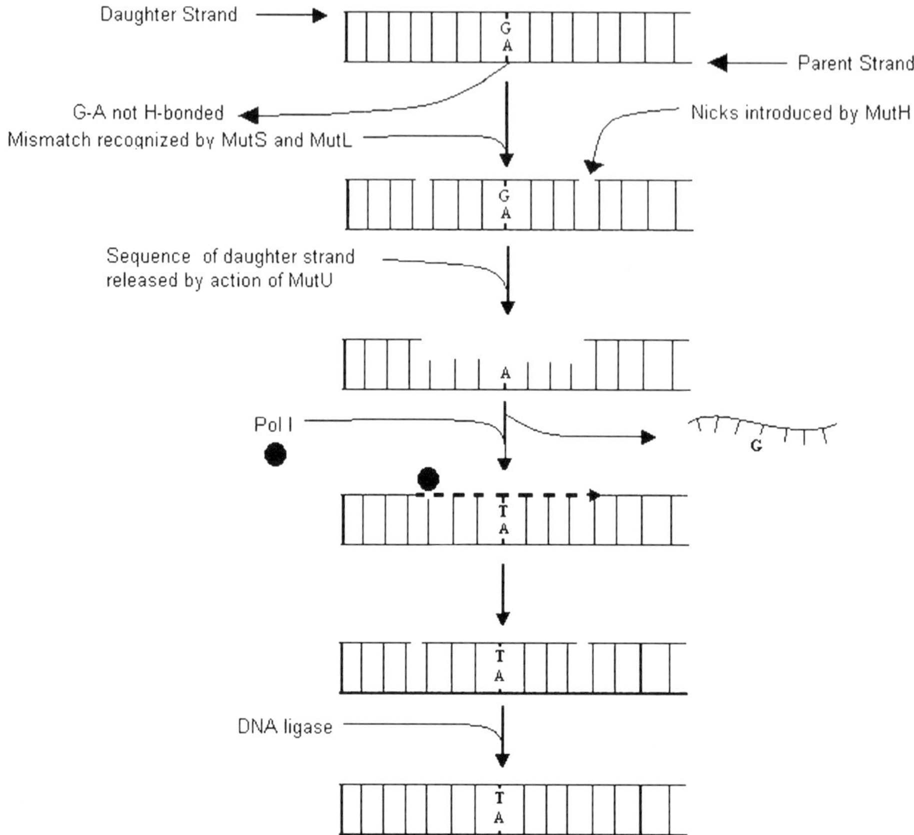

Fig. 2. The lack of a hydrogen bond between two mismatched bases (in this example, G and A) is recognized by an initial MutS dimer. As MutH and MutU are recruited, the strand is nicked and released. Resynthesis and ligation are performed by DNA polymerases and ligases with the assistance of single-strand-binding proteins.

beyond the mismatch. Resynthesis and ligation are performed by DNA polymerases and ligases with the assistance of single-strand-binding proteins (Fig. 2). Defects in one allele of MSH2 and MLH1 are associated with predisposition to hereditary nonpolyposis colorectal cancer, indicating the significant role of MMR in preventing an increased mutation rate in the genome.

Loss of MMR has now been associated with drug resistance by impairing the ability of tumor cells to detect mismatch-induced DNA adducts and trigger proapoptotic signaling *(10,13)*. This has been demonstrated *in vitro* with many DNA-damaging agents, including methylating compounds such as *N*-methyl-*N*-nitrosoguanidine, temozolomide, or platinum-based drugs (e.g., cisplatin and carboplatin) *(11,12,32,33)*. The role of MMR in chemosensitivity to cisplatin is addressed in Subheading 4.3.2.

4. MODERN STRATEGIES TO SENSITIZE TUMOR CELLS TO THE CYTOCIDAL EFFECTS OF PLATINATED ADDUCTS

4.1 Modulation by Pyrimidine Nucleosides

The repair mechanism of lesions induced by cisplatin involving new DNA synthesis, its combination with agents capable of interfering with the latter process appeared as a logical approach to the enhancement of chemosensitivity to cisplatin DNA adducts (Table 1). Indeed, the combination of cisplatin with the nucleoside analog gemcitabine has proven highly synergistic and is increasingly applied in the combination therapy of many solid tumors, including non-small-cell lung carcinoma and urothelial cancer *(4)*. The mechanisms underlying the synergistic interactions between cisplatin and gemcitabine is still under investigation. However, recent results demonstrated that it is based on the modulation of repair of cisplatin-induced lesions by gemcitabine *(34–36)*.

Following cell penetration, gemcitabine is anabolized to its triphosphate form and is incorporated into DNA where it induces termination of DNA synthesis by inhibiting DNA polymerase activity *(37)*. Of all of the mechanisms of repair of DNA adducts induced by genotoxic drugs, NER and HR require the longest nucleotide repair patches. It is now known that only short stretches of DNA, of less than 30 bp are synthesized during NER and much longer stretches (kilobases) are required for HR *(35)*. Thus, the latter process is most likely to be perturbed by gemcitabine incorporation. Indeed, recent studies by Crul et al. *(35)*, using a panel of isogenic CHO cells with varied levels of base excision repair (BER), NER, NHEJ, and HR activities demonstrated that inhibitions of HR and partly NER by gemcitabine were the most probable mechanism accounting for the synergistic interaction observed with the cisplatin+gemcitabine combination. This was further corroborated by the sequence specificity of drug administration required for synergy. Although contradictory results were reported by previous studies, Crul et al. *(35)* observed that the strongest synergistic interactions were obtained when cisplatin was administered before gemcitabine, which is in agreement with a mechanism whereby the nucleotide-incorporation-dependent HR is significantly perturbed by subsequent gemcitabine incorporation. More importantly, gemcitabine has the ability to inhibit ribonucleotide reductase, thereby depleting the intracellular deoxynucleotide pools available for HR-mediated gap-filling DNA synthesis.

Although no studies have yet been published on the role of MMR in cisplatin and gemcitabine synergistic interactions, it is now generally agreed that the potentiation of cisplatin by pyrimidine nucleoside analogs is primarily based on the ability of the latter to interfere with the mechanisms of DNA repair elicited by cisplatin. HR, because of its strong requirement for long patch new DNA synthesis, is perhaps the most affected mechanism. Other

Table 1
Structures of Platinated Agents and Their Experimental or Clinical Chemosensitizers

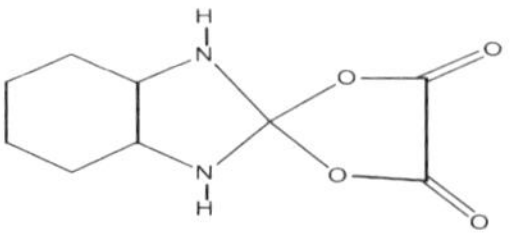 Oxaliplatin	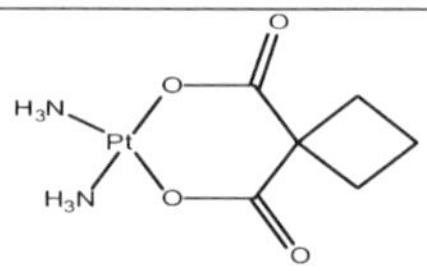 Carboplatin
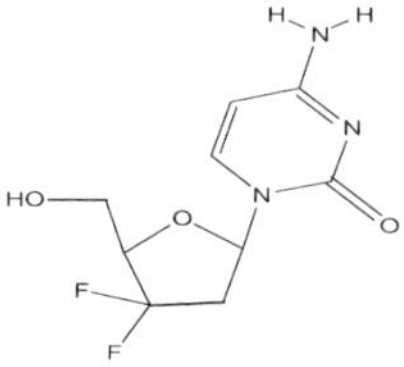 Gemcitabine	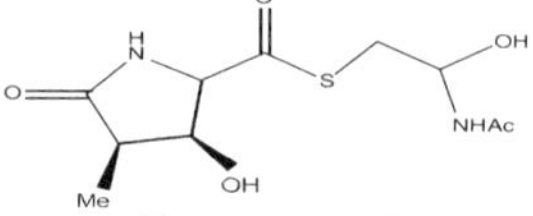 Lactacystin
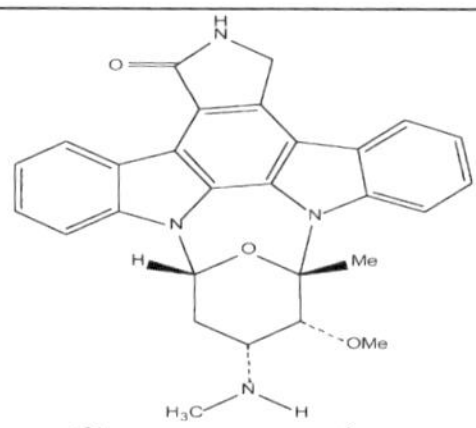 Staurosporine	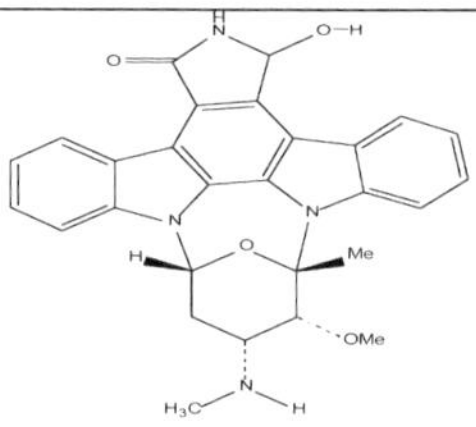 UCN-01
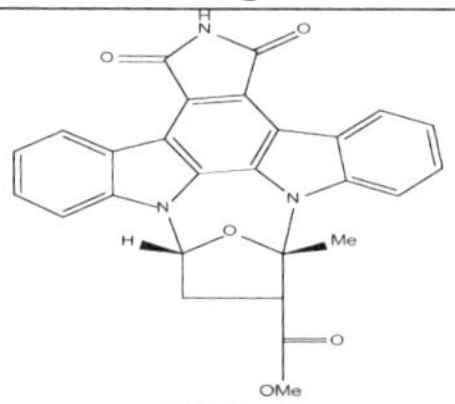 ICP-1	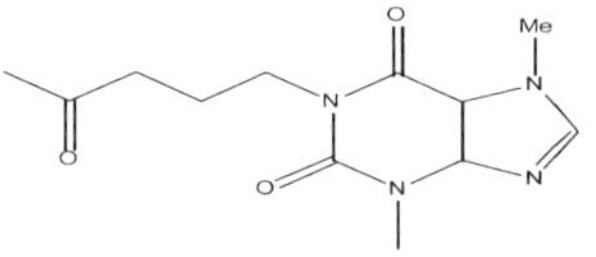 Pentoxifylline
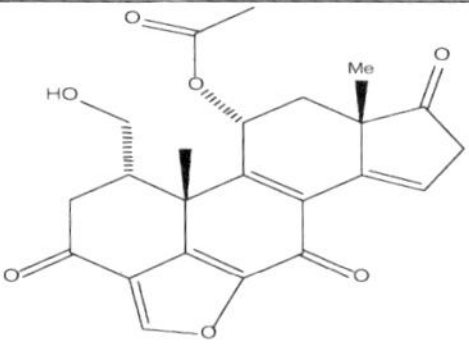 Wortmannin	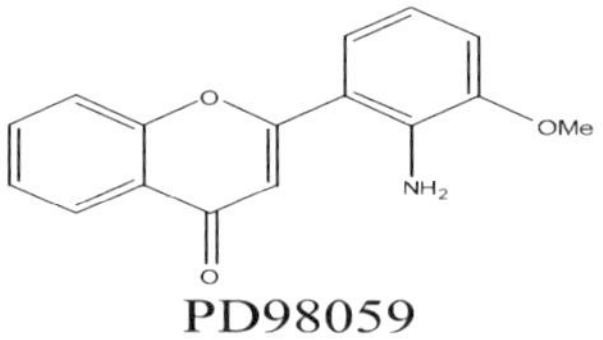 PD98059
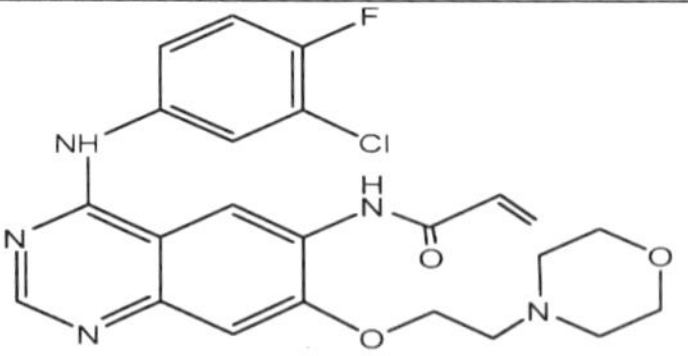 CI-1033	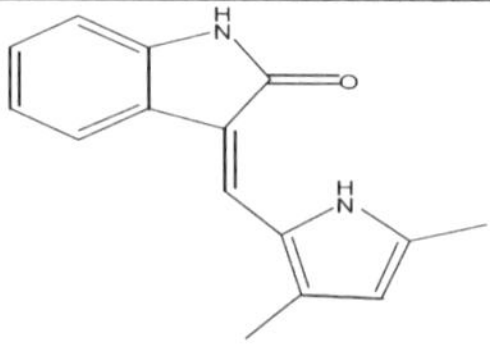 SU-5416

nucleoside analogs such as fludarabine synergize with cisplatin in a similar fashion *(4)*.

4.2. Modulation by Proteasome Inhibitors

Protein degradation is essential to the supply of amino acids for new protein synthesis and the removal of excess enzymes and transcription factors that are no longer needed. This process is maintained by two complex intracellular particles: the lysosomes and the proteasomes. Whereas lysosomes deal with extracellular proteins taken into the cells by endocytosis, proteasomes primarily digest endogenous proteins such as transcription factors, cyclins, encoded viral proteins, intracellular parasites, or proteins encoded by faulty genes *(38)*. The proteasome core particle consists of 2 copies of 14 different proteins assembled in groups of 7 in a ring structure. The central ring stacks are flanked by two complexes of multiple proteins termed "regulatory particles" (RP). More importantly, some of the subunits have recognition sites for small 76-amino-acid proteins known as ubiquitins. Proteins to be destroyed are conjugated to a molecule of ubiquitin through a lysine residue. This ubiquitin tag has a high affinity for the ubiquitin recognition site of the RP. Through a number of ATP hydrolysis reactions that releases energy, the captured protein is unfolded and introduced into the central cavity of the proteasome for digestion. At the end of the cleavage process, the bound ubiquitin is released for reuse. Therefore, agents that block or inhibit proteasome functions decrease the pool of ubiquitin available for further cellular processes such as chromatin conformation or remodeling that avidly requires ubiquitination. Histone ubiquitination–deubiquitination cycles are considered tightly regulated processes that are catalyzed by a number of families of unknown ubiquitin-conjugating and ubiquitin-removing enzymes. Histone ubiquitination is believed to shift the chromatin from a highly ordered structure to a more relaxed conformation, hence allowing greater accessibility to DNA repair enzymes.

Baxter and Smerdon *(39)* showed that nucleosomes transiently unfold during NER in both normal and cancer cells. Thus, blockage of histone ubiquitination and hence the subsequent nucleosome unfolding may indirectly affect DNA repair activity. Indeed, Mimnaugh et al. *(40)* recently demonstrated that proteasome inhibitors lactacystin (LC) and *N*-acetyl-leucyl-leucyl-norleucimil (AL LNL) are capable of indirectly inducing deubiquitination of ubiquitinated histone H2AA (uH2H) and concomitantly promoting chromatin condensation. This significantly decreased NER-dependent repair of cisplatin–DNA adducts. More importantly, these inhibitors downregulated the expression of ERCC1 and sensitized cells to cisplatin. Because of the significant interconnection between histone ubiquitination/deubiqitination with pathways involved in apoptosis, this novel chemosensitization approach may well represent a new model to selectively sensitize tumor cells to platinum-based drugs.

4.3. Modulation by Abrogation of Cell Cycle Checkpoints

4.3.1. p53, CELL CYCLE CHECKPOINT, AND THE CELLULAR RESPONSE TO DNA DAMAGE

The term "checkpoint" was introduced by Weinert and Hartwell *(41)* to define the process by which dependence and regulation in the cell cycle are maintained. Checkpoints are present in the G_1- and G_2-phases to prevent cell cycle progression in the presence of DNA damage, in the S-phase to inhibit mitosis until completion of DNA replication, and in mitosis to block chromosome segregation until mitotic spindles have been properly assembled. It is now commonly accepted that p53 is the major controller of the G_1 checkpoint. Exposure of cells to radiation or DNA-damaging agents leads to increased nuclear levels of p53 and activation of subsequent p53-dependent processes *(42–46)*. These events include the p53-dependent transcription of p21/WAF1 that inhibits the G_1 cyclin-dependent kinase, leading then to cell cycle arrest in the late G_1 stage *(47)*. As a result, the cells fail to enter the S-phase and DNA synthesis is inhibited *(48–53)*. This cell cycle arrest permits the repair of DNA lesions, and in wild-type p53 cells, this may lead to apoptosis. Thus, p53 can play either a cytoprotective or a cytotoxic role. If during the cell cycle arrest the drug- or radiation-induced DNA lesions are repaired, p53 will promote cell survival and will, therefore, play a cytoprotective role. On the other hand, if the induction of p53 leads to cell death through apoptotic pathways, p53 plays a cytotoxic role and this causes an increased sensitivity to DNA-damaging agents. Cogent evidence of the cytoprotective role of p53 in drug response has been reported by Fan et al. *(50)*, who showed that disruption of normal p53 function in MCF-7 breast cancer cells (by transfection with human papillomavirus type-16 E6 gene or a dominant-negative mutant p53 gene) sensitize these cells to cisplatin. More recently, a similar observation was reported by Pestell et al. *(64)*, who demonstrated that A2780 human ovarian tumor cells stably transfected with HPV-16 E6 were threefold to fourfold more sensitive to cisplatin than their wild-type counterparts.

Further sensitization of p53-disrupted MCF-7 cells was induced when they were pre-exposed to pentoxifylline *(50,55–59)*, an agent that abrogated their G_2 arrest *(50,60)*. Pentoxifylline increased the sensitivity of p53-disrupted MCF-7 cells by 30-fold, whereas no potentiation was seen in the control MCF-7 cells expressing wild-type p53. Because the G_1 checkpoint remains intact in normal cells, this strategy was believed to have the potential to selectively potentiate the action of DNA damaging drugs in p53-deficient cells. However, pentoxifylline is active at high-millimolar concentrations and its intracellular targets remain elusive. Similarly, caffeine is also known to abrogate the G_2 checkpoint at high-millimolar doses.

Recently, more potent agents have been developed that abrogate the G_2 checkpoint at submicromolar levels. One such compound, UCN01, an analog of staurosporine, is now in phase I clinical trials. UCN01 a 100,000-fold more potent abrogator of the S and G_2 checkpoint than caffeine, has been shown to inhibit Chk1 and Chk2, two essential components of the DNA-damage checkpoint. When the presence of DNA damage is signaled, ATM or ATM-related proteins (ATR) phosphorylate Chk1 *(61,62)* that, in turn, regulates the G_2 checkpoint by phosphorylating and inactivating the Cdc25 protein phosphatase *(63)*. Blockade of Chk1 by UCN01 results in an absence of inhibitory signal for Cdc25 that mediates cell transition from G_2 to M.

The combination of cisplatin with UCN01 and G_2 abrogators is highly and selectively synergistic in p53-deficient cells *(64–67)*. Unfortunately, the phase I trial of this combination revealed a significantly high plasma-binding property for UCN01 that increases its pharmacokinetic half-life. Recently, a novel analog of UCN01, ICP-1 *(68)* has been shown to possess lower human plasma-binding properties and potency almost equal to that of UCN01. The synergy observed between ICP-1 and cisplatin set the premise for a successful clinical development of this approach for the selective therapy of malignancies in which p53 is dysfunctional. It should be remembered that loss of p53 function is a characteristic common to at least 50% of all known malignancies.

4.3.2. The MMR System As a Potential Target for Selective Chemosensitization to Platinated Agents

Loss of MMR is now known to induce chemoresistance to a great variety of genotoxic agents, including mustards, triazenes and platinum-based drugs *(10–14,69–71)*. This is believed to be the result of a possible link between MMR and the activation of apoptosis. A recent study by Shimodaira et al. *(72)* demonstrated that the link between MMR and apoptosis may be through PMS2. It has been shown that the levels of p73 protein increased with increasing amounts of coexpressed PMS2, suggesting that the latter protein may promote stabilization of p73, a p53-related transcription factor that participates in the regulation of the apoptotic response to DNA damage. To our knowledge, approaches to chemosensitize cells by targeting elements of the MMR system have not yet been reported. However, a recent study by Fedier et al. *(69)* showed that p53 deficiency with loss of PMS2 exhibited increased sensitivity to cisplatin. Based on these results, the authors suggested that tumor-targeted functional inhibition of PMS2 may be a valuable strategy to selectively sensitize p53-mutant cancers to cytotoxic DNA-damaging agents.

4.3.3. Modulation by Abrogation of DNA-PK

Many attempts have been made to disrupt the action of DNA-PK with the purpose of sensitizing tumor cells to the action of ionizing radiation and chemotherapeutic drugs *(73–75)*. Although the use of wortmannin has shown some chemosensitization, the lack of specificity of this inhibitor that also targets phosphatidylinositol 3-kinase (PI3k) has hampered a clear assessment of the role of DNA-PK inhibition in chemosensitization. The most recent and specific approach was reported by Kim et al. *(74)*, who designed a peptide sequence representing part of the C-terminus of Ku80 (HNI-38) to selectively and competitively target and disrupt interaction between the Ku complex and DNA-PKcs. This compound, HNI-38, was found to significantly reduce DNA double-strand break and to sensitize the cells to cisplatin. More importantly, HNI-38 was inactive in the absence of double-strand break, which validates the concept that Ku80 is required for binding DNA-PK to DNA double-strand breaks. Thus, blockade of the NHEJ repair mechanism through selective inhibition of DNA-PK appears to be an efficient strategy to sensitize cells to platinum-based drugs.

4.3.4. MODULATION BY KINASE-RELATED PATHWAYS

The Jun/stress-activated protein kinase (JNK/SAPK) pathway *(see* Fig. 3) is involved in a kinase cascade that phosphorylates the transcription factor c-jun at serine residues 63 and 73. C-jun phosphorylation enhances the transactivation potential of the AP-1 complex. In addition to its significant role in cellular transformation, inflammation, and apoptosis, JNK/SAPK is involved in a pathway that is strongly stimulated by DNA-damaging treatments, including alkylating agents and ionizing radiation. Recently, Potapova et al. *(76)* demonstrated that treatment of T98G glioblastoma cells with cisplatin activated JNK/SAPK by about 10-fold when compared with untreated control. When these cells were modified to express a nonphosphorylatable dominant-negative c-jun, they became sixfold more sensitive to the cytotoxic action of cisplatin. More importantly, the c-jun mutant cell line showed reduced DNA repair activity and exhibited increased apoptosis and elevated bax : bcl2 ratios upon transfection with wild-type p53. This suggests that inhibition of JNK kinase activity may reduce DNA repair activity, thereby sensitizing cells to cisplatin.

More recent studies correlated the activation of MAP kinases (signaling proteins upstream from JNK and c-jun) with chemosensitivity to cisplatin *(76)*. It has been found that mitogen-activating protein (MAP) kinase or ERK1/2 was significantly activated by cell exposure to cisplatin. In contrast to growth-factor-receptor-stimulated response, this activation was rather slow. Blockade of phosphorylation and activation of MEK1, the immediate upstream kinase of ERK1/2 using the small molecule inhibitor PD98059 *(76)*, significantly sensitized ovarian cancer cells to the action of cisplatin. More importantly, although cisplatin exposure induced activation of both ERK1/2 and JNK in these cells, PD98059-mediated inhibition did not affect the phosphorylation status of JNK, indicating that this potentiation occurred via a specific MAP kinase-related pathway, independent of c-jun phosphorylation. The implication of DNA repair in the mechanism of cell sensitization to cisplatin by ERK1/2 inhibition remains to be demonstrated.

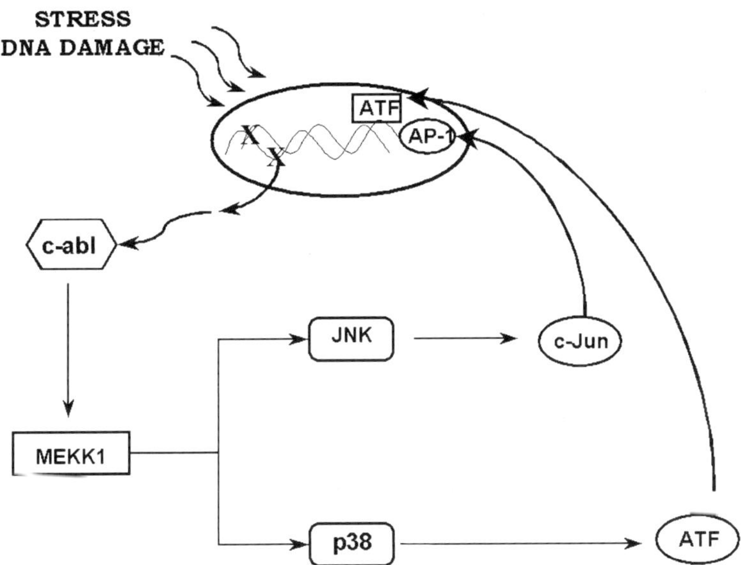

Fig. 3. The JNK/SAPK stress response pathway.

4.3.5. RECEPTOR-MEDIATED CHEMOSENSITIVITY TO PLATINATED DRUGS

Over the past 6 yr, significant effort has been deployed to elucidate the relationship between growth-factor-receptor expression or function and chemosensitivity. Because of their implication in tumor progression and aggressiveness, the epidermal growth factor receptor (EGFR) gene HER1 *(77–79)* and its closest homolog, HER2 *(80–83)*, have received much attention. The relationship between EGFR function or expression and chemosensitivity to cisplatin remains controversial. Studies with a panel of cervical carcinoma squamous cells demonstrated that M180 cells expressing the highest levels of EGFR exhibited higher S-phase fractions and were the most sensitive to cisplatin *(84)*. Dixit et al. *(85)* demonstrated that stably transfected *p*-chloroamphenicol acetyl transferase (pact[*p*]-CAT) vector containing a 4.1-kb full-length antisense EGFR complementary DNA decreased chemosensitivity of the host cells and enhanced the repair of cisplatin-induced intrastrand crosslinks. In contrast, other studies demonstrated that C225, an anti-EGFR antibody, was capable of significantly sensitizing squamous cancer cells to cisplatin *(86,87)*. More recently, Gleseg et al. *(88)* showed that an irreversible inhibitor of the quinazoline class CI-1033 was capable of strongly synergizing with cisplatin. However, the effects were found to be independent of cellular DNA repair activity. These results are in agreement with a previous study by Tsai et al. *(89)*, who showed little interrelationship among EGFR, chemosensitivity, and NER activity in small-cell lung carcinoma cells.

The debate is further complicated by a recent study demonstrating that upon exposure to ionizing radiation, EGFR is activated in a ligand-independent fashion and induces, through the RAS-MAP kinase pathway, the expression of XRRC1 and ERCC1 *(90)*, two critical DNA repair enzymes involved in the repair of cisplatin-induced lesions. A similar result was reported a year earlier by Benhar et al. *(91)*, who demonstrated that cell exposure to cisplatin resulted in a ligand-independent activation of EGFR and that PP1, an inhibitor of Src tyrosine kinase, could block this activation. This indicates that Src tyrosine activity may be involved in cell response to cisplatin-induced DNA lesions. If EGFR activation by Src in response to cisplatin exposure induces XRRC1 and ERCC1, this will be inconsistent with the recent observation by Gleseg et al. *(88)*, who reported a complete absence of a link between DNA repair and the synergistic interaction of EGFR inhibitor CI-1033 with cisplatin. The interrelation between EGFR and the repair of cisplatin-induced lesions may be cell-specific and other factors playing a *sine qua non* role in this process remained to be identified.

Although the relationship between EGFR and chemosensitivity to cisplatin remains controversial, the implication of its closest homolog HER2 in DNA-repair-mediated chemoresistance to cisplatin has been consistently demonstrated by many laboratories. Alaoui-Jamali et al. *(92)* demonstrated that inhibition of p185^{neu}, the HER2 gene product, was associated with downregulation of DNA repair in non-small-cell lung carcinoma. Pietras et al. *(93)* showed that treatment of breast and ovarian carcinoma cells with an anti-HER2 antibody resulted in 35–40% reduction in repair of cisplatin DNA adducts. This chemosensization approach is now referred to as "receptor-mediated chemosensitivity." Similarly, Tsai et al. *(94)* in a panel of 16 human NSLC cells demonstrated that high levels of DNA repair activity correlated with increased HER2 levels and activity. To date, a significant body of work has accumulated to suggest that the HER2 gene product p185^{neu} is a valid target for modulation of DNA repair of cisplatin adducts. However, the type of repair mechanisms (e.g., NER, HR, or NHEJ) most affected by HER2 modulation remains to be determined.

Recently, the implication of the vascular endothelial growth factor (VEGF) in cell sensitivity to cisplatin has been studied by Zhong et al. *(95)*. Binding of VEGF to its cognate receptor FLK1/KDR *(96)* activates phosphorylation of downstream signaling proteins that ultimately induce expression of genes associated with angiogenesis. Blockade of the FLK1/KDR tyrosine kinase activity with the novel inhibitor SU5416 has been shown to be associated with significant antiangiogenic activity. SU5416 *(97)* has been shown to inhibit proliferation of endothelial cells simulated with VEGF and exhibited a broad range of activity in inhibiting xenografts in athymic mice. Zhong et al. *(95)* were the first to assess the benefits of combining an antiangiogenic agent with a cytotoxic DNA-damaging drug in ovarian cancer cells. It should be remembered that cisplatin and carboplatin are widely used in the therapy of ovarian cancers *(98)*. It was found that at nontoxic concentrations, SU5416 enhances cisplatin cytotoxicity in resistant ovarian cancer cells. More importantly, this was associated with reduced expression of ERCC-1 protein and c-jun mRNA as well as a decrease in c-jun and JNK activities. The increased chemosensitivity induced by the cisplatin/SU5416 combination is believed to be mediated by direct or indirect reduction of AP-1 and DNA repair activities. Thus, this is an innovative strategy that may be exploited to perhaps induce a tandem blockade of angiogenesis and enhancement of chemosensitivity in refractory solid tumors.

5. FUTURE PROSPECTS

Cisplatin is one of the most potent cytotoxic drugs used in the clinic in the treatment of an increasing number of malignancies. Over the past three decades, the lack of selectivity of platinum-based drugs for tumor tissues has been considered a major deterrent for their use in the therapy of many cancers. However, despite being a "nontargeted" cytotoxin, cisplatin shows significantly selective potency in specific types of tumors in the clinic. As an example, it is now common knowledge that cisplatin is the most potent drug used in the clinical management of testicular cancers. This is explained by the high bax : bcl2 ratio that characterizes rapid triggering of apoptosis in response to cisplatin adducts in these tumors. Thus, the bax : bcl2 ratio of these tumors is therefore an important signature that underlies the selective potency of cisplatin in testicular cancers. The most common marker for selective action of cisplatin is the specific deficiency of some tumors in one or many elements of the complex DNA repair machinery required for the repair of the rich number of adducted DNA bases induced by cisplatin. However, despite the significant potency of platinum-based drugs in these tumors, acquired resistance mediated by DNA repair enzymes is the major cause of relapse following chemotherapy. To circumvent this problem, the current trend is to identify and develop potent inhibitors of DNA repair proteins involved in both NER and recombinational repair. Because of the cooperative mechanism of action of these proteins, screening assays for small-molecule inhibitors are scarce and minor progress is achieved by the use of pepidomimetics or antisense. As an example, DNA-PK, the critical complex of NHEJ, can only be inhibited by an unspecific inhibitor (wortmannin) and, more specifically, by peptide sequences capable of competitively blocking the binding of Ku80 to the double-strand-break sites. Although these peptides may not translate into useful chemotherapeutic agents, they have been successfully used to demonstrate that selective abrogation of DNA-PK sensitizes tumor cells to cisplatin. Thus, the proof of the concept demonstrating the usefulness of blocking the action of many elements of the DNA repair machinery, including DNA-PK, Rad51, and XRCC1 to sensitize cells to platinum-based drugs, is practically made. Thus, there remains to develop novel and adaptable small-molecule inhibitors to modulate platinated DNA adduct repair in order to improve its selectivity and its potency in the clinic.

Platinum-based drugs are indicated in many tumors including lung, ovarian, and testicular carcinoma, which are often marked by disordered receptor- and nonreceptor-mediated signaling. The recent discovery that JNK and EGFR are phosphorylated following cell exposure to cisplatin and that these events are related to DNA repair has opened the way to novel tumor-selective approaches to the chemosensitization of cisplatin. In contrast to DNA repair enzymes that require complex assembly for their functions, many of the signaling phosphoproteins are readily isolated for direct in vitro screening, and their activity can be monitored by immunodetection of phosphorylated substrates. Inhibitors of EGFR are already in phase III clinical trials; inhibitors of MAP kinases are already available and other kinase inhibitors are currently under investigation. Therefore, the development of specific inhibitors for many signaling kinases could indirectly translate into the development of potent agents designed to selectively enhance chemosensitivity to platinum-based drugs in refractory tumors.

Recently, in order to ameliorate the selectivity and potency of classical DNA-damaging agents, a novel receptor-mediated chemosensitization strategy has been developed in our laboratory. This strategy termed "combi-targeting" seeks to synthesize single molecules termed "combi-molecules" designed to simultaneously block tyrosine-kinase-mediated signaling while inflicting DNA damage to tumor cells *(63,99–104,107)*. The feasibility of this novel tumor-targeting strategy was first demonstrated with two combi-molecules (SMA41 and BJ2000) that showed ability to strongly block EGFR-mediated signaling and to damage DNA in cells overexpressing this receptor *(99,100,107)*. Further, we also reported that the combi-molecules could induce irreversible inhibition of the EGFR tyrosine kinase activity *(100)*. More importantly, these drugs showed superior potency when compared with two-drug combinations involving an individual EGFR inhibitor and a classical DNA damaging agent *(99)*. Given the strong interrelations between the EGFR/HER2 activation and signaling associated with platinum-induced DNA adducts, the development of platinum-based combi-molecules may well represent a promising strategy to enhance the selectivity and potency of platinum-based therapy in refractory ovarian, prostate, head and neck, lung, breast, or other solid tumors with disordered expression of the erb oncogenes.

REFERENCES

1. Loehrer S, Einhorn ADJ. Diagnosis and treatment drugs five years later: cisplatin. *Ann Int Med* 1984;100:704–713.
2. Calsou P, Salles B. Role of DNA repair in the mechanisms of cell resistance to alkylating agents and cisplatin. *Cancer Chemother Pharmacol* 1993;32:85–89.

3. McClay EF, Albright KD, Jones JA, et al. Modulation of cisplatin resistance in human malignant melanoma cells. *Cancer Res* 1992;52:6790–6796.

4. Crul M, van Waardenburg RCAM, Beijen JH, et al. DNA-based drug interactions of cisplatin. *Cancer Treat Rev* 2002;28:291–303.

5. Mistry P, Kelland LR, Abel G, et al. Cellular accumulation of the anticancer agent cisplatin: a review. *Br J Cancer* 1993;67:1171–1176.

6. Pepponi R, Marra G, Fuggetta MP, et al. The effect of O6-alkylguanine-DNA-alkyltransferase and mismatch repair activities on the sensitivity of human melanoma cells to temozolomide, 1,3-bis(2-chloroethyl)1-nitrosourea and cisplatin. *J Pharm Exp Ther* 2003;304:661–668.

7. Chaney SG, Vaisman A. Specificity of platinum–DNA adduct repair. *J Inorg Biochem* 1999;77:71–81.

8. Raymond E, Faivre S, Chaney S, et al. Cellular and molecular pharmacology of oxaliplatin. *Mol Cancer Ther* 2002;1:227–235.

9. Eastman A. The formation, isolation and characterization of adducts produced by anticancer platinum complexes. *Pharmacol Ther* 1987;34:155–156.

10. Li CM. The role of mismatch repair in DNA-damage-induced apoptosis. *Oncol Res* 1999;11:393–400.

11. Lage II, Dictel M. Involvement of the DNA mismatch repair system in antineoplastic drug resistance. *J Cancer Res Clin Oncol* 1999;125:156–165.

12. Manic S, Gatti L, Carenini N, et al. Mechanisms controlling sensitivity to platinum complexes: role of p53 and DNA mismatch repair. *Curr Cancer Drug Targ* 2003;3:21–29.

13. Fink D, Nebel S, Aebi S, et al. The role of mismatch repair in platinum drug resistance. *Cancer Res* 1996;56:4881–4886.

14. Fink D, Nebel S, Norris PS, et al. Enrichment of DNA mismatch repair-deficient cells during treatment with cisplatin. *Int J Cancer* 1998;77:741–746.

15. You JS, Wang M, Lee SH. Biochemical analysis of the damage recognition process in nucleotide excision repair. *J Biol Chem* 2003;278:7476–7485.

16. Khanna KK, Jackson SP. DNA double-strand breaks: signaling, repair and the cancer connection. *Nature Genet* 2001;27:247–254.

17. Dabholkar M, Bostick-Bruton F, Weber C, et al. ERCC1 and ERCC2 expression in malignant tissues from ovarian cancer patients. *J Natl Cancer Inst* 1992;84:1512–1517.

18. Dabholkar M, Vionnet J, Bostick-Bruton F, et al. Messenger RNA levels of XPA and ERCC1 in ovarian cancer tissue correlate with response to platinum-based chemotherapy. *J Clin Invest* 1994;94:703–708.

19. Selvakumaran M, Pisarcik DA, Bao R, et al. Enhanced cisplatin cytotoxicity by disturbing the nucleotide excision repair pathway in ovarian cancer cell lines. *Cancer Res* 2003;63:1311–1316.

20. Furuta T, Ueda T, Aune G, et al. Transcription-coupled nucleotide excision repair as a determinant of cisplatin sensitivity of human cells. *Cancer Res* 2002;62:4899–4902.

21. Wang D, Hara R, Sancar A, et al. Nucleotide excision repair from site-specifically platinum-modified nucleosomes. *Biochemistry* 2003;42:6747–6753.

22. Johnson RD, Jasin M. Sister chromatid gene conversion is a prominent double-strand break repair pathway in mammalian cells. *Nature Genet* 2001;27:247–253.

23. Caslou P, Delteil C, Frit P, et al. Coordinated assembly of Ku and p460 subunits of the DNA-dependent protein kinase on DNA ends is necessary for XRCC4–ligase IV recruitment. *J Mol Biol* 2003;326:93–103.

24. Frit P, Li RY, Arzel D, et al. Ku entry into DNA inhibits inward DNA transactions in vitro. *J Biol Chem* 2000;275:35,684–35,691.

25. Muller C, Caslou P, Frit P, et al. Regulation of the DNA-dependent protein kinase (DNA-PK) activity in eukaryotic cells. *Biochemistry* 1999;8:117–125.

26. Hoeijmakers JH. Genome maintenance mechanisms for preventing cancer. *Nature* 2001;411:366–374.

27. Caldecott K, Jeggo P. Cross-sensitivity of gamma ray sensitive hamster mutants to cross-linking agents. *Mutat Res* 1991;255:111–121.
28. Tanaka K, Yamagami T, Oka Y, et al. The SCID mutation in mice causes defects in the repair system for both the double-strand DNA breaks and DNA cross-links. *Mutat Res* 1993;288:277–280.
29. Sancar A. Excision repair invades the territory of mismatch repair. *Nature Genet* 1999;21:247–249.
30. Obmolova G, Ban C, Hsieh P, et al. Crystal structures of mismatch repair proteins MutS and its complex with a substrate DNA. *Nature* 2000;407:703–710.
31. Lamers MH, Perrakis A, Enzlin JH, et al. The crystal structure of DNA mismatch repair protein MutS binding to a G × T mismatch. *Nature* 2000;407:711–717.
32. D'Atri S, tentori L, Lacal PM, et al. Involvement of the mismatch repair system in temozolomide-induced apoptosis. *Mol Pharmacol* 1998;54:334–341.
33. Hickman MJ, Samson LD. Role of DNA mismatch repair and p53 in signaling induction of apoptosis by alkylating agents. *Proc Natl Acad Sci USA* 1999;96:10,764–10,769.
34. Moufarij A, Phillips DR, Cullinane C. Gemcitabine potentiates cisplatin cytotoxicity and inhibits repair of cisplatin-DNA damage in ovarian cancer cell lines. *Mol Pharmacol* 2003;63:862–869.
35. Crul M, Waardenburg RC, Bocxe S, et al. DNA repair mechanisms involved in gemcitabine cytotoxicity and in the interaction between gemcitabine and cisplatin. *Biochem Pharmacol* 2003;65:275–282.
36. Peters G, Wilt vdL, Moorsel vCJA, et al. Basis for effective combination cancer chemotherapy with antimetabolites. *Pharmacol Ther* 2000;87:227–253.
37. Plunkett W, Huang P, Ganghi V. Preclinical characteristics of gemcitabine. *Anticancer Drugs* 2003;6:7–13.
38. Pietras RJ, Poen JC, Gallardo D, et al. Monoclonal antibody to HER-2/neureceptor modulates repair of radiation-induced DNA damage and enhances radiosensitivity of human breast cancer cells overexpressing this oncogene. *Cancer Res* 1999;59:1347–1355.
39. Baxter BK, Smerdon MJ. Nucleosome unfolding during DNA repair in normal and xerodermal pigmentosum (group C) human cells. *J Biol Chem* 1998;273:17,517–17,524.
40. Mimnaugh EG, Yunmbam MK, Li Q, et al. Prevention of cisplatin–DNA adduct repair and potentiation of cisplatin-induced apoptosis in ovarian carcinoma cells by proteasome inhibitors. *Biochem Pharmacol* 2000;60:1343–1354.
41. Weinert T, Hartwell L. Control of G2 delay by the rad9 of *Saccharomyces cerevisiae. J Cell Sci* 1989;12(Suppl):145–148.
42. Lowe SW, Bodis S, McClatchey A, et al. p53 status and the efficacy of cancer therapy in vivo. *Science* 1994;266:807–810.
43. Murray A, Hunt T. *The Cell Cycle: An Introduction.* Oxford University Press, Oxford, 1993:42–89.
44. Sidransky HM, Vogelstein B. p53 mutations in human cancers. *Science* 1991;253:49–53.
45. Ramet M, Castren K, Jarvinen K, et al. p53 protein expression is correlated with benzo[a]pyrene-DNA adducts in carcinoma cell lines. *Carcinogenesis* 1995;16:2117–2124.
46. Fu LM, Benchimol S. Participation of the human p53 3'UTR in translational repression and activation following gamma-irradiation. *EMBO J* 1997;16:4117–4125.
47. Li Y, Jenkins CW, Nichols MA, et al. Cell cycle expression and p53 regulation of the cyclin-dependent kinase inhibitor p21. *Oncogene* 1994;9:2261–2268.
48. Kastan MB, Onyekwere O, Sidransky D, et al. Participation of p53 protein in the cellular response to DNA damage. *Cancer Res* 1991;51:6304–6311.
49. Nagasawa H, Li CY, Inrich AC, Little JB. Relationship between radiation-induced G1 phase arrest and p53 function in human tumor cells. *Cancer Res* 1995;55:1842–1846.
50. Fan S, Smith ML, Rivet DJ, et al. Disruption of p53 function sensitizes breast cancer MCF-7 cells to cisplatin and pentoxifylline. *Cancer Res* 1995;55:1649–1654.

51. Harris CC, Holstein M. Clinical implications of the p53 function and dysfunction. *Cell* 1992;70:523–526.
52. Wyllie FS, Haughton MF, Bond JA, et al. S phase cell-cycle arrest following DNA damage is independent of the p53/p21WAF1 signalling pathway. *Oncogene* 1996;12:1077–1082.
53. Guilouf C, Rosselli F, Krishnaraju K, et al. p53 involvement in control of G2 exit of the cell cycle: role in DNA damage-induced apoptosis. *Oncogene* 1995;10:2263–2270.
54. Pestell KE, Hobbs SM, Titley JC, et al. Effects of p53 status on sensitivity to platinum complexes in a human ovarian cancer cell line. *Mol Pharmacol* 1999;57:503–511.
55. Bunz F, Dutriaux C, Lengauer T, et al. Requirement for p53 and p21 to sustain G2 arrest after DNA damage. *Science* 1998;282:1497–1500.
56. Zamble DB, Jacks T, Lippard SJ. p53-dependent and -independent responses to cisplatin in mouse testicular teratocarcinoma cells. *Proc Natl Acad Sci USA* 1998;95:6163–6168.
57. Hain J, Crompton EAN, Burkart W, et al. Caffeine release of radiation induced S and G2 phase arrest in V79 hamster cells: increase of histone messenger RNA levels and p34cdc2 activation. *Cancer Res* 1993;53:1507–1510.
58. O'Connor PM, Ferris DK, Pagano M, et al. G2 delay induced by nitrogen mustard in human cells affects cyclin A/cdk2 and cyclin B1/cdc20kinase complexes differently. *J Biol Chem* 1993;268:8298–8308.
59. Powell SN, DeFrank JS, Connell P, et al. Differential sensitivity of p53(–) and p53(+) cells to caffeine-induced radiosensitivity and override of G2 delay. *Cancer Res* 1995;55:1643–1648.
60. Smith ML, Chen IT, Zhan Q, et al. Interaction of the p53-regulated protein GADD45 with proliferating cell nuclear antigen. *Science* 1994;266:1376–1380.
61. Hui Z, Piwnica-Worms H. ATR-mediated checkpoint pathways regulate phosphorylation and activation of human Chk1. *Mol Cell Biol* 2001;21:4129–4139.
62. Damia G, Sanchez Y, Erba E, et al. DNA Damage Induces p53-dependent down-regulation of hCHK1. *J Biol Chem* 2001;276:10,641–10,645.
63. Stewart ZA, Westfall MD, Pietenpol JA. Cell-cycle dysregulation and anticancer therapy. *Trends Phamacol Sci* 2003;24:139–145.
64. Wang Q, Asijun F, Easstman A, et al. UCN-01: a potent abrogator of G2 checkpoint function in cancer cells with disrupted p53. *J Natl Cancer Inst* 1996;88:956–965.
65. Yu L, Orlandi L, Wang P, et al. UCN-01 abrogates G2 arrest through a cdc2-dependent pathway that is associated with inactivation of the Wee1Hu kinase and activation of the cdc25C phosphatase. *J Biol Chem* 1998;273:33,455–33,464.
66. Jiang H, Yang LY. Cell cycle checkpoint abrogator UCN-01 inhibits DNA repair: association with attenuation of the interaction of XPA and ERCC1 nucleotide Eexcision repair proteins. *Cancer Res* 1999;59:4529–4534.
67. Bunch RT, Eastman A. Enhancement of cisplatin-induced cytotoxicity by 7-hydroxy-staurosporine (UCN-01), a new G2-checkpoint inhibitor. *Clin Cancer Res* 1996;2:791–797.
68. Eastman A, Kohn EA, Brown MK, et al. A novel indolocarbazole, ICP-1, abrogates DNA damage-induced cell cycle arrest and enhances cytotoxicity: Checkpoint Abrogator UCN-01. *Mol Cancer Ther* 2002;1:1067–1078.
69. Fedier A, Ruefenacht UB, Schwarz VA, et al. Increased sensitivity of p53-deficient cells to anticancer agents due to loss of Pms2. *Br J Cancer* 2002;87:1027–1033.
70. Jordan P, Carmo-Fonesca M. Molecular mechanisms involved in cisplatin cytotoxicity. *Cancer Res* 2000;57:1229–1235.
71. Colella G, Marchini S, D'incalci M, et al. Mismatch repair deficiency is associated with resistance to DNA minor groove alkylating agents. *Br J Cancer* 1999;80:338–343.
72. Shimodaira H, Yoshioka-Yamashita A, Kolodner RD, et al. Interaction of mismatch repair protein PMS2 and the p53 related transcription factor p73 in apoptosis response to cisplatin. *Proc Natl Acad Sci USA* 2003;100:2420–2425.

73. Sak A, Stuschke M, Wurm R, et al. Selective inactivation of DNA-dependent protein kinase with antisense oligodeoxynucleotides: consequences for the rejoining of radiation-induced DNA double-strand breaks and radiosensitivity of human cancer cell lines. *Cancer Res* 2002;62:6621–6624.

74. Kim CH, Park SJ, Lee SH. A targeted inhibition of DNA-dependent protein kinase sensitizes breast cancer cells following ionizing radiation. *J Pharmacol Exp Ther* 2002;303:753–759.

75. Muller C, Rodrigo G, Caslou P, et al. The DNA dependent protein kinase: a major protein involved in the cellular response to ionizing radiation. *Bull Cancer* 1999;86:977–983.

76. Persons DL, Yazlovitskaya EM, Cui W, et al. Cisplatin-induced activation of mitogen-activated protein kinases in ovarian carcinoma cells: inhibition of extracellular signal-regulated kinase activity increases sensitivity to cisplatin. *Clin Cancer Res* 1999;5:1007–1014.

77. Ilekis JV, Conner JP, Prins GS, et al. Expression of epidermal growth factor and androgen receptors in ovarian cancer. *Gynecol Oncol* 1997;66:250–254.

78. Modjtahedi H, Dean C. The receptor for EGF and its ligands: expression, prognostic value and target for tumour therapy. *Int J Oncol* 1998;4:277–296.

79. Turner T, Chen P, Goodly LJ, et al. EGF receptor signaling enhances in vivo invasiveness of DU-145 human prostate carcinoma cells. *Clin Exp Metastasis* 1996;14:409–418.

80. Meden H, Kuhn W. Overexpression of the oncogene c-erbB-2 (HER2neu) in ovarian cancer: a new prognostic factor. *Eur J Obstet Gynecol Reprod Biol* 1997;71:173–179.

81. Alapetite C, Thirion P, De la Rochefediere A, et al. Analysis by alkaline comet assay of cancer patients with severe reactions to radiotherapy: defective rejoining of radioinduced DNA strand breaks in lymphocytes of breast cancer patients. *Mol Cell Endocrinol* 1999;117:553–558.

82. Scoccia B, Lee YM, Niederberger C, et al. Expression of the ErbB family of receptors in ovarian cancer. *J Soc Gynecol Invest* 1998;5:161–165.

83. Saito Y, Haendeler J, Hojo Y, et al. Receptor heterodimerization: essential mechanism for platelet-derived growth factor-induced epidermal growth factor receptor transactivation. *Mol Cell Biol* 2001;21:6387–6394.

84. Donato NJ, Perez M, Kang H, et al. EGF Receptor and p21WAF1 Expression are reciprocally altered as ME-180 cervical carcinoma cells progress from high to low cisplatin sensitivity. *Clin Cancer Res* 2000;6:193–202.

85. Dixit M, Yang JL, Poirier MC, et al. Abrogation of cisplatin-induced programmed cell death in human breast cancer cells by epidermal growth factor antisense. *J Natl Cancer Inst* 1997;89:365–373.

86. Brown D, Wang R, Russell P. Anti-epidermal growth factor receptor antibodies augment cytotoxicity of chemotherapeutic agents on squamous carcinoma cell lines. *Otolaryngol–Head Neck* 2000;122:75–83.

87. Ciardello F, Bianchi R, Damiano V, et al. Antitumor activity of sequential treatment with topotecan and anti-epidermal growth factor receptor monoclonal antibody C225. *Clin Cancer Res* 1999;5:909–916.

88. Gleseg MA, de Bock C, Ferguson LR, et al. Evidence for epidermal growth factor receptor-enhanced chemosensitivity in combinations of cisplatin and the new irreversible tyrosine kinase inhibitor CI-1033. *Anti-Cancer Drugs* 2001;12:683–690.

89. Tsai CM, Levitzki A, Wu LH, et al. Enhancement of chemosensitivity by tyrphostin AG825 in high-p185 expressing non-small cell lung cancer cells. *Cancer Res* 1996;56:1068–1074.

90. Yacoub A, McKinstry R, Hinman D, et al. Epidermal growth factor and ionizing radiation up-regulate the DNA repair genes XRCC1 and ERCC1 in DU145 and LNCaP prostate carcinoma through MAPK signaling. *Radiat Res* 2003;159:439–452.

91. Benhar M, Engelberg D, Levitzki A. Cisplatin-induced activation of the EGF receptor. *Oncogene* 2002;21:8723–8731.

92. Yen L, Zengrong N, You XL, et al. Regulation of cellular response to cisplatin-induced DNA damage and DNA repair in cells overexpressing p185erbB-2 is dependent on the ras singaling pathway. *Oncogene* 1997;14:1827–1835.

93. Peitras RJ, Poen JC, Gallardo D, et al. Monoclonal antibody to HER2/*neu* receptor modulates repair of radiation-induced DNA damage and enhances radiosensitivity of human breast cancer cells overexpressing this oncogene. *Cancer Res* 1999;59:1347–1355.

94. Tsai C, Chang K, Perng R, et al. Correlation of instrinsic chemoresistance of non-small-cell lung cancer cell line with HER-2/neu gene expression but not with ras gene mutations. *J Natl Cancer Inst* 1993;85:897–901.

95. Zhang X, Li QQ, Reed E. SU5416 sensitizes ovarian cancer cells to cisplatin through inhibition of nucleotide excision repair. *Cell Mol Life Sci* 2003;60:794–802.

96. Quinn TP, Peters KG, De Vries C, et al. Fetal liver kinase 1 is a receptor for vascular endothelial growth factor and is selectively expressed in vascular endothelium. *Proc Natl Acad Sci USA* 1993;90:7533–7537.

97. Strawn LM, Shawver LK. Tyrosine kinases in disease: overview of kinase inhibitors as therapeutic agents and current drugs in clinical trials. *Curr Opin Invest Drugs* 1998;7:553–557.

98. Reed E. Platinum analogs, anticancer drugs. In: DeVita VT, Hellman S, Rosenberg SA, eds. *Principles and Practice of Oncology.* Lippincott–Raven, Philadelphia, 1993:390–400.

99. Matheson S, McNamee J, Jean-Claude BJ. Design of a chimeric 3-methyl-1,2,3-triazene with mixed receptor tyrosine kinase and DNA damaging properties: a novel tumour targeting strategy. *J Pharm Exp Ther* 2001;296:832–840.

100. Brahimi F, Matheson S, McNamee J, et al. Inhibition of epidermal growth factor receptor-mediated signaling by "combi-triazene" BJ2000, a new probe for the combi-targeting postulates. *J Pharm Exp Ther* 2002;303:238–246.

101. Matheson SL, McNamee J, Jean-Claude BJ. Differential responses of EGFR–/–AGT-expressing cells to the "combi-triazene" SMA41. *Cancer Chemother Pharmacol* 2003;51:11–20.

102. Qiu Q, Dudouit F, Matheson SL, McNamee J, et al. The combi-targeting concept: a novel 3,3-disubstituted nitrosourea with EGFR tyrosine kinase inhibitory properties. *Cancer Chemother Pharmacol* 2003;51:1–10.

103. Rachid Z, Katsoulas A, Brahimi F, et al. Synthesis of pyrimidinopyridine–triazene conjugates targeted to the bcr-abl oncogene. *Bioorg Med Chem Lett* 2003;13:3297–3300.

104. Rachid Z, Brahimi F, Teoh N, et al. The combi-targeting concept: chemical dissection of the dual targeting properties of a series of "combi-triazenes." *J Med Chem* 2003;46:4313–4321.

105. Kastan MB, Radin AI, Kuerbitz SJ, et al. Levels of p53 protein increase with maturation in human hematopoietic cells. *Cancer Res* 1991;51:4279–4286.

106. Agarwal ML, Agarwal A, Taylor WR, et al. p53 controls both the G2/M and the G1 cell cycle checkpoints and mediates reversible growth arrest in human fibroblasts. *Proc Natl Acad Sci USA* 1996;92:8493–8497.

107. Banerjee R, Rachid Z, McNamee J, et al. Synthesis of a prodrug designed to release multiple inhibitors of the epidermal growth factor receptor (EGFR) tyrosine kinase and an alkylating agent: a novel tumor targeting concept. *J Med Chem* 2003;46:5546–5551.

5 Regulation of DNA Repair and Apoptosis by p53 and Its Impact on Alkylating Drug Resistance of Tumor Cells

Istvan Boldogh, PhD, *Kishor K. Bhakat*, PhD, *Dora Bocangel*, PhD, *Gokul C. Das*, PhD, *and Sankar Mitra*, PhD

CONTENTS

1. INTRODUCTION

Eukarytotic cells have developed a highly conserved network of processes to ensure that the damaged genome is repaired before replication or mitotic segregation. This protective mechanism is essential for maintaining genomic integrity and stability, cell viability, and prevention of mutations. The drugs used in the treatment of human malignancies are invariably genotoxic, and their effectiveness is limited by a variety of factors. The most important factor is the tumor

From: *Cancer Drug Discovery and Development: DNA Repair in Cancer Therapy*
Edited by: L. C. Panasci and M. A. Alaoui-Jamali © Humana Press Inc., Totowa, NJ

cells' resistance to drugs. Most tumor cells could develop resistance to chemotherapeutic agents, whereas some malignant cells, including prostate carcinomas, melanomas, and gliomas, are intrinsically resistant to most antitumor drugs. For instance, ovarian and small-cell lung cancers are known to respond well to drug treatment initially, only to relapse with the appearance of drug-resistant cells. This acquired drug resistance resulting from genomic instability and consequent plasticity of the genome is encountered in approx 40% of all cancer patients undergoing chemotherapy.

Alkylating drugs are among the oldest and most commonly used therapeutic agents, because of their potent cytotoxic activity toward proliferating cells. This family of drugs includes nitrogen mustards, ethyleneamines, alkylsulfonates, N-chloroethyl-N-nitrosoureas (CNU), and triazenes. Some of these act directly, whereas others require enzymatic or spontaneous activation to form active species that can bind covalently to nucleophilic sites of all cellular molecules. Ring nitrogen, exocyclic oxygen, and phosphates are the common targets for alkylation of DNA, although the extent of reaction at any particular site may vary widely for different agents (reviewed in refs. *1–3*).

The formation of alkyl base monoadducts and/or crosslinks in DNA is the primary mechanism of genotoxicity of alkylating agents. The chemical structure of these adducts varies widely with different alkylating drugs. The most common electrophilic attack site of N-mustards is the N-7 position of guanine (G), whereas CNUs commonly react with the O-6 and N-7 positions of G, in addition to phosphate residues. Some of the monoalkyl adducts are mutagenic, particularly O^6-alkylguanine, and could also be toxic because of their ability to inhibit transcription and replication *(1–3)*. However, most alkylating agents currently used in chemotherapy produce intrastrand and interstrand DNA crosslinks that activate the DNA repair system or signaling cascades for apoptosis *(4)*.

Cellular and pharmacological factors determine the extent of both the intrinsic and acquired drug resistance of tumor cells. Acquired drug resistance could result from enhanced drug detoxification, enhanced export, increased metabolism of reactive oxygen species, and increased repair of drug-induced DNA lesions. DNA double-strand breaks (DSBs) are eliminated via nonhomologous end-joining (NHEJ) and homologous recombination (HR). O^6-Alkylguanine adducts are induced by monofunctional temozolomide (TMZ) and procarbazine and also by bifunctional CNU derivatives. These are repaired via direct reversal by O^6-methylguanine-DNA methyltransferase (MGMT) *(5–7)*. Intrinsic or acquired drug resistance may be developed as a result of dysfunction of apoptotic pathways. A change in the levels of proapoptotic and antiapoptotic proteins could affect the sensitivity of cells to drugs *(8–10)*.

The tumor suppressor protein p53 is a trans-acting factor and a critical mediator of cellular responses to DNA damage induced by genotoxic agents, including radiation and drugs. The p53 level is transiently enhanced as a result of increased

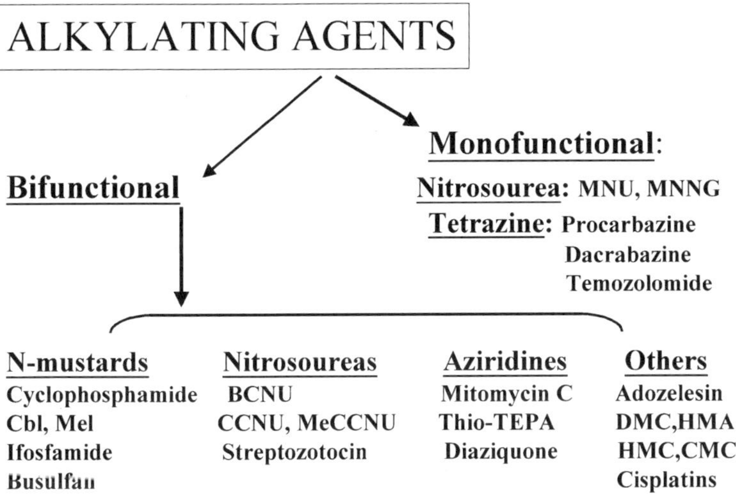

Fig. 1. Classification of alkylating agents. MNU, N-methyl-N-nitrozourea; MNNG, N'-nitro-N-nitrosoguanidine; BCNU, 1,3-bis(2-chloroethyl)-1-nitrosoureas; Cbl, chlorambucil; Mel, melphalan; CCNU, 1-(2-chloroethyl)-3-cyclohexyl-1-nitrosourea; MeCCNU, methyl,1-(2-chloroethyl)-3-cyclohexyl-1-nitrosourea; thio-TEPA, N, N', N', triethylenethiophosphoramide; CMC, 1-(2-chloroethyl)-3-methyl-3-carbethoxytriazene; DMC, 1,3-dimethyl-3-carbethoxytriazene; HMA, 1-(2-hydroxyethyl)-3-methyl-3-acetyltriazen; HMC, 1-(2-hydroxyethyl)-3-methyl-3-carbethoxytriazene.

stability of its mRNA and the protein itself via complex regulatory mechanisms *(11)*. p53 may act directly or function via a series of downstream genes (e.g., $p21^{WAF1}$, GADD45, Bax, Fas/Apo-1, and other p53-inducible genes). In more than 50% of tumor cells, p53 is inactivated because of point mutations, deletion, increased degradation, sequestration, or interaction with cellular and viral proteins *(12)*. Although the presence of wild-type p53 should favor drug-induced apoptosis, increased repair of damaged DNA may still limit the therapeutic index after drug treatment.

In the following sections of this review, we provide a brief overview of the pharmacological actions of different types of alkylating agent used in cancer chemotherapy and of possible mechanisms of drug resistance, and we review the role of p53 in DNA repair and apoptosis.

2. PHARMACOLOGICAL ACTION OF ALKYLATING DRUGS

Based on their genotoxic action, the alkylating drugs are classified broadly into two categories: monofunctional alkylating agents induce DNA base monoadducts, while bifunctional alkylating drugs induce DNA intra- or inter-strand crosslinks (Fig. 1) *(1,13)*. Among DNA base monoadducts, O^6-alkylguanine (O^6-alkyl G) may be most critical because it is highly mutagenic *(2,14)*. O^6-Alkyl G can pair with thymine (T) during DNA replication and produce mismatches in DNA *(15,16)*. Whereas subsequent segregation of replicated genomes results in G→A transition mutations, an O^6-alkyl G:T mismatch is a substrate for the mismatch repair (MMR) process, in which the T-containing nascent strand is cleaved in an attempt to insert the correct base *(17)*. In the absence of O^6-alkyl G repair, these O^6-alkyl G:T mismatches produce persistent DNA single-strand breaks (SSBs) as a result of futile repetition of the DNA MMR cycle. The monofunctional alkylating drugs TMZ and procarbazine are imidazotetrazine derivatives effective against human glioma with minimal side effects *(18)*. First, they require metabolic dealkylation to produce an unstable intermediate, which decomposes to release a methyldiazonium (CH_3-N^+—N) group. TMZ then becomes a monofunctional methylating agent, which produces a high-level O^6-methyl G adduct in DNA *(13)*.

Agents that produce DNA crosslinks include CNUs and nitrogen mustard (NM)-type alkylating agents, as well as cisplatin compounds and mitomycin C. CNU derivates developed in the early 1960s as potent antitumor agents include 1,3-bis(2-chloroethyl)-1-nitrosoureas (BCNU) or 1-(2-chloroethyl)-3-cyclohexyl-1-nitrosourea (CCNU) and methyl CCNU. These bifunctional alkylating agents react within the same or opposite DNA strands to generate intrastrand or interstrand crosslinks, respectively *(1,13)*. These compounds spontaneously decompose in vivo to form a chloroethylcarbonium ion, which alkylates DNA at multiple positions, including the O^6 position of guanine. The O^6-alkyl G adduct undergoes slow intramolecular rearrangement to yield initially an O^6,N^1-guanine cyclized intermediate and, ultimately, a cytotoxic N^1-guanine, N^3-cytosine DNA interstrand crosslinks with the C residue in the complementary strand opposite G *(1,13)*.

The NMs form an important class of bifunctional alkylating agents that initially alkylate the *N*-7 position of G, followed by interstrand crosslinks with a C residue in the complementary strand *(13)*. Cyclophosphamides, the most commonly used NM, requires metabolic activation, whereas chlorambucil (Cbl), a phenylbutyric acid derivative of NM, spontaneously decomposes into active species and has been used in a variety of human malignancies, including chronic lymphocytic leukemia *(19)* and ovarian cell carcinoma *(20)*. Other alkylating agents include mitomycin C, which primarily produces DNA intrastrand crosslinks. The aziridine and carbamate groups on the mitomycin C molecule are necessary for reductive activation of the molecule to a bifunctional alkylating agent. The molecular sites of DNA alkylation by mitomycin C are preferentially at the N^2 and O^6 positions of adjacent guanines in the minor groove of cellular DNA. Chlorambucil, mitomycin C, and cisplatin (a platinum compound) also produce thermolabile glycosylic bonds, yielding apurinic sites that, in turn, can cause DNA DSBs and/or SSBs and are generated as intermediates in DNA repair

(1,13). Although there are several hypothesis, the mechanism of DSB generation during the repair of interstrand crosslinks (ICL) in mammalian cells is not clear.

3. TUMOR CELL RESISTANCE RESULTING FROM ENHANCED LESION REPAIR, EFFECT OF p53

Essentially, two systems are responsible for genome-integrity drug resistance: DNA repair and programmed cell death or apoptosis. Cells that have increased DNA repair and/or are defective in some DNA repair pathways lead to increased survival of tumor cells. It has also become obvious that key proteins, which contribute to DNA repair, are often involved in apoptosis when excessive DNA damage is induced. Many of these proteins, p53, breast-cancer-associated gene (BRCA1), ataxia telagiectasia (ATM), AT-related protein (ATR), Werner syndrome helicase (WRN), and double-stranded DNA (dsDNA)-dependent protein kinase (DNA-PK) have dual functions as sensors of DNA damage and are components of major DNA repair pathways including HR, NHEJ, nucleotide excision repair (NER), base excision repair (BER), and mismatch repair (MMR). Overexpression or mutations in these proteins often leads to increased drug resistance of tumor cells.

In response to DNA damage, the level of activated p53 is elevated, which causes cell cycle arrest both at the G1/S- and G2/M-phases. The cell cycle arrest may also induce apoptosis. In addition to its regulatory role in the cell cycle, p53 has a direct role in repair of DNA damage induced by crosslinking alkylating agents (e.g., cyclophosphamide, platinum [e.g., cisplatin, oxaliplatin]), via MMR *(21)*, NER *(22–24)*, as well as BER *(25)* pathways (Fig. 2). Although p53 protein has no apparent specificity to a particular repair process, it interacts with such protein as xeroderma pigmentosum (XP) complementation group proteins XPB and XPD involved in NER, DNA β polymerase (polβ) involved in BER, and Rad51 required for recombination-mediated DNA DSB repair *(9,26)*.

In vitro studies have shown interaction of p53 with various DNA structural modifications, including single-stranded DNA (ssDNA) or DNA duplex with single-stranded gaps, nicks, and free ends. The p53 polypeptide has been shown to enhance DNA reannealing and to possess 3'→5' exonuclease activity. It binds strongly and with high specificity to Holliday junctions and facilitates their cleavage, an important step in the HR pathway *(27,28)*. In addition, p53 interacts physically and functionally with BRCA2 *(29)* and RAD51 *(30)*, consistent with a role of p53 in HR. It appears that BRCA2 and Rad51 cooperate to downregulate p53 *(29)* and reduce p53 transactivation activity and to limit the extent of p53-mediated cell cycle arrest after DNA damage. Thus, p53 participates not only in HR but also in cell cycle arrest to allow time for DNA repair. Apoptosis could ensue if repair is insufficient.

There is an unusual relationship between defects in the DNA repair machinery and suppression of p53-induced apoptosis *(9)*. Apoptosis mediated by p53 is

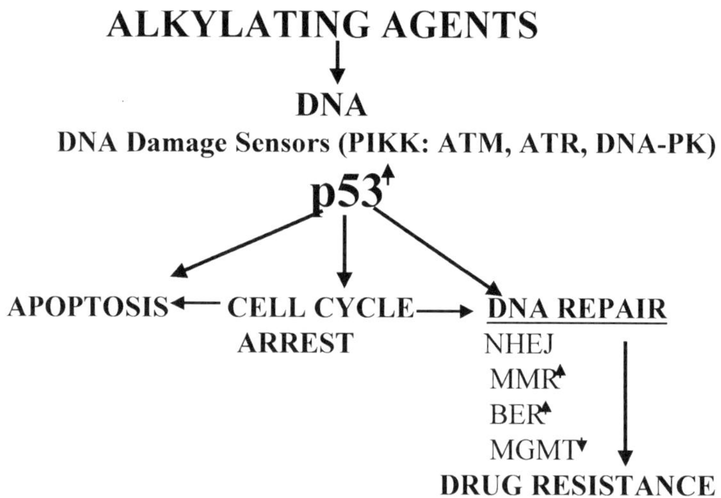

Fig. 2. Role of p53 in cellular response to DNA damage induced by alkylating agents. DNA damage/lesions act as the signals and activates signaling kinases involving the phosphatidylinositol 3-kinase-like (PIKK) family (e.g., ATM, DNA-PK). ATM and ATR then amplify the signal by activating downstream kinases (Chk1 and Chk2) *(211)* and these, in turn, posttranslationally modify p53. p53 induces arrest of the cell cycle to allow cells sufficient time for DNA repair (MGMT, NHEJ, MMR) or to initiate apoptosis when excessive DNA damage has been induced. NHEJ, nonhomologous end-joining; MMR, mismatch repair; BER, base excision repair; MGMT (O^6-methylguanine-DNA methyltransferase).

abrogated in cells containing mutations in XPB or XPD, components of the NER pathway. NER is involved in the repair of bulky lesions *(31)*, whereas the mutant Cockayne syndrome B protein (CSB/XRCC6) induces p53-dependent apoptosis *(32)*. The apoptotic response is restored by the introduction of wild-type XPB and XPD genes *(22)*. Attenuation of p53-mediated apoptosis was also observed in cells with a defect in the Werner protein, a DNA helicase *(33)*. The $3' \rightarrow 5'$ exonuclease activity intrinsic to p53 may be required to ensure sequence fidelity during DNA replication and repair. This activity could provide a molecular basis for p53's involvement not only in BER but also in MMR, where proofreading is necessary *(9)*.

3.1. DNA Crosslink Repair

DNA ICLs are believed to be the most toxic DNA lesions. The unrepaired damage in surviving cells may cause mutations and chromosomal rearrangements, leading to cell transformation and uncontrolled cell growth, tumor formation, increased metastatic capacity, and/or drug resistance during chemotherapy *(34–36)*.

Tumor cell resistance to alkylating drugs is associated with enhanced repair of DNA DSBs via NHEJ *(37)* or HR pathways *(34,38)*. DNA DSBs are likely generated at the site of crosslinks as repair or replication intermediates, which are rejoined by NHEJ and HR processes. NHEJ does not require significant sequence homology for joining two DNA molecules. The broken ends can thus be rejoined even where there is little or no homology at the site of DSBs and without a template. NHEJ requires a conserved set of core proteins to process DNA DSBs. DNA-PK, a critical component of the NHEJ complex, is a nuclear serine–threonine kinase, consisting of a catalytic subunit of 460,000 Da (DNA-PKcs) and the DNA-binding subunit, a heterodimer of Ku70 and Ku86 *(39)*. The Ku complex has high affinity for duplex DNA ends and other discontinuities in the DNA, and it recruits DNA-PKcs to the damage site *(40)*. DNA-PKcs, activated by serine/threonine phosphatases *(41)*, is capable of phosphorylating p53 and other DNA-bound proteins and form a repair complex that includes DNA ligase IV and XRCC4 *(8,42)*. Studies of the NHEJ reaction suggest that the DNA ends are held together precisely in the repair complex, allowing polymerases to fill in gaps and nucleases to trim excess ends as needed before the ligation step. As few as one to two complementary basepairs, if present, can position the junction. It has been inferred that repair by NHEJ is enhanced by alignment factors that help align DNA ends to maintain basepairing and stacking interactions *(34)*.

Mutations in DNA-PKcs and Ku cause increased sensitivity to ionizing radiation (IR) and bifunctional alkylating agents *(43)*, although deficiency of DNA-PK causes milder phenotype than the defects in Ku. This suggests that DNA crosslinks induced by alkylating agents are repaired by NHEJ. Consistent with this, cells containing a variant form of Ku80 have the same level of kinase activity of DNA-PKcs and higher sensitivity to alkylating agents *(35)*. Wortmannin, an inhibitor of the phosphatidylinositol 3-kinase family, inhibits DNA-PKcs and sensitizes chronic lymphoid leukemia cells to NMs *(44)*. Thus, it is likely that increased kinase activity of DNA-PKcs is the result of an increased level of phosphorylated Ku proteins that enhances NHEJ activity and tumor cells' resistance to alkylating drugs *(19)*.

The linkage of p53 to NHEJ is controversial *(45)*. Both in vitro and in vivo studies have shown that p53 is not involved in the rejoining of DNA DSBs *(46,47)*. Furthermore, p53 inhibits both interchromosomal and intrachromosomal recombination and thereby triggers apoptosis. Supporting this observation, embryonic lethality and increased apoptosis associated with lack of functional ligase IV and XRCC4 are rescued by p53 deficiency *(48)*.

DNA DSBs can also be repaired by the HR pathway, in which the DNA terminal sequences of one strand are first removed in the $5' \rightarrow 3'$ direction by exonucleases *(49)*. The resulting single-strand tails then invade the DNA double helix of a homologous, sister chromatide in G2 cells or other chromosome in G1 cells and are extended by the action of a DNA polymerase *(50)*. After branch migration, the resulting DNA crossovers (Holliday junctions) are resolved to yield two intact DNA molecules. Strand invasion requires participation of the

Rad51 strand-exchange protein and a number of associated proteins, including Rad51B, Rad51D, Rad52, Rad54, MRE11, and NBS1 *(49)*. Mammalian Rad51 also interacts with the p53 and with BRCA1 and BRCA2 which have been shown to act as suppressors of breast cancer in the humans. p53 may be directly or indirectly involved in HR, apparently independent of its transcriptional activation properties. However, with RAD51, homozygous loss of BRCA1 or BRCA2 result in early embryonic lethality in mice, which is attenuated by p53 deficiency *(49,51)*. Notably, BRCA1- and BRCA2-deficient cells have enhanced sensitivity to DNA-damaging agents and develop spontaneous chromosomal aberrations. Initial studies suggested that NHEJ was the predominant mechanism of DSB repair in mammalian cells, but it is now established that HR also has a crucial role *(52,53)*. Thus, enhancement of DSB repair via both HR and NHEJ processes may contribute to the resistance of the tumor cells to alkylating agents *(34,49)*.

3.2. Reversal of O^6-Alkylguanine Adduct

The O^6-methylguanine-DNA methyltransferase is a ubiquitous protein, but its expression level is highly variable in different cell types *(5,6,54,55)*. Because the level of MGMT could be directly correlated with tumor cell resistance to both CNU-type and monofunctional alkylating agents, cellular factors controlling MGMT expression are clinically important *(56,57)*. The human MGMT gene spans more than 150 kb and contains five exons, the first of which is non-coding. As with most housekeeping genes, the MGMT gene lacks TATA and CAAT boxes in its promoter and contains a CpG island with multiple CpG sequences in six putative Sp1 recognition sites, two glucocorticoid-receptor-binding (GRE) elements, and two putative AP-1 and AP-2 binding cis elements *(58,59)*. AP-1 and GRE sequences are involved in the activation of MGMT expression *(58,59)*. Most recently, the chromatin structure of the MGMT promoter has also been suggested to play a role in MGMT expression. Furthermore, we have provided the first evidence that histone acetylation plays a role in MGMT expression; both endogenous and MGMT promoter-driven reporter expression was enhanced by TSA, a histone deacetylase inhibitor *(60)*. We speculate that targeted acetylation of histones leads to the loosening of the nucleosome structure, which may, in turn, facilitate the binding of AP-1 and Sp1 transcription factors to the promoter region and modulate the expression of MGMT.

O^6-methylguanine-DNA methyltransferase overexpression renders tumor cells resistant to CNU-type drugs, whereas tumor cells with a low level of MGMT are highly sensitive to these drugs *(54,55,61–63)*. However, some tumor cells do not express MGMT, despite the presence of an intact gene, and were named Mex⁻/Mer⁻ *(5,6)*. DNA alkylation at the O^6 position of guanine represent a major lesion in the carcinogenic, mutagenic, and clastogenic actions of methylating (dacarbazine, temozolomide, streptozotocin) and chloroethylating (e.g., 1,3-bis[2-chloroethyl]-1-nitrosourea [BCNU] and 1-[2-chloroethyl]-3-cyclohexyl-1-

nitrosourea) agents. The O^6-alkyl G and N-alkylpurines are removed from DNA by MGMT *(64–66)* and by N-methylpurine-DNA glycosylase (MPG) *(67,68)*, respectively, which are ubiquitous DNA repair proteins. Furthermore, MGMT removes not only methyl groups but also 2-chloroethyl, benzyl, and pyridyloxobutyl adducts. MGMT carries out a stoichiometric and suicidal reaction in which the alkyl groups bound to the O^6 position of guanine are transferred to the active-site cysteine, resulting in the direct restoration of the normal base (G) in DNA and self-inactivation of MGMT *(69,70)*. In the case of bifunctional alkylators like BCNU, removal of O^6-chloroethyl G, the primary adduct, by MGMT prevents subsequent production of cytotoxic ICLs *(1)*.

The relationship between the p53 status and MGMT expression is complex, and conflicting results have been published *(71–75)*. Expression of p53 renders neonatal mouse astrocytes resistant to BCNU, independent of the protein's ability to regulate cell cycle and induce apoptosis. The drug-resistant astrocytes express some fivefold higher MGMT activity than either heterozygous or p53 null cells *(76)*. A similar effect was observed in glioblastoma (GM47.23) and lung tumor cell lines (H1299, H460) *(77)*. On the other hand, p53 activation strongly represses the MGMT gene, but the molecular basis for this effect is not yet clear *(78)*. Our results indicate that the interaction of p53 with Sp1, which binds to the basal MGMT promoter, reduces its expression (Bocangel et al., unpublished). In any event, a detailed analysis of the effects of p53 on MGMT regulation may help design chemotherapeutic protocols involving CNU-type drugs for tumors with different p53 status.

3.3. Impact of MMR on Tumor Cell Sensitivity to Alkylating Agents

In human cells (as well as in bacteria or yeast), repair of basepair mismatch (MMR) is nascent strand-specific and directed by a nick located 5' or 3' to the mismatch *(79)*. The mismatch is recognized in mammalian cells by MutSα (MSH2/MSH6) and MutSβ (MSH2/MSH3) complexes, which bind to the DNA mismatches (single-basepair mismatches and large mismatched loops, respectively). Binding of an hMLHI–hPMS2 heterodimer to the DNA mismatch MutS complex is dependent on ATP hydrolysis and requires proliferating cell nuclear antigen (PCNA). In vitro studies show that after the formation of 5' or 3' nicks, the DNA strand containing the incorrect base is removed by FEN1, which is stimulated by PCNA after helicase-mediated unwinding of the DNA helix. Using the 3'-OH-containing single-strand end as a primer, a replicative DNA polymerase fills the gap, with the aid of replication protein A (RPA) and DNA ligase seals DNA strand for restoring genome integrity *(80)*.

Resistance of tumor cells to methylating drugs (e.g., TMZ and procarbazine), which induce O^6-methyl G depends not only on the level of O^6-methylguanine methyltransferase (MGMT), but also on the status of the MMR pathway, particularly when MGMT expression is low *(7)*. In contrast to other DNA repair machineries, no increased activity of the MMR pathway was observed in drug-resistant cells, but a deficiency in MMR activity causes resistance to a number of clinically important drugs that include busulfan, *N*-nitroso-*N*-methylurea (MNU), procarbazine, and TMZ as well as antimetabolites such as mercaptopurine, 6-thioguanine, and platinum compounds (carboplatin and cisplatin) *(79,81)*. It is possible that because impaired MMR causes a strong mutator phenotype, and overexpression of DNA repair genes that mediate resistance to antineoplastic agents may result from mutations in the regulatory pathway *(82)*. Because monofunctional alkylating agents are generally mutagenic, the drug itself may produce genomic instability in MMR-deficient cells and induce generalized drug resistance in a clinical situation *(83)*.

Because the genotoxic effect of O^6-methyl G is largely alleviated in cells deficient in DNA MMR, it was proposed that MMR is actively involved in O^6-methyl G-induced genotoxicity *(57,75)*. It appears that the O^6-methyl G adduct induced by monomethylating alkylating agents triggers apoptosis only following cell proliferation and DNA replication. It is important to note that O^6-methyl G itself does not block DNA replication, but during replication, it can pair with thymine and thus produce O^6-methyl G:T mispairs *(15,84)*. Such mispairs are subject to MMR in which T is removed from O^6-methyl G:T in the nascent strand via a process initiated by binding of MutSα heterodimer to the mismatch. It has been postulated that repeated cycles of MMR of this basepair mismatch becomes futile and result in persistent DNA SSBs. A subsequent cycle of replication will convert this SSB into a DSB, a trigger for p53-mediated apoptosis *(85)*. Thus MMR-deficient tumor cells develop resistance to alkylating drugs even when O^6-alkylG is not repaired because of the low level of MGMT in these cells. On the other hand, tumors expressing a low level of MGMT but proficient in MMR have a good response to chemotherapy involving methylating drugs *(85)*.

Although there is no evidence that p53 stimulates repair of DNA mismatches, it is noteworthy that p53 protein exhibits intrinsic 3'→5' exonuclease activity, which maps to the core DNA-binding domain of p53. The observed specificity of mismatch excision shows that p53 exonucleolytic proofreading preferentially repairs transversions *(85)*. p53 exhibits mispair excision with a preference of A>G>C opposite A in the template adenine and with a preference of G>G>T opposite the template guanine residue. The observations that p53 is colocalized with the DNA replication machinery and that DNA polymerase α (polα), unlike polδ and polε, lacks the 3'→5' exonuclease activity for proofreading, providing hints that link p53 with fidelity of DNA replication. Thus, it is possible that the 3'→5' exonuclease activity of p53 complements the function of polα during DNA replication to ensure the excision of misincorporated nucleotides. This hypothesis was supported by the observation that 3'→5' exonuclease activity of p53 significantly reduces the number of mismatched nucleotides incorporated into DNA by polα and thus enhances in vitro DNA replication fidelity *(21,85)*.

4. MODULATION OF CELLULAR REDOX STATE AND ALKYLATING DRUG RESISTANCE

Reactive oxygen species (ROS) are generated when oxygen is partially reduced (e.g., in mitochondria) *(86)*. In addition to the mitochondria, other sources of ROS generation include endogenous enzyme systems (e.g., plasma membrane NADPH-oxidase and cytoplasmic xanthine oxidase), as well as organellar sources (e.g., peroxisomal cytochrome P450 oxidases) *(87)*. When cells are exposed to IR *(88)* and such chemotherapeutic agents as bleomycin *(88)*, anthracyclins (e.g., daunorubicin) *(89–91)*, and DNA crosslinking alkylating agents, cellular ROS levels are significantly increased (*see* Fig. 4).

Reactive oxygen species including superoxide anions ($O_2{}^{-}$), singlet oxygen *(102)*, hydrogen peroxide (H_2O_2), hydroxyl radicals (·OH), lipid peroxides (LOO–) and nitric oxide (NO) rapidly react with all cellular components, including DNA, proteins, and lipids. The cellular redox balance is maintained by antioxidants that inactivate these species; when this balance is disrupted, oxidative stress develops *(87)*. Such stress alters the cellular redox state (e.g., depletion of nucleotide coenzymes and disturbance of sulfhydryl-containing enzymes and other proteins). Several cellular processes are induced, ranging from changes in the activity of signal transduction pathways to gene expression, cell proliferation, mutagenesis, and apoptosis *(87,90,92)*.

4.1. ROS-Induced Signaling and Apoptosis

Reactive oxygen species are associated with cytotoxicity, because they carry out widespread and random oxidation. This attack by ROS on cellular components under conditions where oxidative stress is the initiating stimulus for apoptosis trigger cell death as a result of cumulative oxidative damage. Accumulating evidence now suggests that ROS also act as signaling molecules for the initiation and execution of the apoptotic death program in many, if not all, cases. The paradigm that oxidative stress is a mediator of apoptosis is based on the following observations: (1) apoptosis is associated with the generation of ROS; (2) a wide range of oxidants (H_2O_2, lipid- and thiol-oxidizing compounds, and redox-cycling quinones) have been found to be proapoptotic; (3) depletion of cellular antioxidants accelerates apoptosis; and (4) apoptosis can be inhibited by addition of antioxidant compounds.

Signaling by ROS appears not to be random, but targeted at components of specific metabolic and signal transduction pathways. There is also evidence that the enzymatic generation of ROS is not an unwanted byproduct of the primary reaction, but that ROS are used as signaling molecules rather than simply toxic metabolites to regulate cellular processes, including apoptosis. This view was further supported by the finding that cellular antioxidants not only serve to regulate ROS levels but also act as reversible redox modifiers of enzyme functions and apoptosis, as reviewed in ref. *93*.

Reactive oxygen species activate p53 and distinct mitogen-activated protein kinases (MAPK), including extracellular signal-regulated kinase (ERK), Jun N-terminal-kinase (JNK)/stress-activated protein kinase (SAPK), and p38-MAPK along with translocation/activation of protein kinase C (PKC) isoforms and inhibition of serine–threonine protein phosphatases *(94–96)*. ROS induce lipid peroxidation products (e.g., 4-hydroxynon-2-enal or 4-HNE) *(97)* and activate sphingomyelinase that generates ceramides *(94,98)*. Ceramides and lipid peroxides induce damage to mitochondria, leading to an increase in ROS levels *(99)*, which, in turn, activates proapoptotic processes and induces apoptosis *(99,100)*. These effects of ROS may be inhibited by antioxidants such as *N*-acetyl-L-cysteine (NAC) or butylated hydroxianisole (BHA) *(91)* as well as by the PKC isozyme (PKC-δ and -ε) inhibitor staurosporine *(94)*.

The glutathione-*S*-transferase (GST) family of enzymes play an important role in alkylating drug resistance by catalyzing glutathione (GSH)-dependent sequestration of the drug and ROS, thereby decreasing damage to cellular macromolecules and also the signals for apoptosis *(101–103)*. GSTs reduce intracellular concentration of lipid peroxidation products (e.g., 4-HNE, malondialdehyde). These reactive molecules may be involved in apoptosis by directly damaging cellular components such as mitochondria and activating cell signaling *(102–104)*. Additionally, recent studies show that the monomeric form of GSTπ is a negative regulator of JNK/SAPK signaling in nonstressed cells *(105)*. For example, H_2O_2 treatment causes GSTπ dimerization and dissociation of the GSTπ–JNK/SAPK complex, resulting in JNK/SAPK activation in cells. Supporting this idea, the addition of γ-glutamyl-*S*-(benzyl)cysteinyl-*R*-phenylglycine, a specific inhibitor of GSTπ, caused a dose-dependent activation of JNK/SAPK. Furthermore, overexpression of GSTπ decreased JNK/SAPK activity *(105)*. Thus, in addition to GST's drug-detoxification effect, GSTπ and possibly other GSTs may affect ROS-induced apoptosis by modulating the activity of redox-sensitive stress kinases (*see* Fig. 3).

4.2. Generation of ROS by Genotoxic Agents

In addition to direct induction of DNA damage, most genotoxic agents generate ROS *(90,91)*. To study the role of ROS in alkylating agent-induced apoptosis, we developed a variant (A2780/100) of ovarian carcinoma cell line (A2780) by repeated exposure to escalating concentrations of the drug *(106,107)* that is more than 10-fold resistant to Cbl than the parent cells. A2780/100 shows crossresistance to melphalan and cisplatin, and to a lesser degree to BCNU and etoposide (Eto), a topoisomerase inhibitor *(107)*. Because NAC pretreatment of cells (A2780 and A2780/100) increased survival and BSO pretreatment decreased

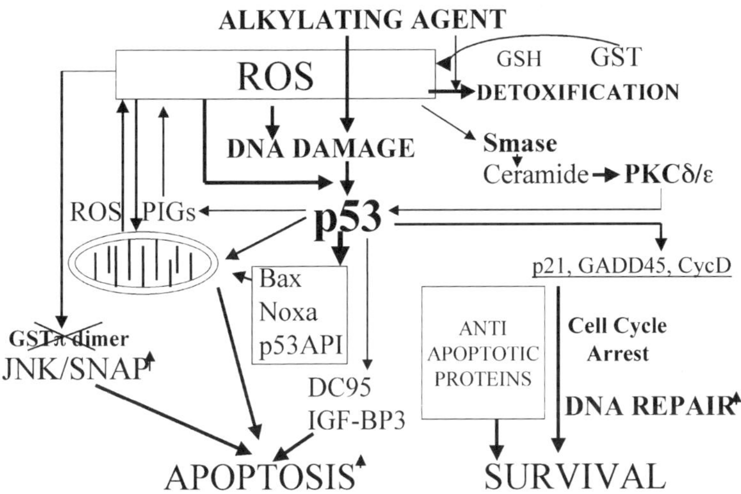

Fig. 3. Implications of ROS in p53-driven apoptosis. DNA damage and/or ROS-induced signaling activate p53, which transactivates regulators of cyclin kinase (e.g., p21^{Waf1}), DNA-damage-response genes (e.g., GADD45), and cyclin G (e.g., CycD) to induce cell cycle arrest to allow DNA repair. In the case of excessive DNA damage, p53 transactivates proapoptotic (e.g., Bax, NOXA, and p53AIP) and cell death proteins, including CD95, DR5, IGF-BP3. The death signal may be initiated at the cell surface by direct ligand–receptor interactions or by accumulation of Bax, Noxa, and p53API (trigger apoptosis by cyt-C release and activation of the Apaf-1/caspase-9 apoptosome) along with p53 in the mitochondria. In addition to drug-induced mitochondrial ROS, p53-inducible proteins (e.g., PIG3, PIG11) augment ROS production, which leads to chronic oxidative stress. ROS-mediated activation by diassociation of GSTπ from JNK/SNAP promotes death receptor (CD95, DR5)-dependent apoptotic signaling leading to promoter activation of CDC95-L, TNF-α, and also p53. ROS-activated sphingomyclinase generates ceramides, which, in turn, can activate p53 via PKC-δ or PKC-ε.

it after treatment with DNA crosslinking Cbl, or Eto (*see* Fig. 4B, right panel), we investigated whether ROS was generated in these cells. We have shown that Cbl (or Eto) exposure of cells induced ROS generation (as measured by 2'7'-dihydro-dichlorofluorescein [H$_2$DCF] assay) from 15 min onward. Remarkably, by 4–6 h after treatment, the ROS levels returned to normal in drug-resistant cells, whereas in drug-sensitive A2780 cells, ROS were continuously produced (*see* Fig. 4B, left panel), ultimately killing cells via apoptosis. Allopurinol (an inhibitor of xanthine oxidoreductase) and diphenylene iodinium chloride (which inhibits NADPH oxidoreductase) did not significantly affect drug-induced ROS production, suggesting that ROS production is associated with mitochondria in treated cells. To test this possibility directly, we used a redox-sensitive probe, dihydrorhodamine-123 (H$_2$RD123), which is primarily oxidized in the mito-chondria (*108*). As expected, the kinetics and extent of ROS levels determined

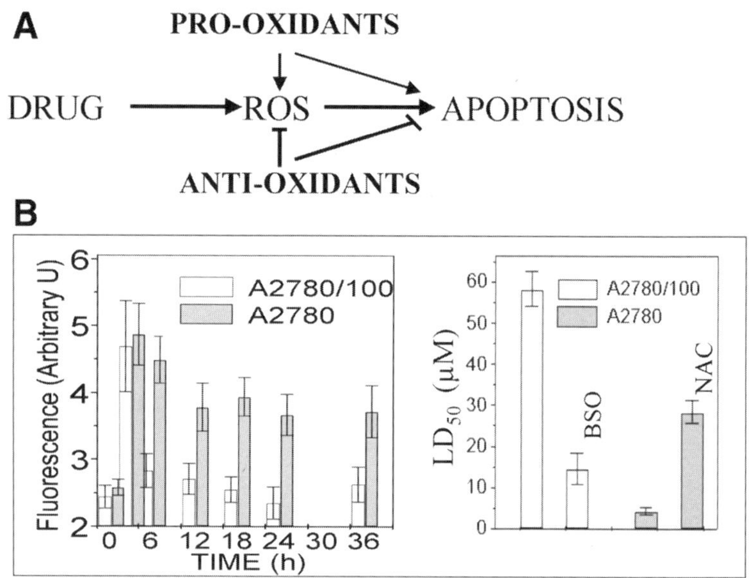

Fig. 4. Cellular redox state is critical for the efficacy of drug treatment. (**A**) Antioxidants inhibit while pro-oxidants enhance drug-induced apoptosis. (**B**) Drug-resistant cells are capable of coping with changes in cellular redox state. **Left panel:** ROS is chronically produced in drug-sensitive (A2780) cells, whereas drug resistant (A2780/100) cells normalize ROS production after Cbl treatment, as shown by the change in fluorescence of redox-sensitive 2'7'-dihydro-dichlorofluorescein. **Right panel:** BSO depletes GSH, increases ROS levels, sensitizes A2780/100 cells to Cbl (decreased LD_{50} from 58.4 ± 4 to 14.1 ± 2.4 μM), whereas NAC, an antioxidant, increases LD_{50} for Cbl (from 4.2 ± 0.4 to 28.4 μM) of drug-sensitive A2780 cells.

by H_2DCF or H_2RD132 were similar, indicating that chronic ROS production in treated drug-sensitive A2780 cells is the result of Cbl-induced damage to mitochondria. This finding is consistent with reactivity of Cbl with cellular macromolecules including DNA, RNA, and proteins *(3)*. We showed that chronic ROS-induced damage culminated in a mitochondria membrane permeability transition, collapse of mitochrondria membrane potential (m$\Delta\Psi$), and release of cytochrome-c (cyt-c) *(106)*. The mechanisms by which chronic ROS production in Cbl-sensitive cells develops and ROS generation terminates in drug-resistant cells warrant further investigation.

5. ROLE OF ROS IN p53-INDUCED APOPTOSIS

The sensitivity of tumor cells to drugs is markedly influenced both positively and negatively by a variety of gene products, many of which are mutated and/or dysfunctionally regulated in human cancers (e.g., p53, DNA repair proteins, and members of the Bcl-2 gene family) involved in regulation of the cellular redox state *(109)*. Redox sensitivity may be one of the biochemical mechanisms by which p53 (or Bcl-2 family members) act as a "sensor" of multiple forms of stress *(110,111)*. It appears that p53 is at the center of a network of complex redox interactions and that p53 regulates production of ROS induced by genotoxic agents *(112)*. Thus, a feedback loop exist between p53 expression and ROS production, which appears to amplify stress signals. Such a process could induce apoptosis, especially when the genotoxic damage is sensed to be beyond repair (*see* Fig. 3). ROS activates p53 and are also downstream mediators of p53-dependent apoptosis *(113)*. Thus, p53 may promote apoptosis via induction of redox-related genes *(114,115)*, some of which cause generation of ROS. The ROS cascade eventually precipitates apoptosis by targeting mitochondria *(116)*.

5.1. Posttranslational Modification of p53 by ROS-Induced Signaling

The p53 polypeptide is constitutively present in cells in an inactive form and at a very low level. Its steady-state level increases rapidly after DNA damage and other cellular insults associated with its posttranslational modifications *(110,111)*. Such modification sites are located in approx 100 amino acid (aa) residues at the N-terminus and 90 residues at the C-terminus of the protein. The N-terminal region becomes heavily phosphorylated, whereas specific Ser/Thr residues at the C-terminus are phosphorylated and Lys residues acetylated. Recent biochemical and genetic studies show that N-terminal phosphorylation(s) are important for stabilizing p53 and are crucial for acetylation of C-terminal Lys residues. Modifications at the C-terminus of p53 regulate stability, oligomerization state, nuclear import/export processes, and ubiquitination as a prelude to its degradation (reviewed in ref. *117*).

Oxidative stress induces posttranslational modification of p53 via several potentially interacting but distinct pathways. Although activation of p53 by oxidants is well documented in some studies, there is evidence to suggest that oxidizing agents could also inhibit p53's trans-activation activity *(112)*. For example, exposing cells to H_2O_2 reduced p53-dependent activation of a target reporter, whereas simultaneous treatment of cells with NAC prevented this inhibition *(118)*. A similar paradoxical situation was reported with nitric oxide (NO). It activates p53 by being an indirect inducer of DNA strand breaks *(119,120)*, but there is also evidence that NO could modify the p53 protein (via tyrosine nitrosation) and thus inhibit its activity *(121)*.

Reactive oxygen species-induced posttranslational modification of p53 appears to be different from that with other types of agents and involves specific redox-sensitive kinases, including DNA-PK, JNK, p38-MAPK, cyclin-dependent protein kinase activating kinase (CAK), and casein kinase 1 (CK1) *(11,122)*. Recent studies have shown involvement of polo-like-kinase-3 (Plk3) in H_2O_2-induced phosphorylation of Ser[20] in p53 *(123)*, in concert with cell cycle checkpoint kinases (Chk)-1 and Chk-2, which are downstream to ATM or ATR *(11,124)*. Whereas Plk3 may preferentially transduce signals generated by oxidative stress, Chk1 and Chk2 are differentially activated by ultraviolet (UV) radiation and IR, respectively. It was also suggested that Plk3 integrates the signals from ATM-Chk-2 and ATR-Chk-1 and induces cell cycle arrest or apoptosis by phosphorylating either Cdc25 at Ser^{216} or p53 at Ser^{20} *(124,125)*. Consistent with the latter scenario, Plk3 is also activated by the IR-mimetic drugs, such as adriamycin or bleomycin, and by H_2O_2. As outlined in Fig. 3, when p53 is activated, a significant increase in expression of $p21^{WAF}$ as well as other p53-inducible target genes, including Bax, p53-inducible genes (PIGs), CD95, DR5 (a receptor for the death ligand TRAIL), IGF-BP3, Rpr, Cdc42 (a Ras-like GTPase), Noxa (a Bcl-2 family protein), and p53AIP1 was observed *(110,124–126)*.

5.2. Redox Regulation of p53 Protein

The redox state of p53 itself affects its binding to the target DNA in such a way that cognate cis elements in DNA are recognized only by the reduced p53. However, damaged DNA is recognized equally well by both oxidized and reduced p53. Thus, specific binding of p53 to DNA requires the presence of reducing agents such as 2-mercaptoethanol or dithiothreitol and is abrogated by thiol oxidants such as diamide *(112)*.

p53 has a Zn-finger domain with Cys residues, and its sequence-specific DNA binding is dependent on metal and redox regulation. Cellular redox status was shown to preferentially affect Cys residues (Cys^{124}, Cys^{135}, Cys^{141}, Cys^{275}, and Cys^{277}) present in the DNA-binding domain *(127)* among which Cys^{275} and Cys^{277} form a C–X–C motif located within a loop binds to the major groove of DNA. Oxidation of Cys^{277} on the protein surface induces conformational and steric changes and prevents the formation of an H bond with bases and, thus, its interaction with DNA *(127)*.

Oxidation–reduction of p53 by thioredoxin (TRX) and APE1/Ref-1 (AP endonuclease 1/redox factor 1) affect p53's conformation and DNA-binding activity in vivo *(27,128)*. Deletion of 30 C-terminal residues of p53 (the regulatory domain for DNA-binding activity) allows p53 to escape its dependence on TRX reductase for activity *(27,129)*.

APE1/Ref-1 not only plays a key role in BER *(130,131)* but is also a key regulator of p53 and, hence, of cell cycle arrest and apoptosis. APE1/Ref-1 interacts with p53 *(128)* and enhances p53's ability to trans-activate a number of p53 target genes *(132)*. Accordingly, APE1/Ref-1 overexpression increases p53-mediated upregulation of p21 and cyclin G, which, in turn, leads to cell cycle arrest by inhibiting cyclin-dependent kinase(s) *(133)*. Downregulation of APE1/Ref-1 decreases the ability of p53 to *trans*-activate the promoter of Bax a key pro-apoptotic protein. APE1/Ref-1 is induced by oxidative stress *(131,134)* and, in turn, stimulates the DNA-binding activity of several transcription factors, includ-

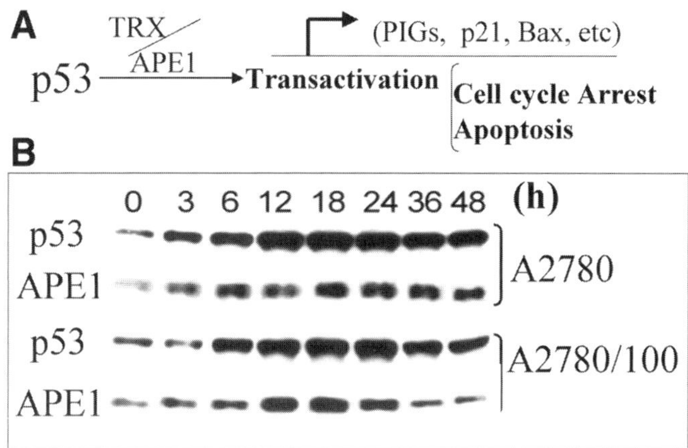

Fig. 5. p53's transactivating activity is redox-regulated. (**A**) AP-endonuclease 1 (APE1/ Ref1) and thioredoxin (TRX) regulate p53's intrinsic redox state, affecting its transactivation functions. (**B**) Kinetics of change in APE1/Ref1 and p53 levels in drug-sensitive (A2780) and drug-resistant (A2780/100) cells. Equal amounts of proteins from treated cells were fractionated by sodium dodecyl sulfate polyacrylamide gel electrophoresis blotted, and probed with anti-APE1 (rabbit polyclonal) and anti-p53 (DO1) antibodies.

ing NFkB, AP1, Myb, ATF/cAMP-responsive-element-binding protein family, HIF-1α, HLF, and PAX *(128,129,135,136)*. Thus, APE1/Ref1 is a novel component of signal transduction processes for regulating gene expression in cooperation with p53.

In view of the direct role of APE1 in p53 regulation, we examined whether APE1/Ref-1 is differentially activated in drug-sensitive (A2780) and drug-resistant (A2780/100) cells after treatment with Cbl. As shown in Fig. 5B, the APE1/Ref-1 level was increased 8- to 10-fold both in A2780 and A2780/100 cells. Similar results were obtained after treatment with the DNA topoisomerase inhibitor etoposide (data not shown). In drug-resistant cells, the increase in APE1/ Ref-1 level was lower and the basal level was restored, starting at 24–28 h posttreatment. In contrast, the APE1/Ref-1 level remained elevated in sensitive cells, which eventually underwent apoptosis. The increase in p53 level paralleled that of APE1/Ref-1 (*see* Fig. 5B) in sensitive but not in drug-resistant cells. Because APE1/Ref-1 regulates p53 activities for the p53-dependent genes *(114)* (e.g., Bax or PIG family genes), it is possible that the decreased APE1/Ref-1 level from 18 h onward was causally linked to the increased survival of drug-resistant A2780/100 cells. Because the APE1/Ref1 level did not decrease during apoptosis in the sensitive cells, it is likely that the elevated level of APE1/Ref-1 was necessary for locking these cells in the G_1-phase of the cell cycle (by p53-dependent upregulation of p21) and ensuring an elevated Bax level *(106)*. Bax

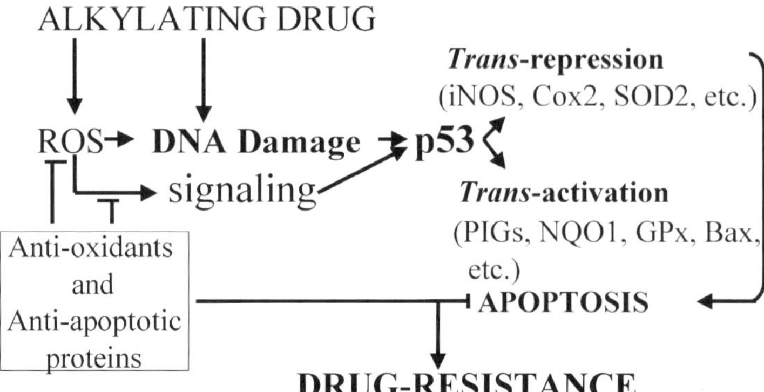

Fig. 6. Regulation of ROS by p53-mediated gene expression. ROS from intracellular sources induce DNA damage, and signaling cascades leading to posttranslational modifications. In turn, p53 trans-activates (e.g., PIGs, GPx) and trans-represses (e.g., iNOS, Cox2) genes involved in ROS production and apoptosis. iNOS, inducible nitric oxide synthetase; SOD2, superoxide dismutase 2; COX2, cyclooxygenase 2; PIGs, p53 inducible genes; GPx, glutathione peroxidase.

forms channels in lipid membranes, and the proapoptotic effect of Bax appears to be elicited through an intrinsic pore-forming activity *(137)*, which causes leakage of cyt-*c* from the mitochondrial intermembrane space into the cytosol. Cyt-*c* is critical in activating caspases, leading to degradation of survival proteins. It is noteworthy that in A2780/100 cells, the ROS levels are normalized while ROS production is chronic in sensitive cells after Cbl treatment (*see* Fig. 4B). Based on these results, we speculate that ROS-generated signals enhance the APE1/Ref1 level/activity, which, in turn, increases the proapoptotic activity of p53. Indeed, APE1/Ref1 is induced by oxidative stress *(134)* and contains several potential sites for phosphorylation by casein kinase I and II and PKC kinases *(138)*, which, in turn, is activated by ROS *(134,139)*.

It is interesting to note that IR sensitivity and chemosensitivity of a wide variety of cancers, such as ovarian, cervical, and prostate, and rhabdomyosarcomas and cancer of the germ cells correlated well with elevated levels of APE1/Ref-1 *(136,140,141)*. Thus, APE1/Ref-1 may provide a molecular bridge between DNA-damage sensing and redox modulation of p53, two events critical for initiation of apoptotic processes by p53, at least in tumor cells containing wild-type p53.

5.3. p53 Regulation of Redox Effector Genes

There is compelling evidence for the role of p53 as a trans-activator or trans-repressor of genes involved in the production and control of ROS (Fig. 6). Transgenic overexpression of p53 in HeLa cells results in a transient increase in

ROS level, causing alterations in the mitochondrial membrane potential and subsequent apoptosis *(93)*. Transfection with mutant p53 abrogated both ROS production and apoptosis. Cotransfection of these cells with p53 and peroxiredoxin V (encoding the antioxidant thioredoxin peroxidase) lowered the oxidative stress level and reduced apoptosis *(142)*. Other genes specifically upregulated in a p53-dependent manner include glutathione peroxidase (GPx) in etoposide-treated or irradiated cells *(143)*. GPx, an antioxidant enzyme, degrades H_2O_2 and detoxifies lipid peroxides. Thus, these results support the notion that a reducing environment decreases p53's proapoptotic activity.

p53 regulates expression of many redox-related genes *(93,115)*, which could induce oxidative stress, leading to apoptosis *(114,115)*. The p53-inducible gene 3 (PIG3) encodes a protein with significant homology to NADH-quinone oxidoreductases (NQO) *(144)*, one of the enzymes that control cellular responses to chemotherapeutic agents and IR *(145)*. This raises the possibility that the cellular redox status, controlled by such enzymes, also regulates the level of p53. Indeed, it has been shown that ectopic overexpression of NQO1 increases the stability of p53 and its inhibition decreased IR-induced level of p53 *(144)* in colon carcinoma and myloid leukemia, which could be prevented by treatment with proteasome inhibitors. These observations suggest that p53 degradation is accelerated in the absence of ROS *(144)*. As expected, wild-type NQO1, but not mutant NQO1, stabilized endogenous as well as transgenically overexpressed wild-type p53. Thus, genetic and pharmacological regulation of p53 have clinical implications for tumor chemotherapy.

In addition to PIG3, which contains p53-binding sequences in its promoter, other PIGs have redox functions, although the cis elements for p53 have not yet been identified. These include PIG1, a member of the galectin family involved in superoxide production *(146)*, PIG6, a homolog of proline oxidoreductase, a mitochondrial enzyme involved in the conversion of proline to glutamate *(147)*, and PIG12, a member of the microsomal glutathione-*S*-transferase gene family *(148)*. The family of redox-regulated genes induced by p53 also include PIG4, which encodes a serum amyloid protein, and PIG7, a gene induced by tumor necrosis factor-α (TNF-α) *(149)*. Although the function of PIG8 is not known, it is interesting that its ectopic expression inhibits cell cycling and induces morphological features of apoptosis *(114)*.

p53 also acts as a repressor for some genes involved in ROS metabolism, including the inducible forms of cyclooxygenase 2 (Cox2) and nitric oxide synthetase (iNOS) *(150)*. Interestingly, Cox2 which catalyzes synthesis of prostaglandins from arachidonic acid, is often upregulated in cancer cells *(151)*. The mitochondrial superoxide dismutase (Mn-SOD or SOD2), provides major cellular defense against oxidative stress *(152)*. Because p53 was shown to inhibit the SOD2 promoter, it is possible that p53 and SOD2 genes are reciprocally coregulated and thus modulate ROS levels, which, in turn, affect various cellular processes, including apoptosis *(153)*. A change in intracellular balance of these proteins by ectopic overexpression can trigger apoptosis in drug-treated cells *(153,154)*. We have shown that Mn-SOD overexpression increases cytotoxicity of Cbl in A2780 cells by increasing mitochondrial ROS (H_2O_2) levels (Boldogh and Mitra, unpublished results). Similarly, transgenic overexpression of Mn-SOD sensitized various tumor cells to apoptosis induced by BCNU *(155)*. These data suggest that increased H_2O_2 level caused by overexpression of Mn-SOD enhances the release of proapoptotic factors from the mitochondria.

It may appear paradoxical that p53 is able to trans-activate both pro-oxidant (e.g., PIGs) *(114)* and antioxidant (e.g., GPx) enzymes *(143)* and represses iNOS and Cox2 genes *(150)*. How these various and sometimes opposite activities are coordinately regulated in vivo is not well understood. One hypothesis is that these genes are regulated both temporally and spatially. Indeed, there is evidence showing that whereas PIG3 is induced rather late in response to p53 induction, GPx activation is an early response. It may be important that p53 plays a dual role within the cell, both as a regulator of cell cycle arrest under conditions of mild genotoxic damage and as a trigger for apoptosis under conditions of severe genotoxic damage *(8,111,112)*. The primary antioxidant response may occur preferentially in cells undergoing cell cycle arrest, whereas the secondary pro-oxidant response may be required for the induction of ROS that may act as second messengers of the apoptotic cascade *(12,112)*.

5.4. Redox Regulation of p53's Organelle Targeting

Although some 40% of human tumors retain the wild-type p53 gene *(26,110)*, they frequently show defects in apoptosis, because of the lack of nuclear p53 import *(156–158)*, which could cause resistance to IR and various drugs *(159)*. Nuclear import of p53 is regulated by cellular redox status *(160)*. For example, treating cells with pyrrolidine dithiocarbamate (PDTC), a thiolantioxidant, reduces p53's nuclear accumulation and prevents transactivation of p53-dependent HDM2 (human homolog of MDM2) *(161)*. It was suggested that oxidation of specific Cys residues in p53 are required for its nuclear accumulation and trans-activation function *(127)*. On the other hand, the nuclear level of p53 could also be affected indirectly by ROS-dependent activation of nuclear import complexes *(162)*.

5.4.1. NUCLEAR IMPORT AND EXPORT OF P53

Nuclear localization of p53 is governed by complex mechanisms and tightly regulated according to cell cycle stages and cellular environment (Fig. 7). p53 has three nuclear localization signals (NLSs) in its C-terminal region *(163)*. Nuclear import of p53 requires posttranslational modification (e.g., phosphorylation by kinases and acetylation by histone acetyltransferases (HAT), an intrinsic activity of transcription coactivators such as p300/CBP- and p300/CBP-associated factor (pCAF) *(164)*. These processes require energy (generated

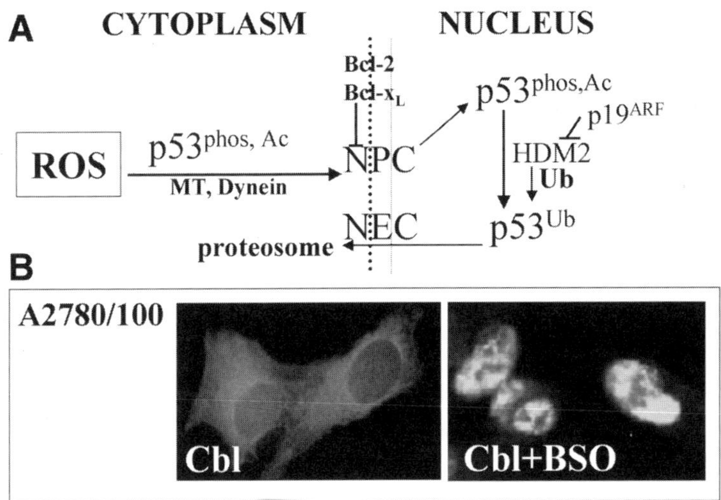

Fig. 7. Effect of ROS on p53's subcellular localization. (**A**) ROS-induced signaling activates p53 and targets it into the nucleus. Signals generated by ROS stabilize p53 by phosphorylation (phos) and acetylation (Ac) that are required for nuclear import processes. Nuclear import of p53 is dependent on its interaction with the microtubule network (MT) and dynein and the activity of nuclear pore complex (NPC). Antiapoptotic Bcl-2/Bcl-xL can inhibit p53's nuclear import. In the nucleus, one of the key regulators of p53 level is the MDM2 protein, which can inhibit p53's transcriptional activity and target p53 for degradation. HDM2 functions as E3 ligase, and conjugates ubiquitin (Ub) to p53, targeting to nuclear export complexes (NEC) and degradation by the proteasome. p19[ARF] binds to and directly inhibits HDM2 ligase activity. (**B**) Function of p53 is abrogated by exclusion from the nucleus in drug-resistant cells (left panel). Application of mild chronic oxidative stress (e.g., subtoxic dose of H_2O_2, BSO treatment) increases nuclear accumulation of p53. NPC, nuclear pore complex, NEC, nuclear export complex; BSO, L-buthionine (*SR*)-sulfoximine; Ac, acetylation; Phos, phosphorylation.

by GTP hydrolysis) and several transport factors (RAN/TC4, karyopherins, and Ran-interacting proteins) for translocation of the complex through a gated NPC *(164)*. Mutations in nuclear localization sequences could lead to accumulation of p53 in the cytoplasm. Nuclear import of p53 is also dependent on its interaction with the microtubule network and the molecular motor dynein, which translocates p53 to the nuclear import machinery *(164,165)*. The nuclear level of p53 is regulated by the level of phosphorylated (Ser17) HDM2. HDM2, an E3 ubiquitin ligase, mediates p53 degradation via the ubiquitin–26S proteasome pathway. The p53 level is also affected by direct binding of p53 to deacetylase complexes *(166,167)* or the level of p19[ARF], which reduces HDM2's interaction with p53 *(168–170)*. However, how these events are affected by ROS is not clear at present.

p53 in the nucleus colocalizes with the ROS-inducible promyelocytic leukemia protein (PML) *(171,172)* and with p300/CBP *(173)*. PML stimulates

N-terminal phosphorylation by SUMO-1 *(174)* and C-terminal acetylation of p53 by HATs *(173)*. It is noteworthy that PML3 specifically enhances p53-dependent activation of the PIG3 promoter *(171)*. In turn, PIG3 increases ROS levels *(115)*, inducing PML3-dependent relocalization of transcriptionally active p53 in the nuclear bodies *(171)*. Loss of PML3 impairs the cellular ability to mount a p53-dependent cell death response both in lymphoid and solid tumors after exposure to genotoxicant(s) *(171)*.

Export of p53 from the nucleus plays an equally important role in the regulation of its function *(175)*. The nuclear export signal of p53 is located at the C-terminus, and efficient export of p53 to the cytoplasm depends on the activity of HDM2. The ubiquitination of p53 by HDM2 occurs within the C-terminus of the p53 protein and mutation of these Lys residues inhibits p53 modification *(167)*. HDM2 also inhibits p300/CBP-mediated p53 acetylation and activation by forming a ternary complex with the two proteins *(176)*. However, significant HDM2-mediated p53 degradation takes place in the presence of leptomycin B, an inhibitor of nuclear export complexes, indicating that endogenous p53 degradation could also occur in the nucleus *(177)*. It may not be surprising that nuclear degradation of p53 occurs during the poststress recovery phase of p53 response, especially after DNA damage *(178)*. Thus, the capability of cells to degrade p53 within the nucleus provides a tighter control and causes a prompt turn-off of an active p53 program. This mechanism of p53 inactivation could be part of a signaling network responsible for the drug resistance of tumor cells.

Intracellular localization of p53 is likely to play a role in controlling its redox state, and thereby its function, because the levels of such redox effectors as TRX and APE1/Ref-1 are higher in the nucleus than in the cytoplasm of stressed cells. In our model system, we have shown that both A2780 and A2780/100 cells express wild-type p53, and the extent of p53 induction by Cbl (or Eto) is similar in both cell lines after drug treatment (*see* Fig. 5B). An increase in p53 level is inhibited by cyclohexamide, but not by actinomycin D. These results are consistent with typical increase in the half-lives of p53 mRNA and polypeptide *(179)*. After drug treatment, p53 continuously accumulated in the nuclei of A2780 cells, which then underwent apoptosis. In A2780/100 cells, p53 was present in the nucleus up to 6 h after Cbl treatment; however, at later time-points, p53 was localized predominantly in the cytoplasm.

Because we showed a close correlation between chronic ROS generation and cellular sensitivity to Cbl (*see* Fig. 4), we investigated the effects of a GSH precursor, NAC *(180)* and ROS scavenger BHA *(181)* nuclear accumulation of p53. We found that both NAC and BHA decreased Cbl-induced ROS levels by at least 80%. More significantly, NAC and BHA reduced p53 nuclear accumulation and, consequently, drug-induced cytotoxicity in A2780 cells. In a complementary study, we observed that treating A2780/100 cells with diethylmaleate (DEM), which inactivates GSH, or buthionine-(*S,R*)-sulfoximine (BSO), an inhibitor of GSH synthesis *(182)*, increased the p53 level in the nucleus and lowered the LD50 of Cbl and Eto in A2780/100 cells several-fold (*see* Fig. 4B). These data indicate that oxidative-stress-mediated signaling plays a key role in the nuclear import of p53 and that nuclear accumulation of p53 may be important for continuous ROS production, possibly the result of transactivation of the PIG genes *(114)*. In support of this hypothesis, we showed increased RNA levels for PIG3 and PIG11 in A2780 cells after drug treatment (Boldogh et al., unpublished observation). Thus, lowering the cellular redox state may sensitize tumor cells expressing wild-type p53 to specific treatments and increase the therapeutic index of drugs. Given the importance of ROS-induced signaling, it seems likely that application of ROS could increase the sensitivity of tumor cells to chemotherapeutic agents.

5.4.2. Targeting of p53 to the Mitochondria

The mitochondria are the focal points in the apoptosis cascade, where death stimuli translate from initiation to execution, irrespective of whether apoptosis is triggered by chemothrapeutic agents, IR, or other types of death stimulus *(100,116,183)*. Mitochondrial changes include membrane permeability transition pore opening, disruption of the electron transport chains, collapse of the inner $m\Delta\Psi$, reduced ATP production, generation of ROS, and swelling that often coincides with outer-membrane rupture. These alterations precede rapid activation of caspase 3 (and caspases 6 and 7) by cyt-*c*, apoptotic protein activating factor 1, and caspase-9. Antiapoptotic Bcl-2 family members counteract the ion movements, thereby stabilizing the $m\Delta\Psi$ and mitochondrial volume and preventing ROS production and the release of apoptogenic protease activators, protecting cells from death *(51,100,116)*.

In response to various types of stress, including oxidative stress and/or DNA damage, p53 is targeted to the mitochondria, as shown both microscopically and biochemically *(184,185)*. This translocation occurs at the onset of p53-dependent apoptosis, but is absent during p53-mediated cell cycle arrest or p53-independent apoptosis. Mitochondrial translocation of p53 is rapid and precedes changes in $m\Delta\Psi$ and cyt-*c* release. A significant amount of p53 is located at the surface of the mitochondria, and a subfraction appears to be intraorganellar, as determined by limited trypsin digestion *(184,186)*. Importantly, the presence of p53 in the mitochondria is sufficient to induce p53-dependent apoptosis. Mitochondria-targeted truncated p53 protein, lacking tetramerization domain, retains its apoptotic activity, suggesting that the C-terminus is dispensable for mitochondrial action *(187)*. This is in contrast to the nuclear functions of p53, which requires tetramerization in order to optimally function as a trans-acting factor *(184)*.

We have investigated the possibility that the lack of p53's mitochondrial targeting may be associated with drug resistance in Cbl-treated cells. A2780 and

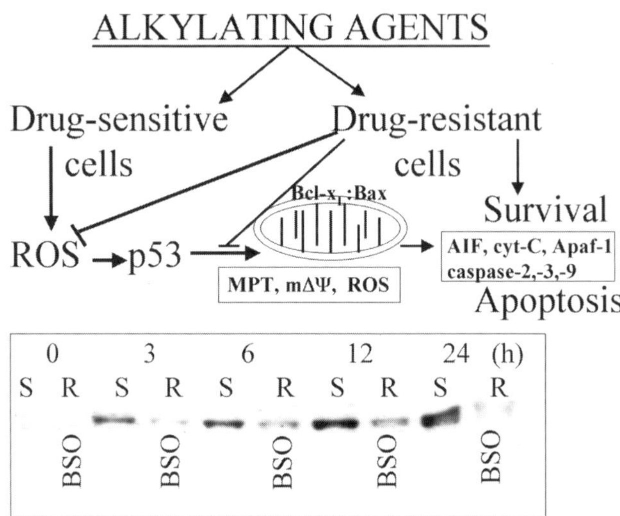

Fig. 8. Targeting of p53 to mitochondria after drug treatment. (**A**) Drug-resistant cells efficiently cope with oxidative stress and abrogate targeting of p53 into the mitochondria. Mitochondria plays a central role in drug-induced apoptosis by releasing apoptogenic factors (AIF, cyt-*c*, Apaf-1) and cysteine proteases (e.g., caspases-2, -3, and -9) into the cytosol. Alteration in mitochondria, such as mitochondria permeability pore transition (MPPT) and mΔΨ, depend on the ratio between proapoptotic (e.g., Bax) and antiapoptotic proteins (e.g., Bcl-x$_L$). In addition, mitochondria can generate reactive oxygen species, following uncoupling and/or inhibition of the respiratory chain. (**B**) Cbl treatment of drug-sensitive A2780 (*S*) cells result in translocation of p53 in mitochondria. Rapid accumulation of p53 in the mitochondria is associated with chronic ROS production (*see* Fig. 3) and cells killed by apoptosis. Drug-resistant, Cbl-treated A2780/100 cells (R) required reduction in the GSH level by BSO pretreatment to transport p53 into the mitochondria.

A2780/100 cells were treated with Cbl and the mitochondria and nuclei were isolated. Western blot analysis showed that a part of the p53 protein is incorporated into the mitochondria soon after Cbl treatment (*see* Fig. 8B), much earlier than the release of cyt-*c* from the mitochondria or the degradation of poly(ADP-ribose) polymerase 1 (PARP) *(106)*. These data are consistent with the observation that overexpression of antiapoptotic Bcl-xL in A2780/100 *(106)* abrogates drug-induced mitochondrial p53 accumulation and apoptosis. Although mitochondrial p53 constitutes only a small fraction of total p53 in Cbl-treated cells, its presence in the mitochondria is highly reproducible. Treating resistant cells with BSO (*see* Fig. 8B) or with multiple additions of nontoxic doses of H$_2$O$_2$ caused mitochondrial accumulation of p53 (data not shown). Mitochondrial accumulation was specific for p53, because other nuclear proteins such as PCNA were not detectable in the mitochondria. We thus conclude that mitochondrial p53 accumulation may be associated with chronic ROS production in drug-treated cells regardless of whether the cells are sensitive or have acquired resistance to drugs and that this transcription-independent pathway is likely to be synergistic with other p53's action.

6. ROS DETOXIFICATION AND INACTIVATION OF p53

As previously discussed, ROS produced by ionizing radiation and chemotherapeutic agents *(90)* appears to be critical in p53-induced apoptosis. Conversely, increased ROS detoxification was shown to inhibit proapoptotic effects of p53, as do inactivating mutations in p53 *(26,110,111)*, overexpression of HDM2 *(188)*, or p53 degradation by E6 protein of human papilloma virus *(11,189–191)*.

To prevent ROS-induced signaling, antioxidant systems act as ROS detoxifiers. The most effective scavengers of ROS are thiol-containing moieties, such as Cys residues of proteins and small molecules such as GSH *(93,192)*. The intracellular GSH level varies in the range 5–10 mM, depending on cell type and cellular compartment, and is maintained in the reduced state by GSSG reductase, a NADPH-dependent enzyme *(193)*. GSH not only defends cells from ROS but is also a substrate for GSTs and inactivates alkylating drugs *(101,102,194)*. Thus, GSH reduces ROS at multiple levels and also inhibits drug-induced DNA damage, resulting in reduced trans-activation by p53 and apoptosis.

Moreover, several enzymatic systems detoxify ROS and may act against drug/IR-induced apoptosis *(195,196)*. These enzymes include catalase, which eliminates H_2O_2, and SOD, which eliminates $O_2^{·-}$ while generating H_2O_2 *(155)* and which catalyses the reduction of peroxides, using the reducing potential of GSH (for review, *see* ref. *197*). The cysteine-rich metallothionein proteins *(198)*, quinone reductase, and heme oxygenase also provide an endogenous protective mechanism against drug-induced oxygen-derived radicals *(199)*. Tumor cells can also protect themselves with antioxidant systems involving a cascade of functional redox molecules, such as TRX, the radical-scavenging ascorbic acid in the cytosol, and the membrane-associated vitamin tocopherol *(200)*. Expression of antioxidant proteins and enzymes that regenerate them (such as glutathione reductase and TRX reductase) is transcriptionaly activated by oxidative stress *(201)* in drug-treated cells.

Antiapoptotic Bcl-2 family proteins (e.g., Bcl-2, Bcl-x$_L$, Mcl-1) are unique among cellular proteins for their indirect ability to prevent p53-mediated cell death after cellular exposure to pro-oxidants (e.g, H_2O_2, menadione) and antitumor agents (NMs, aziridines, cisplatin and bleomycin derivates, and IR) *(92)* as well as after GSH depletion, cytokine withdrawal, or hypoxia *(202,203)*. The mechanism by which antiapoptotic Bcl-2 or Bcl-x$_L$ prevents ROS production and increase their detoxification is not well understood. However, these processes may involve proapoptotic Bak, Bcl-x$_S$, Bax, or BH3 (e.g., Bik, Bad), which form pores in the mitochondrial membrane. Thus, antiapoptotic proteins prevent mitochondria membrane permeability transition pore opening, a change in $m\Delta\Psi$, generation of ROS, and release of cyt-c *(204,205)*. Interestingly, Bcl-2 or Bcl-x$_L$ are as effective antioxidants as GPx, which specifically eliminates H_2O_2 and other peroxides *(197)*. In other studies, the antioxidant effect of Bcl-2 has been shown to be mediated via an increase in mitochondrial glutathione (GSH) levels *(92)* or through increased activities of Mn-SOD and GPx *(206–208)*. It is unlikely that antiapoptotic proteins directly block ROS production or functions in ROS detoxification reactions or act as free-radical scavengers; however, they may regulate the GSH level in the nucleus and mitochondria *(205,209)*.

Finally, Bcl-2 is a membrane protein present not only in the mitochondria and endoplasmatic reticulum but also in the nuclear envelope *(92,100,203,205)*. It has been shown that although p53 was activated to a similar extent in control versus Bcl-2-overexpressing prostate carcinoma cells, nuclear import of p53 after IR treatment was significantly inhibited only in the Bcl-2-positive cells, which caused reduced apoptosis *(210)*. Thus, inhibition of nuclear import and lack of sequential activation of p53-dependent genes (e.g., PIGs) involved in ROS generation by antiapoptotic proteins may provide an additional mechanism for inactivation of p53 function.

7. CONCLUSIONS

Programmed cell killing induced by alkylating therapeutic agents is often disrupted, contributing to such conditions as drug resistance, increased metastic ability, and mutations resulting in altered phenotypes. The key elements of the cell death pathway are closely linked to complex signaling systems that affect multiple processes, including repair of DNA damage, alteration in gene expression, and increased levels of proteins responsible for ROS production and their detoxification. The key cell cycle arrest and proapoptotic transcriptional regulator p53 is subject to redox modulation, and its redox state determines its function. Therefore, p53 belongs to the family of oxidative stress response factors, and is at the center of a network of redox interactions. It is paradoxically dispensable for survival of those cells that are capable of coping with chemotherapy-induced DNA damage and oxidative stress. However, it is critical for those cells that are incapable of scavenging ROS. Understanding the complex interactions between the cellular redox state and the function of p53 could provide insights into mechanisms of chemosensitivity of tumor cells and may provide rational approaches for therapeutic interventions.

ACKNOWLEDGMENTS

The authors' research cited in this chapter was supported by NIH/NCI grants NIH RO1-CA84461 (IB) and CA EOS 08457 (SM) and NIEHS Center at University of Texas Medical Branch at Galveston, TX (grant no. ES06676). We also thank Dr. David A. Konkel (Department of Human Biological Chemistry and

Genetics at UTMB) for scientific/editorial help and comments and members of Mitra's and Boldogh's laboratories for helpful discussions.

REFERENCES

1. Lawley PD, Phillips DH. DNA adducts from chemotherapeutic agents. *Mutat Res* 1996; 355:13–40.
2. Sanderson BJ, Shield AJ. Mutagenic damage to mammalian cells by therapeutic alkylating agents. *Mutat Res* 1996;355:41–57.
3. Sanderson BJ, Ferguson LR, Denny WA. Mutagenic and carcinogenic properties of platinum-based anticancer drugs. *Mutat Res* 1996;355:59–70.
4. Coultas L, Strasser A. The molecular control of DNA damage-induced cell death. *Apoptosis* 2000;5:491–507.
5. Mitra S, Kaina B. Regulation of repair of alkylation damage in mammalian genomes. *Prog Nucleic Acid Res Mol Biol* 1993;44:109–142.
6. Fritz G, Kaina B. Genomic differences between O^6-methylguanine-DNA methyltransferase proficient (Mex+) and deficient (Mex–) cell lines: possible role of genetic and epigenetic changes in conversion of Mex+ into Mex. *Biochem Biophys Res Commun* 1992;183;1184–1190.
7. Gerson SL. Clinical relevance of MGMT in the treatment of cancer. *J Clin Oncol* 2002;20:2388–2399.
8. Norbury CJ, Hickson ID. Cellular responses to DNA damage. *Annu Rev Pharmacol Toxicol* 2001;41:367–401.
9. Bernstein C, Bernstein H, Payne CM, et al. DNA repair/pro-apoptotic dual-role proteins in five major DNA repair pathways: fail-safe protection against carcinogenesis. *Mutat Res* 2002;511:145–178.
10. Christodoulopoulos G, Fotouhi N, Krajewski S, et al. Relationship between nitrogen mustard drug resistance in B-cell chronic lymphocytic leukemia (B-CLL) and protein expression of Bcl-2, Bax, Bcl-X and p53. *Cancer Lett* 1997;121:59–67.
11. Woods DB, Vousden KH. Regulation of p53 function. *Exp Cell Res* 2001;264:56–66.
12. Vogelstein B, Lane D, Levine AJ. Surfing the p53 network. *Nature* 2000;408:307–310.
13. Teicher BA. Antitumor alkylating agents. In: *Cancer: Principles and Practice of Oncology*. deVita VTH, Rosenberg SA, eds. Lippincott–Raven, Philadelphia, 1997:405 418.,
14. Rajewsky MF, Engelbergs J, Thomale J, et al. Relevance of DNA repair to carcinogenesis and cancer therapy. *Recent Results Cancer Res* 1998;154:127–146.
15. Snow ET, Foote RS, Mitra S. Kinetics of incorporation of O^6-methyldeoxyguanosine monophosphate during in vitro DNA synthesis. *Biochemistry* 1984;23:4289–4294.
16. Snow ET, Foote RS, Mitra S. Base-pairing properties of O^6-methylguanine in template DNA during in vitro DNA replication. *J Biol Chem* 1984;259:8095–8100.
17. Lu AL, Welsh K, Clark S, Su SS, et al. Repair of DNA base-pair mismatches in extracts of *Escherichia coli*. *Cold Spring Harb Symp Quant Biol* 1984;49:589–596.
18. Kokkinakis DM, Bocangel DB, Schold SC, et al. Thresholds of O^6-alkylguanine-DNA alkyltransferase which confer significant resistance of human glial tumor xenografts to treatment with 1,3-bis(2-chloroethyl)-1-nitrosourea or temozolomide. *Clin Cancer Res* 2001;7:421–428.
19. Panasci L, Paiement JP, Christodoulopoulos G, et al. Chlorambucil drug resistance in chronic lymphocytic leukemia: the emerging role of DNA repair. *Clin Cancer Res* 2001;7:454–461.
20. Johnson SW, Swiggard PA, Handel LM, et al. Relationship between platinum-DNA adduct formation and removal and cisplatin cytotoxicity in cisplatin-sensitive and -resistant human ovarian cancer cells. *Cancer Res* 1994;54:5911–5916.
21. Lin X, Ramamurthi K, Mishima M, Kondo A, et al. p53 interacts with the DNA mismatch repair system to modulate the cytotoxicity and mutagenicity of hydrogen peroxide. *Mol Pharmacol* 2000;58:1222–1229.

22. Wang XW, Vermeulen W, Coursen JD, et al. The XPB and XPD DNA helicases are components of the p53-mediated apoptosis pathway. *Genes Dev* 1996;10:1219–1232.
23. Xu GW, Nutt CL, Zlatescu MC, et al. Inactivation of p53 sensitizes U87MG glioma cells to 1,3-bis(2-chloroethyl)-1-nitrosourea. *Cancer Res* 2001;61:4155–4159.
24. Xu Z, Chen ZP, Malapetsa A, et al. DNA repair protein levels vis-a-vis anticancer drug resistance in the human tumor cell lines of the National Cancer Institute drug screening program. *Anticancer Drugs* 2002;13:511–519.
25. Seo YR, Fishel ML, Amundson S, et al. Implication of p53 in base excision DNA repair: in vivo evidence. *Oncogene* 2002;21:731–737.
26. Morris SM. A role for p53 in the frequency and mechanism of mutation. *Mutat Res* 2002;511:45–62.
27. Jayaraman L, Prives C. Covalent and noncovalent modifiers of the p53 protein. *Cell Mol Life Sci* 1999;55:76–87.
28. Liu Y, Kulesz-Martin M. p53 protein at the hub of cellular DNA damage response pathways through sequence-specific and non-sequence-specific DNA binding. *Carcinogenesis* 2001;22:851–860.
29. Welcsh PL, Owens KN, King MC. Insights into the functions of BRCA1 and BRCA2. *Trends Genet* 2000;16:69–74.
30. Sturzbecher HW, Donzelmann B, Henning W, et al. p53 is linked directly to homologous recombination processes via RAD51/RecA protein interaction. *EMBO J* 1996;15:1992–2002.
31. Lu Y, Lian H, Sharma P, Schreiber-Agus N, et al. Disruption of the Cockayne syndrome B gene impairs spontaneous tumorigenesis in cancer-predisposed Ink4a/ARF knockout mice. *Mol Cell Biol* 2001;21:1810–1818.
32. Orren DK, Dianov GL, Bohr VA. The human CSB (ERCC6) gene corrects the transcription-coupled repair defect in the CHO cell mutant UV61. *Nucleic Acids Res* 1996;24:3317–3322.
33. Spillare EA, Robles AI, Wang XW, et al. p53-mediated apoptosis is attenuated in Werner syndrome cells. *Genes Dev* 1999;13:1355–1360.
34. Panasci L, Xu ZY, Bello V, et al. The role of DNA repair in nitrogen mustard drug resistance. *Anticancer Drugs* 2002;13:211–220.
35. Muller C, Christodoulopoulos G, Salles B, et al. DNA-dependent protein kinase activity correlates with clinical and in vitro sensitivity of chronic lymphocytic leukemia lymphocytes to nitrogen mustards. *Blood* 1998;92:2213–2219.
36. Sunters A, Springer CJ, Bagshawe KD, et al. The cytotoxicity, DNA crosslinking ability and DNA sequence selectivity of the aniline mustards melphalan, chlorambucil and 4-[bis(2-chloroethyl)amino] benzoic acid. *Biochem Pharmacol* 1992;44:59–64.
37. Jackson SP. Sensing and repairing DNA double-strand breaks. *Carcinogenesis* 2002;23:687–696.
38. Wang ZM, Chen ZP, Xu ZY, et al. In vitro evidence for homologous recombinational repair in resistance to melphalan. *J Natl Cancer Inst* 2001;93:1473–1478.
39. Walker JR, Corpina RA, Goldberg J. Structure of the Ku heterodimer bound to DNA and its implications for double-strand break repair. *Nature* 2001;412:607–614.
40. Dynan WS, Yoo S. Interaction of Ku protein and DNA-dependent protein kinase catalytic subunit with nucleic acids. *Nucleic Acids Res* 1998;26:1551–1559.
41. Douglas P, Moorhead GB, Ye R, et al. Protein phosphatases regulate DNA-dependent protein kinase activity. *J Biol Chem* 2001;276:18992–18998.
42. Smith GC, Jackson SP. The DNA-dependent protein kinase. *Genes Dev* 1999;13:916–934.
43. Okayasu R, Suetomi K, Yu Y, Silver A, et al. A deficiency in DNA repair and DNA-PKcs expression in the radiosensitive BALB/c mouse. *Cancer Res* 2000;60:4342–4345.
44. Christodoulopoulos G, Muller C, Salles B, et al. Potentiation of chlorambucil cytotoxicity in B-cell chronic lymphocytic leukemia by inhibition of DNA-dependent protein kinase activity using wortmannin. *Cancer Res* 1998;58:1789–1792.

45. Jhappan C, Yusufzai TM, Anderson S, et al. The p53 response to DNA damage in vivo is independent of DNA-dependent protein kinase. *Mol Cell Biol* 2000;20:4075–4083.

46. Tang W, Willers H, Powell SN. p53 directly enhances rejoining of DNA double-strand breaks with cohesive ends in gamma-irradiated mouse fibroblasts. *Cancer Res* 1999;59:2562–2565.

47. Bill CA, Yu Y, Miselis NR, et al. A role for p53 in DNA end rejoining by human cell extracts. *Mutat Res* 1997;385:21–29.

48. Gao Y, Ferguson DO, Xie W, et al. Interplay of p53 and DNA-repair protein XRCC4 in tumorigenesis, genomic stability and development. *Nature* 2000;404:897–900.

49. Khanna KK, Jackson SP. DNA double-strand breaks: signaling, repair and the cancer connection. *Nature Genet* 2001;27:247–254.

50. Moynahan ME, Pierce AJ, Jasin M. BRCA2 is required for homology-directed repair of chromosomal breaks. *Mol Cell* 2001;7:263–272.

51. Hakem R, de la Pompa JL, Elia A, et al. Partial rescue of Brca1 (5–6) early embryonic lethality by p53 or p21 null mutation. *Nat Genet* 1997;16:298–302.

52. Jasin M, Berg P. Homologous integration in mammalian cells without target gene selection. *Genes Dev* 1988;2:1353–1363.

53. Jasin M. Chromosome breaks and genomic instability. *Cancer Invest* 2000;18:78–86.

54. Fritz G, Tano K, Mitra S, et al. Inducibility of the DNA repair gene encoding O^6-methylguanine-DNA methyltransferase in mammalian cells by DNA-damaging treatments. *Mol Cell Biol* 1991;11:4660–4668.

55. Tano K, Dunn WC, Darroudi F, et al. Amplification of the DNA repair gene O^6-methylguanine-DNA methyltransferase associated with resistance to alkylating drugs in a mammalian cell line. *J Biol Chem* 1997;272:13,250–13,254.

56. Kokkinakis DM, Hoffman RM, Frenkel EP, et al. Synergy between methionine stress and chemotherapy in the treatment of brain tumor xenografts in athymic mice. *Cancer Res* 2001;61:4017–4023.

57. Kokkinakis DM, Ahmed MM, Delgado R, et al. Role of O^6-methylguanine-DNA methyltransferase in the resistance of pancreatic tumors to DNA alkylating agents. *Cancer Res* 1997;57:5360–5368.

58. Biswas T, Ramana CV, Srinivasan G, et al. Activation of human O^6-methylguanine-DNA methyltransferase gene by glucocorticoid hormone. *Oncogene* 1999;18:525–532.

59. Boldogh I, Ramana CV, Chen Z, et al. Regulation of expression of the DNA repair gene O^6-methylguanine-DNA methyltransferase via protein kinase C-mediated signaling. *Cancer Res* 1998;58:3950–3956.

60. Bhakat KK, Mitra S. Regulation of the human O(6)-methylguanine-DNA methyltransferase gene by transcriptional coactivators cAMP response element-binding protein-binding protein and p300. *J Biol Chem* 2000;275:34197–34204.

61. Grombacher T, Mitra S, Kaina B. Induction of the alkyltransferase (MGMT) gene by DNA damaging agents and the glucocorticoid dexamethasone and comparison with the response of base excision repair genes. *Carcinogenesis* 1996;17:2329–2336.

62. Ostrowski LE, von Wronski MA, et al. Expression of O^6-methylguanine-DNA methyltransferase in malignant human glioma cell lines. *Carcinogenesis* 1991;12:1739–1744.

63. Wang Y, Kato T, Ayaki H, et al. Correlation between DNA methylation and expression of O^6-methylguanine-DNA methyltransferase gene in cultured human tumor cells. *Mutat Res* 1992;273:221–230.

64. Tano K, Shiota S, Collier J, et al. Isolation and structural characterization of a cDNA clone encoding the human DNA repair protein for O^6-alkylguanine. *Proc Natl Acad Sci USA* 1990;87:686–690.

65. Kaina B, Fritz G, Mitra S, et al. Transfection and expression of human O^6-methylguanine-DNA methyltransferase (MGMT) cDNA in Chinese hamster cells: the role of MGMT in protection against the genotoxic effects of alkylating agents. *Carcinogenesis* 1991;12:1857–1867.

66. Natarajan AT, Vermeulen S, Darroudi F, et al. Chromosomal localization of human O^6-methylguanine-DNA methyltransferase (MGMT) gene by in situ hybridization. *Mutagenesis* 1992;7:83–85.

67. Tatsuka M, Ibeanu GC, Izumi T, et al. Structural organization of the mouse DNA repair gene, *N*-methylpurine-DNA glycosylase. *DNA Cell Biol* 1995;14:37–45.

68. Roy R, Kennel SJ, Mitra S. Distinct substrate preference of human and mouse *N*-methylpurine-DNA glycosylases. *Carcinogenesis* 1996;17:2177–2182.

69. Olsson M, Lindahl T. Repair of alkylated DNA in *Escherichia coli*. Methyl group transfer from O^6-methylguanine to a protein cysteine residue. *J Biol Chem* 1980;255:10569–10571.

70. Hazra TK, Roy R, Biswas T, et al. Specific recognition of O^6-methylguanine in DNA by active site mutants of human O^6-methylguanine-DNA methyltransferase. *Biochemistry* 1997;36:5769–5776.

71. Srivenugopal KS, Shou J, Mullapudi SR, et al. Enforced expression of wild-type p53 curtails the transcription of the $O(6)$-methylguanine-DNA methyltransferase gene in human tumor cells and enhances their sensitivity to alkylating agents. *Clin Cancer Res* 2001;7:1398–1409.

72. Guo W, Liu X, Lee S, Park NH. High O^6-methylguanine methyl transferase activity is frequently found in human oral cancer cells with p53 inactivation. *Int J Oncol* 1999;15:817–821.

73. Rolhion C, Penault-Llorca F, Kemeny JL, et al. $O(6)$-Methylguanine-DNA methyltransferase gene (MGMT) expression in human glioblastomas in relation to patient characteristics and p53 accumulation. *Int J Cancer* 1999;84:416–420.

74. Grombacher T, Eichhorn U, Kaina B. p53 is involved in regulation of the DNA repair gene O^6-methylguanine-DNA methyltransferase (MGMT) by DNA damaging agents. *Oncogene* 1998;17:845–851.

75. Bocangel DB, Finkelstein S, Schold SC, et al. Multifaceted resistance of gliomas to temozolomide. *Clin Cancer Res* 2002;8:2725–2734.

76. Nutt CL, Loktionova NA, Pegg AE, et al. $O(6)$-methylguanine-DNA methyltransferase activity, p53 gene status and BCNU resistance in mouse astrocytes. *Carcinogenesis* 1999;20:2361–2365.

77. Mullapudi SR, Ali-Osman F, Shou J, et al. DNA repair protein O^6-alkylguanine-DNA alkyltransferase is phosphorylated by two distinct and novel protein kinases in human brain tumour cells. *Biochem J* 2000;351 Pt 2:393–402.

78. Srivenugopal KS, Mullapudi SR, Shou J, et al. Protein phosphorylation is a regulatory mechanism for O^6-alkylguanine-DNA alkyltransferase in human brain tumor cells. *Cancer Res* 2000;60:282–287.

79. Lage H, Dietel M. Involvement of the DNA mismatch repair system in antineoplastic drug resistance. *J Cancer Res Clin Oncol* 1999;125:156–165.

80. Marti TM, Kunz C, Fleck O. DNA mismatch repair and mutation avoidance pathways. *J Cell Physiol* 2002;191:28–41.

81. Fink D, Nebel S, Aebi S, et al. The role of DNA mismatch repair in platinum drug resistance. *Cancer Res* 1996;56:4881–4886.

82. Aebi S, Kurdi-Haidar B, Gordon R, et al. Loss of DNA mismatch repair in acquired resistance to cisplatin. *Cancer Res* 1996;56:3087–3090.

83. de las Alas MM, Aebi S, Fink D, et al. Loss of DNA mismatch repair: effects on the rate of mutation to drug resistance. *J Natl Cancer Inst* 1997;89:1537–1541.

84. Esteller M, Risques RA, Toyota M, et al. Promoter hypermethylation of the DNA repair gene $O(6)$-methylguanine-DNA methyltransferase is associated with the presence of G : C to A : T transition mutations in p53 in human colorectal tumorigenesis. *Cancer Res* 2001;61:4689–4692.

85. Hickman MJ, Samson LD. Role of DNA mismatch repair and p53 in signaling induction of apoptosis by alkylating agents. *Proc Natl Acad Sci USA* 1999;96:10,764–10,769.

86. Cadenas E, Davies KJ. Mitochondrial free radical generation, oxidative stress, and aging. *Free Radical Biol Med* 2000;29:222–230.

87. Gamaley IA, Klyubin IV. Roles of reactive oxygen species: signaling and regulation of cellular functions. *Int Rev Cytol* 1999;188:203–255.

88. Breen AP, Murphy JA. Reactions of oxyl radicals with DNA. *Free Radical Biol Med* 1995;18:1033–1077.

89. Muller I, Niethammer D, Bruchelt G. Anthracycline-derived chemotherapeutics in apoptosis and free radical cytotoxicity. [review]. *Int J Mol Med* 1998;1:491–494.

90. Mates JM, Sanchez-Jimenez FM. Role of reactive oxygen species in apoptosis: implications for cancer therapy. *Int J Biochem Cell Biol* 2000;32:157–170.

91. Lotem J, Peled-Kamar M, Groner Y, et al. Cellular oxidative stress and the control of apoptosis by wild-type p53, cytotoxic compounds, and cytokines. *Proc Natl Acad Sci USA* 1996;93:9166–9171.

92. Voehringer DW. BCL-2 and glutathione: alterations in cellular redox state that regulate apoptosis sensitivity. *Free Radical Biol Med* 1999;27:945–950.

93. Carmody RJ, Cotter TG. Signalling apoptosis: a radical approach. *Redox Rep* 2001; 6:77–90.

94. Keyse SM. Protein phosphatases and the regulation of mitogen-activated protein kinase signalling. *Curr Opin Cell Biol* 2000;12:186–192.

95. Buschmann T, Yin Z, Bhoumik A, et al. Amino-terminal-derived JNK fragment alters expression and activity of c-Jun, ATF2, and p53 and increases H_2O_2-induced cell death. *J Biol Chem* 2000;275:16590–16596.

96. Rusnak F, Reiter T. Sensing electrons: protein phosphatase redox regulation. *Trends Biochem Sci* 2000;25:527–529.

97. Leonarduzzi G, Arkan MC, Basaga H, et al. Lipid oxidation products in cell signaling. *Free Radical Biol Med* 2000;28:1370–1378.

98. Fanger GR, Johnson NL, Johnson GL. MEK kinases are regulated by EGF and selectively interact with Rac/Cdc42. *EMBO J* 1997;16:4961–4972.

99. Fiers W, Beyaert R, Declercq W, et al. More than one way to die: apoptosis, necrosis and reactive oxygen damage. *Oncogene* 1999;18:7719–7730.

100. Green DR, Reed JC. Mitochondria and apoptosis. *Science* 1998;281:1309–1312.

101. Tew KD. Glutathione-associated enzymes in anticancer drug resistance. *Cancer Res* 1994;54:4313–4320.

102. Tew KD, Dutta S, Schultz M. Inhibitors of glutathione *S* transferases as therapeutic agents. *Adv Drug Deliv Rev* 1997;26:91–104.

103. Cheng JZ, Singhal SS, Saini M, et al. Effects of mGST A4 transfection on 4-hydroxynonenal-mediated apoptosis and differentiation of K562 human erythroleukemia cells. *Arch Biochem Biophys* 1999;372:29–36.

104. Yang Y, Sharma R, Cheng JZ, et al. Protection of HLE B-3 cells against hydrogen peroxide- and naphthalene-induced lipid peroxidation and apoptosis by transfection with hGSTA1 and hGSTA2. *Invest Ophthalmol Vis Sci* 2002;43:434–445.

105. Yin Z, Ivanov VN, Habelhah H, et al. Glutathione *S*-transferase p elicits protection against H_2O_2-induced cell death via coordinated regulation of stress kinases. *Cancer Res* 2000;60:4053–4057.

106. Roy G, Horton JK, Roy R, et al. Acquired alkylating drug resistance of a human ovarian carcinoma cell line is unaffected by altered levels of pro- and anti-apoptotic proteins. *Oncogene* 2000;19:141–150.

107. Horton JK, Roy G, Piper JT, et al. Characterization of a chlorambucil-resistant human ovarian carcinoma cell line overexpressing glutathione S-transferase mu. *Biochem Pharmacol* 1999;58:693–702.

108. Crow JP. Dichlorodihydrofluorescein and dihydrorhodamine 123 are sensitive indicators of peroxynitrite in vitro: implications for intracellular measurement of reactive nitrogen and oxygen species. *Nitric Oxide* 1997;1:145–157.

109. Brown JM, Wouters BG. Apoptosis, p53, and tumor cell sensitivity to anticancer agents. *Cancer Res* 1999;59:1391–1399.
110. Levine AJ. p53, the cellular gatekeeper for growth and division. *Cell* 1997;88:323–331.
111. Prives C, Hall PA. The p53 pathway. *J Pathol* 1999;187:112–126.
112. Parks D, Bolinger R, Mann K. Redox state regulates binding of p53 to sequence-specific DNA, but not to non-specific or mismatched DNA. *Nucleic Acids Res* 1997;25:1289–1295.
113. Li PF, Dietz R, von Harsdorf R. p53 regulates mitochondrial membrane potential through reactive oxygen species and induces cytochrome c-independent apoptosis blocked by Bcl-2. *EMBO J* 1999;18:6027–6036.
114. Polyak K, Xia Y, Zweier JL, et al. A model for p53-induced apoptosis. *Nature* 1997;389:300–305.
115. Flatt PM, Polyak K, Tang LJ, et al. p53-dependent expression of PIG3 during proliferation, genotoxic stress, and reversible growth arrest. *Cancer Lett* 2000;156:63–72.
116. Evan G, Littlewood T. A matter of life and cell death. *Science* 1998;281:1317–1322.
117. Appella E, Anderson CW. Post-translational modifications and activation of p53 by genotoxic stresses. *Eur J Biochem* 2001;268:2764–2772.
118. Datta K, Babbar P, Srivastava T, et al. p53 dependent apoptosis in glioma cell lines in response to hydrogen peroxide induced oxidative stress. *Int J Biochem Cell Biol* 2002;34:148–157.
119. Phoa N, Epe B. Influence of nitric oxide on the generation and repair of oxidative DNA damage in mammalian cells. *Carcinogenesis* 2002;23:469–475.
120. Wang X, Michael D, de Murcia G, et al. p53 Activation by nitric oxide involves down-regulation of Mdm2. *J Biol Chem* 2002;277:15,697–15,702.
121. Cobbs CS, Samanta M, Harkins LE, et al. Evidence for peroxynitrite-mediated modifications to p53 in human gliomas: possible functional consequences. *Arch Biochem Biophys* 2001;394:167–172.
122. Ljungman M. Dial 9-1-1 for p53: mechanisms of p53 activation by cellular stress. *Neoplasia* 2000;2:208–225.
123. Xie S, Wang Q, Wu H, et al. Reactive oxygen species-induced phosphorylation of p53 on serine 20 is mediated in part by polo-like kinase-3. *J Biol Chem* 2001;276:36,194–36,199.
124. Cox LS, Lane DP. Tumour suppressors, kinases and clamps: how p53 regulates the cell cycle in response to DNA damage. *Bioessays* 1995;17:501–508.
125. Yu K, Ravera CP, Chen YN, et al. Regulation of Myc-dependent apoptosis by p53, c-Jun N-terminal kinases/stress-activated protein kinases, and Mdm-2. *Cell Growth Differ* 1997;8:731–742.
126. Ko LJ, Prives C. p53: puzzle and paradigm. *Genes Dev* 1996;10:1054–1072.
127. Rainwater R, Parks D, Anderson ME, et al. Role of cysteine residues in regulation of p53 function. *Mol Cell Biol* 1995;15:3892–3903.
128. Jayaraman L, Murthy KG, Zhu C, et al. Identification of redox/repair protein Ref-1 as a potent activator of p53. *Genes Dev* 1997;11:558–570.
129. Evans AR, Limp-Foster M, Kelley MR. Going APE over ref-1. *Mutat Res* 2000;461:83–108.
130. Mitra S, Hazra TK, Roy R, et al. Complexities of DNA base excision repair in mammalian cells. *Mol Cells* 1997;7:305–312.
131. Ramana CV, Boldogh I, Izumi T, et al. Activation of apurinic/apyrimidinic endonuclease in human cells by reactive oxygen species and its correlation with their adaptive response to genotoxicity of free radicals. *Proc Natl Acad Sci USA* 1998;95:5061–5066.
132. Gaiddon C, Moorthy NC, Prives C. Ref-1 regulates the transactivation and pro-apoptotic functions of p53 in vivo. *EMBO J* 1999;18:5609–5621.
133. Okamoto K, Prives C. A role of cyclin G in the process of apoptosis. *Oncogene* 1999;18:4606–4615.
134. Hsieh MM, Hegde V, Kelley MR, et al. Activation of APE/Ref-1 redox activity is mediated by reactive oxygen species and PKC phosphorylation. *Nucleic Acids Res* 2001;29:3116–3122.

135. Xanthoudakis S, Curran T. Identification and characterization of Ref-1, a nuclear protein that facilitates AP-1 DNA-binding activity. *EMBO J* 1992;11:653–665.

136. Kelley MR, Cheng L, Foster R, et al. Elevated and altered expression of the multifunctional DNA base excision repair and redox enzyme Ape1/ref-1 in prostate cancer. *Clin Cancer Res* 2001;7:824–830.

137. Antonsson B, Conti F, Ciavatta A, et al. Inhibition of Bax channel-forming activity by Bcl-2. *Science* 1997;277:370–372.

138. Yacoub A, Kelley MR, Deutsch WA. The DNA repair activity of human redox/repair protein APE/Ref-1 is inactivated by phosphorylation. *Cancer Res* 1997;57:5457–5459.

139. Mandlekar S, Kong AN. Mechanisms of tamoxifen-induced apoptosis. *Apoptosis* 2001;6:469–477.

140. Schindl M, Oberhuber G, Pichlbauer EG, et al. DNA repair-redox enzyme apurinic endonuclease in cervical cancer: evaluation of redox control of HIF-1alpha and prognostic significance. *Int J Oncol* 2001;19:799–802.

141. Puglisi F, Aprile G, Minisini AM, et al. Prognostic significance of Ape1/ref-1 subcellular localization in non-small cell lung carcinomas. *Anticancer Res* 2001;21:4041–4049.

142. Zhou Y, Kok KH, Chun AC, et al. Mouse peroxiredoxin V is a thioredoxin peroxidase that inhibits p53-induced apoptosis. *Biochem Biophys Res Commun* 2000;268:921–927.

143. Tan M, Li S, Swaroop M, Guan K, et al. Transcriptional activation of the human glutathione peroxidase promoter by p53. *J Biol Chem* 1999;274:12,061–12,066.

144. Asher G, Lotem J, Kama R, et al. NQO1 stabilizes p53 through a distinct pathway. *Proc Natl Acad Sci USA* 2002;99:3099–3104.

145. Phillips RM, Burger AM, Fiebig HH, et al. Genotyping of NAD(P)H:quinone oxidoreductase (NQO1) in a panel of human tumor xenografts: relationship between genotype status, NQO1 activity and the response of xenografts to mitomycin C chemotherapy in vivo(1). *Biochem Pharmacol* 2001;62:1371–1377.

146. Yamaoka A, Kuwabara I, Frigeri LG, et al. A human lectin, galectin-3 (epsilon bp/Mac-2), stimulates superoxide production by neutrophils. *J Immunol* 1995;154:3479–3487.

147. Donald SP, Sun XY, Hu CA, et al. Proline oxidase, encoded by p53-induced gene-6, catalyzes the generation of proline-dependent reactive oxygen species. *Cancer Res* 2001;61:1810–1815.

148. Ji Y, Toader V, Bennett BM. Regulation of microsomal and cytosolic glutathione *S*-transferase activities by S-nitrosylation. *Biochem Pharmacol* 2002;63:1397–1404.

149. Souza DG, Soares AC, Pinho V, et al. Increased mortality and inflammation in tumor necrosis factor-stimulated gene-14 transgenic mice after ischemia and reperfusion injury. *Am J Pathol* 2002;160:1755–1765.

150. Chiarugi V, Magnelli L, Gallo O. Cox-2, iNOS and p53 as play-makers of tumor angiogenesis.review]. *Int J Mol Med* 1998;2:715–719.

151. Bianchi A, Becuwe P, Franck P, et al. Induction of MnSOD gene by arachidonic acid is mediated by reactive oxygen species and p38 MAPK signaling pathway in human HepG2 hepatoma cells. *Free Radical Biol Med* 2002;32:1132–1142.

152. Macmillan-Crow LA, Cruthirds DL. Invited review: manganese superoxide dismutase in disease. *Free Radic Res* 2001;34:325–336.

153. Drane P, Bravard A, Bouvard V, et al. Reciprocal down-regulation of p53 and SOD2 gene expression-implication in p53 mediated apoptosis. *Oncogene* 2001;20:430–439.

154. Pani G, Bedogni B, Anzevino R, et al. Deregulated manganese superoxide dismutase expression and resistance to oxidative injury in p53-deficient cells. *Cancer Res* 2000;60:4654–4660.

155. Oberley LW. Anticancer therapy by overexpression of superoxide dismutase. *Antioxid Redox Signal* 2001;3:461–472.

156. Beham A, Marin MC, Fernandez A, et al. Bcl-2 inhibits p53 nuclear import following DNA damage. *Oncogene* 1997;15:2767–2772.

157. Fritsche M, Haessler C, Brandner G. Induction of nuclear accumulation of the tumor-suppressor protein p53 by DNA-damaging agents. *Oncogene* 1993;8:307–318.
158. Schott AF, Apel IJ, Nunez G, et al. Bcl-XL protects cancer cells from p53-mediated apoptosis. *Oncogene* 1995;11:1389–1394.
159. Wosikowski K, Regis JT, Robey RW, et al. Normal p53 status and function despite the development of drug resistance in human breast cancer cells. *Cell Growth Differ* 1995;6:1395–1403.
160. Martinez JD, Pennington ME, Craven MT, et al. Free radicals generated by ionizing radiation signal nuclear translocation of p53. *Cell Growth Differ* 1997;8:941–949.
161. Wu HH, Thomas JA, Momand J. p53 protein oxidation in cultured cells in response to pyrrolidine dithiocarbamate: a novel method for relating the amount of p53 oxidation in vivo to the regulation of p53-responsive genes. *Biochem J* 2000;351:87–93.
162. Liang SH, Clarke MF. Regulation of p53 localization. *Eur J Biochem* 2001;268:2779–2783.
163. Shaulsky G, Goldfinger N, Ben-Ze'ev A, et al. Nuclear accumulation of p53 protein is mediated by several nuclear localization signals and plays a role in tumorigenesis. *Mol Cell Biol* 1990;10:6565–6577.
164. Steggerda SM, Paschal BM. Regulation of nuclear import and export by the GTPase Ran. *Int Rev Cytol* 2002;217:41–91.
165. Giannakakou P, Sackett DL, Ward Y, et al. p53 is associated with cellular microtubules and is transported to the nucleus by dynein. *Nature Cell Biol* 2000;2:709–717.
166. Manteuffel-Cymborowska M. Nuclear receptors, their coactivators and modulation of transcription. *Acta Biochim Pol* 1999;46:77–89.
167. Ito A, Lai CH, Zhao X, et al. p300/CBP-mediated p53 acetylation is commonly induced by p53-activating agents and inhibited by MDM2. *EMBO J* 2001;20:1331–1340.
168. Pomerantz J, Schreiber-Agus N, Liegeois NJ, et al. The Ink4a tumor suppressor gene product, p19Arf, interacts with MDM2 and neutralizes MDM2's inhibition of p53. *Cell* 1998;92:713–723.
169. Zhang Y, Xiong Y, Yarbrough WG. ARF promotes MDM2 degradation and stabilizes p53: ARF-INK4a locus deletion impairs both the Rb and p53 tumor suppression pathways. *Cell* 1998;92:725–734.
170. Momand J, Wu HH, Dasgupta G. MDM2—master regulator of the p53 tumor suppressor protein. *Gene* 2000;242:15–29.
171. Fogal V, Gostissa M, Sandy P, et al. Regulation of p53 activity in nuclear bodies by a specific PML isoform. *EMBO J* 2000;19:6185–6195.
172. Seeler JS, Dejean A. The PML nuclear bodies: actors or extras? *Curr Opin Genet Dev* 1999;9:362–367.
173. Pearson M, Carbone R, Sebastiani C, et al. PML regulates p53 acetylation and premature senescence induced by oncogenic Ras. *Nature* 2000;406:207–210.
174. Rodriguez MS, Desterro JM, Lain S, et al. SUMO-1 modification activates the transcriptional response of p53. *EMBO J* 1999;18:6455–6461.
175. Gottifredi V, Prives C. Molecular biology. Getting p53 out of the nucleus. *Science* 2001;292:1851–1852.
176. Kobet E, Zeng X, Zhu Y, et al. MDM2 inhibits p300-mediated p53 acetylation and activation by forming a ternary complex with the two proteins. *Proc Natl Acad Sci USA* 2000;97:12,547–12,552.
177. Stommel JM, Marchenko ND, Jimenez GS, et al. A leucine-rich nuclear export signal in the p53 tetramerization domain: regulation of subcellular localization and p53 activity by NES masking. *EMBO J* 1999;18:1660–1672.
178. Shirangi TR, Zaika A, Moll UM. Nuclear degradation of p53 occurs during down-regulation of the p53 response after DNA damage. *FASEB J* 2002;16:420–422.
179. Kastan MB, Onyekwere O, Sidransky D, et al. Participation of p53 protein in the cellular response to DNA damage. *Cancer Res* 1991;51:6304–6311.

180. van Zandwijk N. N-acetylcysteine (NAC) and glutathione (GSH): antioxidant and chemopreventive properties, with special reference to lung cancer. *J Cell Biochem* 1995;22(Suppl):24–32.

181. Whysner J, Williams GM. Butylated hydroxyanisole mechanistic data and risk assessment: conditional species-specific cytotoxicity, enhanced cell proliferation, and tumor promotion. *Pharmacol Ther* 1996;71:137–151.

182. Biaglow JE, Varnes ME, Tuttle SW, et al. The effect of L-buthionine sulfoximine on the aerobic radiation response of A549 human lung carcinoma cells. *Int J Radiat Oncol Biol Phys* 1986;12:1139–1142.

183. Decaudin D, Geley S, Hirsch T, et al. Bcl-2 and Bcl-XL antagonize the mitochondrial dysfunction preceding nuclear apoptosis induced by chemotherapeutic agents. *Cancer Res* 1997;57:62–67.

184. Marchenko ND, Zaika A, Moll UM. Death signal-induced localization of p53 protein to mitochondria. A potential role in apoptotic signaling. *J Biol Chem* 2000;275: 16,202–16,212.

185. Sansome C, Zaika A, Marchenko ND, et al. Hypoxia death stimulus induces translocation of p53 protein to mitochondria. Detection by immunofluorescence on whole cells. *FEBS Lett* 2001;488:110–115.

186. Moll UM, Zaika A. Nuclear and mitochondrial apoptotic pathways of p53. *FEBS Lett* 2001;493:65–69.

187. Donahue RJ, Razmara M, Hoek JB, et al. Direct influence of the p53 tumor suppressor on mitochondrial biogenesis and function. *FASEB J* 2001;15:635–644.

188. Chen L, Agrawal S, Zhou W, et al. Synergistic activation of p53 by inhibition of MDM2 expression and DNA damage. *Proc Natl Acad Sci USA* 1998;95:195–200.

189. Ryan KM, Phillips AC, Vousden KH. Regulation and function of the p53 tumor suppressor protein. *Curr Opin Cell Biol* 2001;13:332–337.

190. Prives C. Signaling to p53: breaking the MDM2-p53 circuit. *Cell* 1998;95:5–8.

191. Jayaraman L, Freulich E, Prives C. Functional dissection of p53 tumor suppressor protein. *Methods Enzymol* 1997;283:245–256.

192. Locigno R, Castronovo V. Reduced glutathione system: role in cancer development, prevention and treatment. [review]. *Int J Oncol* 2001;19:221–236.

193. Kirsch M, De Groot H. NAD(P)H, a directly operating antioxidant? *FASEB J* 2001; 15:1569–1574.

194. Gulick AM, Fahl WE. Mammalian glutathione *S*-transferase: regulation of an enzyme system to achieve chemotherapeutic efficacy. *Pharmacol Ther* 1995;66:237–257.

195. Mantymaa P, Siitonen T, Guttorm T, et al. Induction of mitochondrial manganese superoxide dismutase confers resistance to apoptosis in acute myeloblastic leukaemia cells exposed to etoposide. *Br J Haematol* 2000;108:574–581.

196. Izutani R, Asano S, Imano M, Kuroda D, Kato M, Ohyanagi H. Expression of manganese superoxide dismutase in esophageal and gastric cancers. *J Gastroenterol* 1998;33:816–822.

197. Morel Y, Barouki R. Repression of gene expression by oxidative stress. *Biochem J* 1999;342(Pt 3):481–496.

198. Lazo JS, Kuo SM, Woo ES, et al. The protein thiol metallothionein as an antioxidant and protectant against antineoplastic drugs. *Chem Biol Interact* 1998;111–112:255–262.

199. Hayashi S, Takamiya R, Yamaguchi T, et al. Induction of heme oxygenase-1 suppresses venular leukocyte adhesion elicited by oxidative stress: role of bilirubin generated by the enzyme. *Circ Res* 1999;85:663–671.

200. Sies H, Stahl W, Sundquist AR. Antioxidant functions of vitamins. Vitamins E and C, beta-carotene, and other carotenoids. *Ann NY Acad Sci* 1992;669:7–20.

201. Crawford DR, Davies KJ. Adaptive response and oxidative stress. *Environ Health Perspect* 1994;102(Suppl 10):25–28.

202. Merad-Saidoune M, Boitier E, Nicole A, et al. Overproduction of Cu/Zn-superoxide dismutase or Bcl-2 prevents the brain mitochondrial respiratory dysfunction induced by glutathione depletion. *Exp Neurol* 1999;158:428–436.
203. Lee YJ, Chen JC, Amoscato AA, et al. Protective role of Bcl2 in metabolic oxidative stress-induced cell death. *J Cell Sci* 2001;114:677–684.
204. Reed JC. Cytochrome c: can't live with it—can't live without it. *Cell* 1997;91:559–562.
205. Vander Heiden MG, Chandel NS, Williamson EK, et al. Bcl-x$_L$ regulates the membrane potential and volume homeostasis of mitochondria. *Cell* 1997;91:627–637.
206. Ellerby LM, Ellerby HM, Park SM, et al. Shift of the cellular oxidation-reduction potential in neural cells expressing Bcl-2. *J Neurochem* 1996;67:1259–1267.
207. Papadopoulos MC, Koumenis IL, Xu L, et al. Potentiation of murine astrocyte antioxidant defence by bcl-2: protection in part reflects elevated glutathione levels. *Eur J Neurosci* 1998;10:1252–1260.
208. Lee M, Hyun DH, Marshall KA, Ellerby LM, et al. Effect of overexpression of BCL-2 on cellular oxidative damage, nitric oxide production, antioxidant defenses, and the proteasome. *Free Radical Biol Med* 2001;31:1550–1559.
209. Zamzami N, Marzo I, Susin SA, et al. The thiol crosslinking agent diamide overcomes the apoptosis-inhibitory effect of Bcl-2 by enforcing mitochondrial permeability transition. *Oncogene* 1998;16:1055–1063.
210. Zhan Q, Kontny U, Iglesias M, et al. Inhibitory effect of Bcl-2 on p53-mediated transactivation following genotoxic stress. *Oncogene* 1999;18:297–304.
211. Durocher D, Jackson SP. DNA-PK, ATM and ATR as sensors of DNA damage: variations on a theme? *Curr Opin Cell Biol* 2001;13:225–231.

6

Stress-Activated Signal Transduction Pathways in DNA Damage Response

Implications for Repair, Arrest, and Therapeutic Interventions

Moulay A. Alaoui-Jamali, P. James Scrivens, and Martin Loignon

CONTENTS

1. INTRODUCTION

Tumor cell resistance to chemotherapy is a common clinical problem, which is often a primary cause of treatment failure. In particular, drug-resistant cells develop an impressive arsenal of constitutive and/or inducible DNA-damage response mechanisms, which can deregulate the cell cycle checkpoints and DNA repair and allow cells to escape from apoptotic cell death.

In many instances, stress signals can originate at the plasma membrane or in the cytoplasm as a result of growth-factor-receptor activation; the signals are then propagated via signal transduction cascades, ultimately resulting in nuclear responses that dictate the fate of tumor cells. Such fundamental mechanisms are

From: *Cancer Drug Discovery and Development: DNA Repair in Cancer Therapy*
Edited by: L. C. Panasci and M. A. Alaoui-Jamali © Humana Press Inc., Totowa, NJ

regulated by a multitude of feedback loops and a high level of crosstalk between distinct transduction pathways. It is, therefore, predictable that the alteration of growth factor receptors frequently seen in cancer has an impact not only on the proliferation state of tumor cells but also on a variety of stress response and survival pathways contributing to the malignant phenotype.

How the signaling network directly modulates DNA repair is also beginning to take shape with the discovery that most of the cell's major repair processes, including base excision repair (BER), nucleotide excision repair (NER), double-strand break (DSB) repair, and recombination repair have protein kinase components directly modulating their activity. In this chapter, we will explore the links between stress-activated signaling cascades and DNA-repair-associated mechanisms, with particular emphasis given to stress-activated kinases of the mitogen-activated protein kinase (Mapk) family, but with discussion of other kinases recently shown to play a role in DNA repair. We will discuss a framework for the development of alternative strategies to modulate stress-regulated mechanisms and DNA repair in the context of therapeutics. Given the complex nomenclature of the proteins involved, we will attempt to follow that of Kyriakis and Avruch *(1)*. This article and others *(2)* are highly recommended for readers seeking further information on the molecular biology of stress-activated signaling pathways.

2. CELL SIGNAL TRANSDUCTION PATHWAYS THAT REGULATE EFFECTORS OF THE DNA-DAMAGE RESPONSE

The stress response comprises a network of integrated signaling pathways that regulate a multifaceted response, and its components can be broadly divided as sensors, transducers, and effectors (*see* Fig. 1). With respect to drug-induced genotoxic stress, sensors are believed to sense aberrant DNA structures and initiate the global DNA-damage response. Unlike yeast, the identity of sensors in mammalian cells has not been well established, but several speculative models implicate Atm, Brca1, the Nbs1–Mre11–Rad50 complex, and some mismatch proteins as potential DNA-damage sensors in mammalian cells (reviewed in ref. *3*). The transducers and effectors, which represent the cellular reaction to stress, include a variety of kinases and substrates involved in the regulation of DNA repair, transcription, and cell cycle checkpoints, which together constitute the core of the DNA-damage response network.

The protein kinases that act at the forefront of a stress response to phosphorylate DNA repair and cell-cycle-arrest effectors can be divided in two groups. The first group includes kinases activated by damaged DNA and has been associated with genetic repair and cell-cycle-arrest disorders. As such, deficiencies in the gene products of ataxia telangiectasia mutated (Atm), its homolog Atr, or the DNA-dependent protein kinase (Dna-pk) predispose to cancer and correlate with high radiosensitivity and abnormal cell cycle arrest. The role of these kinases

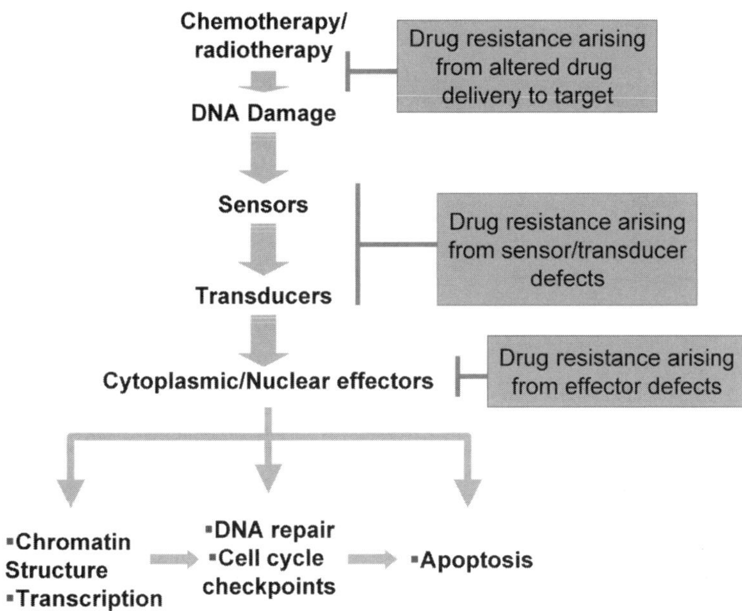

Fig 1. Major cellular factors that can affect tumor cell response to genotoxic stress and account for drug resistance. Listed are changes in the bioavailability of a drug or its active metabolites at the target site, inability of cells to recognize DNA damage, which can lead to increased tolerance, defects in cell ability to signal genotoxic stress to downstream targets, or aberrant function of DNA damage response effector(s). Such defects can impact on cell ability to arrest and repair DNA-damage and the drug's ability to induce tumor cell death.

in the DNA-damage-induced checkpoints has been extensively reviewed *(3)*, but, as described in some of the accompanying chapters, there is a growing body of information expanding our understanding of the roles of these kinases in the regulation of DNA repair processes. A second and very broad group includes the kinases of stress-activated signal transduction pathways (*see* Figs. 2A and 3A) that are activated by several stress stimuli both physiological and exogenous (e.g., chemotherapy) and play roles in a multitude of other cellular processes. The kinases activated during the genotoxic stress response include the p38 and stress-activated, protein kinase (Sapk) serine–threonine kinases, as well as extracellular signal-regulated kinase (Erks) (*see* Fig. 4A). Depending on the stress, activation of the Sapks of the Mapk superfamily can be the result of growth-factor-receptor activation, cytoskeletal alterations, or the signals emanating from the damage-activated kinases (e.g., via the Atm → Abl → Map3k pathway). It is therefore evident that the cellular response to stress depends on a multitude of factors, including the unique characteristics of the stress itself as well as the

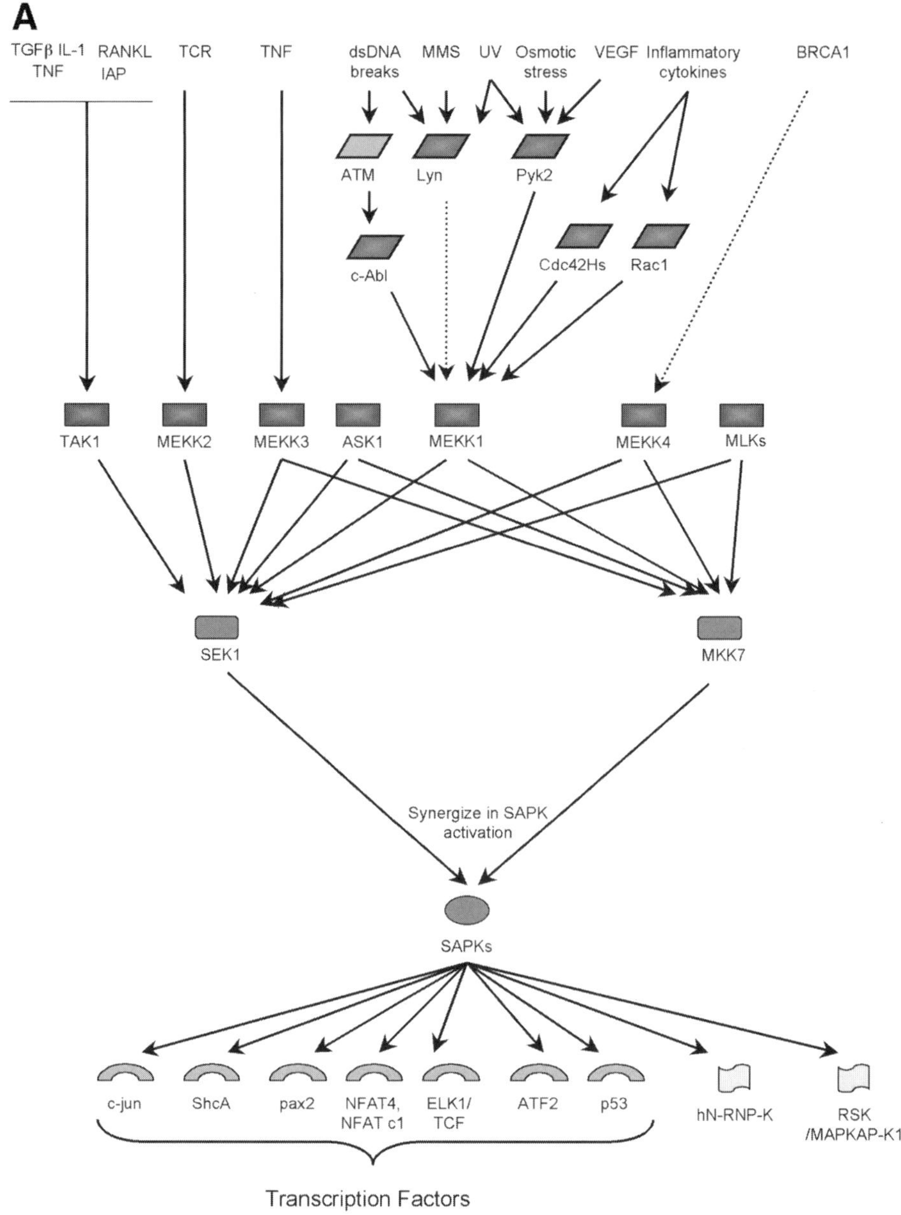

Fig. 2. (A) Major activators of SAPK pathways. Activators stimulate Map3ks either directly or via kinases upstream of the prototypical Map3k>MEK>MAPK cassette. A given genotoxic/chemothereapeutic agent may activate one or several Map3ks. MEKK1 represents a major point of convergence for signals arising from genotoxic agents. Signals

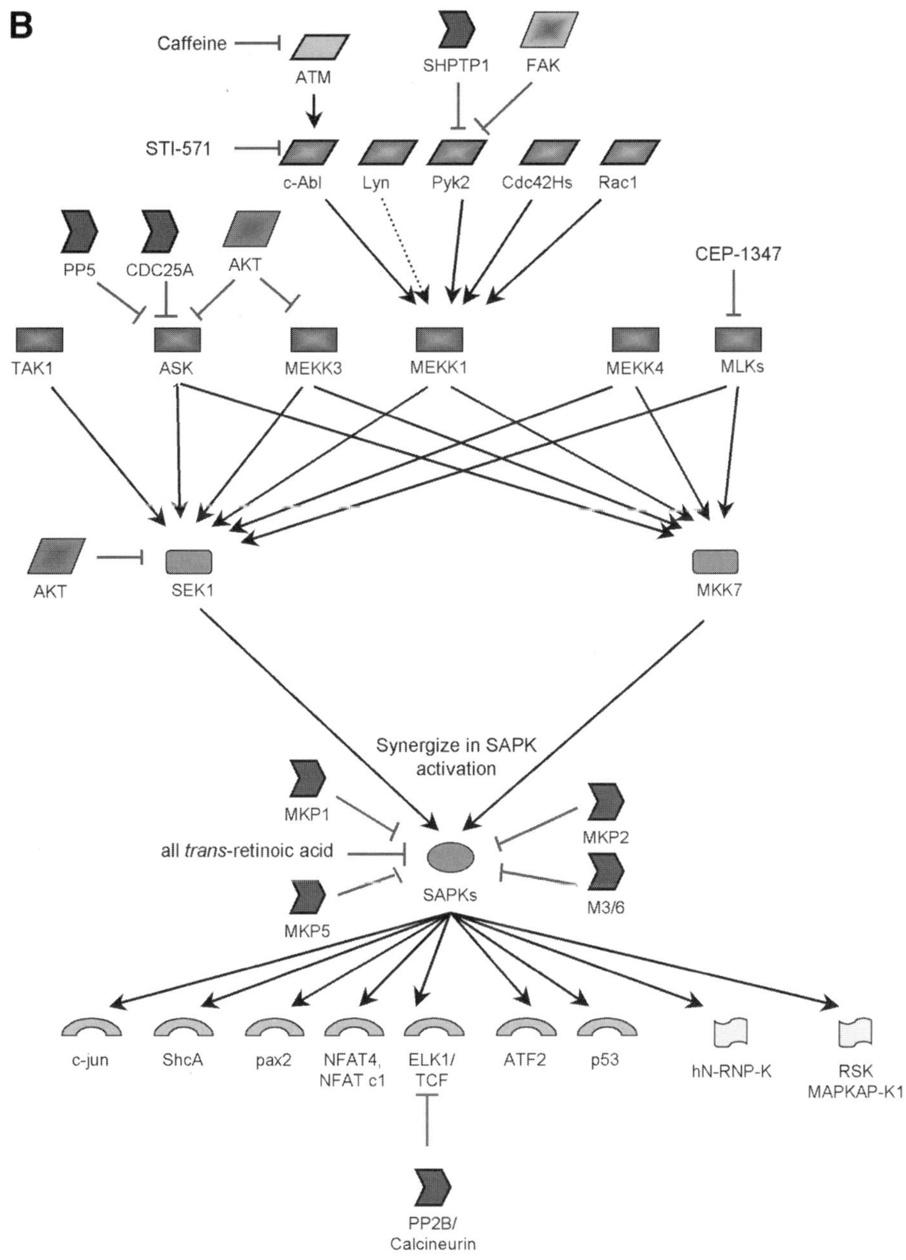

are then transmitted to SEK1 and MKK7, which synergize in the activation of the SAPKs, resulting in effects on transcriptional activity via SAPK-target transcription factors. **(B)** Inhibitors of the SAPK pathway and upstream kinases include physiological inhibitors, such as phosphatases and kinases, as well as pharmacological inhibitors.

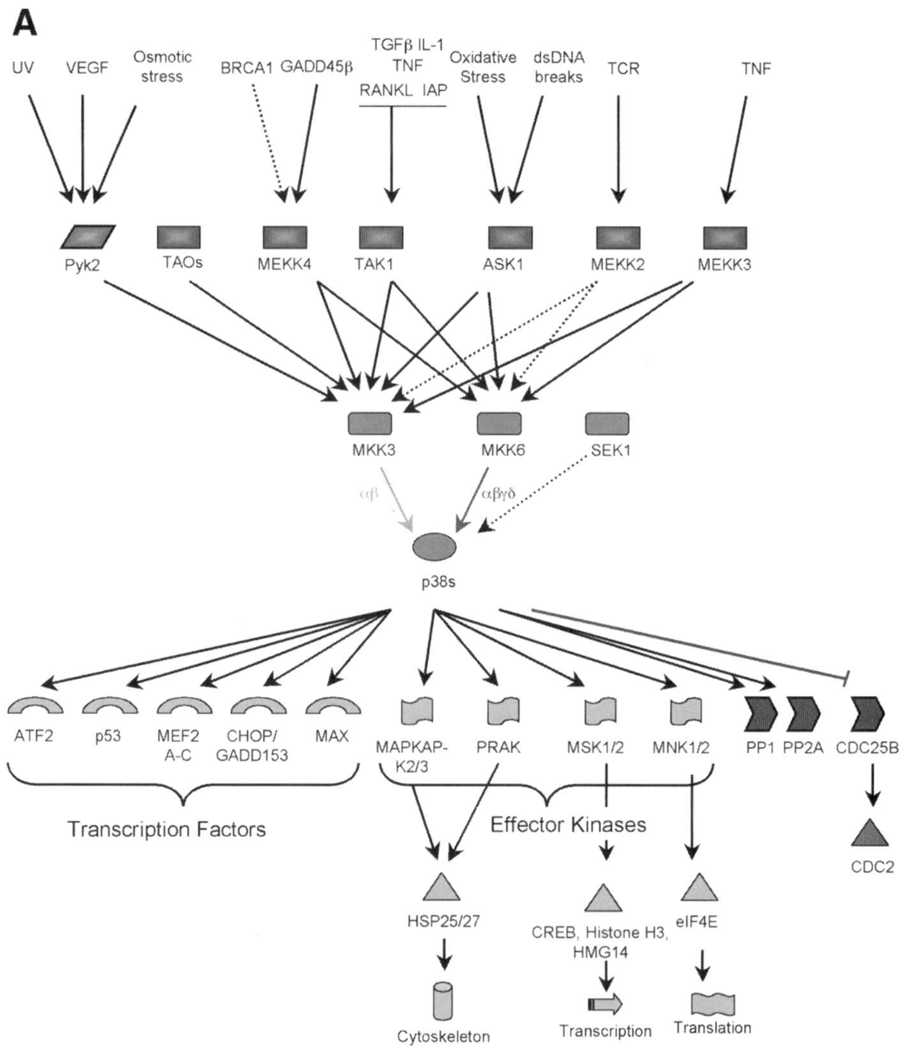

Fig. 3. (A) Many of the Map3ks that stimulate SAPK activation appear to be shared with the p38 pathway (e.g., ASK1, MEKK3, MEKK4, TAK1). Each of these activates MKK3 or MKK6 (or both). MKK3 appears to activate p38α and p38β, whereas MKK6 activates all four p38 isoforms. SEK1 has also been reported to activate p38s; its upstream activators are shown in Fig. 2A. p38 kinases exert their impact via several mechanisms, including activation of transcription factors, effector kinases, and phosphatases. In the case of Cdc25, phosphorylation results in inactivation and degradation. **(B)** Inhibitors of the p38 pathway include a wide array of MAPK phosphatases, kinases, and pharmacological inhibitors such as the cytokine-suppressive anti-inflamatory drugs (CSAIDs). A notable characteristic of the CSAIDs is that they inhibit p38α and p38β, but have little activity toward p38γ or p38δ (i.e., they inhibit the same subset of isoforms activated by MKK3).

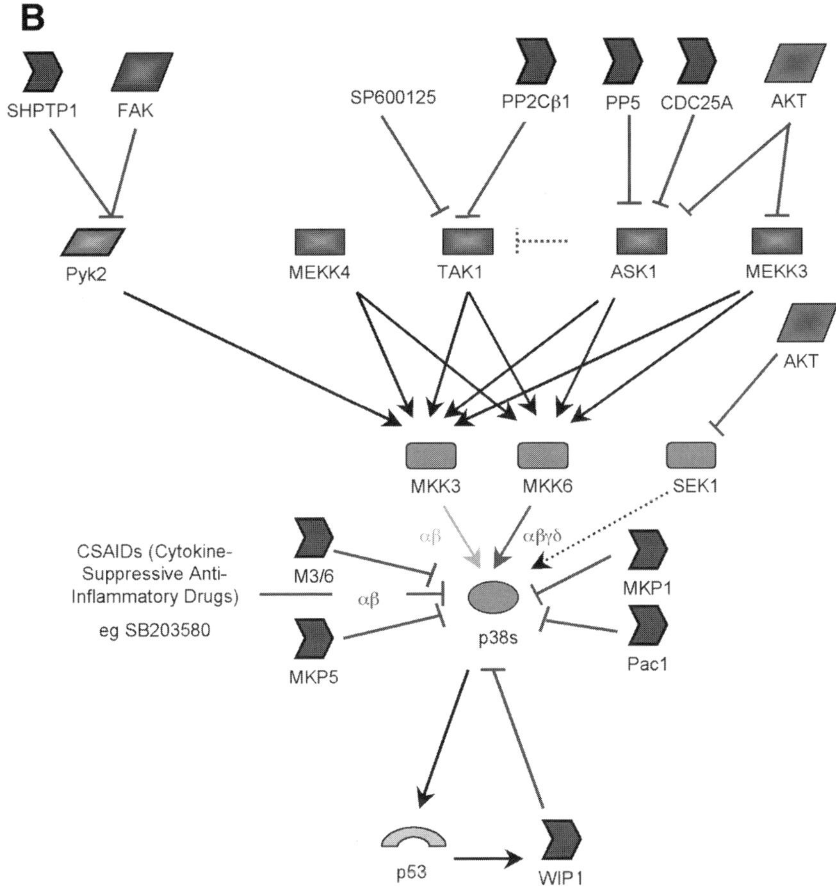

expression patterns of a vast number of proteins with highly integrated yet often opposing functions. The pathways described herein reflect the complexity and the diversity of phosphorylation-dependent mechanisms that mammalian cells use to deal with a stress response.

2.1. SAPK/MAPK Transduction Pathways and the Stress Response

The term "stress-activated protein kinase" (Sapk) has been generally applied to the c-jun N-terminal kinases (JNKs) and p38. The Jnk (henceforth referred to as SAPK) and p38 kinases are members of the Mapk superfamily. The hierarchical nature of signaling through the superfamily is illustrated by the receptor tyrosine kinase (RTK) → Erk pathway (*see* Fig. 4A). In the prototypical cascade, ligand binding stimulates receptor activation (e.g., through dimerization), which results in autophosphorylation and transphosphorylation of multiple tyrosine

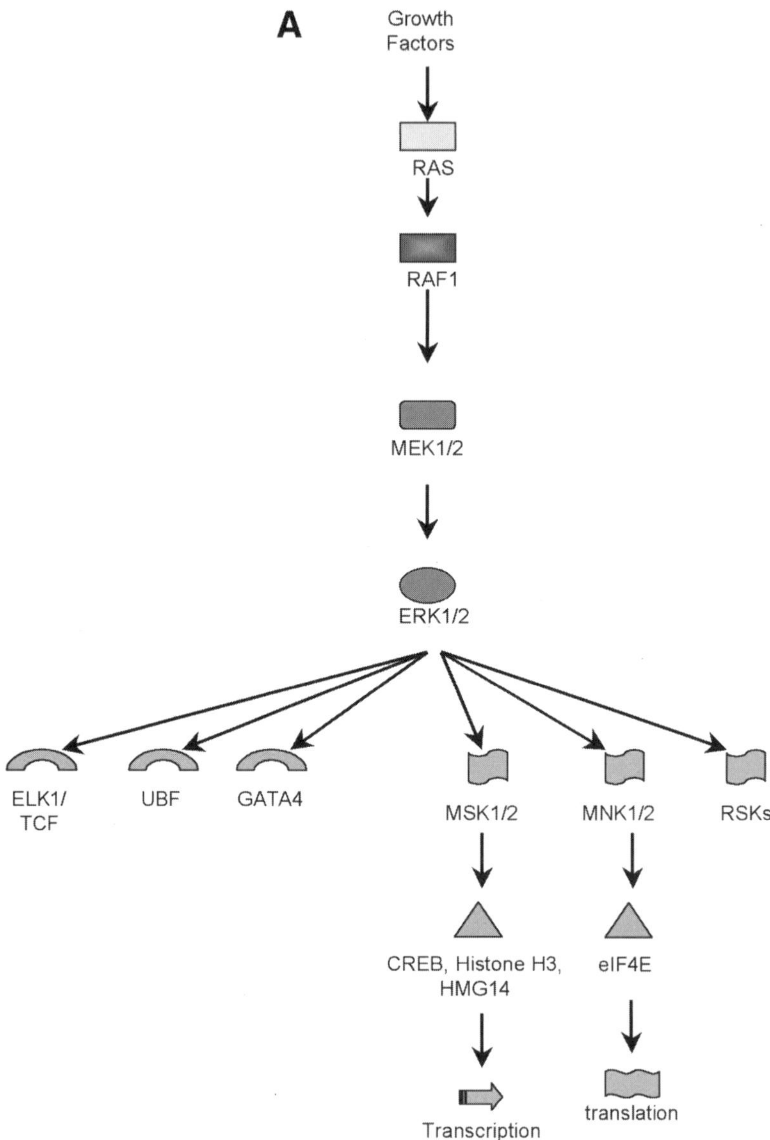

Fig. 4. (A) A pared-down representation of the Map3k>MEK>MAPK cassette. **(B)** Inhibitors of the ERK pathway include a number of MAPK phosphatases, the commonly used MEK1 inhibitor PD98059, and several effector kinases activated by the p38 pathways, such as PRAK and MAPKAP-K2. Note also that PP1 and PP2A have been reported as targets of p38 (*see* Fig. 3A).

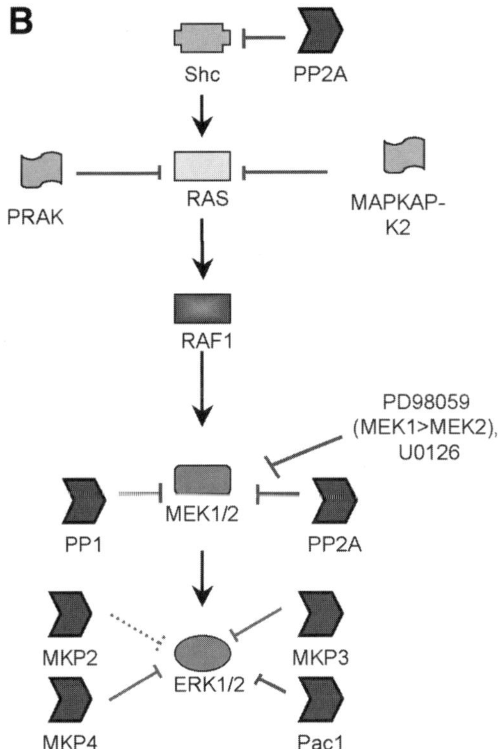

residues. These residues are bound by adapter proteins such as the SH2-containing growth-factor-receptor-bound protein-2 (Grb2). Grb2 is bound to son of sevenless (Sos) via SH3 domains, and the latter acts as a guanine nucleotide exchange factor (GEF), stimulating the exchange of GDP for GTP bound by Ras, resulting in activation of this kinase. Ras is a farnesylated protein and, therefore, membrane bound; through its activation, it becomes an adaptor itself, recruiting Raf to the membrane. The precise mechanism of Raf-1 activation by Ras is not yet clear, but localization to the membrane as well as other signals emanating from activated Ras are necessary. Raf-1 represents the first component of the Mapk core pathway, a signaling module that is reiterated in several parallel forms responding to a variety of stimuli. The highest-level component of this module is variously labeled Mapkkk, Map3k, or Mekk, and, in the case of Raf-1, its activation results in the amplification of the extracellular signal through Mek1/2 and Erk1/2. The use of such a cascade of signaling molecules results not only in signal amplification but also provides additional control points to modulate both the duration and specificity signaling.

2.1.1. Stress Response and DNA Damage Specificity

The Sapks and p38 are strongly activated by cellular stresses. These can include oxidative stresses, DNA damage by chemotherapy drugs, hyperosmolarity and hypo-osmolarity, heat shock, anisomycin, heavy metals, and other insults. Indeed, the c-jun N-terminal kinase, Jnk1, was cloned and identified as a kinase phosphorylating c-jun on Ser-63 and Ser-73 *(4)* following ultraviolet (UV) irradiation. A naïve impression is that Erk activation results from growth factor stimulation and promotes survival/proliferation, whereas cytotoxic agents activate the Sapks, leading to "damage control" or apoptotic responses. Unfortunately, reality is not so simple, and there is a great deal of reiteration between the various pathways, each making distinct (although not yet fully resolved) contributions to survival in response to various stress conditions. Furthermore, an individual "stress," such as chemotherapy, is often multifactorial. These stimuli, for instance, all have concomitant elements of oxidative stress, which modulate Sapk activation.

Common laboratory models provide a second example of the complexity of these pathways and must also be examined carefully. UV radiation is a convenient method to study the response of cancer cell lines to DNA-damaging agents. Although these agents have distinct characteristics, it has been suggested that the response to UV radiation is correlated to that seen with the more therapeutically relevant ionizing radiation *(5)*. UV can potentially activate Sapks through at least three mechanisms: direct DNA/protein damage, as a consequence of RTK oligomerization, or via inactivation of phosphatases or other effects of oxidative stress. These phenomena were examined in Rat1 fibroblasts *(6)* treated with UVB (λ 280–320 nm), with the goal of discerning the contribution of oxidative stress to Sapk activation. Interestingly, the induction of Sapkγ activity was strong and rapid (<15 min) in response to a dose of 1200 J/m^2, and the kinetics of this response were similar to those seen with treatment with the ribotoxin anisomycin, but differed from the slower and more gradual induction seen upon arsenite treatment (an inducer of oxidative stress). These kinetics were mimicked by those of Sek1 activation. Further, the activation of Sapkγ in response to UVB, anisomycin, and interleukin (IL)-1α was not blunted by 30 mM n-acetyl cysteine (NAC) pretreatment, although this was sufficient to completely ablate the arsenite and cadmium chloride stimulation of the enzyme. Thus, it appears that oxidative stress does not play a substantial role in Sapkγ activation by UVB, but that the activation may be via another mechanism, such as direct ribotoxicity.

2.1.2. The Case of Cisplatin

A survey of the literature indicates a broad and varied activation of Mapk/Sapks in response to different chemotherapeutics and DNA-damaging agents. A prolonged activation of c-jun N-terminal kinase activity was reported in mouse keratinocytes treated with cisplatin, in contrast to the response of these cells to transplatin (a therapeutically inactive isomer of cisplatin) *(7)*. Whereas transplatin produced a rapid and transient increase in c-jun phosphorylation, cisplatin stimulated a more prolonged increase. Furthermore, transplatin was a very effective inducer of MKP-1 (a dual-specificity phosphatase that inactivates p38 and Sapk), whereas cisplatin induced only a marginal increase in MKP-1 protein levels. A similar correlation was reported between the level of c-jun phosphorylation and UV radiation resistance in a number of lung cancer cell lines *(5)*.

The amplitude of Sapk activation by cisplatin may vary depending on cell lines used, but most reports indicate a similarly prolonged Sapk activation in response to this drug *(7–9)*. In contrast to the c-jun kinases, there are discrepancies with respect to the reported effects of cisplatin on p38 activity. Whereas some studies reported no effect *(10)*, others observed a strong induction of p38 *(8)*, notably p38γ *(11)*. This induction is significant with respect to analyses of Sapk pathways because, although it is often considered a minor isoform, p38γ has been suggested to be more efficient in phosphorylation of ATF2 than p38α *(12,13)*. More importantly, the cytokine-suppressive anti-inflammatory drugs (CSAIDs) such as SB203580, which are commonly used to inhibit the p38 kinases, do not inhibit p38γ or p38δ, but exert their effect exclusively through the predominant isoforms p38α and p38β *(13)*. Furthermore, the emerging role for p38 as a principal mediator of UV-induced G_2/M arrest deserves particular attention when it comes to DNA-damaging agents *(14)*.

2.1.3. Integration of Signaling Via Common Pathways

Another step in the pathway from DNA damage to Sapk activation was recently filled in with the finding that cells from c-Abl –/– mice are defective in Sapk activation in response to ionizing radiation (IR) and cisplatin, but not inflammatory cytokines *(15)*. c-Abl was shown to physically associate with Mekk1, a Map3k upstream of Sek1 and Sapk. This association was inducible in the nucleus upon treatment with DNA-damaging agents. Furthermore, a cellular inhibitor of Sapk activation, Jip-1, can inhibit Bcr/Abl-induced transformation *(16)*. Rac1 and Cdc42Hs are kinases upstream of Mekk1, and the expression of dominant-negative mutants of these enzymes ablate Sapk activation in response to cytokines *(17)*. They do not, however, affect Sapk activation upon treatment with IR *(15)*, indicating that the Mekk1 sits at a point of convergence in the regulation of Sapk responses to various stresses *(see* Fig. 2A). This is further supported by the observation that UV stimulates Sapk activation through a Pyk2 → Mekk1 pathway *(18)*.

Phosphorylation of c-Abl by the DNA damage-signaling kinase Atm has also been proposed, with the further suggestion that c-Abl may be involved in the downregulation of Dna-pk activity *(19)*. This would indicate that double-stranded DNA breaks induce Sapks via the pathway Atm → c-Abl → Mekk1 → Sek1 → Sapk.

A final example of Sapk pathway activation by DNA-damaging agents can be found in the apoptosis signal regulated kinase-1 (Ask1). Ask1 lies upstream of both the Sapk and p38 kinases and is a Map3k. It is strongly induced by cisplatin treatment of Ovcar3 ovarian carcinoma cells with kinetics similar to those observed for Sapk activation in response to cisplatin *(20)*. Interestingly, Ask1 has also been shown to associate with and phosphorylate Cdc25A *(21)* (*see* Fig. 2B), a proto-oncogene that is overexpressed in several cancers. In this case, however, the interaction between Cdc25A and Ask1 appears to be independent of the former enzyme's phosphatase activity. Overexpression of Cdc25A or phosphatase-deficient Cdc25A (C430S) resulted in decreased activation of Ask1 in response to the oxidant H_2O_2. This also led to suppression of Sapk and p38 activation in response to this stress. This apparent inhibition of Ask1 by Cdc25A may be a negative-feedback mechanism for p38 or Sapks. Although Cdc25A is predominantly nuclear, it has previously been shown to associate with cytoplasmic Raf1 *(22)*, and all three Cdc25 isoforms conditionally associating with 14-3-3 proteins, with the phosphorylation and nuclear export of Cdc25A being a mechanism of Chk1 and Chk2 regulation of its activity in response to UV and IR, respectively *(23,24)*. Furthermore, p38 phosphorylates Cdc25B in the cytoplasm at the G_2/M checkpoint in response to UV *(14)*; thus, all three Cdc25 isoforms could potentially participate in cytoplasmic interactions to regulate Sapk signaling.

In summary, the activation of multiple Sapk/Mapks in response to genotoxic stress can trigger multiple signals whose specificity is often cell-type dependent. The precise role of each kinase in the DNA-damage response is, however, somewhat more difficult to discern. This is further complicated by the broad range of DNA-damage types induced by anticancer drugs, the high degree of crosstalk between the mitogen- and stress-activated protein kinase pathways as well as by the cell heterogeneity observed in cancer tissue.

2.1.4. ACTIVATION OF SAPKS, TO WHAT END?

As seen earlier, DNA-damaging agents, including chemotherapy drugs, produce strong activation of Sapk/Mapks through a number of different mechanisms, but what of their impact on cell survival? Again, the result is not black and white. The activation of these kinases can be associated with either cell death or survival, depending on the context. Expression of a dominant-negative (dn) (nonphosphorylatable) c-jun construct was shown to sensitize a cisplatin-resistant cell line, and the authors suggest that this sensitization is the result of a repair defect in the dn-c-jun-expressing cells *(25)*. The authors also conclude that DNA damage is required for c-jun N-terminal kinase activation, as they did not observe an increase in activity upon incubation with transplatin. Note that this reasoning may be questionable because transplatin does create DNA lesions *(26)*.

The lack of induction of c-jun kinase activity by transplatin conflicts with the result of Sanchez-Perez et al. *(7)*, although this is in agreement with Hayakawa et al. *(9)*, suggesting that differences in cell lines and assay conditions may be responsible for some of the discrepancy. Hayakawa et al. also found that dn-c-jun sensitizes cell lines to cisplatin, but not transplatin treatment. Further, treatment with the Mek1/2 inhibitor PD98059 resulted in similar sensitization, suggesting that both Erks and a c-jun kinase are involved in resistance to cisplatin-induced cellular stress. It is also possible that these kinases function through a similar mechanism, because in this study, the use of both dn-c-jun and PD98059 does not enhance cell killing over that seen with either agent alone.

In contrast to the suggested protective role of c-jun in response to cisplatin, an article by Sanchez-Perez et al. *(27)* indicates a proapoptotic role for c-jun in response to cisplatin. Using a knockout mouse embryonic fibroblast cell model, the authors show that c-jun –/– cells are resistant to cisplatin, but they can be sensitized by restoration of c-jun by transfection. Clearly, some of the effects of Sapk/Mapk activation in response to DNA-damaging agents remain to be established, with particular attention given to choice of cell line, method of measuring kinase activation, and mechanism of pathway inhibition. The importance of the latter issue is emphasized by the disparate results of studies using dn mutants *(9,25)* or knockout cell lines *(27)* to study the function of jun in response to cisplatin. Nevertheless, although the global effects of Sapk activation are proving more complicated to dissect than was perhaps originally assumed, some of the direct interactions between these pathway members and cell cycle or DNA repair proteins are being worked out. Some results are conflicting even on this scale, but careful consideration of the data provided suggests they may be reconciled.

2.2. Modulation of Sapk/Mapk Activation by DNA Damage: The Case of p53

p53 regulation (via phosphorylation) by Sapks has broad implications for the regulation for DNA-damage response, including DNA repair. The multifunctional tumor suppressor p53 is involved in both DNA repair and cell cycle arrest *(28)*. Transcriptional control of gene expression by the tumor suppressor p53 *(29)* is essential for the cellular response after DNA damage and phosphorylation is limiting to this regulation. In DNA-damaged cells, p53 is phosphorylated on many serine/threonine residues, resulting in modulation of its affinity for different transcriptional targets. For example, phosphorylation of Ser15 is increased following UV-induced DNA damage and correlates with nuclear shuttling of p53 *(30)*. Phosphorylation on Ser residues enhances transcription of the Cdk inhibitor p21^{waf1}, which contributes to cell cycle arrest *(31)*. This implies that this is one mechanism by which deficiencies in kinases upstream of p53, such as Atm and Chk2, result in impairment of DNA-damage-induced cell cycle arrest *(32)*. Also, loss of p53 function can compromise induction of apoptosis and DNA-damage repair resulting in drug resistance, increased mutation, and neoplastic progression. During genotoxic stress, p53 is subject to multiple phosphorylations. Sapk phosphorylation of p53 on Thr-81 is important for p53 stabilization and transcriptional activities in response to stress *(33)*. Both Erk1/2 and p38 have been implicated in the regulation of p53 function in response to NO *(34)*. However, the phosphorylation of p53 by Pka, Sapks, and CKII is conformation dependent *(35)*. The mutations affecting the p53 tumor suppressor genes in the Li–Fraumeni syndrome and more than 50% of all sporadic cancers are clustered in the DNA-binding domain and affect the transcriptional activity and conformation, which, in turn, is likely to affect its phosphorylation, resulting in inactive forms of p53 *(35)*. Furthermore, viral oncoproteins functionally inactivate p53 in a large proportion of tumors with genetically intact p53 locus *(36–38)*. As such, p53-dependent pathways are attractive targets to manipulate cancer cell response to chemotherapy drugs.

2.2.1. Direct Interaction of MAPK/SAPKs With p53

The association of p38 and Erks with p53 in untreated UVB- and UVC-treated cell lysates has been reported *(39,40)*, whereas the dissociation of p38 from p53 following UV or cisplatin *(41)* has also been observed. As is often the case, some of these differences may be the reslt of cell lines or the types of UV (UVB vs UVC) used. The most significant difference, however, is that one report suggests p38 and Erk phosphorylation of p53 Ser15 in response to UV and cisplatin *(39)*, whereas another states that the phosphorylation is primarily on Ser33, not Ser15 *(40)*. Whereas the former study shows that p38 and Erk can coprecipitate p53 and that their inhibition blocks phosphorylation of p53 on Ser15, the latter shows a similar coprecipitation and an absence of kinase activity toward an artificial p53 substrate consisting of the first 25 amino acids (aa) of this protein. The solution to the conflict seems to lie in the phosphorylation of Ser33 by p38, which appears to be required for phosphorylation at surrounding sites. In a similar vein, another study examined the effect of Erk inhibition on p53 Ser15 phosphorylation in response to cisplatin *(42)*. This report suggests that the MAPK/ERK inhibitor PD98059 is more effective than wortmannin (DNA-PK, Atm inhibitor), caffeine (Atr inhibitor). or the p38 inhibitor SB202190 at inhibiting phosphorylation of p53 at Ser15 *(see* Figs. 2B, 3B, and 4B). Further, PD98059 completely ablates both p21waf1 and Mdm2 induction following cisplatin treatment, suggesting a strong effect on p53 transactivation. Again, however, these extensive inhibitor studies involve mostly whole-cell treatments with inhibitors, suggesting that the basic conclusion of Bulavin et al. *(40)*, namely that Mapk phosphorylation of p53 Ser33 coordinates further N-terminal phosphorylations, may be correct.

Bulavin's study demonstrates that in A549 and RKO cells (p53wt), p38 inhibition prevents apoptosis after UV exposure. There is no such protection in the p53-deficient H1299 cells or RKO cells expressing the E6 protein of HPV. This finding is in line with the dogma suggesting proapoptotic roles for p38 and Sapks, with an antiapoptotic role for Erks. However, this dogma must be immediately questioned, given the observation that PD98059 did not affect cell survival after

UV *(40)*. Furthermore, certain data suggest that these proapoptotic and antiapoptotic roles for Mapk family members may be dependent on the specific cellular stress *(8)*. In response to cisplatin treatment of HeLa cells, a delayed but sustained activation of Erk is observed. This correlates with a similar induction of the upstream kinases MEK1/2. A 30-min pretreatment with either of two MEK inhibitors (PD980159 or U0126) dramatically reduced both Erk1/2 activity *and* apoptosis in these cells, whereas similar, although less pronounced, results were seen in A549. This contrasts with the inability of PD980159 to affect apoptosis resulting from UV in the latter cell line *(40)*.

The involvement of other Sapks in the response to cisplatin was further investigated using dn-Sek1 to inhibit the Sapk pathway and using SB202190 or SB203580 to inhibit p38α/β *(8)*. Neither of these treatments had an observable effect on apoptosis, indicating that Erk, but not Sapk or p38, plays a role in inducing apoptosis in response to cisplatin. As is often the case, however, these results do not apply to all cell lines, as PC3 cells show no effect of Erk inhibition on apoptosis and, in fact, these results conflict with the above studies using dn-c-jun and c-jun knockouts *(9,25,27)*. Interestingly, the fact that PC3 cells are p53 mutant may suggest that the mechanism of Erk-dependent apoptosis is via p53, as suggested for p38 and as would be expected given the results of Persons et al. *(42)*, as noted earlier. This, too, must be appraised cautiously, however, given the contrasting findings that PD980159 *sensitizes* Caov-3 (p53 mutant) and A2780 (p53 wt) ovarian carcinoma cells to cisplatin *(9)*, as well as C8161 melanoma cells (p53 wt) *(43)*.

2.2.2. GADD45/MAPK INTERACTIONS

Growth arrest and DNA damage 45 (Gadd45) is a stress-inducible protein, which has recently been shown to be regulated by Sapk/Mapks through several mechanisms *(44)*. Its regulation following ionizing radiation is p53 dependent, but following other stresses, such as UV and methyl methanesulfonate (MMS), is more complicated, involving both p53-dependent and independent mechanisms. Gadd45 is thought to play a role in mediating the G_2/M checkpoint and may directly affect repair by modifying chromatin structure. It has also been shown to associate with Mekk4, activating this Map3k upstream of p38 and Sapks *(45)*. Tong et al. *(44)* examined the activation of Gadd45 in response to UV in p53-proficient colon carcinoma cells (HCT116). In these cells, Erk2 induction is seen as rapidly as 30 min following UV treatment, with Erk1 following, and both peaking at 12–24 h posttreatment. Strong induction of Sapkγ, and to a lesser extent Sapkα, is also seen by 1 h posttreatment, with a second peak at 8 h. Cotransfection of Raf-1, Sapkγ, or Mekk1 leads to an increase in the activity of Gadd45 promoter. This effect is observed in the absence of p53. Expression of Mkp1 with the kinases can abrogate Gadd45 promoter activity and also ablate the increase in promoter activity observed upon UV treatment, indicating that Sapks play a major role in the transcriptional activation of the Gadd45 following stress response.

2.3. Sapk Regulation and Cell Cycle: The Case of Cdc25

Eukaryotic cell cycles revolve around the Rb pathway. In order for progression through the G_1-phase and the S-phase to occur, Rb must be phosphorylated by Cdks, resulting in the release of E2F transcription factors. This central pathway is regulated at many levels in order to prevent untimely cell division and the propagation of cells harboring mutations. In the absence of Cdc25 activity, Cdks are kept inactive by Wee1 phosphorylation of inhibitory tyrosine and threonine residues. Their activity is also regulated by the presence or absence of their cyclin partners, as well as CKIs (cyclin-dependent kinase inhibitors), such as p21$^{waf1/cip1}$, p16^{ink4a}, and p27^{kip}.

Cell cycle arrest has been described to involve the induction of cell cycle inhibitors and/or the inhibition of cell cycle activators. Phosphorylation of key substrates plays a determining role in both processes. Recent advances in our understanding of DNA-damage-induced cell cycle checkpoints support a two-step mechanism: the p53-independent initiation of cell cycle arrest and further maintenance of the arrest, requiring p53-dependent transcription of the CKI p21^{waf1} *(46)*.

Consequent to UV exposure, the dual-specificity phosphatase Cdc25A is rapidly degraded, resulting in the maintenance of inhibitory phosphorylation on Cdks and delayed transition from G_1-phase to the S-phase. In the absence of functional p53, the cell cycle resumes concomitant with restoration of Cdc25A expression and arrest can be completely avoided by Cdc25A overexpression *(23)*. This arrest is therefore reinforced by p53-dependent p21^{waf1} expression, which similarly targets the Cdks, resulting in Rb hypophosphorylation.

Ultraviolet-induced G_1 arrest does not always correlate with the degradation of Cdc25A. For example, in early melanoma cells, Cdc25A is not degraded after-UV treatment but does undergo nuclear exclusion *(47)*. Phosphorylation of Cdc25 phosphatases creates 14-3-3 binding sites, leading to their sequestration in the cytosol following various stresses *(48)*. Chk1, Chk2, and p38 have been shown to phosphorylate the various Cdc25 isoforms in response to several cellular stresses (reviewed in ref. *49*). In response to ionizing radiation, Cdc25A is phosphorylated by Chk2, whereas Chk1 phosphorylates Cdc25C *(50)*. In contrast, upon UV irradiation, Cdc25A is phosphorylated in a Chk1-dependent fashion, representing the first wave of a bipartite G_1/S checkpoint *(23)*. Cdc25B is phosphorylated by p38 following UV treatment, initiating the G_2/M checkpoint *(14)*. As mentioned earlier, UV-induced checkpoints are reinforced by p38 phosphorylation of p53, coordinating subsequent phosphorylations around the N-terminus of p53. Similarly, p38, Chk1, and Chk2 play a dual role in phosphorylating both p53 and Cdc25s.

In a negative-feedback loop, p53 downregulates Chk1 transcription *(51)*, whereas p38 is inactivated by the p53-inducible Wip1 *(52)*. Repression of Chk1 by p53 requires p21waf1, because p21waf1 alone is sufficient for this to occur and cells lacking p21waf1 cannot downregulate Chk1 *(51)*. Interestingly, pRb is also required for Chk1 downregulation. p53 and Chk1 play interdependent and complementary roles in regulating both the arrest and resumption of G_2 after DNA damage *(51)*. p53/p21waf1/pRb is also required for maintenance of G_2 arrest *(53,54)*. Another transcriptional target of p53, the 14-3-3 phospho-binding proteins, is involved in the initiation and maintenance of the G_2 arrest by sequestering Cdc25C in the cytoplasm *(55)*. Although 14-3-3 proteins are not kinases, their cell-cycle-arrest function relies mostly on kinase activities because they bind phosphorylated proteins with much greater affinity. Thus, the Mapk/Sapks are emerging as regulators of this pathway at several points. Erk and p38 can phosphorylate p53 *(39–42)*, resulting in induction of p21 as well as Gadd45, which is proposed to regulate the G_2/M checkpoint by disruption of the Cdc2–cyclinB1 complex *(56)*. Gadd45 induction following UV is also proposed to be directly mediated by Sapks and Erks, although not p38, in a p53-independent manner *(44)*.

We therefore see each of the major Mapk/Sapk family members playing an important role in checkpoint regulation: Sapk and Erks through Gadd45 induction, p38 and Erks through p53 phosphorylation, and p38 through Cdc25B phosphorylation. Additional effects of these kinases on apoptosis (terminal cell cycle exit) are also apparent, but are beyond the scope of this chapter.

2.4. Kinases Involved in Phosphorylation of DNA Repair Proteins

In general, mammalian DNA repair proteins are not thought to be transcriptionally inducible, although some show minor induction in specific circumstances. This may be because basal levels of genomic insult are sufficient to require a constantly functioning repair system. It is also logical that DNA repair proteins are primarily regulated posttranscriptionally, because DNA lesions would impede their expression. ERCC1, for example, is induced following cisplatin treatment of A2780 ovarian carcinoma cells, reportedly by a combination of increased transcription and mRNA stabilization *(57)*. Additionally, some of the enzymes providing the basic building blocks required for repair may be induced following UV irradiation *(58)*. However, it is likely safe to assume that the major part of repair activity modulation derives from posttranslational modification or association with proteins (e.g., p53), which are stabilized in response to genotoxic insult. Indeed, posttranslational modification of cell cycle checkpoint and DNA repair proteins can stimulate arrest and repair via several mechanisms *(59,60)*. The phosphorylation status of these proteins can modulate their stability *(61)*, complex formation, subcellular localization *(62,63)*, catalytic activity *(64)*, DNA-binding affinity, and transcriptional activity *(65)*, as well as structural remodeling affecting both the protein and chromatin structure *(66,67)*. As such, the kinases of the signal transduction pathways activated by genotoxic stress will directly or indirectly modulate DNA repair and cell cycle.

2.4.1. NER Is Modulated by Phosphorylation and Kinase Inhibitors

Activation of p53 by phosphorylation is important for efficient DNA repair. To date, however, little is know regarding the modulation of DNA repair activities resulting from phosphorylation of DNA repair proteins *per se*. It has been shown that NEW is inhibited by phosphorylation (via CAK phosphorylation of repair components) and that the inhibition of CAK by the cyclic nucleotide protein kinase inhibitor H-8 restores the NER activity to original levels *(68)*, suggesting that the activity of the NER can be downregulated by phosphorylation. This is an important finding to understand the controversial role of the p53-regulatory pathway and specifically its downstream effector p21^{waf1} in the regulation of NER. The role of CAK in NER provides a link between p21^{waf1} and NER because high levels of p21^{waf1} can inhibit CAK in vivo *(54)*, which, in turn, should increase NER activity. This model would support studies describing the contribution of p21^{waf1} in NER *(69,70)*. This is in apparent contrast with other studies showing that p21^{waf1} has little effect *(70–72)* or an inhibitory role in NER *(73)*. In addition, a recent study showed by ligation-mediated PCR (LM-PCR) that basal levels of p21^{waf1} inhibited NER in a p53-deficient background *(74)*. This deficiency in NER may be interpreted on the basis of CAK–NER complex interaction.

The inhibition of NER is an inviting avenue by which to improve cisplatin-based chemotherapy. Unfortunately, treatment with cisplatin and other bulky adduct-inducing drugs (e.g., alkylating agents), is inconsistently successful despite frequent low NER capacity in tumor cells because of p53 deficiencies (50% of all cancers), which impairs both global genomic NER *(75)* and transcription–coupled NER *(76)*. Other repair pathways such as BER can act on damage preferentially repaired by NER, perhaps representing a mechanism by which to overcome NER deficiency. It should be noted, however, that p53 plays a direct role in BER by stabilizing the interaction between DNA polymerase beta (DNApolβ) and abasic DNA *(77,78)*. Therefore, p53 deficiencies would compromise BER as well, and this repair mechanism is unlikely to compensate for NER lost in a p53-deficient background.

2.4.2. p53-Dependent Repair Proteins

Gadd45 and p21^{waf1} are two DNA-damage-inducible genes that can be induced via both p53-dependent and p53-independent pathways. Many studies implicate those two stress-inducible proteins in NER and apoptosis *(79)*. In vivo p21^{waf1} can be phosphorylated by protein kinase B (Akt/Pkb) *(80)*, an antiapoptotic kinase. Both Gadd45 and p21^{waf1} interact with PCNA *(81,82)*, which is known to affect cell cycle progression by supporting DNA repair and, indirectly, survival. An additional function of Gadd45 is to bind to UV-damaged chromatin, which affect lesion accessibility *(79)*. A direct role for p21^{waf1} phosphorylation in NER has not yet been addressed; however phosphorylation by mitogen-activated protein (MAP) kinases is involved in the induction of the Gadd45 promoter after DNA damage *(44)*. Similarly, inhibition of Sapkγ and Erk kinase activities either by expression of a dominant negative mutant Sapkγ or by treatment with a selective chemical inhibitor of Erk (PD098059) substantially abrogates the UV induction of the Gadd45 promoter *(44)*. p53-independent induction of Gadd45 *(83)* and p21^{waf1} *(84)* has been described following DNA damage, including treatment with cisplatin *(85)*.

Colon carcinoma is characterized by frequent p53 and mismatch repair deficiencies. The p53-dependent upregulation of human mismatch repair gene MSH2 in UV-irradiated colon carcinoma cells depends on a functional interaction with c-jun *(86)*. (Although UV light is not a therapeutic agent, some of its properties may reflect those of more relevant chemotherapeutic agents.) As described earlier, the c-jun kinases (Sapks) are activated by many cellular stresses, including cisplatin.

2.4.3. REPLICATION PROTEIN A

Among the many proteins involved in NER, replication protein A (RPA) is one factor known to be phosphorylated after DNA damage, although the kinase(s) responsible have not yet been determined. The single-stranded-DNA-binding protein RPA is a multifunctional heterotrimer involved in NER *(68,87)* replication and repair of strand breaks *(88,89)*. RPA is modified by phosphorylation during replication *(90)* and the DNA-damage response *(91)*. In particular, the 32-kDa subunit is phosphorylated following UVC treatment *(92)*. Hyperphosphorylation of RPA has been observed in cells from patients with either GGR or transcription-coupled repair (TCR) deficiency (A, C, and G complementation groups of xeroderma pigmentosum and A and B groups of Cockayne syndrome, respectively). This excludes both intermediates in the NER pathway and signals from stalled transcription as essential signals for RPA hyperphosphorylation. However, UV-sensitive cells deficient in NER and TCR require lower doses of UV irradiation to induce RPA32 hyperphosphorylation than normal cells, suggesting that persistent unrepaired lesions contribute to RPA phosphorylation. UVC irradiation experiments on nonreplicating cells and S-phase-synchronized cells emphasize a role for DNA replication arrest in the presence of UV-induced lesions in RPA UV-induced hyperphosphorylation in mammalian cells *(92)*. One might, therefore, speculate that inhibition of RPA phosphorylation could improve treatments, inducing NER-substrate lesions.

2.4.4. O^6-ALKYLGUANINE-DNA ALKYLTRANSFERASE

The expression of O^6-alkylguanine-DNA alkyltransferase (AGT) *(93)*, a DNA repair protein that confers tumor resistance to many anticancer alkylating agents (described in detail in Chapter 7) is upregulated in the absence of p53 *(94)* and frequently overexpressed in oral cancer cells genetically and functionally deficient for p53 *(95)*. Thus, p53 acts as a repressor of AGT expression, whereas the activators of Pkc, phorbol-12-myristate-13-acetate (PMA), and 1,2-diacyl-*sn*-glycerol (DAG), as well as the protein phosphatase inhibitor okadaic acid (OA), increase the transcriptional level of AGT *(96)*. The activity of AGT is inhibited by phosphorylation that can be catalyzed by Pka, Pkc, and/or CKII *(97)*. Thus, the activation of these kinases may impair the elimination of akylated DNA lesions.

2.4.5. BLM HELICASE

Bloom's syndrome (BS), a rare genetic disease, arises through mutations in both alleles of the Blm gene, which encodes a 3'–5' DNA helicase. BS patients exhibit a high predisposition to development of all types of cancer affecting the general population and Blm-deficient cells display a strong genetic instability. Blm participates in the cellular response to ionizing radiation. Blm defect is associated with a partial escape of cells from the γ-irradiation-induced G_2/M cell cycle checkpoint. In response to ionizing radiation, Blm protein is phosphorylated and accumulates through an Atm-dependent pathway *(98)*. Caffeine, by inhibiting Atm and its homolog Atr *(99,100)*, enhances the radiosensitivity of cells in part by altering the phosphorylation of Blm, in addition to its effects on Chk1 and Chk2, as outlined earlier.

3. LINKS BETWEEN GROWTH FACTOR RECEPTORS, Sapk/Mapk ACTIVATION, DNA-DAMAGE RESPONSE, AND DRUG RESISTANCE

A biochemical feature of human solid tumors is the frequent occurrence of aberrant expression (most commonly overexpression) of growth factors and growth factor receptors. In some cases, overexpression of a specific growth factor receptor has been correlated with poor response to first-line chemotherapy/radiotherapy, a phenomenon broadly defined as intrinsic drug resistance (in contrast to acquired resistance, which is secondary to drug treatment). Based on the involvement of downstream components of growth-factor-receptor-coupled cell signaling in the regulation of the stress response, it is logical to predict that altered growth-factor-receptor expression can impact on tumor response to chemotherapy-induced genotoxic stress.

3.1. The Case of the ErbB Tyrosine Kinase Receptors

Overexpression of the ErbB2 tyrosine kinase receptor in cancer cells has been shown to result in a prolonged activation of the Ras-Mapk and JNK pathways (reviewed in ref. *101*). Overexpression of this same receptor has been reported to protect against drug-induced apoptosis in some cancer cell types, possibly via impaired regulation of DNA repair and the associated cell cycle checkpoints *(102–108)*. Interestingly, selective inhibition of ErbB-2 by Herceptin (trastuzumab, a humanized monoclonal antibody that selectively inhibit ErbB2) sensitizes tumor cells to chemotherapy both in experimental systems and in patients *(106–109)*. Another ErbB receptor family member, EGFR, can be activated by UV damage via a mechanism involving reactive oxygen intermediates *(110)*. Importantly, the amplitude of receptor-associated response will dictate the physiological outcome of a given stimulus. In the case of EGFR signaling for example, stimulation with EGF normally produces proliferative signals, notably through the activation of the transcription factor AP-1. However, conditional activation of STAT1 by EGF correlates with p21^{waf1} induction and cell growth inhibition *(111,112)*. In this case, one important question is how do chemotherapeutic agents mimic the natural ligand of receptors and how does this contribute to chemo-resistance? The following experiments have provided some clues to this question and show a prominent role for cell cycle regulators.

p21^{waf1} is involved in the resistance to bulky adducts induced by cisplatin; p21^{waf1} expression is generally associated with a positive prognosis in patients after adjuvant chemoradiation for pancreatic cancer *(113)* and in the treatment of colon cancer *(114)* but conflicting results are reported in breast cancer *(115)*. The resistance to cisplatin has been correlated with increased p21^{waf1} expression and a decrease in EGFR expression *(116)*. Conversely, p21^{waf1} disruption preferentially sensitizes some cell types to the DNA crosslinking agents cisplatin and nitrogen mustard *(117)*. Cisplatin-induced intrastrand DNA adducts accumulate to higher levels in cells overexpressing EGFR than in low-EGFR-expressing cells. Moreover, antisense directed against EGFR provides resistance to cisplatin. This indicates that a critical level of EGFR signaling, which is amplified in several common human cancers, is necessary for cisplatin-mediated apoptosis in tumor cells and suggests an inhibitory effect of this pathway on the repair of cisplatin adducts *(118)*.

In contrast, the inhibition of the EGFR with a dominant-negative mutant EGFR sensitizes cells to radiotherapy as effectively as antisense *(119)*. The inhibition of the Mapk cascade downstream of EGFR by the inhibitor PD98059 recapitulates the sensitivity to ionizing radiation, providing further evidence that the EGFR-mediated signaling is involved in resistance to ionizing radiation *(120)*. Paradoxically, sustained stimulation of cells with EGF downregulates Atm protein levels *(121)*. The Atm kinase is involved in the response to ionizing radiation, and Atm-deficient cells are extremely sensitive to ionizing radiation. Therefore, inhibition of Atm would be expected to increase the radiosensitivity. EGFR-neutralizing monoclonal antibodies also increase the radiosensitivity of tumor cells *(122)*.

3.2. The Case of the Hepatocyte Growth Factor/Scatter Factor

Hepatocyte growth factor/scatter factor (HGF/SF) is a pleiotropic mediator of epithelial cell motility, morphogenesis, angiogenesis, and tumorigenesis. HGF/SF protects cells against DNA damage as well as apoptosis induced by adriamycin, X-rays, ultraviolet radiation, and other agents *(123,124)* by a pathway from its receptor c-Met to PI3'K to c-Akt *(125)*, an antiapoptotic kinase. This protection against the TopoII-inhibitor adriamycin requires the Grb2-binding site of c-Met, and overexpression of the Grb2-associated binder Gab1 inhibits the ability of HGF/SF to mediate adriamycin resistance *(125)*. In contrast to Gab1 and its homolog Gab2, overexpression of c-Cb1, another multisubstrate adapter that associates with c-Met, does not afford protection. Gab1 blocks the ability of HGF/SF to cause the sustained activation of c-Akt and c-Akt signaling. The Gab1 inhibition of sustained c-Akt activation and of cell protection does not require the Gab1 pleckstrin homology or Shp2 phosphatase-binding domain, but it does require the PI3'K-binding domain *(126)*. The protection conferred by HGF/SF can be blocked by wortmannin, expression of Pten, and dominant negative mutants of p85 (a regulatory subunit of PI3'K), Akt, and Pak1 *(126)*. In vivo, the tumorigenic properties of HGF/SF in tumors depending on this pathway (glioblastoma) can be blocked by a combination of neutralizing antibodies *(127)*.

4. TARGETING DNA-DAMAGE AND DNA REPAIR RESPONSES: POTENTIAL AND LIMITATIONS

The concerted roles of stress kinase signaling in the regulation of the DNA-damage response (i.e., in cell cycle checkpoints, DNA repair, and apoptosis), make these components potentially useful targets to modulate chemotherapy/ radiotherapy response. This is becoming increasingly attractive given the discovery of novel Mapk/Sapk inhibitors, many of which have entered clinical trials (reviewed in ref. *128*). It can be anticipated that inhibitors that modulate specific stress-activated signaling pathways can impact on the outcome of treatment, hopefully enhancing the therapeutic efficacy of chemotherapy/radiotherapy. The dilemma remains that, like most DNA damage-based strategies, this may also increase levels of genetic instability and result in tumor cell heterogeneity and drug resistance. Some promising anticancer therapies target both genomic integrity and cell cycle checkpoints simultaneously. Hence, the activity of DNA-damaging agent-based chemotherapy may be potentiated when combined with inhibitors of Sapk/p38/Mapks kinases, which have been implicated in the phosphorylation of p53 as well as the cell cycle protein CDC25B, thus affecting both DNA repair activity and cell cycle checkpoints. Because p53-deficient cells lack a complete G_1 checkpoint in response to genotoxic agents, a high proportion of tumor cells rely on the G_2 checkpoint to survive chemotherapeutic DNA damage. Thus, it is possible that p38 inhibitors, which would impede the induction of the G_2/M checkpoint in response to certain treatments, might specifically sensitize tumor cells while having a reduced effect on normal cells, given their intact G_1/S checkpoint. Treatments that induce a p53-dependent G_1 checkpoint, which prevails in normal cells, would cause selective killing of tumor cells when used in combination with a G_2/M checkpoint inhibitor. Examples of this strategy are presented next.

The protein kinase C (Pkc) inhibitor UCN-01 (7-hydroxystaurosporine) *(129)* abrogates G_2 arrest through a Cdc2-dependent pathway *(130–136)*. Clinical trials using UCN-01 have revealed several side effects, likely the result of the limited specificity of this drug, which inhibits Chk1 and other kinases in addition to PKC *(131)*. UCN-01 appears to activate the Cdc25C phosphatase via inactivation of the Wee1 kinase *(130)* or, *via* the inactivation Chk1 *(132)*. UCN-01 thus abrogates the γ-radiation-induced G_2 checkpoint and radiosensitizes tumor cells *(132)*. UCN-01 similarly enhances the anti-tumor activity of the interstrand crosslinking agent mitomycin C (MMC) in vitro and in vivo through the abrogation of S and/or G_2 arrest induced by MMC *(137)*, consistent with its Chk1 inhibitory function. This treatment is effective in p53-deficient cells but shows very little toxicity in p53-proficient cells, probably the result of the intact p53-dependent G_1 checkpoint. This may, therefore, represent a means to increase the therapeutic index, sparing normal cells while eliminating tumor cells, because more than 50% of tumors harbor p53 mutations and a large proportion of tumors with intact p53 gene are functionally deficient for p53 because of mutations or the expression of inactivating viral oncoproteins (e.g., HPV E6) There are several variations on this theme, but the choice of agent must be made carefully, given that certain agents appear to promote G_2 arrest in an Atm/Atr-independent fashion *(138,139)*, whereas the G_1 arrest is not always p53-dependent *(140–142)*.

In addition to its cell-cycle-checkpoint inhibitory function, UCN-01 inhibits NER by disrupting the Ercc1–Xpa interaction, resulting in low repair incision activity *(143)*. This mechanism may contribute to the positive effect of UCN-01 on cisplatin-based chemotherapeutic regimens *(144)*. UCN-01 may also regulate the response to alkylating agents via Pkc inhibition. Another action of UCN-01 is to increase the activity of deoxynucleoside analogs by downregulating the expression of the thymidylate synthase *(145)*. Thymidylate-synthase-deficient tumor cells are particularly sensitive to ionizing radiation, and the inhibition of thymidylate synthase by fluoropyrimidine induces radiosensitization *(146)*. Thus, this may contribute to the radiosensitization observed with UCN-01. UCN-01 is not only a kinase inhibitor, but it also stimulates the p42/p44 Mapks, most probably as part of a defense pathway activated to counteract the cytotoxic effects of UCN-01. Coadministration of the MEK inhibitor PD184352 blocks UCN-01-induced Mapk activation, causing marked mitochondrial damage and increased cell death in leukemia cell lines *(147)*.

The Atm and Atr kinases and their downstream effectors Chk1 and Chk2 are also appealing targets to enhance chemotherapy. Atm and its homolog Atr are inhibited by caffeine, rendering cells sensitive to ionizing radiation *(99,100)*. Caffeine also inhibits the Atm-dependent G_2 cell cycle arrest *(148)*. Its mechanism of action may also involve the inhibition of homologous recombination, a process required for the repair of strand breaks *(149)*. However, the doses efficient for radiosensitization are relatively toxic *(150,151)*.

A novel G_2 checkpoint inhibitor has been shown to radiosensitize tumor cells. This compound, PD0166285, in the pyridopyrimidine class of molecules inhibits Wee1 and Myt1 at nanomolar concentrations. Through the induction of premature mitosis, it enhances radiosensitivity slightly in p53-proficient cells but shows enhanced efficacy in p53-deficient cells, again showing the bipartite (i.e., Cdc25B and p53-dependent) nature of the G_2 checkpoint.

The semisynthetic flavonoid flavopiridol induces cell cycle arrest in G_1 or G_2/M corresponding to its ability to inhibit both Cdk2 and Cdk1, respectively *(152)*. The biological activity of flavopiridol is not restricted to the direct inhibition of Cdks; it also downregulates the expression of cyclins D1 and D3 *(152,153)*, two cyclins involved in G_1 progression *(154)*. Flavopiridol also inhibits Cdk7 of the CAK complex. This compound might, therefore, reduce cisplatin efficacy because inhibition of CAK stimulates nucleotide excision repair, which is responsible for the removal of cisplatin adducts.

STI571 (Gleevec) inhibits Abl *(155)* and could therefore diminish Mekk1 activation following IR. Because Abl is proposed to lie upstream of Mekk1 in the IR induction of Sapks, STI571 might be expected to modulate the response of tumor cells to IR. Indeed, Abl knockout fibroblasts have a decreased sensitivity to radiation *(156)*. Both in vivo and in vitro experiments suggest, however, that STI571 does not affect sensitivity to radiation *(157)*.

A common strategy for molecular targeting is to inhibit specific kinases, whereas the phosphatases that inactivate them have occasionally been overlooked. This is unfortunate, because many of these phosphatases are themselves enticing targets, whereas overactivation of the pathways they regulate can be highly desirable. In the case of the Abl→Sapk cascade, it may be possible that activation of the pathway would result in increased apoptosis or sensitivity to ionizing radiation and other stresses. The activation of Mapk pathways is negated by Mapk phosphatases (Mkps). In accord with the results of Mapk inhibitor studies, enforced expression of Mkp1 ablates p38 and Sapk induction following UV and is cytoprotective *(158)*; similar results were described with respect to Mkp1 and hVH5 expression following cisplatin *(159)*. Mkp1 induction follows that of Sapk and serves to inactivate these kinases. Because persistent activation of Sapks has been associated with cisplatin-induced cell death, it is possible that Mkp1 inhibition would enhance the cytotoxicity of stressors that activate Sapks.

As discussed earlier, p38 has emerged as a key component of DNA-damage checkpoints via its phosphorylation of both Cdc25 and p53. p38 is itself regulated by a p53-inducible phosphatase, Wip1, a member of the PP2C family. In p53-proficient cells, Wip1 expression inactivates p38 and suppresses UV-induced apoptosis *(52)*. Thus, for that portion of tumors expressing a functional p53, inhibition of Wip1 should provide a means to sensitize those cells to apoptosis arising from prolonged p38 activation. Although the human Wip1 has been reported to be inducible by IR in a p53-dependent manner, basal levels of Wip1 are detectable in the absence of irradiation both in human *(160)* and mouse *(161)*, suggesting that inhibition of Wip1 might be a useful strategy for chemosensitization even in p53-deficient cells.

Another phosphatase target of interest is Cdc25. Cdc25 isoforms are thought to be essential for progression through various stages of the cell cycle, as evidenced by the fact that they are the targets of the checkpoint kinases Chk1 and Chk2 as well as p38. Their inhibition would be the equivalent of Cdk inhibition and, thus, they, like the Cdks, are the subjects of intense efforts in drug development.

5. CONCLUDING REMARKS AND PERSPECTIVES

As highlighted in this chapter, the progress in developing specific DNA repair inhibitors has been slow in part because of the complexity of these mechanisms and the multitude of proteins involved. The future is, however, very promising with the rapid progress on the characterization of crystal structures of many DNA repair proteins. Modulating DNA damage response via cell signaling is very attractive, in particular because specific DNA-damaging agents can trigger selective receptor-activated signal transduction cascades. This is becoming feasible with the development of several specific kinase inhibitors that have proven to be well tolerated by humans *(128)*. An alternative strategy is based on the evidence that receptor-signaling–mediated chemoresistance involves, at least in part, impaired cell cycle checkpoints, increased DNA repair, and/or downregulation of the apoptotic threshold. Growth-factor-receptor upregulation is a frequent cause of innate resistance in many types of cancers. Inhibition of the upstream receptors not only can interfere with the proliferative signals but can also render cells more susceptible to drug-induced apoptosis. Finally, inhibitors of cell cycle checkpoints are certainly an important Achille's heel of tumor resistance to genotoxic chemotherapy drugs and, therefore, represent a promising avenue for future therapies. It is important to keep in mind that the efficiency of single or multiple adjuvant therapies depends on the genetic background of individual tumors. Defining genetic alterations is prerequisite to maximizing therapeutic efficacy.

ACKNOWLEDGMENTS

Work from this laboratory on the regulation of chemotherapy stress response has been supported by grants from the Canadian Breast Cancer Research Initiative (CBCRI) of the National Cancer Institute of Canada (CBCRI), the Canadian Institutes for Health Research (CIHR), and the Cancer Research Society Inc. (CRS). M.A.A-J. is supported by a Senior Scientist Award from the "Fonds de la Recherche en Santé du Quebec" (FRSQ), M.L. is supported by a CIHR postdoctoral training award, and P.J.S is supported by a US Army CDMRP BCRP Predoctoral Fellowship.

REFERENCES

1. Kyriakis JM, Avruch J. Mammalian mitogen-activated protein kinase signal transduction pathways activated by stress and inflammation. *Physiol Rev* 2001;81(2):807–869.
2. Shaulian E, Karin M. AP-1 in cell proliferation and survival. *Oncogene* 2001;20(19): 2390–2400.
3. Shiloh Y. ATM and ATR: networking cellular responses to DNA damage. *Curr Opin Genet Dev* 2001;11(1):71–77.
4. Derijard B, Hibi M, Wu IH, et al. JNK1: a protein kinase stimulated by UV light and Ha-Ras that binds and phosphorylates the c-Jun activation domain. *Cell* 1994;76(6):1025–1037.
5. Butterfield L, Storey B, Maas L, et al. c-Jun NH2-terminal kinase regulation of the apoptotic response of small cell lung cancer cells to ultraviolet radiation. *J Biol Chem* 1997;272(15): 10,110–10,116.
6. Iordanov MS, Magun BE. Different mechanisms of c-Jun NH(2)-terminal kinase-1 (JNK1) activation by ultraviolet-B radiation and by oxidative stressors. *J Biol Chem* 1999;274(36): 25,801–25,806.
7. Sanchez-Perez I, Murguia JR, Perona R. Cisplatin induces a persistent activation of JNK that is related to cell death. *Oncogene* 1998;16(4):533–540.
8. Wang X, Martindale JL, Holbrook NJ. Requirement for ERK activation in cisplatin-induced apoptosis. *J Biol Chem* 2000;275(50):39435–39443.
9. Hayakawa J, Ohmichi M, Kurachi H, et al. Inhibition of extracellular signal-regulated protein kinase or c-Jun N-terminal protein kinase cascade, differentially activated by cisplatin, sensitizes human ovarian cancer cell line. *J Biol Chem* 1999;274(44):31648–31654.
10. Persons DL, Yazlovitskaya EM, Cui W, et al. Cisplatin-induced activation of mitogen-activated protein kinases in ovarian carcinoma cells: inhibition of extracellular signal-regulated kinase activity increases sensitivity to cisplatin. *Clin Cancer Res* 1999;5(5): 1007–1014.
11. Pillaire MJ, Nebreda AR, Darbon JM. Cisplatin and UV radiation induce activation of the stress-activated protein kinase p38gamma in human melanoma cells. *Biochem Biophys Res Commun* 2000;278(3):724–728.
12. Kumar S, McDonnell PC, Gum RJ, et al. Novel homologues of CSBP/p38 MAP kinase: activation, substrate specificity and sensitivity to inhibition by pyridinyl imidazoles. *Biochem Biophys Res Commun* 1997;235(3):533–538.
13. Goedert M, Cuenda A, Craxton M, Jakes R, Cohen P. Activation of the novel stress-activated protein kinase SAPK4 by cytokines and cellular stresses is mediated by SKK3 (MKK6); comparison of its substrate specificity with that of other SAP kinases. *EMBO J* 1997; 16(12):3563–3571.
14. Bulavin DV, Higashimoto Y, Popoff IJ, et al. Initiation of a G2/M checkpoint after ultraviolet radiation requires p38 kinase. *Nature* 2001;411(6833):102–107.
15. Kharbanda S, Pandey P, Yamauchi T, et al. Activation of MEK kinase 1 by the c-Abl protein tyrosine kinase in response to DNA damage. *Mol Cell Biol* 2000;20(14):4979–4989.
16. Dickens M, Rogers JS, Cavanagh J, et al. A cytoplasmic inhibitor of the JNK signal transduction pathway. *Science* 1997;277(5326):693–696.
17. Coso OA, Chiariello M, Yu JC, et al. The small GTP-binding proteins Rac1 and Cdc42 regulate the activity of the JNK/SAPK signaling pathway. *Cell* 1995;81(7):1137–1146.
18. Tokiwa G, Dikic I, Lev S, et al. Activation of Pyk2 by stress signals and coupling with JNK signaling pathway. *Science* 1996;273(5276):792–794.

19. Shangary S, Brown KD, Adamson AW, et al. Regulation of DNA-dependent protein kinase activity by ionizing radiation-activated abl kinase is an ATM-dependent process. *J Biol Chem* 2000;275(39):30,163–30,168.

20. Chen Z, Seimiya H, Naito M, et al. ASK1 mediates apoptotic cell death induced by genotoxic stress. *Oncogene* 1999;18(1):173–180.

21. Zou X, Tsutsui T, Ray D, et al. The cell cycle-regulatory CDC25A phosphatase inhibits apoptosis signal-regulating kinase 1. *Mol Cell Biol* 2001;21(14):4818–4828.

22. Galaktionov K, Jessus C, Beach D. Raf1 interaction with Cdc25 phosphatase ties mitogenic signal transduction to cell cycle activation. *Genes Dev* 1995;9(9):1046–1058.

23. Mailand N, Falck J, Lukas C, et al. Rapid destruction of human Cdc25A in response to DNA damage. *Science* 2000;288(5470):1425–1429.

24. Falck J, Mailand N, Syljuasen RG, et al. The ATM-Chk2-Cdc25A checkpoint pathway guards against radioresistant DNA synthesis. *Nature* 2001;410(6830):842–847.

25. Potapova O, Haghighi A, Bost F, et al. The Jun kinase/stress-activated protein kinase pathway functions to regulate DNA repair and inhibition of the pathway sensitizes tumor cells to cisplatin. *J Biol Chem* 1997;272(22):14041–14044.

26. Kartalou M, Essigmann JM. Recognition of cisplatin adducts by cellular proteins. *Mutat Res* 2001;478(1–2):1–21.

27. Sanchez-Perez I, Perona R. Lack of c-Jun activity increases survival to cisplatin. *FEBS Lett* 1999;453(1–2):151–158.

28. Sanchez Y, Elledge SJ. Stopped for repairs. *BioEssays* 1995;17(6):545–548.

29. Chao C, Saito S, Kang J, et al. p53 transcriptional activity is essential for p53-dependent apoptosis following DNA damage. *EMBO J* 2000;19(18):4967–4975.

30. Ljungman M, O'Hagan HM, Paulsen MT. Induction of ser15 and lys382 modifications of p53 by blockage of transcription elongation. *Oncogene* 2001;20(42):5964–5971.

31. Cuddihy AR, Li S, Tam NW, et al. Double-stranded-RNA-activated protein kinase PKR enhances transcriptional activation by tumor suppressor p53. *Mol Cell Biol* 1999;19(4):2475–2484.

32. Waterman MJ, Stavridi ES, Waterman JL, et al. ATM-dependent activation of p53 involves dephosphorylation and association with 14-3-3 proteins. *Nat Genet* 1998;19(2):175–178.

33. Buschmann T, Potapova O, Bar-Shira A, et al. Jun NH2-terminal kinase phosphorylation of p53 on Thr-81 is important for p53 stabilization and transcriptional activities in response to stress. *Mol Cell Biol* 2001;21(8):2743–2754.

34. Kim SJ, Ju JW, Oh CD, et al. ERK-1/2 and p38 kinase oppositely regulate nitric oxide-induced apoptosis of chondrocytes in association with p53, caspase-3, and differentiation status. *J Biol Chem* 2002;277(2):1332–1339.

35. Adler V, Pincus MR, Minamoto T, et al. Conformation-dependent phosphorylation of p53. *Proc Natl Acad Sci USA* 1997;94(5):1686–1691.

36. Barbosa MS. The oncogenic role of human papillomavirus proteins. *Crit Rev Oncol* 1996; 7(1–2):1–18.

37. Malkin D, Chilton-MacNeill S, Meister LA, et al. Tissue-specific expression of SV40 in tumors associated with the Li-Fraumeni syndrome. *Oncogene* 2001;20(33):4441–4449.

38. zur HH. Papillomavirus infections—a major cause of human cancers. *Biochim Biophys Acta* 1996;1288(2):F55-F78.

39. She QB, Chen N, Dong Z. ERKs and p38 kinase phosphorylate p53 protein at serine 15 in response to UV radiation. *J Biol Chem* 2000;275(27):20,444–20,449.

40. Bulavin DV, Saito S, Hollander MC, et al. Phosphorylation of human p53 by p38 kinase coordinates N-terminal phosphorylation and apoptosis in response to UV radiation. *EMBO J* 1999;18(23):6845–6854.

41. Sanchez-Prieto R, Rojas JM, Taya Y, et al. A role for the p38 mitogen-acitvated protein kinase pathway in the transcriptional activation of p53 on genotoxic stress by chemotherapeutic agents. *Cancer Res* 2000;60(9):2464–2472.

42. Persons DL, Yazlovitskaya EM, Pelling JC. Effect of extracellular signal-regulated kinase on p53 accumulation in response to cisplatin. *J Biol Chem* 2000;275(46):35,778–35,785.

43. Mandic A, Viktorsson K, Heiden T, et al. The MEK1 inhibitor PD98059 sensitizes C8161 melanoma cells to cisplatin-induced apoptosis. *Melanoma Res* 2001;11(1):11–19.

44. Tong T, Fan W, Zhao H, et al. Involvement of the map kinase pathways in induction of gadd45 following UV radiation. *Exp Cell Res* 2001;269(1):64–72.

45. Takekawa M, Saito H. A family of stress-inducible GADD45-like proteins mediate activation of the stress-responsive MTK1/MEKK4 MAPKKK. *Cell* 1998;95(4):521–530.

46. Agami R, Bernards R. Distinct initiation and maintenance mechanisms cooperate to induce G1 cell cycle arrest in response to DNA damage. *Cell* 2000;102(1):55–66.

47. Petrocelli T, Slingerland J. UVB induced cell cycle checkpoints in an early stage human melanoma line, WM35. *Oncogene* 2000;19(39):4480–4490.

48. Fu H, Subramanian RR, Masters SC. 14-3-3 proteins: structure, function, and regulation. *Annu Rev Pharmacol Toxicol* 2000;40:617–647.

49. Walworth NC. DNA damage: Chk1 and Cdc25, more than meets the eye. *Curr Opin Genet Dev* 2001;11(1):78–82.

50. Hirose Y, Berger MS, Pieper RO. Abrogation of the Chk1-mediated G(2) checkpoint pathway potentiates temozolomide-induced toxicity in a p53-independent manner in human glioblastoma cells. *Cancer Res* 2001;61(15):5843–5849.

51. Gottifredi V, Karni-Schmidt O, Shieh SS, et al. p53 down-regulates CHK1 through p21 and the retinoblastoma protein. *Mol Cell Biol* 2001;21(4):1066–1076.

52. Takekawa M, Adachi M, Nakahata A, et al. p53-inducible wip1 phosphatase mediates a negative feedback regulation of p38 MAPK-p53 signaling in response to UV radiation. *EMBO J* 2000;19(23):6517–6526.

53. Bunz F, Dutriaux A, Lengauer C, et al. Requirement for p53 and p21 to sustain G2 arrest after DNA damage. *Science* 1998;282(5393):1497–1501.

54. Smits VA, Klompmaker R, Vallenius T, et al. p21 inhibits Thr161 phosphorylation of Cdc2 to enforce the G2 DNA damage checkpoint. *J Biol Chem* 2000;275(39):30,638–30,643.

55. Chan TA, Hwang PM, Hermeking H, et al. Cooperative effects of genes controlling the G(2)/M checkpoint. *Genes Dev* 2000;14(13):1584–1588.

56. Jin S, Antinore MJ, Lung FD, et al. The GADD45 inhibition of Cdc2 kinase correlates with GADD45-mediated growth suppression. *J Biol Chem* 2000;275(22):16,602–16,608.

57. Ferry KV, Hamilton TC, Johnson SW. Increased nucleotide excision repair in cisplatin-resistant ovarian cancer cells: role of ERCC1-XPF. *Biochem Pharmacol* 2000;60(9):1305–1313.

58. Li D, Turi TG, Schuck A, et al. Rays and arrays: the transcriptional program in the response of human epidermal keratinocytes to UVB illumination. *FASEB J* 2001;15(13):2533–2535.

59. Aguda BD. Instabilities in phosphorylation-dephosphorylation cascades and cell cycle checkpoints. *Oncogene* 1999;18(18):2846–2851.

60. Steegenga WT, van der Eb AJ, Jochemsen AG. How phosphorylation regulates the activity of p53. *J Mol Biol* 1996;263(2):103–113.

61. Fuchs SY, Fried VA, Ronai Z. Stress-activated kinases regulate protein stability. *Oncogene* 1998;17(11):1483–1490.

62. Huang HC, Nguyen T, Pickett CB. Regulation of the antioxidant response element by protein kinase C-mediated phosphorylation of NF-E2-related factor 2. *Proc Natl Acad Sci USA* 2000;97(23):12,475–12,480.

63. Ben Levy R, Hooper S, Wilson R, et al. Nuclear export of the stress-activated protein kinase p38 mediated by its substrate MAPKAP kinase-2. *Curr Biol* 1998;8(19):1049–1057.

64. Lawler S, Fleming Y, Goedert M, et al. Synergistic activation of SAPK1/JNK1 by two MAP kinase kinases in vitro. *Curr Biol* 1998;8(25):1387–1390.

65. Wang XZ, Ron D. Stress-induced phosphorylation and activation of the transcription factor CHOP (GADD153) by p38 MAP Kinase. *Science* 1996;272(5266):1347–1349.

66. Modesti M, Kanaar R. DNA repair: spot(light)s on chromatin. *Curr Biol* 2001;11(6): R229-R232.

67. Zhong S, Jansen C, She QB, et al. Ultraviolet B-induced phosphorylation of histone H3 at serine 28 is mediated by MSK1. *J Biol Chem* 2001;276(35):33213–33219.

68. Araujo SJ, Tirode F, Coin F, et al. Nucleotide excision repair of DNA with recombinant human proteins: definition of the minimal set of factors, active forms of TFIIH, and modulation by CAK. *Genes Dev* 2000;14(3):349–359.

69. Stivala LA, Riva F, Cazzalini O, et al. p21(waf1/cip1)-null human fibroblasts are deficient in nucleotide excision repair downstream the recruitment of PCNA to DNA repair sites. *Oncogene* 2001;20(5):563–570.

70. Sheikh MS, Chen YQ, Smith ML, et al. Role of p21Waf1/Cip1/Sdi1 in cell death and DNA repair as studied using a tetracycline-inducible system in p53-deficient cells. *Oncogene* 1997;14(15):1875–1882.

71. Adimoolam S, Lin CX, Ford JM. The p53-regulated cyclin-dependent kinase inhibitor, p21 (cip1, waf1, sdi1), is not required for global genomic and transcription-coupled nucleotide excision repair of UV-induced DNA photoproducts. *J Biol Chem* 2001;276(28):25,813–25,822.

72. Shivji MK, Grey SJ, Strausfeld UP, et al. Cip1 inhibits DNA replication but not PCNA-dependent nucleotide excision-repair. *Curr Biol* 1994;4(12):1062–1068.

73. Cooper MP, Balajee AS, Bohr VA. The C-terminal domain of p21 inhibits nucleotide excision repair In vitro and In vivo. *Mol Biol Cell* 1999;10(7):2119–2129.

74. Therrien JP, Loignon M, Drouin R, et al. Ablation of p21waf1cip1 expression enhances the capacity of p53-deficient human tumor cells to repair UVB-induced DNA damage. *Cancer Res* 2001;61(9):3781–3786.

75. Ford JM, Hanawalt PC. Expression of wild-type p53 is required for efficient global genomic nucleotide excision repair in UV-irradiated human fibroblasts. *J Biol Chem* 1997;272(44): 28,073–28,080.

76. Therrien JP, Drouin R, Baril C, et al. Human cells compromised for p53 function exhibit defective global and transcription-coupled nucleotide excision repair, whereas cells compromised for pRb function are defective only in global repair. *Proc Natl Acad Sci USA* 1999;96(26):15,038–15,043.

77. Zhou J, Ahn J, Wilson SH, et al. A role for p53 in base excision repair. *EMBO J* 2001;20(4): 914–923.

78. Offer H, Milyavsky M, Erez N, et al. Structural and functional involvement of p53 in BER in vitro and in vivo. *Oncogene* 2001;20(5):581–589.

79. Smith ML, Ford JM, Hollander MC, et al. p53-mediated DNA repair responses to UV radiation: studies of mouse cells lacking p53, p21, and/or gadd45 genes. *Mol Cell Biol* 2000;20(10):3705–3714.

80. Rossig L, Jadidi AS, Urbich C, et al. Akt-dependent phosphorylation of p21(Cip1) regulates PCNA binding and proliferation of endothelial cells. *Mol Cell Biol* 2001;21(16):5644–5657.

81. Smith ML, Chen IT, Zhan Q, et al. Interaction of the p53-regulated protein Gadd45 with proliferating cell nuclear antigen. *Science* 1994;266(5189):1376–1380.

82. Waga S, Hannon GJ, Beach D, et al. The p21 inhibitor of cyclin-dependent kinases controls DNA replication by interaction with PCNA. *Nature* 1994;369(6481):574–578.

83. O'Reilly MA, Staversky RJ, Watkins RH, et al. p53-independent induction of GADD45 and GADD153 in mouse lungs exposed to hyperoxia. *Am J Physiol Lung Cell Mol Physiol* 2000;278(3):L552-L559.

84. Russo T, Zambrano N, Esposito F, et al. A p53-independent pathway for activation of WAF1/CIP1 expression following oxidative stress. *J Biol Chem* 1995;270(49):29,386–29,391.

85. Zamble DB, Jacks T, Lippard SJ. p53-dependent and -independent responses to cisplatin in mouse testicular teratocarcinoma cells. *Proc Natl Acad Sci USA* 1998;95(11): 6163–6168.

86. Scherer SJ, Maier SM, Seifert M, et al. p53 and c-Jun functionally synergize in the regulation of the DNA repair gene hMSH2 in response to UV. *J Biol Chem* 2000;275(48):37,469–37,473.

87. Aboussekhra A, Biggerstaff M, Shivji MK, et al. Mammalian DNA nucleotide excision repair reconstituted with purified protein components. *Cell* 1995;80(6):859–868.

88. Wold MS. Replication protein A: a heterotrimeric, single-stranded DNA-binding protein required for eukaryotic DNA metabolism. *Annu Rev Biochem* 1997;66:61–92.

89. Iftode C, Daniely Y, Borowiec JA. Replication protein A (RPA): the eukaryotic SSB. *Crit Rev Biochem Mol Biol* 1999;34(3):141–180.

90. Dutta A, Stillman B. cdc2 family kinases phosphorylate a human cell DNA replication factor, RPA, and activate DNA replication. *EMBO J* 1992;11(6):2189–2199.

91. Oakley GG, Loberg LI, Yao J, et al. UV-induced hyperphosphorylation of replication protein a depends on DNA replication and expression of ATM protein. *Mol Biol Cell* 2001;12(5):1199–1213.

92. Rodrigo G, Roumagnac S, Wold MS, et al. DNA replication but not nucleotide excision repair is required for UVC-induced replication protein A phosphorylation in mammalian cells. *Mol Cell Biol* 2000;20(8):2696–2705.

93. Christmann M, Pick M, Lage H, et al. Acquired resistance of melanoma cells to the antineoplastic agent fotemustine is caused by reactivation of the DNA repair gene MGMT. *Int J Cancer* 2001;92(1):123–129.

94. Srivenugopal KS, Shou J, Mullapudi SR, et al. Enforced expression of wild-type p53 curtails the transcription of the $O(6)$-methylguanine-DNA methyltransferase gene in human tumor cells and enhances their sensitivity to alkylating agents. *Clin Cancer Res* 2001;7(5):1398–1409.

95. Guo W, Liu X, Lee S, et al. High O^6-methylguanine methyl transferase activity is frequently found in human oral cancer cells with p53 inactivation. *Int J Oncol* 1999;15(4):817–821.

96. Boldogh I, Ramana CV, Chen Z, et al. Regulation of expression of the DNA repair gene O^6-methylguanine-DNA methyltransferase via protein kinase C-mediated signaling. *Cancer Res* 1998;58(17):3950–3956.

97. Srivenugopal KS, Mullapudi SR, Shou J, et al. Protein phosphorylation is a regulatory mechanism for O^6-alkylguanine-DNA alkyltransferase in human brain tumor cells. *Cancer Res* 2000;60(2):282–287.

98. Ababou M, Dutertre S, Lecluse Y, et al. ATM-dependent phosphorylation and accumulation of endogenous BLM protein in response to ionizing radiation. *Oncogene* 2000;19(52):5955–5963.

99. Blasina A, Price BD, Turenne GA, et al. Caffeine inhibits the checkpoint kinase ATM. *Curr Biol* 1999;9(19):1135–1138.

100. Sarkaria JN, Busby EC, Tibbetts RS, et al. Inhibition of ATM and ATR kinase activities by the radiosensitizing agent, caffeine. *Cancer Res* 1999;59(17):4375–4382.

101. Olayioye MA, Neve RM, Lane HA, et al. The ErbB signaling network: receptor heterodimerization in development and cancer. *EMBO J* 2000;19(13):3159–3167.

102. Arteaga CL, Winnier AR, Poirier MC, et al. p185c–erbB-2 signal enhances cisplatin-induced cytotoxicity in human breast carcinoma cells: association between an oncogenic receptor tyrosine kinase and drug-induced DNA repair. *Cancer Res* 1994;54(14):3758–3765.

103. Yen L, Nie ZR, You XL, et al. Regulation of cellular response to cisplatin-induced DNA damage and DNA repair in cells overexpressing p185(erbB-2) is dependent on the ras signaling pathway. *Oncogene* 1997;14(15):1827–1835.

104. You XL, Yen L, Zeng-Rong N, et al. Dual effect of erbB-2 depletion on the regulation of DNA repair and cell cycle mechanisms in non-small cell lung cancer cells. *Oncogene* 1998;17(24):3177–3186.

105. Pietras RJ, Fendly BM, Chazin VR, et al. Antibody to HER-2/neu receptor blocks DNA repair after cisplatin in human breast and ovarian cancer cells. *Oncogene* 1994;9(7):1829–1838.

106. Pietras RJ, Pegram MD, Finn RS, et al. Remission of human breast cancer xenografts on therapy with humanized monoclonal antibody to HER-2 receptor and DNA-reactive drugs. *Oncogene* 1998;17(17):2235–2249.

107. Pietras RJ, Poen JC, Gallardo D, et al. Monoclonal antibody to HER-2/neureceptor modulates repair of radiation-induced DNA damage and enhances radiosensitivity of human breast cancer cells overexpressing this oncogene. *Cancer Res* 1999;59(6):1347–1355.
108. Tsai CM, Chang KT, Li L, et al. Interrelationships between cellular nucleotide excision repair, cisplatin cytotoxicity, HER-2/neu gene expression, and epidermal growth factor receptor level in non-small cell lung cancer cells. *Jpn J Cancer Res* 2000;91(2):213–222.
109. Pegram MD, Slamon DJ. Combination therapy with trastuzumab (Herceptin) and cisplatin for chemoresistant metastatic breast cancer: evidence for receptor-enhanced chemosensitivity. *Semin Oncol* 1999;26(4 Suppl 12):89–95.
110. Huang RP, Wu JX, Fan Y, et al. UV activates growth factor receptors via reactive oxygen intermediates. *J Cell Biol* 1996;133(1):211–220.
111. Chin YE, Kitagawa M, Su WC, et al. Cell growth arrest and induction of cyclin-dependent kinase inhibitor p21 WAF1/CIP1 mediated by STAT1. *Science* 1996;272(5262):719–722.
112. Bromberg JF, Fan Z, Brown C, et al. Epidermal growth factor-induced growth inhibition requires Stat1 activation. *Cell Growth Differ* 1998;9(7):505–512.
113. Ahrendt SA, Brown HM, Komorowski RA, et al. p21WAF1 expression is associated with improved survival after adjuvant chemoradiation for pancreatic cancer. *Surgery* 2000;128(4):520–530.
114. Yang W, Velcich A, Mariadason J, et al. p21(WAF1/cip1) is an important determinant of intestinal cell response to sulindac in vitro and in vivo. *Cancer Res* 2001;61(16):6297–6302.
115. Tsihlias J, Kapusta L, Slingerland J. The prognostic significance of altered cyclin-dependent kinase inhibitors in human cancer. *Annu Rev Med* 1999;50:401–423.
116. Donato NJ, Perez M, Kang H, et al. EGF receptor and p21WAF1 expression are reciprocally altered as ME-180 cervical carcinoma cells progress from high to low cisplatin sensitivity. *Clin Cancer Res* 2000;6(1):193–202.
117. Fan S, Chang JK, Smith ML, et al. Cells lacking CIP1/WAF1 genes exhibit preferential sensitivity to cisplatin and nitrogen mustard. *Oncogene* 1997;14(18):2127–2136.
118. Dixit M, Yang JL, Poirier MC, et al. Abrogation of cisplatin-induced programmed cell death in human breast cancer cells by epidermal growth factor antisense RNA. *J Natl Cancer Inst* 1997;89(5):365–373.
119. Lammering G, Hewit TH, Hawkins WT, et al. Epidermal growth factor receptor as a genetic therapy target for carcinoma cell radiosensitization. *J Natl Cancer Inst* 2001;93(12):921–929.
120. Reardon DB, Contessa JN, Mikkelsen RB, et al. Dominant negative EGFR-CD533 and inhibition of MAPK modify JNK1 activation and enhance radiation toxicity of human mammary carcinoma cells. *Oncogene* 1999;18(33):4756–4766.
121. Keating KE, Gueven N, Watters D, et al. Transcriptional downregulation of ATM by EGF is defective in ataxia-telangiectasia cells expressing mutant protein. *Oncogene* 2001;20(32):4281–4290.
122. Harari PM, Huang SM. Head and neck cancer as a clinical model for molecular targeting of therapy: combining EGFR blockade with radiation. *Int J Radiat Oncol Biol Phys* 2001;49(2):427–433.
123. Fan S, Wang JA, Yuan RQ, et al. Scatter factor protects epithelial and carcinoma cells against apoptosis induced by DNA-damaging agents. *Oncogene* 1998;17(2):131–141.
124. Gao M, Fan S, Goldberg ID, et al. Hepatocyte growth factor/scatter factor blocks the mitochondrial pathway of apoptosis signaling in breast cancer cells. *J Biol Chem* 2001;276(50):47257–47265.
125. Fan S, Ma YX, Wang JA, et al. The cytokine hepatocyte growth factor/scatter factor inhibits apoptosis and enhances DNA repair by a common mechanism involving signaling through phosphatidyl inositol 3' kinase. *Oncogene* 2000;19(18):2212–2223.
126. Fan S, Ma YX, Gao M, et al. The multisubstrate adapter Gab1 regulates hepatocyte growth factor (scatter factor)–c-Met signaling for cell survival and DNA repair. *Mol Cell Biol* 2001;21(15):4968–4984.

127. Cao B, Su Y, Oskarsson M, Zhao P, et al. Neutralizing monoclonal antibodies to hepatocyte growth factor/scatter factor (HGF/SF) display antitumor activity in animal models. *Proc Natl Acad Sci USA* 2001;98(13):7443–7448.

128. English JM, Cobb MH. Pharmacological inhibitors of MAPK pathways. *Trends Pharmacol Sci* 2002;23(1):40–45.

129. Mizuno K, Noda K, Ueda Y, et al. UCN-01, an anti-tumor drug, is a selective inhibitor of the conventional PKC subfamily. *FEBS Lett* 1995;359(2–3):259–261.

130. Yu L, Orlandi L, Wang P, et al. UCN-01 abrogates G2 arrest through a Cdc2-dependent pathway that is associated with inactivation of the Wee1Hu kinase and activation of the Cdc25C phosphatase. *J Biol Chem* 1998;273(50):33,455–33,464.

131. Senderowicz AM. Small molecule modulators of cyclin-dependent kinases for cancer therapy. *Oncogene* 2000;19(56):6600–6606.

132. Busby EC, Leistritz DF, Abraham RT, et al. The radiosensitizing agent 7-hydroxystaurosporine (UCN-01) inhibits the DNA damage checkpoint kinase hChk1. *Cancer Res* 2000;60(8):2108–2112.

133. Piwnica-Worms H, Atherton-Fessler S, Lee MS, et al. p107wee1 is a serine/threonine and tyrosine kinase that promotes the tyrosine phosphorylation of the cyclin/p34cdc2 complex. *Cold Spring Harbor Symp Quant Biol* 1991;56:567–576.

134. Parker LL, Atherton-Fessler S, Piwnica-Worms H. p107wee1 is a dual-specificity kinase that phosphorylates p34cdc2 on tyrosine 15. *Proc Natl Acad Sci USA* 1992;89(7):2917–2921.

135. Rowley R, Hudson J, Young PG. The wee1 protein kinase is required for radiation-induced mitotic delay. *Nature* 1992;356(6367):353–355.

136. Sanchez Y, Wong C, Thoma RS, et al. Conservation of the Chk1 checkpoint pathway in mammals: linkage of DNA damage to Cdk regulation through Cdc25. *Science* 1997;277(5331):1497–1501.

137. Sugiyama K, Shimizu M, Akiyama T, et al. UCN-01 selectively enhances mitomycin C cytotoxicity in p53 defective cells which is mediated through S and/or G(2) checkpoint abrogation. *Int J Cancer* 2000;85(5):703–709.

138. Theard D, Coisy M, Ducommun B, et al. Etoposide and adriamycin but not genistein can activate the checkpoint kinase Chk2 independently of ATM/ATR. *Biochem Biophys Res Commun* 2001;289(5):1199–1204.

139. Xu B, Kim ST, Lim DS, et al. Two molecularly distinct g(2)/m checkpoints are induced by ionizing irradiation. *Mol Cell Biol* 2002;22(4):1049–1059.

140. Loignon M, Fetni R, Gordon AJ, et al. A p53-independent pathway for induction of p21waf1cip1 and concomitant G1 arrest in UV-irradiated human skin fibroblasts. *Cancer Res* 1997;57(16):3390–3394.

141. Haapajarvi T, Kivinen L, Heiskanen A, et al. UV radiation is a transcriptional inducer of p21(Cip1/Waf1) cyclin-kinase inhibitor in a p53-independent manner. *Exp Cell Res* 1999;248(1):272–279.

142. Zeng YX, El Deiry WS. Regulation of p21WAF1/CIP1 expression by p53-independent pathways. *Oncogene* 1996;12(7):1557–1564.

143. Jiang H, Yang LY. Cell cycle checkpoint abrogator UCN-01 inhibits DNA repair: association with attenuation of the interaction of XPA and ERCC1 nucleotide excision repair proteins. *Cancer Res* 1999;59(18):4529–4534.

144. Bunch RT, Eastman A. Enhancement of cisplatin-induced cytotoxicity by 7-hydroxystaurosporine (UCN-01), a new G2-checkpoint inhibitor. *Clin Cancer Res* 1996;2(5):791–797.

145. Hsueh CT, Kelsen D, Schwartz GK. UCN-01 suppresses thymidylate synthase gene expression and enhances 5-fluorouracil-induced apoptosis in a sequence-dependent manner. *Clin Cancer Res* 1998;4(9):2201–2206.

146. Hwang HS, Davis TW, Houghton JA, et al. Radiosensitivity of thymidylate synthase-deficient human tumor cells is affected by progression through the G1 restriction point

into S-phase: implications for fluoropyrimidine radiosensitization. *Cancer Res* 2000; 60(1):92–100.

147. Dai Y, Yu C, Singh V, et al. Pharmacological inhibitors of the mitogen-activated protein kinase (MAPK) kinase/MAPK cascade interact synergistically with UCN-01 to induce mitochondrial dysfunction and apoptosis in human leukemia cells. *Cancer Res* 2001;61(13):5106–5115.

148. Zhou BB, Chaturvedi P, Spring K, et al. Caffeine abolishes the mammalian G(2)/M DNA damage checkpoint by inhibiting ataxia-telangiectasia-mutated kinase activity. *J Biol Chem* 2000;275(14):10,342–10,348.

149. Asaad NA, Zeng ZC, Guan J, et al. Homologous recombination as a potential target for caffeine radiosensitization in mammalian cells: reduced caffeine radiosensitization in XRCC2 and XRCC3 mutants. *Oncogene* 2000;19(50):5788–5800.

150. Boonkitticharoen V, Laohathai K, Puribhat S. Differential radiosensitization of radioresistant human cancer cells by caffeine. *J Med Assoc Thai* 1993;76(5):271–277.

151. Jackson JR, Gilmartin A, Imburgia C, et al. An indolocarbazole inhibitor of human checkpoint kinase (Chk1) abrogates cell cycle arrest caused by DNA damage. *Cancer Res* 2000;60(3):566–572.

152. Carlson BA, Dubay MM, Sausville EA, et al. Flavopiridol induces G1 arrest with inhibition of cyclin-dependent kinase (CDK) 2 and CDK4 in human breast carcinoma cells. *Cancer Res* 1996;56(13):2973–2978.

153. Carlson B, Lahusen T, Singh S, et al. Down-regulation of cyclin D1 by transcriptional repression in MCF-7 human breast carcinoma cells induced by flavopiridol. *Cancer Res* 1999;59(18):4634–4641.

154. Sherr CJ. The Pezcoller lecture: cancer cell cycles revisited. *Cancer Res* 2000;60(14): 3689–3695.

155. Carroll M, Ohno-Jones S, Tamura S, et al. CGP 57148, a tyrosine kinase inhibitor, inhibits the growth of cells expressing BCR-ABL, TEL-ABL, and TEL-PDGFR fusion proteins. *Blood* 1997;90(12):4947–4952.

156. Yuan ZM, Huang Y, Ishiko T, et al. Regulation of DNA damage-induced apoptosis by the c-Abl tyrosine kinase. *Proc Natl Acad Sci USA* 1997;94(4):1437–1440.

157. Uemura N, Griffin JD. The ABL kinase inhibitor STI571 does not affect survival of hematopoietic cells after ionizing radiation. *Blood* 2000;96(9):3294–3295.

158. Franklin CC, Srikanth S, Kraft AS. Conditional expression of mitogen-activated protein kinase phosphatase-1, MKP-1, is cytoprotective against UV-induced apoptosis. *Proc Natl Acad Sci USA* 1998;95(6):3014–3019.

159. Sanchez-Perez I, Martinez-Gomariz M, Williams D, et al. CL100/MKP-1 modulates JNK activation and apoptosis in response to cisplatin. *Oncogene* 2000;19(45):5142–5152.

160. Fiscella M, Zhang H, Fan S, et al. Wip1, a novel human protein phosphatase that is induced in response to ionizing radiation in a p53-dependent manner. *Proc Natl Acad Sci USA* 1997;94(12):6048–6053.

161. Choi J, Appella E, Donehower LA. The structure and expression of the murine wildtype p53-induced phosphatase 1 (Wip1) gene. *Genomics* 2000;64(3):298–306.

7

Overcoming Resistance to Alkylating Agents by Inhibitors of O^6-Alkylguanine-DNA Alkyltransferase

Anthony E. Pegg, PhD
and M. Eileen Dolan, PhD

CONTENTS

1. INTRODUCTION

Despite a myriad of toxic side effects, alkylating agents still represent a major class of cancer chemotherapeutic agents. DNA is the primary target for these agents and the correct repair of DNA damage provides protection from them *(1,2)*. Virtually every DNA repair pathway is able to interact with one or more facets of alkylation damage. Alkylated bases are repaired via base excision repair (BER), nucleotide excision repair (NER), and direct reversal. Mechanisms for the repair of single- and double-strand breaks, including poly(ADP)ribose

From: *Cancer Drug Discovery and Development: DNA Repair in Cancer Therapy*
Edited by: L. C. Panasci and M. A. Alaoui-Jamali © Humana Press Inc., Totowa, NJ

polymerase binding, also aid in protection from alkylating agents. NER is also involved in the repair of intrastrand crosslinks and, probably, interstrand crosslinks, although these mechanisms have not been characterized fully. As described in more detail in this chapter, mismatch repair (MMR) also recognizes alkylation damage in the form of the O^6-methylguanine (m⁶G), but, in this case, it causes apoptosis rather than repair. This review is focused on the role played in alkylating agent resistance by the direct reversal reaction brought about by O^6-alkylguanine-DNA alkyltransferase (AGT) and the possibilities for improving therapy by using O^6-benzylguanine (BG), an inactivator of AGT. This field has expanded rapidly in recent years and only a limited number of key and recent citations can be provided in this short review chapter. Further material for the interested reader is contained within these citations and in several recent reviews *(2–11)*.

O^6-Alkylguanine-DNA alkyltransferase repairs adducts on the O^6 position of guanine via a direct transfer of the alkyl group from the DNA to a cysteine acceptor site on the protein. The resulting *S*-alkylcysteine is not converted back to cysteine, so the AGT can act only once. The addition of an alkyl group to human AGT occurs at Cys145. This addition disrupts a hydrogen-bonding network and also causes a steric clash with an adjacent helix, resulting in the opening of the protein structure at an asparagine hinge formed by Asn139 *(12)*. This opening leads to the ubiquitination of the AGT protein and its rapid degradation *(13,14)*.

The AGT-like proteins have been identified in almost all the organisms for which complete genomes are known (except for the notable recent exceptions of *Arabidopsis thaliana* and *Deinococcus radiodurans*) with the sequence surrounding the cysteine acceptor site, –ProCysHisArgVal(Ile)–, providing a signature motif *(15,16)*. Crystal structures for the AGTs from human *Pyrococcus kodakaraensii* and *Escherichia coli* sources have now been solved *(12,17–19)*. These structures show that the AGTs contain a helix–turn–helix DNA-binding domain and that the target substrate base is flipped out of the DNA helix into a binding pocket that contains the cysteine acceptor residue.

O^6-Alkylguanine-DNA alkyltransferase repairs m⁶G in DNA very efficiently. This base is only a minor proportion (<10%) of the total alkylation produced by methylating agents used for cancer chemotherapy, like 5-(3,3-dimethyl-L-triazenyl)imidazole-4-carboximide (dacarbazine, DTIC) and temozolomide, but it is a major factor in causing tumor cell death *(20)* and chromosomal aberrations *(21)*. These effects are mediated via the MMR system *(2,22,23)*. Although m⁶G does provide a minor barrier to DNA replication *(24–26)*, replicative bypass does occur, probably mediated via polymerase η *(27)*, and m⁶G is copied with the frequent incorporation of a thymine instead of cytosine. The T:m⁶G pair (and probably the C:m⁶G pair also) is recognized by the MutSα protein as a mismatch *(22,23,28,29)*. This recognition leads to apoptopic cell death *(30–32)*. Two plausible explanations (which are not mutually exclusive) have been proposed for this *(22,23,33)*. First, the resulting MMR complex can lead to the degradation of a section of the daughter strand. Because the strand containing m⁶G is not degraded, resynthesis produces the same site for recognition by MMR, setting up a futile cycle leading to cell demise. Second, the binding of the MutSα and MutLα complex may initiate cell cycle arrest and programmed cell death directly. Very recently, it has been proposed that the second mechanism may be triggered by collapsed replication forks at lesions leading to double-strand breaks whose repair is aborted by MMR *(23)*. The induction of p53, which also requires a functional MMR system, may play a major role in the onset of apoptosis *(34–36)*. Cyclin B1 has also been implicated in the pathway leading to MMR-induced apoptosis of alkylation lesions *(37)*. The loss of MMR has a much greater protective effect in human cells than in mouse cells and the latter are intrinsically more resistant to methylating agents *(38)*.

Thus, AGT activity, which removes the methyl group from m^6G and restores the original guanine, provides a very effective means of resistance to such methylating agents. However, it should be noted that tumor cells can also become "tolerant" of m^6G in their DNA by the loss of MMR *(23)*. This loss is known to occur (and indeed to be a causative factor) in human tumors associated with the hereditary nonpolyposis colorectal carcinoma (HNPCC) syndrome *(39–41)*. In such tumors, which are not confined to the gastrointestinal (GI) tract, the exposure to methylating agents in the absence of AGT leads to persistent and highly mutagenic lesions in the DNA and is likely to be highly counterproductive for chemotherapy.

Chemotherapeutic chloroethylating agents such as carmustine, 1,3-bis(2-chlorothyl)-1-nitrosourea (BCNU), 1-(4-amino-2-methyl-5-pyrimidinyl)methyl-3-(2-chloroethyl)-3-nitrosourea (ACNU), and 1-(2-chloroethyl)-3-(4-methylcyclohexyl)-1-nitrosourea (MeCCNU) also kill tumor cells by the induction of apoptosis *(30,42,43)*. However, in this case, the principal lesion leading to cell death is a 1-(3-deoxycytidyl)-2-(1-deoxyguanosinyl)ethane interstrand crosslink and the MMR system is not involved in the development of toxicity from such lesions and may actually increase cell killing by unknown mechanisms *(44)*. The crosslink is formed by a rearrangement from an initial O^6-chloroethylguanine adduct *(45)*. This spontaneously forms $1,O^6$-ethanoguanine, which then reacts with the cytosine in the complementary strand of DNA to form the crosslink. AGT repairs O^6-chloroethylguanine and thus prevents the subsequent reactions and the development of interstrand crosslinks. The presence of high levels of AGT therefore provides a powerful protection from BCNU and similar compounds. Other mechanisms of resistance to BCNU have been reported, but AGT is clearly the most predominant means to prevent cell death in response to such chloroethylating agents.

2. RELATIONSHIP BETWEEN AGT AND CLINICAL RESISTANCE TO ALKYLATING AGENTS

2.1. AGT Activity and Resistance

The level of AGT activity and expression varies widely in tumors with some tumors, having high and others having undetectable AGT activity *(7,8)*. For example, lack of AGT activity has been detected in approx 24% of brain tumors and cutaneous T-cell lymphomas *(46)*. Although there is a vast amount of data demonstrating a role for AGT in protecting tumors from alkylnitrosourea toxicity in cells and in animals, more recent studies evaluating AGT levels in clinical samples from patients receiving BCNU therapy have supported this concept. Following an evaluation of 167 primary brain tumors from patients receiving BCNU therapy, Belanich et al. *(47)* reported that AGT content in tumors correlates with high response to treatment and greater survival. Conversely, the presence of a subpopulation of cells in a tumor with elevated AGT is correlated with poor prognosis. A Southwest Oncology Group study evaluated 64 patients with malignant astrocytoma (63% glioblastoma, 37% anaplastic astrocytoma), finding that 64% had high (>60,000 molecules/nucleus) AGT levels. The overall median survival for patients treated with BCNU with high versus low AGT levels was 8 and 29 mo, respectively *(48)*.

2.2. AGT Promoter Methylation and Resistance to Alkylating Agents

The AGT gene is not commonly deleted or mutated; thus, loss of AGT function is most frequently the result of epigenetic changes, specifically 5-methylation of cytosines within the promoter region. Hypermethylation of CpG islands within the promoter of the AGT gene as the cause of *AGT* transcriptional silencing in cell lines defective in repair of O^6-methylguanine lesion has been demonstrated *(49,50)*. Furthermore, in vitro treatment of cancer cells with demethylating agents restores AGT expression. Aberrant AGT gene methylation in cell culture has been firmly correlated with loss of mRNA expression, lack of AGT protein, and loss of enzymatic activity *(49–51)*. Several studies have now established an increased incidence of promoter hypermethylation in human tumors, although these studies have not always been combined with measurement of the AGT activity *(52–56)*.

The first clinical study relating promoter methylation to response to BCNU came from Esteller et al. *(57)* who established that *AGT*-promoter region methylation in brain tumors was a strong predictor of response, overall survival, and time to progression in patients treated with BCNU. It was an independent and stronger prognostic factor than age, stage, tumor grade, or performance status. More recently, the relationship between *AGT*-promoter methylation and clinical outcome in patients with B-lineage diffuse large cell lymphoma (B-DLCL) treated with multiagent chemotherapy, including the alkylating agent cyclophosphamide, was studied *(58)*. *AGT* hypermethylation was a strong predictor of overall survival but not of response. As discussed in Subheading 8, it is unclear that there is a direct role of AGT in resistance to the antitumor effect of cyclophosphamide . Therefore, the favorable outcome of patients with *AGT* hypermethylation may be the result of the possibility that *AGT* hypermethylation is associated with biochemical or epigenetic changes resulting in greater sensitivity to the chemotherapeutic regimen. *AGT* hypermethylation may be a prognostic marker of natural history that identifies a specific pathogenetic subset of lymphomas with a more favorable outcome. In support of this, promoter methylation of *AGT* is likely to relate to the process of tumorigenesis. Loss of *AGT* expression as a result of promoter methylation appears to be associated with increased frequency of TP53 mutations, in particular G:C to A:T transitions in secondary gliomas, non-small-lung cancers, and colorectal tumors *(53–56)*.

3. INACTIVATION OF AGT

3.1. Use of Ribozymes to Inactivate AGT

In an attempt to increase the sensitivity of cells to chloroethylating and methylating agents, efforts have been made to downregulate the AGT protein *(59–62)*. A partial sensitization of cells expressing AGT to alkylating agents has been obtained using oligonucleotides targeted against AGT mRNA and hammerhead ribozymes designed to degrade the long-lived AGT mRNA. The ribozymes designed against eight GUC sites within the AGT mRNA showed the most promising results in terms of enhancing BCNU toxicity *(62)*. In general, this approach has been successful in partial sensitization of cells grown in culture to alkylating agents, but, as of now, it has not been tested in a clinical setting because delivery of the ribozyme is problematic.

3.2. Design of BG

O^6-benzylguanine (BG), a free alkylated base, has received the most attention as a means to sensitize cells to alkylating agents and is presently in phase III clinical trials. BG, a substrate for AGT *(63)*, was designed based on an understanding of the bimolecular displacement reaction between the AGT protein and the leaving group at the O^6 position of guanine *(64)*. Benzyl groups are known to enter more readily into bimolecular reactions compared to alkyl groups because the electron charge stabilizes the benzyl group in the transition state. BG is, indeed, a very effective inhibitor of mammalian AGT and, as designed, it inactivates by acting as a substrate for alkyl transfer and by forming *S*-benzylcysteine at the acceptor site of the protein *(65)*. The reactivity of the benzyl group contributes to this efficient inactivation. However, recent crystallographic studies of the AGT protein, biochemical studies, and models of the binding of BG to AGT show that a major factor in its ability to act as a substrate is the fact that the benzyl group facilitates the binding of the free base to the active site by stacking with Pro140 and interacting with other residues, including Tyr158 *(9,12,19)*. This allows the free base to bind even though it lacks the ability to interact with the residues of the DNA-binding domain of the protein, which are located in a winged helix–turn–helix domain and critical for the rapid repair of DNA adducts.

Administration of micromolar concentrations of BG to human tumor cells for a few minutes resulted in a complete depletion of the AGT protein, rendering cells more sensitive to agents that alkylate at the O^6 position of guanine *(64)*. Transfer of the benzyl group to the AGT protein leads to a very marked decrease in the stability of the protein in HT29 cells *(66)*. The benzylated AGT protein is then degraded via the ubiquitin proteolytic pathway *(14)*. Loss of AGT led to a marked enhancement of the in vitro sensitivity of cultured tumor cells to alkylnitrososureas *(3,7,9)* and to DTIC and temozolomide *(67–71)*. Wedge and Newlands *(70)* found that repeated dosing of BG and/or continuous exposure to micromolar BG in MMR-proficient cells potentiated temozolomide cytotoxicity to a greater extent than a single BG dose. These data suggest that repeat dosing of BG with temozolomide might have a clinical role. However, MMR-deficient cell lines, unlike MMR wild-type cell lines, are more resistant to temozolomide and no synergistic effect is observed when these cells are exposed to BG *(72)*.

3.3. Effect of BG on Alkylating Agent-Induced Antitumor Activity and Toxicity

Using human tumor xenografts as a model, BG in combination with BCNU was shown to inhibit the growth of tumors expressing AGT *(73–77)*. In addition to enhancing BCNU, BG can potentiate the activity of BCNU delivered intracranially via polymers in rats challenged with a lethal brain tumor *(78)*. The advantage to using polymers to deliver BCNU is the lack of evidence of treatment-related toxicity. Recently, studies to sustain AGT inactivation in tumor xenografts for 24 h included a second bolus injection of BG administered 8 h after the first dose *(79)*. Sustained AGT inactivation is particularly important when combining BG with BCNU polymers because these agents are released over several days, requiring prolonged AGT suppression.

BG has also been shown to increase the antitumor activity of temozolomide, a direct acting methylating agent. Several human tumor xenograft studies have shown the enhancement of temozolomide by BG *(10,80,81)*. The beneficial effect is dependent on the schedule of drug administration, with multiple dosing of BG plus temozolomide producing the greatest effect *(70,80–82)*. Repeated administration of the combination can lead to an increase in the therapeutic index of temozolomide.

DTIC is a methylating agent requiring metabolic activation. Although BG was, as expected, able to increase tumor cell killing by MTIC, an activated form of DTIC, in cell culture *(67)*, BG paradoxically inhibited DTIC-induced apoptosis in the small intestine of mice, implying lack of therapeutic benefit of the combination *(83)*. This inhibition is readily explained by the observation that BG is a competitive substrate of cytochrome P450 1A1 and 1A2, which are the same P450 isoforms responsible for conversion of DTIC to its active methylating agent *(84)*. In the presence of therapeutic concentrations of BG, there is little to no conversion of DTIC to its active methylating species *(84)*. Therefore, treatment of patients with BG and DTIC would likely be less beneficial than DTIC alone and it is not recommended.

Coupled with BG-enhanced antitumor activity of alkylating agents is increased toxicity to bone marrow in animals. BG enhances the bone marrow toxicity of BCNU by fourfold to eightfold in mice and the toxicity of temozolomide by onefold to twofold as measured by marrow cellularity, granulocyte macrophage–colony forming cells, and colony-forming-units *(85)*. When the clastogenic effects of BCNU and temozolomide were examined in the mouse bone marrow micronucleus assay, a significantly higher frequency of micronuclei formation was observed in mice that received BG pretreatment compared with mice that received no pretreatment *(85)*. Similar results have been observed in humans with greater dose reductions for BCNU than temozolomide when treated in combination with BG (see below).

3.4. Other AGT Inactivators

3.4.1. O^6-(4-Bromothenyl)Guanine and O^6-(Hetarylmethyl)Guanines

Another AGT inactivator that is currently in clinical development is O^6-(4-bromothenyl)guanine (4-BTG), a member of a group of O^6-(hetarylmethyl)guanines that were designed in order to use a different O^6 substituent from benzyl but which would have similar properties allowing facile inactivation of AGT *(86)*. 4-BTG was reported to be somewhat more potent than BG against isolated AGT in vitro but has a potential disadvantage in its chemical instability *(86)*. Preclinical studies indicated enhanced tumor growth delay combining 4-BTG with temozolomide *(87)*. Although the delay in tumor growth was indistinguishable from that observed with BG and temozolomide, the 4-BTG combination resulted in considerably less toxicity at the doses used *(88)*. No pharmacokinetic measurements or studies of the metabolism of 4-BTG have been reported. Recently, glucoside analogs of 4-BTG with improved water solubility and selective uptake in tumor cells have been reported *(89,90)*.

3.4.2. O^6-Benzyl-2'-Deoxyguanosine

An analog of BG that has been studied extensively in a preclinical setting is O^6-benzyl-2'-deoxyguanosine (B2dG), a water-soluble inhibitor of AGT. B2dG potentiates the efficacy of BCNU against AGT-positive xenografts in athymic mice *(43,81)*. The therapeutic effect of the B2dG and BCNU combination against brain tumor xenografts was optimized without inducing substantial toxicity in the host by adjusting the doses of both compounds. B2dG plus BCNU doses were 133 mg/m^2 plus 25 mg/m^2, 200 mg/m^2 plus 17 mg/m^2, and 300 mg/m^2 plus 11 mg/m^2. The growth delays of 30.2, 38.4, and 22.3 d, respectively, observed for the above regimens suggest that the optimal drug combination did not require the maximum doses of B2dG. In fact, the highest doses of B2dG (300 mg/m^2) contributed to more frequent BCNU-related toxicities, despite the reduced BCNU dosage, and a reduction of the therapeutic effect. Tumors were eradicated without toxicity at 200 mg/m^2 B2dG and 23 mg/m^2 BCNU in 8 of 12 animals *(91)*. Evidence suggests the effectiveness of B2dG is related to its in vivo conversion to BG and 8-oxoBG *(92)*.

3.4.3. DERIVATIVES OF BG WITH SUBSTITUENTS AT THE 8 OR 9 POSITIONS

Based on encouraging animal results with B2dG, additional 9-substituted derivatives of BG (viz. O^6-benzyl-9-cyanomethylguanine and O^6-benzylguanosine) have been tested. However, they do not appear to be better than BG *(92)*. In addition, 8-substituted BG analogs have been studied for their ability to inactivate AGT in vitro *(93)*, in mice *(94)* and their ability to penetrate the cerebrospinal fluid *(95)*. Although these compounds were more potent than BG at inactivating AGT in vitro *(93)*, they were not significantly more potent in animal studies *(94)*. The CSF penetration percentages, based on the ratio $AUC_{(CSF)} : AUC_{(plasma)}$ for BG, 8-aza-BG, 8-oxo-BG, 8-trifluoromethyl-BG, 8-bromo-BG, and B2dG were 3.2%, 0.18%, 4.1%, 1.4%, <0.3%, and 2%, respectively. Therefore, the CSF penetration of BG, and its active metabolite 8-oxoBG, was greater than the penetration of 8-aza-BG, 8-Br-BG, 8-trifluoromethyl-BG, and B2dG.

3.4.4. O^6-ALLYL-, O^6-(2-OXOALKYL)- AND O^6-CYCLOALKENYL-GUANINES

A series of O^6-allyl- and O^6-(2-oxoalkyl)guanines were synthesized and evaluated but were found to be weak inactivators, with IC_{50} values ranging from 100 to 1000 μM *(96)*. O^6-Cycloalkenylguanines proved to be good AGT inactivators, with O^6-cyclobutenylmethylguanine and O^6-cyclopentenylmethylguanine exhibiting potency approaching that of BG. O^6-Cyclopentenylmethylguanine was shown to inactivate AGT in intact HT29 human colorectal carcinoma cells and potentiated the cytotoxicity of the temozolomide in colon tumor cell lines *(96)*.

3.4.5. BENZYLATED PYRIMIDINES

In addition to guanine analogs, several substituted 6(4)-(benzyloxy)pyrimidines were tested for their ability to inactivate AGT. One group of compounds, the 5-substituted 2,4-diamino-6-(benzyloxy)pyrimidines bearing electron-withdrawing groups at the 5-position (e.g., 2,4-diamino-6-(benzyloxy)-5-nitroso- and 2,4-diamino-(benzyloxy)-5-nitro-pyrimidine), was found to be significantly more effective than BG at inactivating AGT in human HT29 colon tumor cell extracts *(93)*. These compounds were also effective in sensitizing HT29 cells to killing by BCNU *(65)*. Unfortunately, neither derivative was as effective as BG at depleting AGT activity in vivo, as measured by loss of activity in mouse liver or spleen, and further in vivo studies have revealed extremely short half-lives for these compounds *(97)*.

The related compound 2,4-diamino-6-([4-fluorobenzyl]oxy)-5-nitrosopyrimidine was synthesized and found to be 50 times more potent than BG *(98)*. It was able to sensitize HeLa S3 cells to ACNU, but its stability in vivo was not investigated. Clearly, more animal studies will need to be completed to determine whether these agents have clinical potential.

3.4.6. OLIGODEOXYRIBONUCLEOTIDES CONTAINING m^{6}G OR BG

Oligodeoxyribonucleotides containing m^{6}G or BG *(99–101)* are excellent inhibitors of isolated AGT in vitro and are much more potent than free-base BG. However, major questions relating to uptake and stability must be addressed before these become candidates for clinical trials. Although much remains to be done to establish the validity of this approach, it is noteworthy that oligodeoxyribonucleotides containing multiple BG residues are very potent inactivators of AGT in vitro *(101,102)* and that the use of terminal methylphosphonate linkages to reduce exonuclease-mediated degradation does not impair the ability to interact with AGT *(100)*.

4. BG CLINICAL INVESTIGATIONS

Phase I studies of BG combined with BCNU are now complete *(103–106)*. Phase II and III studies have been initiated. Because BG is a biochemical modulatory agent as opposed to a cytotoxic, the dose of BG required to deplete AGT activity in tumor or lymphocytes was considered in its clinical development *(103–106)*. Two important lessons were learned in the early clinical development of BG: (1) AGT activity in lymphocytes was not a reliable predictor for tumor tissue depletion despite the logistical convenience of this surrogate marker *(105)*; and (2) measurements of AGT required biochemical assays on fresh-frozen tissue, not antibody-based approaches measuring AGT protein because degradation of BG-inactivated AGT takes place too slowly to lead to a good correlation between protein content and depletion of AGT activity immediately after BG *(107)*. Two phase I studies employing the same biochemical assay and measuring tumor AGT activity resulted in different recommended doses of BG. Spiro et al. *(105)* recommended a dose of 120 mg/m^2 BG based on evaluating AGT activity in needle biopsies of systemic tumors, whereas Friedman et al. *(108)* determined the dose to be 100 mg/m^2 for patients with malignant glioma. The difference in the biochemical modulatory dose between these studies could reflect differences in the site of the tumor, although one might envision a higher dose required for entry into the central nervous system (CNS). Neither study compared both doses; therefore, the possibility arose that 100 mg/m^2 would be an appropriate dose in systemic tumors *(105)*. To resolve this issue, a later study using surgically resected systemic tumors compared 120 mg/m^2 and 100 mg/m^2 for AGT depletion and concluded 120 mg/m^2, not 100 mg/m^2, was optimal for these tumors *(108a)*.

O^6-benzylgrianine alone is not toxic; however, the combination of BG plus BCNU causes bone marrow suppression, which may be cumulative, and is dose limiting. Bone marrow suppression included thrombocytopenia and neutropenia *(106)*. Nonhematologic toxicity was minimal but included fatigue, anorexia, increased bilirubin, and transaminase elevation. Similar to animal studies, BG results in a decrease in the dose of BCNU that can be safely administered. In humans, 120 mg per m^2/40 mg per m^2 is the recommended combined phase II dose of BG/BCNU, respectively *(103–106)*. This dose of BCNU is a fraction of the standard clinical dose of BCNU (200 mg/m^2) and the 80% reduction is a serious obstacle to clinical success that may necessitate the development of regional-based therapies, tumor-specific derivatives of BG, or strategies to protect normal cells that will allow increases in the BCNU dose.

5. BG METABOLISM AND PHARMACOKINETICS

O^6-Benzylguanine is converted to O^6-benzyl-8-oxoguanine (8-oxo-BG) a longer-lived, yet equally potent AGT inactivator and further debenzylated to 8-oxoguanine *(109)*. CYP450 isoforms 1A1 (extrahepatic) and 1A2 (hepatic) are primarily responsible for both BG oxidation and debenzylation. Both CYP3A4 and aldehyde oxidase also contribute to BG oxidation; however, the K_m values are much higher for these enzymes than for CYP1A1/2 *(110,111)*. The catalytic efficiency for CYP1A2 is higher than CYP1A1 for oxidation and lower for debenzylation *(111)*. Prolonged AGT suppression is likely attributable primarily to the effect of 8-oxo-BG.

The disposition and pharmacokinetic parameters of intravenous BG have been evaluated in rats, dogs, nonhuman primates (monkeys), and humans *(109,111–113)*. The plasma profile of BG was best described by a one-compartment model for each species. Although 8-oxoBG is formed in each species, the rate and degree to which it is formed differs with the highest area-under-the-curve ratio 8-oxoBG/BG for humans. In addition, acetylation at the N^2 position is a species-specific reaction because N^2-acetyl-O^6-benzylguanine, N^2-acetylguanine, and N^2-acetyl-O^6-benzyl-8-oxoguanine were found in rats but not mice, monkeys, or humans *(114)*.

6. AGT MUTANT PROTEINS

Current clinical trials with BG are quite encouraging, but a serious potential problem with this approach is the possible generation or selection of tumor cells expressing resistant forms of the AGT protein. Several of the known AGTs from nonhuman species are extremely resistant to inactivation by BG because of the presence of residues in the active-site pocket that disfavor the binding of BG *(12,19,63,115)*. Based on these comparisons and studies in which BG-resistant mutants of human AGT have been selected in an *E. coli* screening procedure, many point mutations have now been shown to cause human AGT to become less sensitive to the inhibitor *(115–122)*. Alterations of the binding pocket that either reduce the size available to exclude BG or disrupt the hydrophobic binding between the BG and the protein are quite readily produced and at least 134 known single-point mutations at 27 distinct codons provide sites at which point mutations can cause resistance *(9,122)*. Some of these mutants, particularly those at positions Pro140, Gly156, Tyr158, and Lys165, render AGT almost totally refractory to BG. When expressed in Mer⁻ cells, these mutant AGT proteins are highly effective in providing protection against BCNU that was not abolished by BG or 8-oxo-BG *(123–125)*.

Although no cases of BG-resistant AGTs have yet been reported from patients treated in clinical trials, mutations imparting resistance to alkylating agent plus BG by alterations at either Lys165 *(126)* or Gly156 (Friedman, personal communication) in tumor cells treated in the laboratory have been observed. These positions in the AGT sequence had been identified previously from the screening of AGT mutant plasmid libraries in *E. coli* as places at which multiple changes can give rise to BG resistance *(118,119)*. Additional AGT inhibitors capable of inactivating these BG resistant forms may, therefore, become necessary. The only current candidates for such inhibitors are the oligodeoxyribonucleotides containing m⁶G or BG *(99–101)*. The BG-resistant AGT mutants are able to react with such substrates because the DNA binding domain of the protein allows binding even though the active-site-pocket interactions with BG are reduced.

It is worth pointing out that although all mammalian AGTs appear to be susceptible to inactivation by BG, the mouse AGT is less sensitive than the human or rat *(127,128)*. This difference is mainly the result of the presence in the substrate-binding pocket of a His residue in place of Asn157 in the human and rat *(122)*. The extent to which this contributes to the success of laboratory studies in which human tumor xenografts in mice are treated with alkylating agents plus BG is not clear. The doses of BG used in these studies are high enough that AGT in both tumor and host is likely to have been completely inactivated.

After clinical treatment with BG, the loss of AGT is not specific for the tumors and occurs in peripheral lymphocytes, bone marrow cells, and presumably other normal tissues. The bone marrow has a low starting level of AGT and myelosuppression is a major limitation in the use of alkylating agents for cancer chemotherapy. This problem is exacerbated with the addition of BG. One solution would be to use gene transfer techniques to express additional AGT protein in normal cells that are the major sites of alkylation damage. Several groups have shown that this approach is feasible in animal models (*see* refs. *129–136* and references therein). An even more promising concept is to express a BG-resistant AGT prior to treatment with BG plus the alkylating agent. Encouraging results with this approach in rodent models have been obtained using constructs containing cDNAs for human AGT modified at Gly156 or Pro140 or with multiple mutations *(125,137–144)*. Based on laboratory studies, the P140K and Y158H single mutants are perhaps the most promising candidates for such an approach because they are both stable, extremely resistant to BG, and less likely to cause an immune response *(116,117)*. An interesting but unproven extension of this concept is the proposal to use a chimeric protein having AGT fused to AP endonuclease to increase repair by both alkyl transfer and the base excision repair pathway *(145)*. One problem with this approach is the rapid degradation that would be imposed by DNA repair leading to the production of the alkylated form of AGT.

7. NATURALLY OCCURRING AGT VARIANTS

Several variants in human AGTs have been reported. The exact incidence of these forms is unclear and they should probably be referred to as variants rather than polymorphisms because highly variable frequencies have been reported in different studies and some of the studies contain only a small number of measurements.

The currently known variants in the coding region of the protein that alter the amino acid sequence are W65C (0.4% incidence) *(146,147)*, L84F (approx 20%) *(147,148)*; I143V/K178R (2–21%) *(148)*, and G160R (0–15%) *(149,150)*. It is possible that these changes could alter the AGT properties in DNA repair and in response to inhibitors like BG. No difference in the ability to protect *Escherichia coli* from MNNG was seen when comparing L84F variant with wild-type AGT, but the W65C alteration was much less effective and appears to render AGT unstable *(151)*. A preliminary study showed a marginally increased risk for lung cancer in individuals with the I143V form *(150)*, but this alteration had little effect on the ability of the protein to repair methylated DNA even though Ile143 is very close to the active-site Cys145 *(148)*. It is possible that the I143V difference could affect the ability to repair larger adducts or O^4-methylthymine, which is a good substrate for the *E. coli* Ogt alkyltransferase, but a very poor substrate for human AGT. The K178R change [which is linked to the I143V alteration *(148)*] is also a conservative substitution that occurs in the carboxyl terminal region of the AGT protein, which is not needed for m^6G repair activity *(152)*.

The G160R variant was as effective as wild type in allowing protection from MNNG and BCNU, but this variant is significantly resistant to inactivation by BG *(123,153)*. Therefore, it might be expected that therapy with BG would be ineffective in individuals who were homozygous or even heterozygous for this alteration because the selection pressure of treatment with BG plus a chemotherapeutic agent would greatly favor its expression. However, although the initial report observed a 15% incidence in a small group of Japanese *(149)*, subsequent studies with much larger numbers of Caucasian and Chinese populations have found no examples or only a very low incidence of the G160R alteration *(148,150,154,155)*.

8. POTENTIAL USE OF BG TO ENHANCE OTHER CHEMOTHERAPEUTIC AGENTS

8.1. Nitrogen Mustards

Evidence has emerged recently from several studies suggesting a role for AGT in the repair of certain cyclophosphamide-induced lesions. Cyclophosphamide is metabolized to acrolein and phosphoramide mustard (PM) *(156)*. The antitumor effect is thought to be mainly mediated by interstrand crosslinks formed from the reaction of PM and DNA *(157)*.

Didechlorocyclophosphamide, which releases acrolein and a nontoxic analog of PM, was not found to have significant antitumor activity, therefore, acrolein is not thought to play a major role in the antitumor activity of cyclophosphamide but may be responsible for some toxic effects *(156)*. Friedman and colleagues *(158)* demonstrated that CHO cells stably transfected with a plasmid that led to AGT expression were significantly less sensitive to the toxic effects of both 4-HC (an activated form of cyclophosphamide) and 4-HDC (an activated form of didechlorocyclophosphamide, which generates acrolein and a nonalkylating form of PM) than CHO cells without detectable AGT. Further studies demonstrated that AGT-expressing cells were also less sensitive to the mutagenic effects of these agents *(159)*. However, neither the toxic nor mutagenic effects of PM were altered in the presence or absence of AGT. Taken together, these results suggest that AGT is likely to contribute to an increase in acrolein-induced lesions in DNA and is therefore unlikely to have a major impact on the antitumor activity produced by phosphoramide mustard.

As described earlier, a recent clinical study showed a correlation between AGT promoter methylation and clinical outcome in patients treated with combination chemotherapy that included cyclophosphamide *(58)*. Another study showed a positive correlation between high tumor AGT activity and poor initial response of ovarian cancer patients to postoperative combination chemotherapy with cyclophosphamide and cisplatin *(160)*. In contrast, no significant association was observed between AGT levels in ovarian carcinomas and the survival of patients treated with cyclophosphamide and carboplatin *(161)*, albeit few patients with low AGT were included in this study. A limitation in the design of studies to date is that patients were treated with multiple chemotherapeutic agents, making it difficult to interpret the contribution of cyclophosphamide alone to response or survival. Low AGT levels may be associated with other biochemical or epigenetic changes resulting in greater sensitivity to cyclophosphamide *(162)*.

Similarly, in laboratory studies of human tumor xenografts, confounding results have been reported. A correlation between AGT activity and sensitivity to cyclophosphamide was observed in a study of 14 lung cancer xenografts *(163)*. In contrast, D'Incalci and colleagues *(164)* reported no correlation between AGT activity in human tumor xenografts and response to cyclophosphamide. A cyclo-phosphamide-resistant tumor was found to have an elevated AGT activity and was rendered more sensitive by exposure to BG *(158)*. However, it is unclear whether the effect was the result of the ability of BG to inactivate AGT. As described next, there is evidence that BG may act in other ways to enhance cyclophosphamide-induced cell killing.

In some cell lines, BG definitely enhances the toxicity of activated forms of cyclophosphamide, ifosfamide, and nitrogen mustards (chlorambucil and melphalan) *(159,165,166)*. Based on the studies described earlier, a plausible explanation might be depletion of AGT activity by BG, resulting in an increase in acrolein-induced DNA lesions. However, there is compelling evidence that enhancement of the activity of nitrogen mustards can involve a mechanism dis-tinct from AGT inactivation *(166)*. Evidence pointing away from the well-known mechanism of BG is twofold: (1) The known toxic DNA lesions associated with nitrogen mustards occur at nucleophilic nitrogens of guanine, not the O^6 position of guanine and there is no evidence to suggest AGT can repair these lesions; (2) enhancement is observed in cell lines such as CHO (Chinese hamster ovary) and SQ20b (squamous cell carcinoma), which are apparently devoid of the AGT protein.

To elucidate whether the ability of BG to increase killing by cyclophospha-mide involved interstrand crosslinks, the effect of BG on acrolein and propyl-PM, an analog of PM with only one alkylating moiety and therefore unable to form interstrand or intrastrand crosslinks, was evaluated. BG did not enhance the toxicity of either compound in cells devoid of AGT, indicating that the effect was specific to alkylating species that can form interstrand or intrastrand crosslinks *(165,166)*. Further evidence for the importance of crosslinks was the enhancement of cisplatin and carboplatin, agents that form interstrand and intrastrand crosslinks and lack of enhancement of radiation by BG. Thus, it is likely that the formation or processing of intrastrand and/or interstrand DNA crosslinks is critical for the interaction. The effect may be cell-type-specific; it has been observed in CHO and several squamous head and neck carcinoma cell lines.

Concomitant with increased toxicity when BG is combined with these alky-lating agents is a greater percentage of cells undergoing apoptosis compared to cells treated with alkylating agent alone *(165)*. One potential mechanism that has been suggested involves inhibition of cyclin-dependent kinase (CDK). Structur-ally similar BG analogs inhibit CDK1 and CDK2 *(167)* and BG has been shown to arrest cells in stationary phase (i.e., G_1) at 16 h posttreatment *(165)*. Thus, an accumulation of PM-induced damage, particularly DNA interstrand crosslinks, may result in greater toxicity and induce apoptosis by virtue of the intermediates formed during repair of these lesions in G_1-phase compared to S-phase. The ability of BG to inhibit CDKs and enhance the cell killing by crosslinks in this way may also explain the observation that BG is able to increase the effectiveness of BCNU, even in tumor xenografts that apparently do not express AGT *(168)*.

8.2. Topoisomerase I Inhibitors

Recent studies have revealed an unexpected and previously unknown link between AGT activity and sensitivity to topoisomerase inhibitors such as irinotecan (CPT-11), a camptothecin derivative. Furthermore, in studies using xenografts, there are clear signs of a schedule-dependent synergistic antitumor activity between irinotecan and either temozolomide *(169,170)* or BCNU *(171)* and an enhancement of the activity of irinotecan and tamozolamide by BG *(172)*. A mechanistic explanation for these results is provided by the observation that there is a 10-fold increase in the formation of topoisomerase I (top1) cleavage complexes when m^6G is incorporated in oligonucleotides at the +1 position relative to a unique top1 cleavage site *(173)*. This top1 poisoning is the result of a decrease of the top1-mediated DNA religation as well as an increase in the enzyme cleavage step. (The enhanced cleavage is probably linked to a change in the hydrogen-bonding pattern, whereas inhibition of religation may be attributed to altered basepairing.) The top1–DNA covalent complexes were readily produced in MNNG-treated CHO cells constitutively lacking AGT but not in CHO cells transfected with the wild-type AGT. Similarly, yeast strains overexpressing human top1 are more sensitive to methylation damage, whereas strains with the top1 gene inactivated knockout top1 strain are resistant *(173,174)*.

These results suggest a role for top1 poisoning by alkylated bases in the antiproliferative activity of alkylating agents. Furthermore, AGT-repairable DNA lesions resulting from endogenous DNA modifications (probably m^6G) may impart sensitivity to top1 inhibitors. The finding that there is an inverse correlation between AGT activity and killing by irinotecan and related top1 inhibitors in the absence of any alkylating agent *(175,176)* supports this hypothesis. The recent observation that high AGT levels reduced, and that inhibition of AGT by either antisense or BG increased, killing by TAS-103, a top1/2 inhibitor is also consistent with this concept *(177)*. Treatment with BG to reduce AGT may, therefore, be useful in enhancing the response to top1 poisons even in the absence of treatment with an exogenous alkylating agent.

9. CONCLUSIONS

O^6-Alkylguanine-DNA alkyltransferase activity is clearly a significant factor causing resistance to therapeutic methylating and chloroethylating agents. The abolition of this activity by the use of BG is a promising means to improve the efficacy of these agents, but there are several obstacles to the success of this strategy. First, AGT inactivation is not likely to be helpful when combined with methylating agents in the treatment of tumors deficient in MMR and may be counterproductive because unrepaired m^{6}G in DNA is highly mutagenic. This limitation does not apply to chloroethylating agents. However, the success of the combination of chloroethylating agents with BG may be limited to situations in which one or the other of the agents can be administered regionally, preferably the alkylating agent because BG has little to no systemic toxicity when administered alone. The maximum tolerated dose of the alkylating agent is reduced significantly when combined with BG compared to that which can be given in the absence of BG. Because of the more limited acute toxic profile of methylating agents such as temozolomide, their combination with BG may be most likely to establish the proof of principle that AGT inactivation can enhance alkylating agent chemotherapy in the clinic.

Another serious consideration is the possibility of genetic damage leading to secondary tumors. Although there is no current animal or human data demonstrating an increase incidence in therapy-related leukemia, it is not unreasonable to suggest that a greater number of mutagenic lesions will be present in the absence of AGT. For this reason, therapy with BG plus alkylating agents should be limited to tumors for which there is currently a very poor long-term prognosis.

Second-generation AGT inhibitors are under development. The only one of these, which is in clinical trials, is 4-BTG and it is unclear what advantage this might have over BG. The detailed knowledge of the metabolism and pharmacokinetics of BG (which shows that much of its activity may be the result of metabolism to the active and longer lasting 8-oxoBG) is not yet available for 4-BTG, which could be subject to other metabolic reactions, particularly involving the bromine substituent, that are not favorable. Other second-generation compounds have not yet entered development but do hold the promise of being more adaptable to modifications that could lead to tumor specificity and/or the ability to inactivate BG-resistant forms of AGT.

The expression of BG-resistant forms of AGT in key tissues, such as bone marrow stem cells, that are the sites of dose-limiting toxicity with alkylating agents is an exciting approach to improving the therapeutic index of BG plus alkylating agent therapy. Suitable BG-resistant AGTs with point mutations are now available; the major unresolved issue, as with many other gene therapy approaches, is the vector for safe and efficient delivery. The facile selection in animal models of marrow cells which contain such BG-resistant AGTs as a result of transduction with retroviral vectors is both an indication of the potential of this technique and a compelling demonstration of the myelosuppresive potential of the treatment with BG plus alkylation. It is worth noting that this selection could also be used to improve the degree of expression of an additional gene of interest, which could be inserted into the same vector.

It is firmly established that many cultured tumor cell lines exhibit the Mer$^-$ phenotype and do not express AGT. Methylation of the promoter region of the AGT gene may not be the initiator, but it is definitively implicated in this lack of expression. Therefore, the rapidly increasing number of recent reports of hypermethylation of the AGT promoter in a variety of primary human tumors is of major interest particularly because such hypermethylation has been linked to sensitivity to alkylating agent therapy. Because a low level of AGT expression in brain tumors is clearly linked in both retrospective and prospective studies to sensitivity to therapy with BCNU, it is tempting to extend this conclusion to tumors with AGT-promoter hypermethylation. However, the link may not be a direct one because it has not been proven that these tumors have reduced AGT levels.

The possibility that AGT and/or BG treatment may influence responsiveness to top1 inhibitors and to other alkylating agents is intriguing and worthy of further investigation, but it is not yet established whether this provides additional therapeutic opportunities. The ability of m^6G to increase the formation of top1 cleavage complexes provides a possible mechanistic explanation for the current experimental results with top1 inhibitors if there is a sufficient endogenous formation of methylation adducts. A low dose of a methylating agent plus AGT inactivation by BG may be a useful means of increasing the activity of irinotecan and related compounds.

Exposure to BG may increase tumor cell killing by cyclophosphamide by multiple mechanisms of which AGT inactivation leading to slower repair of DNA lesions may be only a minor factor. Effects on the cell-cycle influencing repair of lesions that are neither repairable by AGT nor generated by secondary reactions from AGT-repairable lesions may be involved in the sensitivity to cyclophosphamide. It is possible that such effects may be an exploitable facet of the pharmacology of BG and related compounds. They may also contribute to the therapeutic effects of combinations of BG with alkylating agents in which AGT inactivation is more clearly involved such as BCNU. Further laboratory and clinical investigations are needed to follow up on these interesting possibilities.

ACKNOWLEDGMENTS

Research on this topic in the authors' laboratories is supported in part by grants CA-71976 (AEP), CA-71627 (MED), and CA-81485 (MED). This work was completed in February 2002.

The authors are among the holders of patents for the use of O^6-benzylguanine and related compounds to inactivate O^6-alkylguanine-DNA alkyltransferase and improve therapy with alkylating agents and, therefore, have a potential financial interest in the success of this therapy.

REFERENCES

1. Berthet N, Boturyn D, Constant J-F. DNA repair inhibitors. *Exp Opin Ther Patents* 1999;9:401–415.
2. Kaina B, Ochs K, Grösch S, et al. BER, MGMT, and MMR in defense against alkylation-induced genotoxicity and apoptosis. *Prog Nucleic Acid Res Mol Biol.*2001;68:41–54.

3. Dolan ME, Pegg AE. O^6-Benzylguanine and its role in chemotherapy. *Clin Cancer Res* 1997;3:837–847.

4. Dolan ME. Inhibition of DNA repair as a means of increasing the antitumor activity of DNA reactive agents. *Adv Drug Deliv Rev* 1997;26:105–118.

5. Margison GP, Hickson I, Jelinek J, et al. Resistance to alkylating agents: more or less. *Anti-Cancer Drugs* 1996;7:109–116.

6. Pieper RO. Understanding and manipulating O^6-methylguanine-DNA methyltransferase expression. *Pharmacol Ther* 1997;74:285–297.

7. Pegg AE, Dolan ME, Moschel RC. Structure, function and inhibition of O^6-alkylguanine-DNA alkyltransferase. *Prog Nucleic Acid Res Mol Biol.*1995;51:167–223.

8. Pegg AE. Repair of O^6-alkylguanine by alkyltransferases. *Mutat Res* 2000;462:83–100.

9. Pegg AE, Xu-Welliver M, Loktionova NA. The DNA repair protein O^6-alkylguanine-DNA alkyltransferase as a target for cancer chemotherapy. In: Ehrlich E, ed. *DNA Alterations in Cancer: Genetic and Epigenetic Shanges.* Eaton Publishing, Natick, MA, 2000:471–488.

10. Friedman HS, McLendon RE, Dolan ME, et al. The biology of sensitivity and resistance to the molecule (temozolomide). *Proc Am Assoc Cancer Res* 1999;40:754–755.

11. Friedman HS. Can O^6-alkylguanine-DNA alkyltransferase depletion enhance alkylator activity in the clinic? *Clin Cancer Res* 2000;6:2967–2968.

12. Daniels DS, Mol CD, Arval AS, et al. Active and alkylated human AGT structures: a novel zinc site, inhibitor and extrahelical binding. DNA damage reversal revealed by mutants and structures of active and alkylated human AGT. *EMBO J* 2000;19:1719–1730.

13. Srivenugopal KS, Yuan XH, Bigner DD, et al. Ubiquitination of O^6-alkylguanine-DNA alkyltransferase in human tumor cells following inactivation with O^6-benzylguanine or 1,3 bis(2-chloroethyl)-1-nitrosourea. *Biochemistry* 1996;35:1328–1334.

14. Xu-Welliver M, Pegg AE. Ubiquitin-mediated degradation of alkylated O^6-alkylguanine-DNA alkyltransferase. *Carcinogenesis* 2002;23:823–830.

15. Kanugula S, Pegg AE. A Novel DNA repair alkyltransferase from *Caenorhabditis elegans.* *Environ Mole Mutag* 2001;38:235–243.

16. Sekiguchi M, Sakumi K. Roles of DNA repair methyltransferase in mutagenesis and carcinogenesis. *Jpn J Hum Genet* 1997;42:389–399.

17. Moore MH, Gulbus JM, Dodson EJ, et al. Crystal structure of a suicidal DNA repair protein: the Ada O^6-methylguanine-DNA methyltransferase from *E. coli. EMBO J* 1994;13:1495–1501.

18. Hashimoto H, Inoue T, Nishioka M, et al. Hyperthermostable protein structure maintained by intra- and inter-helix ion pairs in archael O^6-methylguanine-DNA methyltransferase. *J Mol Biol* 1999;292:707–716.

19. Wibley JEA, Pegg AE, Moody PCE. Crystal structure of the human O^6-alkylguanine-DNA alkyltransferase. *Nucleic Acid Res* 2000;28:393–401.

20. Bignami M, O'Driscoll M, Aquilina G, et al. Unmasking a killer: DNA O^6-methylguanine and the cytotoxicity of methylating agents. *Mutat Res* 2000;462:71–82.

21. Kaina B. Critical steps in alkylation-induced aberration formation. *Mutat Res* 1998;404:119–124.

22. Buermeyer AB, Deschenes SM, Baker SM, et al. Mammalian DNA mismatch repair. *Annu Rev Genet* 2000;33:533–564.

23. Karran P. Mechanism of tolerance to DNA damaging therapeutic drugs. *Carcinogenesis* 2001;22:1931–1937.

24. Dosanjh MK, Galeros G, Goodman MF, et al. Kinetics of extension of O^6-methylguanine paired with cytosine or thymine in defined oligonucleotide sequences. *Biochemistry* 1991;30:11,595–11,599.

25. Voigt JM, Topal MD. O^6-Methylguanine-induced replication blocks. *Carcinogenesis* 1995;16:1775–1782.

26. Reha-Krantz LJ, Nonay RL, Day RS III, Wilson SL. Replication of O^6-methylguanine-containing DNA by repair and replicative DNA polymerases. *J Biol Chem* 1996;271: 20,088–20,095.

27. Haracska L, Prakash S, Prakash L. Replication past O^6-methylguanine by yeast and human DNA polymerase η. *Mol Cell Biol* 2000;20:8001–8007.

28. Christmann M, Kaina B. Nuclear translocation of mismatch repair proteins MSH2 and MSH6 as a response of cells to alkylating agents. *J Biol Chem* 2000;275:36256–36262.

29. Rasmussen LJ, Samson L. The *Escherichia coli* MutS DNA mismatch binding protein specifically binds O^6-methylguanine DNA lesions. *Carcinogenesis* 1996;17:2085–2088.

30. Meikrantz W, Bergom MA, Memisoglu A, et al. O^6-Alkylguanine DNA lesions trigger apoptosis. *Carcinogenesis* 1998;19:369–372.

31. Tentori L, Turriziani M, Franco D, et al. Treatment with temozolomide and poly(ADP-ribose) polymerase inhibitors induces early apoptosis and increases base excision repair gene transcripts in leukemic cells resistant to triazene compounds. *Leukemia* 1999;13:901–909.

32. Fritz G, Kaina B. Ras-related GTPase RhoB forces alkylation-induced apoptotic cell death. *Biochem Biophys Res Commun* 2000;268:784–789.

33. Li G-M. The role of mismatch repair in DNA damage-induced apoptosis. Oncol. Res. 2000;11:393–400.

34. D'Atri S, Tentori L, Lacal PM, et al. Involvement of the mismatch repair system in temozolomide-induced apoptosis. *Mol Pharmacol* 1998;54:334–341.

35. Tentori L, Lacal PM, Benincasa E, et al. Role of wild-type p53 on the antineoplastic activity of temozolomide alone or combined with inhibitors of poly(ADP-ribose) polymerase. *J Pharmacol Exp Ther* 1998;285:884–893.

36. Toft NJ, Winton DJ, Kelly J, et al. *Msh2* status modulates both apoptosis and mutation frequency in the murine small intestine. *Proc Natl Acad Sci USA* 1999;96:3911–3915.

37. Rasmussen LJ, Rasmussen M, Lützen A, et al. The human cyclin B1 repair protein modulates sensitivity of DNA mismatch repair deficient prostate cancer cell lines to alkylating agents. *Exper Cell Res* 2000;257:127–134.

38. Humbert O, Siumicino S, Aquilina A, et al. Mismatch repair and differential sensitivity of mouse and human cells to methylating agents. *Carcinogenesis* 1999;20:205–214.

39. Markowitz S. DNA repair defects inactivate tumor suppressor genes and induce hereditary and sporadic colon cancers. *J Clin Oncol* 2000;18:75s–80s.

40. Jiricny J, Nyström-Lahti M. Mismatch repair defects in cancer. *Curr Opin Gen Dev* 2000;10:157–161.

41. Peltomäki P. DNA mismatch repair and cancer. *Mutat Res* 2001;488:77–85.

42. Peták I, Mihalik R, Bauer PI, et al. BCNU is a caspase-mediated inhibitor of drug-induced apoptosis. *Cancer Res* 1998;58:614–618.

43. Schold SC, Kokkinakis DM, Rudy JL, et al. Treatment of human brain tumor xenografts with O^6-benzyl-2'-deoxyguanosine and BCNU. *Cancer Res* 1996;56:2076–2081.

44. Aquilina G, Ceccotti S, Martinelli S, et al. *N*-(2-Chloroethyl)-*N'*-cyclohexyl-*N*-nitrosourea sensitivity in mismatch repair-defective human caells. *Cancer Res* 1998;58:135–141.

45. Ludlum DB. The chloroethylnitrosoureas: Sensitivity and resistance to cancer chemotherapy at the molecular level. *Cancer Invest* 1997;15:588–598.

46. Silber JR, Blank A, Bobola MS, et al. O^6-Methylguanine-DNA methyltransferase-deficient phenotype in human gliomas: frequency and time to tumor progression after alkylating agent-based chemotherapy. *Clin Cancer Res* 1999;5:807–814.

47. Belanich M, Pastor M, Randall T, et al. Retrospective study of the correlation between the DNA repair protein alkyltransferase and survival of brain tumor patients treated with carmustine. *Cancer Res* 1995;56:783–788.

48. Jaeckle KA, Eyre HJ, Townsend JJ, et al. Correlation of tumor O^6-methylguanine-DNA methyltransferase levels with survival of malignant astrocytoma patients treated with

bis-chloroethylnitrosourea: a Southwest Oncology Group study. *J Clin Oncol* 1998;16: 3310–3315.

49. Watts GS, Pieper RO, Costello JF, et al. Methylation of discrete regions of the O^6-methylguanine-DNA methyltransferase (MGMT) CpG island is associated with heterochromatinization of the MGMT transcription start site and silencing of the gene. *Mol Cell Biol* 1997;17:5612–5619.

50. Danam RP, Qian XC, Howell SR, et al. Methylation of selected CpGs in the human O^6-methylguanine-DNA methyltransferase promoter region as a marker of gene silencing. *Mol Carcinog* 1999;24:85–89.

51. Qian XC, Brent TP. Methylation hot spots in the 5' flanking region denote silencing of the O^6-methylguanine-DNA methyltransferase gene. *Cancer Res* 1997;57:3672–3677.

52. Esteller M, Hamilton SR, Burger PC, et al. Inactivation of the DNA repair gene O^6-methylguanine-DNA methyltransferase by promoter hypermethylation is a common event in primary neoplasia. *Cancer Res* 1999;59:793–797.

53. Esteller M, Toyota M, Sanchez-Cespedes M, et al. Inactivation of the DNA repair gene O^6-methylguanine-DNA methyltransferase by promoter hypermethylation is associated with G to A mutations in *K-ras* in colorectal tumorigenesis. *Cancer Res* 2000;60:2368–2371.

54. Esteller M, Risques RA, Toyota M, et al. Promoter hypermethylation of the DNA repair gene O^6-Methylguanine-DNA methyltransferase is associated with the presence of G:C to A:T transition mutations in *p53* in human colorectal tumorigenesis. *Cancer Res* 2001;61: 4689–4692.

55. Nakamura M, Watanabe T, Yonekawa Y, et al. Promoter methylation of the DNA repair gene MGMT in astrocytomas is frequently associated with G:C A:T mutation of the *TP53* tumor suppressor gene. *Carcinogenesis* 2001;10:1715–1719.

56. Wolf P, Hu YC, Doffek K, et al. O^6-Methylguanine-DNA methyltransferase promoter hypermethylation shifts the *p53* mutational spectrum in non-small cell lung cancer. *Cancer Res* 2001;61:8113–8117.

57. Esteller M, Garcia-Foncillas J, Andion E, et al. Inactivation of the DNA-repair gene *MGMT* and the clinical response of gliomas to alkylating agents. *N Engl J Med* 2000;343: 1350–1354.

58. Esteller M, Gaidano G, Goodman SN, et al. Hypermethylation of the DNA repair gene O^6-methylguanine DNA methyltransferase and survival of patients with diffuse large B-cell lmyphoma. *J Natl Cancer Inst* 2002;94:26–32.

59. Potter PM, Harris LC, Remack JS, et al. Ribozyme-mediated modulation of human O^6-methylguanine-DNA methyltransferase expression. *Cancer Res* 1993;53:1731–1734.

60. Citti L, Eckstein F, Capecchi B, et al. Transient transfection of a synthetic hammerhead ribozyme targetted against human MGMT gene to cells in culture potentiates the genotoxicity of the alkylation damage induced by mitozolomide. *Antisense Nucleic Acid Drug Dev* 1999;9:125–133.

61. Mariani L, Citti L, Nevischi S, et al. Ribozyme and free alkylated base: a dual approach for sensitizing Mex$^+$ cells to the alkylating antineoplastic drug. Cancer Gene Ther. 2000;7:905–909.

62. Zhang Q, Ohannesian DW, Kreklau EL, Erickson LC. Modulation of 1,3-bis-(2-chloroethyl)-1-nitrosourea resistance in human tumor cells using hammerhead ribozymes designed to degrade O^6-methylguanine DNA methyltransferase mRNA. *J Pharmacol Exp Ther* 2001;298:141–147.

63. Pegg AE, Boosalis M, Samson L, et al. Mechanism of inactivation of human O^6-alkylguanine-DNA alkyltransferase by O^6-benzylguanine. *Biochemistry* 1993;32:11,998–12,006.

64. Dolan ME, Moschel RC, Pegg AE. Depletion of mammalian O^6-alkylguanine-DNA alkyltransferase activity by O^6-benzylguanine provides a means to evaluate the role of this protein in protection against carcinogenic and therapeutic alkylating agents. *Proc Natl Acad Sci USA* 1990;87:5368–5372.

65. Pegg AE, Swenn K, Dolan ME, et al. Increased killing of prostate, breast, colon and lung tumor cells by the combination of inactivators of O^6-alkylguanine-DNA alkyltransferase and N,N-bis(2-chloroethyl)-N-nitrosourea. *Biochem Pharmacol* 1995;50:1141–1148.
66. Pegg AE, Wiest L, Mummert C, et al. Use of antibodies to human O^6-alkylguanine-DNA alkyltransferase to study the content of this protein in cells treated with O^6-benzylguanine or N-methyl-N'-nitro-N-nitrosoguanidine. *Carcinogenesis* 1991;12:1679–1683.
67. Dolan ME, Mitchell RB, Mummert C, et al. Effect of O^6-benzylguanine analogues on sensitivity of human tumor cells to the cytotoxic effects of alkylating agents. *Cancer Res* 1991;51:3367–3372.
68. Baer JC, Freeman AA, Newlands ES, et al. Depletion of O^6-alkylguanine-DNA alkyltransferase correlates with potentiation of temozolomide and CCNU toxicity in human tumour cells. *Br J Cancer* 1993;67:1299–1302.
69. Wedge SR, Porteous JK, May BL, et al. Potentiation of temozolomide and BCNU cytotoxicity by O^6-benzylguanine: a comparative study in vitro. *Br J Cancer* 1996;73:482–490.
70. Wedge SR, Newlands ES. O^6-Benzylguanine enhances the sensitivity of a glioma xenograft with low O^6-alkylguanine-DNA alkyltransferase activity to temozolomide and BCNU. *Br J Cancer* 1996;73:1049–1052.
71. Bobola MS, Tseng SH, Blank A, et al. Role of O^6-methylguanine-DNA methyltransferase in resistance of human brain tumor cell lines to the clinically relevant methylating agents temozolomide and streptozotocin. *Clin Cancer Res* 1996;2:735–741.
72. Liu L, Taverna P, Whitacre CM, et al. Pharmacologic disruption of base excision repair sensitizes mismatch repair-deficient and -proficient colon cancer cells to methylating agents. *Clin Cancer Res* 1999;5:2908–2917.
73. Dolan ME, Pegg AE, Biser ND, et al. Effect of O^6-benzylguanine on the response to 1,3-bis(2-chloroethyl)-1-nitrosourea in the Dunning R3327G model of prostatic cancer. *Cancer Chemother Pharmacol* 1993;32:221–225.
74. Felker GM, Friedman HS, Dolan ME, et al. Treatment of subcutaneous and intracranial brain tumor xenografts with O^6-benzylguanine and 1,3-bis(2-chloroethyl)-1-nitrosourea. *Cancer Chemother Pharmacol* 1993;32:471–476.
75. Friedman HS, Dolan ME, Moschel RC, et al. Enhancement of nitrosourea activity in medulloblastoma and glioblastoma multiforme. *J Natl Cancer Inst* 1992;84:1926–1931.
76. Mitchell RB, Moschel RC, Dolan ME. Effect of O^6-benzylguanine on the sensitivity of human tumor xenografts to 1,3-bis(2-chloroethyl)-1-nitrosourea and on DNA interstrand cross-link formation. *Cancer Res* 1992;52:1171–1175.
77. Gerson SL, Zborowska E, Norton K, et al. Synergistic efficacy of O^6-benzylguanine and BCNU in a human colon cancer xenograft completely resistant to BCNU alone. *Biochem Pharmacol* 1993;45:483–491.
78. Rhines LD, Sampath P, Dolan ME, et al. O^6-Benzylguanine potentiates the antitumor effect of locally delivered carmustine against an intracranial rat glioma. *Cancer Res* 2000;60:6307–6310.
79. Kreklau EL, Liu N, Li Z, et al. Comparison of single- versus double-bolus treatments of O^6-benzylguanine for depletion of O^6-methylguanine DNA methyltransferase (MGMT) activity in vivo: development of a novel fluorometric oligonucleotide assay for measurement of MGMT activity. *J Pharmacol Exp Ther* 2001;297:524–530.
80. Friedman HS, Dolan ME, Pegg AE, et al. Activity of temozolomide in the treatment of central nervous system tumor xenografts. *Cancer Res* 1995;55:2853–2857.
81. Kokkinakis DM, Bocangel DB, Schold SC, et al. Thresholds of O^6-alkylguanine-DNA alkyltransferase which confer significant resistance of human glial tumor xenografts to treatment with 1,3-bis(2-chloroethyl)-1-nitrosourea or temozolomide. *Clin Cancer Res* 2001;7:421–428.

82. Wedge SR, Porteous JK, Newlands ES. Effect of single and multiple administration of an O^6-benzylguanine/temozolomide combination: an evaluation in a human melanoma xenograft model. *Cancer Chemother Pharmacol* 1997;40:266–272.

83. Toft NJ, Sansom OJ, Brookes RA, et al. In vivo administration of O^6-benzylguanine does not influence apoptosis or mutation frequency following DNA damage in the murine intestine, but does inhibit P450-dependent activation of dacarbazine. *Carcinogenesis* 2000;21: 593–598.

84. Long L, Moschel RC, Dolan ME. Debenzylation of O^6-benzyl-8-oxoguanine in human liver: implications for O^6-benzylguanine metabolism. *Biochem Pharmacol* 2001;61:721–726.

85. Chinnasamy N, Rafferty JA, Hickson I, et al. O^6-Benzylguanine potentiates the *in vivo* toxicity and clastogenicity of temozolomide and BCNU in mouse bone marrow. *Blood* 1997;89:1566–1573.

86. McElhinney RS, Donnelly DJ, McCormick JE, et al. Inactivation of O^6-alkylguanine-DNA alkyltransferase 1. Novel O^6-(hetarylmethyl)guanines having basic rings in the side chain. *J Med Chem* 1998;41:5265–5271.

87. Middleton MR, Ranson M, Bowen R, et al. Phase I study of 4-bromothenylguanine (4-BTG) and temozolomide: preliminary results and O^6-alkylguanine-DNA alkyltransferase (ATASE) depletion. *Proc Am Assoc Cancer Res* 2000;41:3878.

88. Middleton MR, Kelly J, Thatcher N, et al. O^6-(4-Bromothenyl)guanine improves the therapeutic index of temozolomide against A375M melanoma xenografts. *Int J Cancer* 2000;85:248–252.

89. Reinhard J, Eichorn U, Wiessler M, Kaina B. Inactivation of O^6-methylguanine-DNA methyltransferase by glucose-conjugated inhibitors. *Int J Cancer* 2001;93:373–379.

90. Reinhard J, Hull WE, von der Lieth CW, et al. Monosaccharide-linked inhibitors of O^6-methylguanine-DNA methyltransferase (MGMT): synthesis, molecular modeling, and structure-activity relationships. *J Med Chem* 2001;44:4050–4061.

91. Kokkinakis DM, Moschel RC, Pegg AE, et al. Eradication of brain tumor xenografts with a combination of O^6-benzyl-2'-deoxyguanosine and 1,3-bis(2-chloroethyl)1- nitrosourea. *Clin Cancer Res* 1999;5:3676–3681.

92. Kokkinakis DM, Moschel RC, Pegg AE, et al. Potentiation of BCNU antitumor efficacy by 9-substituted O^6-benzylguanines. Effect of metabolism. *Cancer Chemother Pharmacol* 2000;45:69–77.

93. Chae M-Y, Swenn K, Kanugula S, et al. 8-Substituted O^6-benzylguanine, substituted 6(4)-benzyloxypyrimidine and related derivatives as inactivators of O^6-alkylguanine-DNA alkyltransferase. *J Med Chem* 1995;38:359–365.

94. Ewesuedo RB, Wilson LR, Friedman HS, et al. Inactivation of O^6-alkylguanine-DNA alkyltransferase by 8-substituted O^6-benzylguanine analogs in mice. *Cancer Chemother Pharmacol* 2001;47:63–69.

95. Long L, Berg SL, Roy SK, et al. Plasma and cerebrospinal fluid pharmacokinetics of O^6-benzylguanine and analogues in nonhuman primates. *Clin Cancer Res* 2000;6:3662–3669.

96. Griffin RJ, Arris CE, Bleasdale C, et al. Resistance-modifying agents. 7. Inhibition of O^6-alkylguanine-DNA alkyltransferase by O^6-alkenyl-, O^6-cycloalkenyl- and O^6-(2-oxoalkyl)-guanines, and potentiation of temozolomide cytotoxicity in vitro by O^6-(1-cyclopentenylmethyl)guanine. *J Med Chem* 2000;43:4071–4083.

97. Roy SK, Moschel RC, Dolan ME. Pharmacokinetics and metabolism in rats of 2,4-diamino-6-benzyloxy-5-nitrosopyrmidine, an inactivator of O^6-alkylguanine-DNA alkyltransferase. *Drug Metab Dispos* 1996;24:1205–1211.

98. Terashima I, Kohda K. Inhibition of human O^6-alkylguanine-DNA alkyltransferase and potentiation of the cytotoxicity of chloroethylnitrosourea by 4(6)-(benzyloxy)-2,6(4)-diamino-5-(nitro or nitroso)pyrimidine derivatives and analogues. *J Med Chem* 1998;41: 503–508.

99. Pegg AE, Kanugula S, Edara S, et al. Reaction of O^6-benzylguanine-resistant mutants of human O^6-alkylguanine-DNA alkyltransferase with O^6-benzylguanine in oligodeoxyribonucleotides. *J Biol Chem* 1998;273:10863–10867.

100. Pegg AE, Goodtzova K, Loktionova NA, et al. Inactivation of human O^6-alkylguanine-DNA alkyltransferase by modified oligodeoxyribonucleotides containing O^6-benzylguanine. *J Pharm Exp Ther* 2001;296:958–965.

101. Luu KX, Kanugula S, Pegg AE, et al. Repair of oligodeoxyribonucleotides by O^6-alkylguanine-DNA alkyltransferase. *Biochemistry* 2002;27:8689–8697.

102. Kanugula S, Liu K, Pegg AE, et al. Oligodeoxyribonucleotides containing O^6-benzylguanine (b6G) are potent inactivators of wild type and b6G-resistant mutant forms of human O^6-alkylguanine-DNA alkyltransferase. *Proc Am Assoc Cancer Res* 2000;41:360.

103. Friedman HS, Kokkinakis DM, Pluda J, et al. Phase 1 trial of O^6-benzylguanine for patients undergoing surgery for malignant glioma. *J Clin Oncol* 1998;16:3570–3575.

104. Friedman HS, Pluda J, Quinn JA, et al. Phase I trial of carmustine plus O^6-benzylguanine for patients with recurrent or progressive malignant glioma. *J Clin Oncol* 2000;18:3522–3528.

105. Spiro TP, Gerson SL, Liu L, et al. O^6-Benzylguanine: a clinical trial establishing the biochemical modulatory dose in tumor tissue for alkyltransferase-directed DNA repair. *Cancer Res* 1999;59:2402–2410.

106. Schilsky RL, Dolan ME, Bertucci D, et al. Phase I clinical and pharmacological study of O^6-benzylguanine followed by carmustine in patients with advanced cancer. *Clin Cancer Res* 2000;6:3025–3031.

107. Liu L, Spiro TP, Qin X, et al. Differential degradation rates of inactivated alkyltransferase in blood mononuclear cells and tumors of patients after treatment with O^6-benzylguanine. *Clin Cancer Res* 2001;7:2318–2324.

108. Friedman HS, McLendon RE, Kerby T, et al. DNA mismatch repair and O^6-alkylguanine-DNA alkyltransferase analysis and response to temodal in newly diagnosed malignant glioma. *J Clin Oncol* 1998;16:3851–3857.

108a. Dolan ME, Posner M, Karrison T, et al. Determination of the optimal modulatory dose of O^6-benzylguanine in patients with surgically respectable tumors. *Clin Cancer Res* 2002; 5:2519–2523.

109. Dolan ME, Roy SK, Fasanmade A, et al. O^6-Benzylguanine in humans: metabolic, pharmacokinetic and pharmacodynamic findings. *J Clin Oncol* 1998;16:1803–1810.

110. Roy SK, Korzekwa KR, Gonzalez FJ, et al. Human liver oxidative metabolism of O^6-benzylguanine. *Biochem Pharmacol* 1995;50:1385–1389.

111. Long L, Dolan ME. Role of cytochrome P450 isoenzymes in metabolism of O(6)-benzylguanine: implications for dacarbazine activation. *Clin Cancer Res* 2001;7: 4239–4244.

112. Rogers TS, Rodman LE, Tomaszewski, E. J, et al. Preclinical toxicology and pharmacokinetic studies of O^6-benzylguanine (NSC-637037) in mice and dogs. *Proc Am Assoc Cancer Res* 1994;35:1953.

113. Berg S, Gerson SL, Godwin K, et al. Plasma and cerebrospinal fluid pharmacokinetics of O^6-benzylguanine and time course of peripheral blood mononuclear cell O^6-methylguanine-DNA methyltransferase inhibition in the nonhuman primate. *Cancer Res* 1995;55: 4606–4610.

114. Dolan ME, Chae MY, Pegg AE, et al. Metabolism of O^6-benzylguanine, an inactivator of O^6-alkylguanine-DNA alkyltransferase. *Cancer Res* 1994;54:5123–5130.

115. Crone TM, Goodtzova K, Edara S, et al. Mutations in O^6-alkylguanine-DNA alkyltransferase imparting resistance to O^6-benzylguanine. *Cancer Res* 1994;54:6221–6227.

116. Xu-Welliver M, Kanugula S, Pegg AE. Isolation of human O^6-alkylguanine-DNA alkyltransferase mutants highly resistant to inactivation by O^6-benzylguanine. *Cancer Res* 1998;59:1936–1945.

117. Xu-Welliver M, Leitao J, Kanugula S, et al. Alteration of the conserved residue tyrosine-158 to histidine renders human O^6-alkylguanine-DNA alkyltransferase insensitive to the inhibitor O^6-benzylguanine. *Cancer Res* 1999;59:1514–1519.

118. Xu-Welliver M, Kanugula S, Loktionova NA, et al. Conserved residue lysine-165 is essential for the ability of O^6-alkylguanine-DNA alkyltransferase to react with O^6-benzylguanine. *Biochem J* 2000;347:527–534.

119. Xu-Welliver M, Pegg AE. Point mutations at multiple sites including highly conserved amino acids maintain activity but render O^6-alkylguanine-DNA alkyltransferase insensitive to O^6-benzylguanine. *Biochem J* 2000;347:519–526.

120. Encell LP, Loeb LA. Redesigning the substrate specificity of human O^6-alkylguanine-DNA alkyltransferase. Mutants with enhanced repair of O^4-methylthymine. *Biochemistry* 1999;38:12,097–12,103.

121. Christians FC, Dawson BJ, Coates MM, Loeb LA. Creation of human alkyltransferases resistant to O^6-benzylguanine. *Cancer Res* 1997;57:2007–2012.

122. Loktionova NA, Pegg AE. Interaction of mammalian O^6-alkylguanine-DNA alkyltransferases with O^6-benzylguanine. *Biochem Pharmacol* 2002;8:1431–1442.

123. Loktionova NA, Xu-Welliver M, Crone T, et al. Mutant forms of O^6-alkylguanine-DNA alkyltransferase protect CHO cells from killing by BCNU plus O^6-benzylguanine or O^6-8-oxo-benzylguanine. *Biochem Pharmacol* 1999;58:237–244.

124. Loktionova NA, Pegg AE. Point mutations in O^6-alkylguanine-DNA alkyltransferase prevent the sensitization by O^6-benzylguanine to killing by N,N'-bis(2-chloroethyl)-N-nitrosourea. *Cancer Res* 1996;56:1578–1583.

125. Maze R, Kurpad C, Pegg AE, et al. Retroviral-mediated expression of the P140A, but not P140A/G156A, mutant form of O^6-methylguanine-DNA methyltransferase protects hematopoietic cells against O^6-benzylguanine sensitization to chloroethylnitrosourea treatment. *J Pharm Exp Ther* 1999;290:1467–1474.

126. Liu L, Gerson SL. Upregulation of anti-apotosis gens and decreased G2 arrest contributes to the acquired resitance to BG and BCNU observed in MMR deficient cells harboring BG-resistant K165 mutant AGT. *Proc Am Cancer Res* 2001;42:3469.

127. Elder RH, Margison GP, Rafferty JA. Differential inactivation of mammalian and E. coli O^6-alkylguanine-DNA alkyltransferase by O^6-benzylguanine. *Biochem J* 1994;298:231–235.

128. Liu L, Lee K, Wasan E, et al. Differential sensitivity of human and mouse alkyltransferase to O^6-benzylguanine using a transgenic model. *Cancer Res* 1996;56:1880–1885.

129. Fairbairn LJ, Rafferty JA, Lashford LS. Engineering drug resistance in human cells. *Bone Marrow Transplant* 2000;25:S110–S113.

130. Roth RB, Samson LD. Gene transfer to suppress bone marrow alkylation sensitivity. *Mutat Res* 2000;462:107–120.

131. Jansen M, Bardenheuer W, Sorg UR, et al. Protection of hematopoietic cells from O^6-alkylation damage by O^6-methylguanine DNA methyltransferase gene transfer: studies with different O^6-alkylating agents and retroviral backbones. *Eur J Hematol* 2001;67:2–13.

132. Wu M, Kelley MR, Hansen WK, et al. Reduction of BCNU toxicity to lung cells by high level expression of O^6-methylguanine-DNA methyltransferase. *Am J Physiol Lung Cell Mol Physiol* 2001;280:L755-L761.

133. Gerson SL. Drug resistance gene transfer: stem cell protection and therapeutic efficacy. *Exp Hematol* 2000;28:1315–1324.

134. Maze R, Kapur R, Kelley MR, et al. Reversal of 1,3-bis(2-chloroethyl)-1-nitrosourea-induced severe immunodeficiency by transduction of murine long-lived hemopoietic progenitor cells using O^6-methylguanine DNA methyltransferase complementary DNA. *J Immunol* 1997;158:1006–1013.

135. Kleibl K, Margison GP. Increasing DNA repair capacity in bone marrow by gene transfer as a prospective tool in cancer therapy. *Neoplasma* 1998;45:181–186.

136. Suzuki M, Sugimoto Y, Tsuruo T. Efficient protection of cells from the genotoxicity of nitrosoureas by the retrovirus-mediated transfer of human O^6-methylguanine-DNA methyltransferase using bicistronic vectors with human multidrug resistance gene 1. *Mutat Res* 1998;401:133–141.

137. Davis BM, Koc ON, Gerson SL. Limiting numbers of G156A O^6-methylguanine-DNA methyltransferase-transduced marrow progenitors repopulate nonmyeloablated mice after drug selection. *Blood* 2000;95:3078–3084.

138. Ragg S, Xu-Welliver M, Bailey J, et al. Direct reversal of DNA damage by mutant methyltransferase protein protects mice against dose-intensified chemotherapy and leads to *in vivo* selection of hematopoietic stem cells. *Cancer Res* 2000;60: 5187–5195.

139. Williams DA, Maze R, Kurpad C, et al. Protection of hematopoietic cells against combined O^6-benzylguanine and chloroethylnitrosourea treatment by mutant forms of O^6-methylguanine DNA methyltransferase. *Bone Marrow Transplant* 2000;25:S105-S109.

140. Lee K, Gerson SL, Maitra B, et al. G156A MGMT-transduced human mesenchymal stem cells can be selectively enriched by O^6-benzylguanine and BCNU. *J Hematother Stem Cell Res* 2001;10:691–701.

141. Sawai N, Zhou S, Vanin EF, et al. Protection and *in vivo* selection of hematopoietic stem cells using temozolomide, O^6-benzylguanine, and an alkyltransferase-expressing retroviral vector. *Mol Ther* 2001;3:78–87.

142. Chinnasamy N, Rafferty JA, Hickson I, et al. Chemoprotective gene transfer II: multilineage in vivo protection of haemopoiesis against the effects of an antitumour agent by expression of a mutant human O^6-alkylguanine-DNA alkyltransferase. *Gene Ther* 1998;5:842–847.

143. Koc ON, Reese JS, Davis BM, et al. DMGMT transduced bone marrow infusion increases tolerance to O^6-benzylguanine & BCNU and allows intensive therapy of BCNU resistant human colon cancer xenografts. *Hum Gene Ther* 1999;10:1021–1030.

144. Reese JS, Davis BM, Liu L, et al. Simultaneous protection of G156A methylguanine DNA methyltransferase gene-transduced hematopoietic progenitors and sensitization of tumor cells using O^6-benzylguanine and temozolomide. *Clin Cancer Res* 1999;5:163–169.

145. Hansen WK, Deutsch WA, Yacoub A, et al. Creation of a fully functional human chimeric DNA repair protein. *J Biol Chem* 1998;273:756–762.

146. Otsuka M, Abe M, Nakabeppu Y, et al. Polymorphism in the human O^6-methylguanine-DNA methyltransferase gene detected by PCR-SSCP analysis. *Pharmacogenetics* 1996;6:361–363.

147. Abe M, Inoue R, Suzuki T. A convenient method for genotyping of human O^6-methylguanine-DNA methyltransferase polymorphism. *Jpn J Hum Genet* 1997;42:425–428.

148. Deng C, Capasso H, Zhao Y, et al. Genetic polymorphism of human O^6-alkylguanine-DNA alkyltransferase: identification of a missense variation in the active site region. *Pharmacogenetics* 1999;9:81–87.

149. Imai Y, Oda H, Nakatsuru Y, et al. A polymorphism at codon 160 of human O^6-methylguanine-DNA methyltransferase gene in young patients with adult type cancers and functional assay. *Carcinogenesis* 1995;16:2441–2445.

150. Kaur TB, Travaline JM, Gaughan JP, et al. Role of polymorphisms in codons 143 and 160 of the O^6-alkylguanine DNA alkyltransferase gene in lung cancer risk. *Cancer Epidemiol Biomarkers Prev* 2000;9:339–342.

151. Inoue R, Abe M, Nakabeppu Y, et al. Characterization of human polymorphic DNA repair methyltransferases. *Pharmacogenetics* 2000;10:59–66.

152. Hazra TK, Roy R, Biswas T, et al. Specific recognition of O^6-methylguanine in DNA by active site mutants of human O^6-methylguanine-DNA methyltransferase. *Biochemistry* 1997;36:5769–5776.

153. Edara S, Kanugula S, Goodtzova K, et al. Resistance of the human O^6-alkylguanine-DNA alkyltransferase containing arginine at codon 160 to inactivation by O^6-benzylguanine. *Cancer Res* 1996;56:5571–5575.

154. Wu MH, Lohrbach KE, Olopade OI, et al. Lack of evidence for a polymorphism at codon 160 of human O^6-alkylguanine-DNA alkyltransferase gene in normal tissue and cancer. *Clin Cancer Res* 1999;55:209–213.

155. Gerson SL, Schupp J, Liu L, et al. Leukocyte O^6-alkylguanine-DNA alkyltransferase from human donors is uniformly sensitive to O^6-benzylguanine. *Clin Cancer Res* 1999;5:521–524.

156. Colvin OM. An overview of cyclophosphamide development and clinical application. *Curr Pharm Des* 1999;5:555–560.

157. Dong Q, Barsky D, Colvin ME, et al. A structural basis for a phosphoramide mustard-induced DNA interstrand cross-link at 5'-d(GAC). *Proc Natl Acad Sci USA* 1995;92: 12,170–12,174.

158. Friedman HS, Pegg AE, Johnson SP, et al. Modulation of cyclophosphamide activity by O^6-alkylguanine-DNA alkyltransferase. *Cancer Chemother Pharmacol* 1999;43:80–85.

159. Cai Y, Wu H, Ludeman SM, et al. Role of O^6-alkylguanine-DNA alkyltransferase in protecting against cyclophosphamide-induced toxicity and mutagenicity. *Cancer Res* 1999;59:3059–3063.

160. Chen SS, Citron M, Spiegel G, et al. O^6-methylguanine-DNA methyltransferase in ovarian malignancy and its correlation with postoperative response to chemotherapy. *Gynecol Oncol* 1994;52:172–174.

161. Hengstler JG, Tanner B, Möller L, et al. Activity of O^6-methylguanine-DNA methyltransferase in relation to p53 status and therapeutic response in ovarian cancer. *Int J Cancer* 1999;84:388–395.

162. Dolan ME, Schilsky RL. Silence is golden: Gene hypermethylation and survival in large-cell lymphoma. *J Natl Cancer Inst* 2002;94:6–7.

163. Mattern J, Eichhorn U, Kaina B, et al. O^6-Methylguanine-DNA methyltransferease activity and sensitivity to cyclophosphamide and cisplatin in human lung tumor xenografts. *Int J Cancer* 1998;77:919–922.

164. D'Incalci M, Bonfanti M, Pifferi A, et al. The antitumour activity of alkylating agents is not correlated with the levels of glutathione, glutathione transferase and O^6-alkylguanine-DNA alkyltransferase of human tumour xenografts. *Eur J Cancer* 1998;34:1749–1755.

165. Cai Y, Ludeman SM, Wilson LR, et al. Effect of O^6-benzylguanine on nitrogen mustard-induced toxicity, apoptosis, and mutagenicity in Chinese hamster ovary cells. *Mol Cancer Thera* 2001;1:21–28.

166. Cai Y, Wu M, Xu-Welliver M, et al. Effect of O^6-benzylguanine on alkylating agent-induced toxicity and mutagenicity in CHO cells expressing wild type and mutant O^6-alkylguanine-DNA alkyltransferase. *Cancer Res* 2000;60:5464–5669.

167. Arris CE, Boyle FT, Calvert AH, et al. Identification of novel purine and pyrimidine cyclin-dependent kinase inhibitors with distinct molecular interactions and tumor cell growth inhibition profiles. *J Med Chem* 2000;43:2797–2804.

168. Keir ST, Dolan ME, Pegg AE, et al. O^6-Benzylguanine-mediated enhancement of nitrosourea activity in Mer minus central nervous system tumor xenografts- implications for clinical trials. *Cancer Chemother Pharmacol* 2000;45:437–440.

169. Houghton PJ, Stewart CF, Cheshire PJ, et al. Antituor activity of temozolomide combined with irinotecan is partly independent of O^6-methylguanine-DNA methyltransferase and mismatch repair phenotypes in xenograft models. *Clin Cancer Res* 2000;6:4110–4118.

170. Patel VJ, Elion GB, Houghton PJ, et al. Schedule-dependent activity of temozolomide plus CPT-11 against a human central nervous system tumor-derived xenograft. *Clin Cancer Res* 2000;6:4154–4157.

171. Friedman HS, Castellino RC, Elion GB, et al. Schedule-dependent activity of irinotecan plus BCNU against malignant glioma xenografts. *Cancer Chemother Pharmacol* 2000;45:345–349.

172. Friedman HS, Keir S, Pegg AE, et al. O^6-Benzylguanine-mediated enhancement of chemotherapy. *Mol Cancer Ther* 2002;11:943–948.

173. Pourquier P, Waltman JL, Urasaki Y, et al. Topoisomerase I-mediated cytoxicity of *N*-methyl-*N*'-nitro-*N*-nitrosoguanidine: trapping of topoisomerase I by the O^6-methylguanine. *Cancer Res* 2001;61:53–58.

174. Nitiss JL, Nitiss KC, Rose A, et al. Overexpression of type I topoisomerases sensitizes yeast cells to DNA damage. *J Biol Chem* 2001;276:26708–26714.

175. Sekikawa T, Takano H, Okamuro T, et al. O^6-Methylguanine-DNA methyltransferase is a critical determinant of cytotoxicity for DNA topoisomerase I inhibitors. *Proc Am Assoc Cancer Res* 2000;41:Abstract 1145.

176. Okamoto R, Takano H, Sekikawa T, et al. Unique action determinants of double acting topoisomerase inhibitor, TAS-103. *Int J Oncol* 2001;19:921–927.

177. Okamoto R, Takano H, Okamura T, et al. O^6-Methylguanine-DNA methyltransferase (MGMT) as a determinant of resistance to camptothecin derivatives. *Jpn J Cancer* 2002;93:93–102.

8   Cellular Protection Against the Antitumor Drug Bleomycin

Dindial Ramotar, PhD, Huijie Wang, PhD, and Chaunhua He, PhD

CONTENTS

1. INTRODUCTION

Bleomycin is a basic hydrosoluble antibiotic originally isolated as a copper complex from the culture medium of *Streptomyces verticillis (1,2)*. Bleomycin comprises a family of 11 isomers differing only in the terminal amine moiety (*see* Fig. 1), and the most abundant form is bleomycin-A2 *(2–5)*. By the late sixties, substantial evidence had accumulated showing that bleomycin can diminish the growth of experimentally induced tumors in mice and rats and dramatically decrease the size of human tumors *(6–10)*. It has been postulated that bleomycin mediates the cell killing by directly attacking the DNA *(11,12)*. This notion rapidly gained support from subsequent independent studies showing that bleomycin triggers the induction of lysogenic phage in bacteria, a result of DNA damage, and induces mitotic recombination and mutations in many model systems, including the budding yeast *Saccharomyces cerevisiae, Aspergillus*, and

From: *Cancer Drug Discovery and Development: DNA Repair in Cancer Therapy*
Edited by: L. C. Panasci and M. A. Alaoui-Jamali © Humana Press Inc., Totowa, NJ

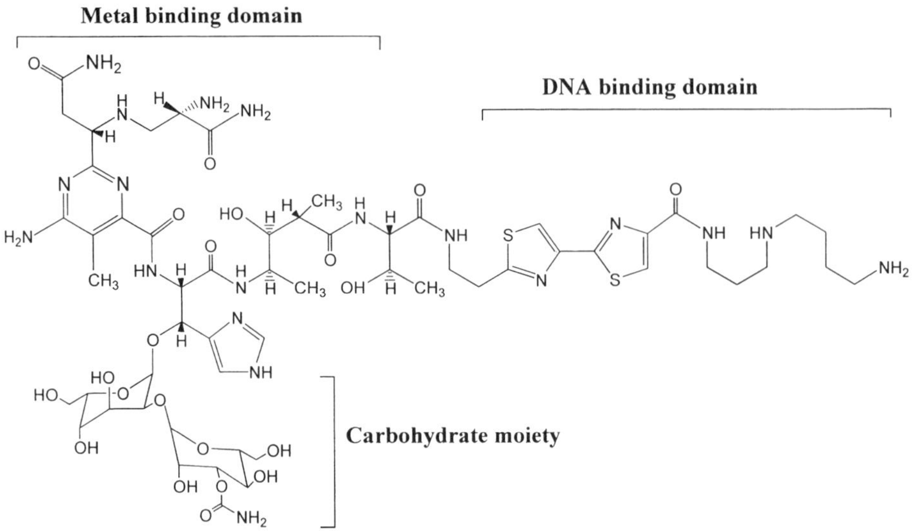

Fig. 1. Structure of the antitumor drug bleomycin depicting the three domains.

Drosophila (13–18). Later studies also showed that bleomycin can induce micronuclei formation and chromosome aberrations in human lymphocytes *(19)*. The accumulated findings strongly suggest that bleomycin may mediate its effect as a chemotherapeutic agent by mutating the DNA *(20–23)*. However, more recent studies showed that RNA is also a target for bleomycin, raising debate about the actual therapeutic cellular target (i.e., DNA vs RNA) *(24)*.

Bleomycin is routinely used in the clinic in many parts of the world as a mixture (blenoxane), consisting primarily of the isomers bleomycin-A2 and bleomycin-B2 *(4)*. It is used only in combination therapy with a number of other antineoplastic agents *(4,25,26)*. Bleomycin is most effective against lymphomas, testicular carcinomas, and squamous cell carcinomas of the cervix, head, and neck *(27,28)*. One useful property of bleomycin is that it does not appear to cause myelosuppression, as compared to other cytotoxic antineoplastic drugs *(4,28,29)*. Moreover, bleomycin is eliminated rapidly from the circulatory system by renal excretion. At least half of the drug is cleared from the blood within 2–4 h, except for patients with impaired renal function *(29)*. Like many other antitumor drugs, bleomycin also manifests clinical limitations. For example, at high doses (i.e., >400 units or approx 235 mg), bleomycin can induce pulmonary fibrosis, a condition characterized as a diffuse disease of the lung parenchyma that can cause pulmonary insufficiency leading to fatal hypoxemia *(30,31)*. The exact mechanisms by which bleomycin induces pulmonary fibrosis is not known, but findings from several experimental animal models suggest that the onset of the disease is triggered by lipid peroxidation *(32,33)*. Another common factor that limits the clinical application of bleomycin is tumor resistance *(28)*. So far, there is no definitive mechanism(s) to explain the development of tumor resistance toward bleomycin, although possible ones could include (1) decreased drug uptake, (2) increased drug extrusion, (3) enhanced repair of bleomycin-induced DNA lesions, and (4) increased inactivation of bleomycin *(34–38)*. Clues regarding the actual bleomycin-resistance mechanism are likely to emanate from a better understanding of how the drug gains entry into tumor cells. Some reports have documented the presence of an uncharacterized receptor-like protein on the surface of mammalian cells, which might effect bleomycin transport *(39–41)*. The transport of the bleomycin–receptor complex proceeds via a receptor-mediated endocytosis, which leads to the accumulation of toxic bleomycin levels in the cell within a few hours at relevant pharmacological concentrations *(42;* Mir, personal communication). It is therefore logical that modification or inactivation of the bleomycin–receptor complex would likely provoke a resistant phenotype. The preliminary evidence that the bleomycin–receptor endocytosis process is gradual prompted a prediction that increasing the rate of bleomycin uptake will result in a more effective anticancer treatment *(43)*. A test of this hypothesis showed that electropermeabilization of cells greatly improves the uptake of bleomycin and increases the cell killing *(42,43)*. These studies also estimated that nearly 500 molecules of bleomycin are sufficient to trigger apoptotic cell death, which is preceded by G_2/M-phase cell cycle arrest in a manner analogous to mitotic death induced by ionizing radiation *(43–45)*.

2. STRUCTURE OF BLEOMYCIN

Bleomycin serves as an experimental model drug because its chemical structure and modes of action are reasonably well established. The structure of bleomycin comprises three functional domains, including the N-terminal metal-binding domain, the C-terminal DNA-binding domain, and the carbohydrate moiety (*see* Fig. 1). The N-terminal domain also binds to molecular oxygen, in addition to the minor groove of DNA. This domain is thus largely responsible for the antitumor activity of bleomycin. It has a flexible requirement for metal ions, as it is capable of binding to both redox-active transition metal ions such as iron and copper and non-redox-active metal ions such as zinc, cobalt, and cadmium *(46–51)*. The metal ion plays two roles in bleomycin-induced genotoxicity: One is to facilitate contact between bleomycin and DNA, and the other is to activate oxygen such that a reactive radical species is generated *(20,47,48,52–54)*. Among the metal ions, cobalt forms the most stable complex with bleomycin. Despite this, iron is the metal ion predominantly used in clinical preparations of bleomycin, as it enhances the production of DNA lesions *(54,55)*.

The function of the two other domains of bleomycin is not clearly established. The C-terminus, encompassing the bithiazole group, is also required to bind

DNA and to engage in sequence-selective DNA cleavage *(22,56,57)*. Blocking the C-terminal end of bleomycin does not alter its ability to cleave DNA *(57)*. The role of the carbohydrate moiety is far more elusive. Recent studies demonstrated that removal of the carbohydrate moiety from bleomycin does not alter the resulting deglycobleomycin ability to cleave DNA, excluding a role for this region in incising DNA *(58)*. No additional studies have been conducted with deglycobleomycin to examine whether it is capable of entering cells or causing cell killing.

To date, several modifications have been introduced to alter the structure of bleomycin, but so far none has led to a more potent antitumor activity without the ability to cause pulmonary fibrosis. Thus, enhancing the antitumor effect of bleomycin will likely rely on the alteration of cellular molecules that would allow, for example, increased uptake of bleomycin.

3. BLEOMYCIN-INDUCED DNA LESIONS

Bleomycin can enter and accumulate in the nucleus of mammalian cells, where it inflicts a narrow set of DNA lesions through a multistep process *(59)*. In the earliest events, bleomycin binds to reduce iron (Fe [II]) and molecular oxygen followed by its conversion into an activated form *(20,60)*. The activated bleomycin $(Blm-Fe[II]-O_2)$ complex then acts as an oxidant, abstracting a hydrogen atom from the 4'-carbon of deoxyribose to produce an unstable sugar carbon radical and a single-electron reduced form of activated bleomycin $(Blm-Fe[III]-OH\cdot)$, which can carry out multiple attacks on DNA *(61–64)*. The unstable sugar generated by activated bleomycin can be rearranged to generate at least four types of oxidative DNA lesion (*see* Fig. 2). These lesions are structurally and chemically related to some of the lesions produced by ionizing radiation, and include the following:

1. Oxidized (ketoaldehyde) apurinic/apyrimidinic (AP) sites, where the entire base is lost, resulting in no template information for DNA polymerase *(59,65)*.
2. DNA single-strand breaks where the 3' ends are terminated with a portion of the deoxyribose ring to form 3'-phosphoglycolate (3'-PG), which effectively blocks DNA synthesis *(59,65,66)*. The remaining portion of the fragmented sugar, left attached to the base, exists in the free-base propenal form and exhibits a high propensity to undergo secondary reactions to form a variety of base adducts *(65,67)*.
3. Pyrimidopurinone of deoxyguanosine (M_1G) is the most abundant base adduct produced when the malondialdehyde moiety of the propenal base reacts with guanine *(67)*. The M_1G lesion is also generated by aerobic metabolism and it is detected at levels of approx 5000 adducts/cell in normal human liver *(68–70)*. This lesion is mutagenic in bacterial test systems *(68–70)*.
4. Bistranded DNA lesions, which are produced at certain sequences, such as CGCC, when the Fe–bleomycin complex induces an AP site on one strand and

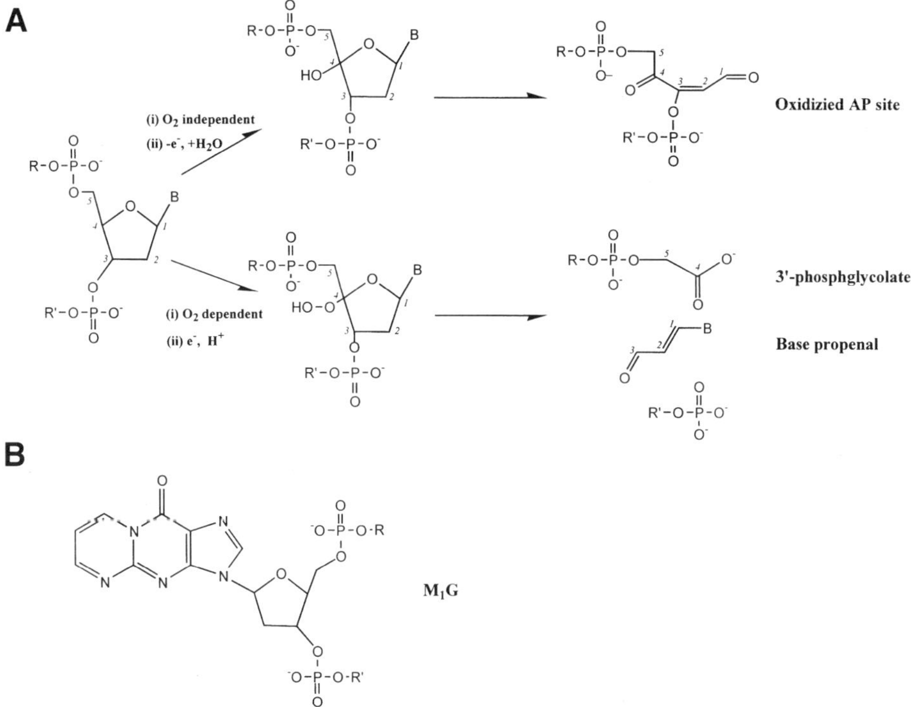

Fig. 2. Structure of bleomycin-induced DNA lesions. Production of the various types of bleomycin-induced lesions is dependent on oxygenation conditions. In the absence of oxygen, bleomycin produces a primarily oxidized apurinic/apyrimidinic (AP) site, whereas in the presence of oxygen, it generates mostly DNA strand breaks, such as 3'-phosphoglycolate. The M_1G lesion, 3-(2'-deoxy-β-D-erythro-pentofuranosyl)-pyrimido[1,2-a]purin-10(3H)-one is produced by reaction of the base propenal with deoxyguanosine.

a directly opposed strand break on the complementary strand *(51,71–73)*. This lesion requires a single activated bleomycin molecule, which binds to both strands of the duplex DNA *(51)*. The bistranded lesions can be converted to double-strand breaks following spontaneous cleavage of the AP site by primary amines (e.g., histone amine) in vivo *(71–73)*.

The extent of formation of the various bleomycin-induced lesions depends on the redox status of the cells (74–77). In the presence of oxygen, bleomycin produces primarily DNA strand breaks, but under low oxygen tension, it forms largely AP sites in the DNA *(59,65,71,78)*. Thus, the redox state of the cells is likely to dictate the types of DNA lesion that are generated by bleomycin. The types of DNA lesions are also influenced by bleomycin concentrations. At high concentrations, bleomycin releases all four bases from DNA in the order of preference thymine > cytosine > adenosine > guanine *(59,62,79)*. At lower concentrations, bleomycin exhibits significant base sequence specificity. Although bleomycin cuts mixed sequence DNAs with a disposition for GC = GT > GA >> GG, it efficiently cleaves regions of (AT)n·(TA)n and hardly at (ATT)n·(TTA)n, (ATT)n·(AAT)n, (AC)n·(GT)n, and (A)n·(T)n *(80–82)*. The structure of DNA also plays a role in the outcome of bleomycin-induced DNA lesions *(83)*. DNA that is pre-exposed to other DNA-damaging agents, such as cisplatin, alters the pattern of lesions produce by bleomycin *(84–86)*. Thus, the clinical application of bleomycin together with other DNA-damaging agents is likely to produce irreparable DNA lesions.

Several studies clearly demonstrate that bleomycin-induced DNA lesions are mutagenic *(87–91)*. For example, the introduction of bleomycin-treated vectors into mammalian cells, followed by recovery, revealed that the vectors contain high levels of base substitutions and single-base deletions *(87,88)*. The base substitutions are claimed to be the result of misincorporation of nucleotides by DNA polymerase at unrepaired oxidized AP sites, whereas the one-base deletions may arise from incorrect repair of bistranded DNA lesions *(87,88)*. Thus, the normal cells of a cancer patient exposed to bleomycin must rely on enzymes to efficiently repair bleomycin-induced DNA lesions to prevent the production of lethal mutations that can lead to toxic side effects and secondary tumors. Likewise, tumor cells are likely to employ even more efficient DNA repair mechanism to evade the genotoxic effects of bleomycin.

4. BLEOMYCIN-INDUCED RNA CLEAVAGE

Several reports demonstrated that bleomycin can also cleave many different RNAs, including human immunodeficiency virus (HIV)-1 reverse transcriptase mRNA, transfer RNAs, ribosomal RNA, and RNA present in the RNA·DNA heteroduplex *(92–97)*. Incision of RNA also occurs via an oxidative pathway reminiscent of the cleavage mechanism of DNA *(98,99)*. Furthermore, RNA cleavage occurs preferentially at 5'-GU-3' sequences reminiscent of the site-specific 5'-GT-3' incision observed in DNA *(92,98)*. In addition to these similarities, there are distinct differences between RNA and DNA with respect to cleavage with bleomycin. A notable difference is that not all RNA molecules (e.g., *E. coli* tRNA^{Tyr} and yeast mitochondrial tRNA^{Asp}) are cleaved by bleomycin *(92,98)*. This observation led to the suggestion that bleomycin-induced cleavage of RNA is structure-specific. Another key difference is that double-stranded RNA is not cleaved by bleomycin *(99)*. Moreover, the extent of RNA cleavage is significantly less that that of DNA *(92)*. Finally, the cleavage of RNA, but not DNA, is impeded with as low as 0.5 mM Mg^{2+} ions *(93)*. Because Mg^{2+} ions are required to maintain most RNA structure and function, it is postulated that the Mg^{2+} ions bind to RNA at the same exact site that coincides with binding of bleomycin *(93)*. The exceptional selectivity for destruction of certain RNAs, even with excess nonsubstrate RNAs, led to the suggestion that at least one unique RNA species could be targeted for destruction by bleomycin during chemotherapy. However, the following findings stand against RNA being a therapeutic target: (1) most RNA molecules exist in multiple copies and that destruction of a few molecules is unlikely to cause cell death, unless bleomycin is able to target a specific essential RNA species present in extremely low abundance, (2) RNA cleavage by bleomycin is inhibited by the physiological concentration of Mg^{2+} (2 mM), and (3) cleavage of RNA is structure-specific and occurs much slower than DNA *(92,93,98)*. As such, it can be inferred that DNA is the most likely target during bleomycin chemotheraphy.

5. OTHER CELLULAR TARGETS

In addition to DNA and RNA, bleomycin can also attack the integrity of the cell wall (a complex structure composed mainly of glucans, mannoproteins, and chitin) of microbes. At high doses, or under prolonged exposure, bleomycin can create small incisions in the cell wall, thereby exposing the protoplast *(100,101)*. The protoplast is osmotically fragile and this can lead to plasma membrane rupturing and cell death *(100,101)*. Because the sugar constituents of the cell wall have a stereochemistry at the C-5 position which is similar to the C-4 position of the deoxyribose moiety of DNA, it is believed that bleomycin destroys the cell wall via oxidative damage to the sugar *(101,102)*. Another relevant target affected by bleomycin is the plasma membrane, which is believed to undergo lipid peroxidation, and this may constitute the initiation process of bleomycin-induced pulmonary fibrosis *(32,103,104)*.

6. PROKARYOTIC AND EUKARYOTIC DEFENSE MECHANISMS AGAINST BLEOMYCIN TOXICITY

6.1. Cell Wall and Membrane Barriers

The highly reactive nature of bleomycin toward various cellular components, particularly DNA, suggests that organisms must employ multiple defense mechanisms to combat the deleterious effects of this drug. Some of these defense mechanisms include the barrier afforded by the cell wall and plasma membrane, proteins that bind and sequester bleomycin, and proteins that repair bleomycin-induced DNA lesions *(39,101,105)*. The contribution of each mechanism's protection against bleomycin toxicity is often determined by measuring the sensitivity of mutants to the drug. This type of analysis revealed that in the budding yeast *S. cerevisiae*, the cell wall appears to play a minor passive role in the protection against bleomycin toxicity. This is supported by the fact that some, not all, cell-wall-defective mutants displayed only a modest sensitivity toward bleomycin (Leduc and Ramotar, unpublished data).

Recent evidence suggests that the plasma membrane may play a more direct (active) role in limiting the entry of bleomycin into cells. Although the exact mechanism is not clear, a receptor protein appears to exist on the plasma membrane of mammalian and yeast cells that may mediate bleomycin internalization *(39,41)*. This putative receptor (approx 250 kDa in size) was initially identified by its specific interaction with a labeled cobalt–bleomycin complex *(39)*. It is reasonable to speculate that the presence of a plasma membrane receptor might provide a rational explanation for why certain tumor types, and not others, can be reduced by bleomycin chemotherapy. Aside from a potential bleomycin–receptor, it is equally plausible that the plasma membrane may harbor a specific efflux pump to limit bleomycin uptake. However, the fact that bleomycin can enter and accumulate in cells excludes the presence of an active plasma membrane efflux pump (Wang and Ramotar, unpublished). In fact, there is no direct evidence that any of the known drug efflux pumps has a role in expelling bleomycin from cells as part of a detoxification process.

6.2. Bleomycin-Binding Proteins

The transposon Tn5, commonly used for insertion mutagenesis in many Gramnegative bacteria, was serendipitously discovered to harbor a gene, *ble,* that renders cells resistant to bleomycin *(106–108)*. Two other genes, *Sa ble* and *Sh ble*, have been characterized and shown to encode proteins that are homologous to *Tn5 ble (106,109–111)*. The *Sh ble* gene from *Streptoalloteichus hindustanus* encodes a 14-kDa protein that confers resistance to bleomycin by sequestering the drug *(112,113)*. The X-ray crystal structure of Sh ble revealed that it consists of two halves that are identically folded despite no sequence similarity *(113)*. The structure further revealed that the Sh ble dimer binds to two molecules of bleomycin *(113)*. In vitro assays demonstrated that this protein prevents the action of bleomycin on DNA. At concentrations as low as 1 μ*M*, bleomycin can completely degrade 0.2 μg of chromosomal, linear, or covalently closed circular DNA within a few minutes at ambient temperature, a process that is completely inhibited in the presence of a fivefold molar excess of the Sh ble protein *(114)*. It is likely that the related ble members may also function to sequester bleomycin and possess no direct role in DNA repair, as previously suggested *(115,116)*. A blerelated protein is also present in the bleomycin-producing strain *Streptomyces verticillus*, raising the possibility that the ble-related protein could have yet another role by sequestering bleomycin in *S. verticillus* for efficient transport to the exterior *(117,118)*. To date, database searches reveal that eukaryotes do not possess the blerelated protein and suggest that higher organisms may have evolved other mechanisms to mount a defense against bleomycin.

6.3. Bleomycin Hydrolase

Earlier studies demonstrated that bleomycin can be metabolically inactivated in normal and tumor tissues by an enzyme called bleomycin hydrolase and that such inactivation may play a role in bleomycin resistance *(119–121)*. This is supported by the correlation that tissues with low levels of bleomycin hydrolase are usually sensitive to bleomycin, and tumor cells that acquire resistance to bleomycin possess higher levels of activity *(35,36,38,122,123)*. To better understand the role of bleomycin hydrolase, the enzyme was characterized and shown to be a thiol protease that hydrolyzes the β-aminoalanine amide moiety at the carboxyl terminus of bleomycin to generate the inactive deamido metabolite *(35,37,122)*. Using a specific thiol protease inhibitor (E64) that blocks bleomycin hydrolase activity, it was further shown that cells become more sensitive to bleomycin *(124)*. This finding quickly led to the isolation of the bleomycin hydrolase-corresponding gene from yeast and mammalian cells *(125–128)*. Expression of the yeast bleomycin hydrolase gene *BLH1* in mammalian cells conferred a nearly eightfold increased resistance to bleomycin, but the effect was blocked by the E64 inhibitor *(129)*. One would expect that removal of the *BLH1* gene from yeast would cause a bleomycin-hypersensitive phenotype. However, this may not be the case, as two independent studies show conflicting data regarding the role of Blh1 in the detoxification of bleomycin *(125,126)*. Whereas one study showed that the *blh1Δ* mutant is mildly sensitive to bleomycin, another clearly established that the mutant is not at all sensitive *(125,126)*. Findings from another laboratory also showed that *blh1Δ* mutants are not sensitive to bleomycin *(130)*. Moreover, overexpression of the *BLH1* gene in yeast cells confers no additional resistance to bleomycin *(130)*. Thus, the role of bleomycin hydrolase in producing tumor resistance is controversial. The situation is further complicated by the fact that the Blh1 protein, also called Gal6, is under the control of the Gal4 transcriptional activator *(131)*. Blh1/Gal6 binds specifically to the Gal4 transcription factor DNA-binding site and acts as a repressor to negatively control the galactose metabolism pathway *(131–133)*. On the basis of the foregoing studies, it would appear that bleomycin hydrolase has a more general role in the cells to degrade proteins or, perhaps, to degrade transcription factors to regulate gene expression *(134–136)*.

6.4. DNA Repair Pathways

Repairing of bleomycin-induced DNA lesions is likely the most crucial mechanism employed by cells to avert bleomycin-induced genotoxicity. Thus, organisms exposed to bleomycin must recruit a variety of enzymes and/or proteins to repair the diverse types of bleomycin-induced DNA lesion. Although such enzymes are poorly characterized in eukaryotic cells, the bacterium *E. coli* has two well-documented enzymes (i.e., endonuclease IV and exonuclease III) that repair bleomycin-induced DNA lesions *(105,137,138)*. Both enzymes possess (1) a 3'-diesterase that removes 3'-blocking groups (such as 3'-phosphoglycolate) at strand breaks, and (2) an AP endonuclease that cleaves AP sites. These enzymatic activities regenerate 3'-hydroxyl groups that allow DNA repair synthesis by DNA polymerase *(138–140)*. *E. coli* mutants lacking both endonuclease IV and exonuclease III are severely impaired in the removal of bleomycin-induced DNA lesion, and, as a consequence, display extreme hypersensitivity to bleomycin *(138)*. Between the two enzymes, endonuclease IV plays a more predominant role in repairing bleomycin-damaged DNA *(141)*. This is supported by two independent studies, the first showing that mutants deficient in endonuclease IV are substantially more sensitive to bleomycin than exonuclease III-deficient mutants *(138,142)*. The second study demonstrated that purified endonuclease IV is more active at processing bleomycin-induced DNA lesions in vitro compared to purified exonuclease III *(141)*.

While the *E. coli* studies were in progress, the first eukaryotic homolog of endonuclease IV, called Apn1, was discovered in *S. cerevisiae (143,144)*. Surprisingly, yeast mutants lacking Apn1 are not sensitive to bleomycin, leading to the prediction that yeast may use alternative enzyme(s) to combat the genotoxic effects of bleomycin *(130,144)*. Consequently, a rigorous search was initiated for a possible auxiliary enzyme(s) in yeast that might repair bleomycin-induced DNA lesions. One approach exploited the power of biochemistry to detect enzymatic activities that would process lesions along defined DNA substrates. In one case, a highly sensitive assay was developed consisting of a double-stranded DNA substrate where one strand ($*[^{32}P]$-labeled) bears a single-strand break terminated with 3'-phosphoglycolate (*see* Fig. 3). This biochemical assay identified an extremely weak 3'-diesterase in total extracts derived from an *apn1Δ* mutant (i.e., lacking the major 3'-diesterase/AP endonuclease activity of Apn1). The weak activity removed the 3'-phosphoglycolate from the labeled DNA strand to produce 3'-OH (*see* Fig. 3) *(145)*. The enzyme, called Pde1, was partially purified and also found to have an AP endonuclease in addition to the 3'-diesterase activity *(145)*. Immediately following this report, the gene (*APN2/ETH1*) encoding Pde1 was quickly isolated by two independent laboratories, and the deduced amino acid sequence shared 19% identity with the *E. coli* exonuclease III *(146,147)*. Thus, Pde1/Apn2/Eth1 is the yeast homolog of *E. coli* exonuclease III. The most surprising finding is that yeast mutants lacking both Apn1 and Pde1 (Apn2/Eth1), if at all, showed very mild sensitivity to bleomycin (Ramotar, unpublished). However, the *apn1Δ pde1Δ* double mutants are exquisitely sensitive to the alkylating agent methyl methane sulfonate, which produces natural AP sites, as opposed to oxidized AP sites generated by bleomycin *(59)*. It is therefore possible that the 3'-phosphoglycolate and the oxidized AP site lesions produced by bleomycin are inaccessible or refractory to cleavage by the 3'-diesterase/AP endonuclease activities of either Apn1 or Pde1 in vivo. If so, yeast may possess yet other "backup" enzymes to initiate the repair of bleomycin-induced DNA lesions. This possibility is supported by the discovery of the hPNKP gene encoding the human polynucleotide kinase, which possesses two enzymatic activities—a kinase that phosphorylates the 5'-hydroxyl group of DNA and a

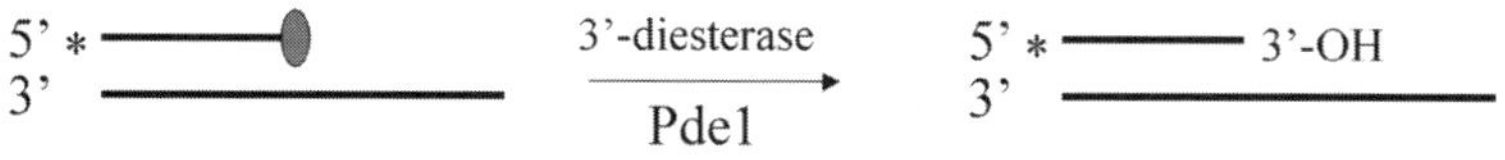

5' * ——————●   3'-diesterase   5' * —————— 3'-OH
3' ———————————   ⟶   3' ———————
                  Pde1

Fig. 3. Depiction of an oligonucleotide DNA substrate bearing a 3'-phophoglycolate terminus. The 3'-phosphoglycolate (oval shape) is produced by bleomycin and requires processing by a 3'-repair diesterase in order to regenerate a 3'-hydroxyl group for DNA polymerase activity. Labeling (*) the 5' end with ^{32}P allows detection of the processed product by polyarcylamide gels.

strong 3'-diesterase activity that repairs oxidative DNA lesions in *E. coli (148)*. hPNPK is unrelated to any of the known 3'-diesterase/AP endonuclease belonging to the endonuclease IV or exonuclease III family, but it may share a related active site *(105,149)*. A gene *(TPP1)* encoding a yeast homolog of the human hPNKP has been recently isolated and it is expected that a gene knockout of *TPP1* should facilitate studies to establish if this enzyme contributes to the repair of bleomycin induced DNA lesions in vivo *(148,150)*.

The repair of bleomycin-induced DNA lesions is not restricted to enzymes with the ability to cleave AP sites or remove 3'-blocking groups, as other DNA repair pathways also participate in the repair process. In yeast, the recombination and the postreplication DNA repair pathways respectively represented by the Rad52 and Rad6 proteins are involved in the repair of bleomycin-induced DNA lesions *(151,152)*. These two pathways also repair a wide spectrum of other DNA lesions, including those generated by the alkylating agent methyl methane sulfonate, 4-nitroquinoline-1-oxide (which forms bulky DNA adducts), and γ-rays. The *rad52Δ* and *rad6Δ* mutants are hypersensitive to a large number of DNA-damaging agents, including bleomycin *(153–156)*. On the basis of cell killing and growth kinetic analyses, both Rad52 and Rad6 showed different contribution to the repair of bleomycin-induced DNA lesions *(152)*. At low bleomycin concentrations (10–15 μg/mL culture), the Rad52 pathway is required to repair bleomycin-damaged DNA, whereas at higher concentrations (15–30 μg/mL culture), the Rad6 pathway plays a more prominent role *(152)*. This disparate response can be explained if distinct lesions are generated in yeast cells in a manner that depends on the bleomycin dose.

A few studies also implicated the involvement of other proteins in the repair of bleomycin-damaged DNA. For example, the Ku proteins, a heterodimer composed of a 70-kDa subunit and a 80-kDa subunit that is involved in nonhomologous end-joining of DNA, is implicated in the repair of bleomycin-induced DNA lesions *(157,158)*. However, a number of laboratories could not convincingly confirm the earlier findings that Ku heterodimer-deficient yeast mutants (*hdf1Δ hdf2Δ*) are sensitive to bleomycin *(157,158)* (Masson, and Ramotar, unpublished data). Perhaps this discrepancy may be related to the yeast strain background used in the initial studies *(157,158)*. Proteins that remodel the chromatin structure are also involved in protecting the genome from the genotoxic effects of bleomycin. In a recent study, it was shown that either bleomycin or methyl methane sulfonate (MMS) can activate the Mec1 kinase in yeast, leading to direct phosphorylation of serine 129 of histone H2A *(159)*. A mutation (S129A) that prevented the phosphorylation of H2A causes cells to be hypersensitive to both bleomycin and MMS *(159)*. The investigators proposed that phosphorylation of H2A is required to relax the chromatin to either allow gene expression to facilitate repair or to permit access of repair proteins and other factors directly to the DNA lesions *(159)*.

Although it is clear that DNA repair plays an important role in the protection against bleomycin-induced DNA lesions, there is no direct evidence that the overproduction of DNA repair proteins can contribute to enhance bleomycin resistance in normal or tumor cells. At least in yeast, the overproduction of some of the above-described DNA repair proteins do not confer additional bleomycin resistance to parental cells *(130,156)*. This is in disaccord with one of the earlier predictions that tumor resistance to bleomycin may be attributed to elevated DNA repair activities *(38)*. Irrespective of whether DNA repair activities are subsequently discovered to be elevated in bleomycin-resistant tumors, any attempts to promote the antitumor potential of bleomycin should take into consideration the possibility of diminishing the DNA repair capacity of tumor cells.

7. YEAST AS A MODEL

The pursuit of cellular defense mechanisms against bleomycin continues to be a central focus and many studies are utilizing yeast as a model system. This organism offers numerous technical advantages, including the relative ease of (1) isolating drug-sensitive mutants, (2) creating gene knockouts, and (3) examining for protein–protein interactions *(160–167)*. Moreover, the current status regarding the nearly 6000 yeast genes can be readily gleaned from the *S. cerevisiae* genomic database. Perhaps the most important benefit of yeast as a model system is the high degree of conservation of many biological processes such as DNA repair, cell cycle control, signal transduction, and drug transport that are shared between yeast and human cells *(168–170)*. In fact, nearly 30% of positionally cloned genes implicated in human diseases have homologs in yeast *(163)*. To date, studies with yeast have directly contributed to the understanding of many human diseases, including those caused by defective DNA repair (e.g., hereditary nonpolyposis colon cancer, Nijmegen syndrome, Bloom syndrome, and xeroderma pigmentosum) *(171–174)*.

7.1. Reverse Genetics

Because biochemical approaches are limited by either the substrate design or by the ability to detect DNA repair enzymes in total extracts, it became apparent that reverse genetics in yeast would be a powerful approach to identify defective genes involved in bleomycin resistance by complementation of bleomycin-hypersensitive mutants. To date, several bleomycin-sensitive yeast variants have been isolated from a parental strain by using either chemical or insertional mutagenesis with a modified form of the *mini-Tn3* transposon (*see* Table 1) *(175–177)*. The transposon strategy is unique because it provides a rapid way to identify the transposon-interrupted gene either by a plasmid rescue or by polymerase chain reaction (PCR) *(176)*. Among the mutants, several show marked hypersensitivities to bleomycin, ranging from 5- to 15-fold, whereas others are less sensitive (less than fivefold), as compared to the parent (*see* Table 1) *(177)*. The hypersensitive mutants were chosen for further analysis, and characterization to date revealed that two are defective in the previously identified DNA repair genes, *RAD52* and *RAD6* (*see* Table 1). This reverse-genetics approach also revealed that altering the RNA polymerase subunit complex, or mitochondrial function, can result in bleomycin sensitivity (*see* Table 1) *(177,178)*. The involvement of RNA polymerase suggests that bleomycin resistance in yeast may be controlled at the transcriptional level, and the requirement for mitochondrial function indicates the need for an energy-dependent process *(177,178)*. The energy dependent process could involve a protein that either directly repairs bleomycin-induced DNA lesions or limits the uptake of bleomycin into the cells *(177)*.

7.2. Imp2 Function and Primary Structure

Among the genes identified to date, *IMP2* is the most intriguing and may uncover yet another mechanism by which cells mount a defense against bleomycin *(130)*. The rest of this chapter is devoted to highlight our current understanding of the role of the Imp2 protein. *IMP2* was initially characterized as a gene permitting yeast cells to utilize maltose, galactose, and raffinose as carbon sources *(179)*. Targeted deletion of the *IMP2* gene in various parental background strains reproduced the same bleomycin-hypersensitive phenotype *(130)*. The *imp2Δ* mutants are also hypersensitive to several other oxidants, including hydrogen peroxide (H_2O_2) and *tert*-butylhydroperoxide, but they are not sensitive to the DNA damaging agents 254-nm ultraviolet light, MMS, 4-nitroquinoline-1-oxide, and ionizing radiation, as compared to the parent strains *(130)*. Reintroduction of the *IMP2* gene in the *imp2Δ* mutant restored full resistance to the oxidants *(130)*. Further analysis revealed that the *imp2Δ* mutants accumulate unrepaired DNA lesions when the cells are challenged with either bleomycin or H_2O_2 *(130)*. No more than 5% of the bleomycin- or H_2O_2-induced DNA lesions in the *imp2Δ* mutant are repaired even after 6 h postincubation, whereas greater than 90% of the DNA lesions are repaired in the parent *(130)*. Consistent with a defect in DNA repair, the *imp2Δ* mutant showed more than a 10-fold higher level of chromosomal translocation when exposed to either bleomycin or H_2O_2, but not to MMS *(180)*. Because H_2O_2 generates predominantly DNA

Table 1
Bleomycin-Hypersensitive Yeast Mutants

Strains	Defective gene	Gene-encoded function	Fold sensitivity	Sensitivities to other DNA-damaging agents
DBY747 (parent)	Normal		1	
DRY212 (*imp2Δ*)	*IMP2*	Transcriptional (co)activator	15	H_2O_2, t-BH
PHL40	*RAD6*	Ubiquitin conjugating enzyme and postreplication DNA repair	10	UV, MMS, 4-NQO
HCY9	*RAD52*	Recombinational DNA repair	8	MMS, 4-NQO, γ-ray
HCY53	*RPB7*	Subunit of RNA polymerase II	5	H_2O_2
HCY69	*OXA1*	Mitochondrial membrane protein required to process Cox II	5	H_2O_2
LMY8	Unknown		7	H_2O_2
HCY55	Unknown		15	H_2O_2, t-BH

Note: t-BH, *tert*-butylhydroperoxide; UV, ultraviolet light; MMS, methyl methane sulfonate; 4-NQO, 4-nitroquinoline-1-oxide.

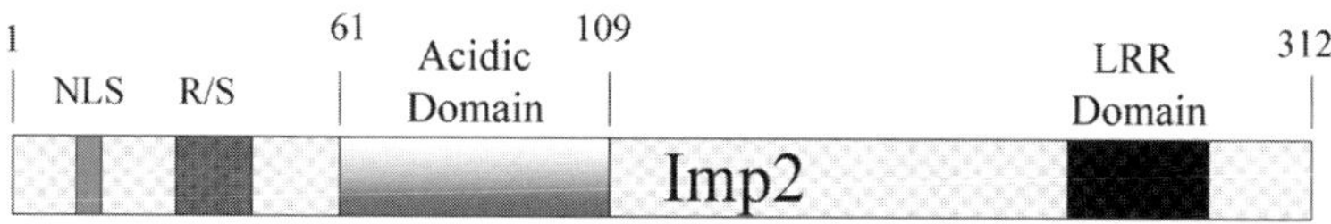

Fig. 4. The primary structural domains deduced from the predicted amino acid sequence of the transcriptional activator Imp2. The N-terminus consists of the putative nuclear localization signal (NLS) and arginine/serine (R/S)-rich domain. The central region bears the activation domain and the C-terminus possesses a leucine-rich repeat (LRR) domain.

single-strand breaks terminated with 3'-phosphate, it is likely that the hypersensitivity of the *imp2Δ* mutant to bleomycin is also the result of a defect in repairing DNA single-strand breaks *(139)*. The fact that *imp2Δ* mutants are not sensitive to ionizing radiation, which produces mostly DNA double-strand breaks and oxidized bases, strongly indicates that the bleomycin hypersensitivity of these mutants is not the result of unrepaired DNA double-strand breaks.

The Imp2 is a small protein, 37 kDa in size, which can activate transcription of a reporter gene by virtue of its acidic domain that is characteristic of many transcriptional activator proteins *(130)*. Confirmation that Imp2 is indeed a transcriptional activator came from an independent study demonstrating that the *MALS* and *MALT* genes, belonging to the maltose utilization pathway, are positively regulated by Imp2 *(181)*. Interestingly, the *imp2Δ* mutants also display hypersensitivity to the monovalent salts Na⁺ and Li⁺ the divalent salts Ca^{2+}, Cu^{2+}, Zn^{2+}, and Mn^{2+}, as well as to the heavy metals arsenite and cadmium *(130,182)*. These latter phenotypes are unrelated to the oxidant hypersensitivity of the mutant *(182)*. The multiple phenotypes displayed by the mutant are consistent with the Imp2 protein being a stress-response transcriptional activator that exerts control on several genes.

The actual mechanism of gene activation by Imp2 is not known, although this mechanism does not involve its own induction. It is possible that pre-existing Imp2 is posttranslationally modified in response to oxidative and salt stress. The protein has an arginine/serine (R/S) rich domain, and proteins with such domain are known to be phosphorylated in vivo, thus influencing their biological function *(see* Fig. 4) *(183,184)*. Imp2 also has a C-terminus leucine-rich repeat (LRR) domain, which shares a high degree of similarity to those present in a diverse group of proteins that function in DNA repair, signal transduction, cell adhesion, cell development, and RNA processing *(185)*. In a few cases, the LRR has been clearly documented to mediate protein–protein interactions (e.g., that of the yeast transcriptional activator Ccr4) *(186,187)*. Because Imp2 has no obvious

DNA-binding domain, it is conceivable that the C-terminus LRR may interact with protein(s) bound to the promoter element of target genes thereby permitting the N-terminus (containing the acidic domain) to efficiently activate gene expression. In this context, Imp2 can be viewed as a coactivator.

Because Imp2 is involved in multiple stress responses, it must be capable of discriminating and activating precisely those gene targets required to counteract the particular stress burden encountered so as not to participate in unnecessary gene activation. For example, Imp2 could distinguish between oxidative and salt stress if it is subjected to different types of posttranslational modification, such as the multiple modifications bequeathed upon p53 in response to different stimuli. Indeed, in the case of p53, the serine amino acid residue Ser15 is phosphorylated by ATM kinase in response to ionizing radiation, and Ser392 is phosphorylated independently of ATM kinase in response to UV light *(188,189)*. Moreover, redox stress affects the sequence-specific DNA-binding capabilities of p53 as well as its ability to bind divalent metal ions which depends on the reduced state of cysteine residues *(188–190)*. However, to examine if Imp2 is differentially modified in response to stress would require structure–function analysis of the protein. This objective would, of course, be greatly simplified if the Imp2 gene targets for both oxidative and salt stress were first established. Hence, the need exists to find Imp2-controlled target genes, especially the one involved in preventing the genotoxic effects of bleomycin. This is likely to be a novel gene, as Imp2 protein does not influence the expression of either Rad6 or Rad52. Moreover, a *rad6Δimp2Δ* or a *rad52Δ imp2Δ* double mutant is exquisitely sensitive to bleomycin, relative to the corresponding single mutants, suggesting that Imp2 performs its function in a separate pathway to repair bleomycin-induced DNA lesions *(180)*. Furthermore, the Imp2 target gene is unlikely to encode a drug efflux pump responsible for expelling bleomycin from the cell, as the *imp2Δ* mutant also exhibits hypersensitivity to H_2O_2, which diffuses freely in and out of cells and apparently requires no extrusion pump for detoxification. Known drug efflux pumps in yeast, such as Snq2, Atr1, Flr1, and Yor1, are not involved in the extrusion of bleomycin, as deficient mutants are not sensitive to the drug *(191–197)*. Additional experimentation excluded the possibility that the hypersensitivity of the *imp2Δ* mutant to bleomycin and H_2O_2 are the result of a deficiency in antioxidant function *(180)*. The null mutants *trx2Δ* and *gsh1Δ* lacking production of the antioxidant thioredoxin and glutathione, respectively, are not sensitive to bleomycin or H_2O_2 *(180)*. In addition, mutants lacking superoxide dismutase that destroys superoxide anion, or catalase T that destroys H_2O_2 are also not sensitive to bleomycin *(180)*. Thus, the hypersensitivity of the *imp2Δ* mutant to bleomycin is not related to lack of expression of any known drug efflux pumps, antioxidants, and/or bleomycin hydrolase *(126,180)*.

The foregoing data clearly indicate that Imp2 is required to protect yeast cells from the genotoxic effect of bleomycin and H_2O_2. Because Imp2 is a transcriptional activator (or perhaps coactivator), it has been postulated that this protein may positively regulate the expression of at least one gene encoding a protein that repairs bleomycin-induced DNA lesions. This protein is evidently not one of the known proteins involved in the major DNA repair pathways, and efforts are currently being directed to identify the Imp2 target gene(s) encoding a protein required to repair bleomycin-induced DNA lesions. Whether the target is defective in the HCY55 mutant strain shown in Table 1 remains to be determined. Perhaps the advent of DNA microarrays might shed some light into possible Imp2 targets.

8. PERSPECTIVE

To date, our greatest understanding of eukaryotic resistance to bleomycin originates from the yeast *S. cerevisiae*, largely because of its powerful genetics. Based on current information, it would appear that cellular defense against bleomycin is a complex process involving multiple pathways. Among these defense pathways, inhibiting DNA repair proteins might be a fruitful approach to selectively enhance the sensitive of tumor cells to bleomycin; otherwise, attention should be focused on increasing the uptake of bleomycin into the cell. One avenue currently being explored is the search for a putative receptor protein that mediates bleomycin uptake. As such, modulating cells to enhance the receptor production should sensitize tumor cells to bleomycin. A further prediction from this bleomycin–receptor model is that mutations altering the receptor function are likely to bring about complete resistance to bleomycin. Such mutations are likely to be discovered in tumors that gradually develop resistance to bleomycin. An intriguing possibility currently being explored relates to the fact that Imp2-deficient yeast mutants sustain an elevated level of a bleomycin–receptor protein on the plasma membrane. If so, the mutant should exhibit an increased intracellular concentration of bleomycin. A further prediction is that once the putative bleomycin–receptor gene becomes available, increasing the gene dosage in parental cells should cause considerable sensitivity to bleomycin. The logical route thereafter would be to isolate the mammalian counterpart of the yeast bleomycin–receptor gene and to test if it is responsible for tumor resistance. Although the corresponding mammalian bleomycin–receptor gene would be difficult to use for sensitizing tumor cells to bleomycin, it would at least provide an immediate screening assay to determine in advance which patients should be subjected to bleomycin chemotherapy.

REFERENCES

1. Umezawa H. Bleomycin and other antitumor antibiotics of high molecular weight. *Antimicrobial Agents Chemother* 1965;5:1079–1085.
2. Umezawa H, Maeda K, Takeuchi T, et al. New antibiotics, bleomycin A and B. *J Antibiot (Tokyo)* 1966;19:200–209.
3. Umezawa H, Ishizuka M, Maeda K, et al. . Studies on bleomycin. *Cancer* 1967;20:891–895.

4. Umezawa H. Natural and artificial bleomycins: chemistry and antitumor activities. *Pure Appl Chem* 1971;28:665–680.

5. Fisher LM, Kuroda R, Sakai TT. Interaction of bleomycin A2 with deoxyribonucleic acid: DNA unwinding and inhibition of bleomycin-induced DNA breakage by cationic thiazole amides related to bleomycin A2. *Biochemistry* 1985;24:3199–31207.

6. Suzuki H, Nagai K, Yamaki H, et al. Mechanism of action of bleomycin. Studies with the growing culture of bacterial and tumor cells. *J Antibiot (Tokyo)* 1968;21:379–386.

7. Kanno T, Nakazawa T, Sugimoto T. Study of bleomycin on brain tumors. 1. Inhibitory effect of bleomycin on cultured brain tumor cells. *Seishin Igaku Kenkyusho Gyosekishu* 1969;16:23–31 (in Japanese).

8. Terasima T, Umezawa H. Lethal effect of bleomycin on cultured mammalian cells. *J Antibiot (Tokyo)* 1970;23:300–304.

9. Ichikawa T, Nakano I, Hirokawa I. Bleomycin treatment of the tumors of penis and scrotum. *J Urol* 1969;102:699–707.

10. Oka S, Sato K, Nakai Y, et al. Treatment of lung cancer with bleomycin. II. *Sci Rep Res Inst Tohoku Univ [Med]* 1970;17:77–91.

11. Terasima T, Yasukawa M, Umezawa H. Breaks and rejoining of DNA in cultured mammalian cells treated with bleomycin. *Gann* 1970;61:513–516.

12. Suzuki H, Nagai K, Akutsu E, et al. On the mechanism of action of bleomycin. Strand scission of DNA caused by bleomycin and its binding to DNA in vitro. *J Antibiot (Tokyo)* 1970;23:473–480.

13. Haidle CW, Weiss KK, Mace ML Jr. Induction of bacteriophage by bleomycin. *Biochem Biophys Res Commun* 1972;48:1179–1184.

14. Ferguson LR, Turner PM. "Petite" mutagenesis by anticancer drugs in *Saccharomyces cerevisiae*. *Eur J Cancer Clin Oncol* 1988;24:591–596.

15. Koy JF, Pleninger P, Wall L, et al. Genetic changes and bioassays in bleomycin- and phleomycin-treated cells, and their relationship to chromosomal breaks. *Mutat Res* 1995;336:19–27.

16. Demopoulos NA, Stamatis ND, Yannopoulos G. Induction of somatic and male crossing-over by bleomycin in *Drosophila melanogaster*. *Mutat Res* 1980;78:347–351.

17. Demopoulos NA, Kappas A, Pelecanos M. Recombinogenic and mutagenic effects of the antitumor antibiotic bleomycin in *Aspergillus nidulans*. *Mutat Res* 1982;102:51–57.

18. Cederberg H, Ramel C. Modifications of the effect of bleomycin in the somatic mutation and recombination test in *Drosophila melanogaster*. *Mutat Res* 1989;214:69–80.

19. Hoffmann GR, Colyer SP, Littlefield LG. Induction of micronuclei by bleomycin in G0 human lymphocytes: I. Dose–response and distribution. *Environ Mol Mutagen* 1993;21:130–135.

20. Burger RM, Peisach J, Horwitz SB. Activated bleomycin. A transient complex of drug, iron, and oxygen that degrades DNA. *J Biol Chem* 1981;256:11,636–11,644.

21. Burger RM, Peisach J, Horwitz SB. Stoichiometry of DNA strand scission and aldehyde formation by bleomycin. *J Biol Chem* 1982;257:8612–8614.

22. Hecht SM. DNA strand scission by activated bleomycin group antibiotics. *Fed Proc* 1986;45:2784–2791.

23. Kane SA, Hecht SM. Polynucleotide recognition and degradation by bleomycin. *Prog Nucleic Acid Res Mol Biol* 1994;49:313–352.

24. Hecht SM. Bleomycin: new perspectives on the mechanism of action. *J Nat Prod* 2000;63:158–168.

25. Wharam MD, Phillips TL, Kane L, et al. Response of a murine solid tumor to in vivo combined chemotherapy and irradiation. *Radiology* 1973;109:451–455.

26. Jani JP, Mistry JS, Morris G, et al. In vivo circumvention of human colon carcinoma resistance to bleomycin. *Cancer Res* 1992;52:2931–2937.

27. Povirk LF, Austin MJ. Genotoxicity of bleomycin. *Mutat Res* 1991;257:127–143.
28. Lazo JS, Sebti SM, Schellens JH. Bleomycin. *Cancer Chemother Biol Response Modif* 1996;16:39–47.
29. Dorr RT. Bleomycin pharmacology: mechanism of action and resistance, and clinical pharmacokinetics. *Semin Oncol* 1992;19:3–8.
30. Sikic BI. Biochemical and cellular determinants of bleomycin cytotoxicity. *Cancer Surveys* 1986;5:81–91.
31. Harrison JH Jr, Lazo JS. High dose continuous infusion of bleomycin in mice: a new model for drug-induced pulmonary fibrosis. *J Pharmacol Exp Ther* 1987;243:1185–1194.
32. Ekimoto H, Takahashi K, Matsuda A, et al. Lipid peroxidation by bleomycin–iron complexes in vitro. *J Antibiot (Tokyo)* 1985;38:1077–1082.
33. Wang Q, Wang Y, Hyde DM, et al. Effect of antibody against integrin alpha4 on bleomycin-induced pulmonary fibrosis in mice. *Biochem Pharmacol* 2000;60:1949–1958.
34. Miyaki M, Ono T, Hori S, et al. Binding of bleomycin to DNA in bleomycin-sensitive and -resistant rat ascites hepatoma cells. *Cancer Res* 1975;35:2015–2019.
35. Akiyama S, Ikezaki K, Kuramochi H, et al. Bleomycin-resistant cells contain increased bleomycin-hydrolase activities. *Biochem Biophys Res Commun* 1981;101:55–60.
36. Morris G, Mistry JS, Jani JP, et al. Neutralization of bleomycin hydrolase by an epitope-specific antibody. *Mol Pharmacol* 1992;42:57–62.
37. Sebti SM, Jani JP, Mistry JS, et al. Metabolic inactivation: a mechanism of human tumor resistance to bleomycin. *Cancer Res* 1991;51:227–232.
38. Urade M, Ogura T, Mima T, et al. Establishment of human squamous carcinoma cell lines highly and minimally sensitive to bleomycin and analysis of factors involved in the sensitivity. *Cancer* 1992;69:2589–2597.
39. Pron G, Belehradek J Jr, Mir LM. Identification of a plasma membrane protein that specifically binds bleomycin. *Biochem Biophys Res Commun* 1993;194:333–337.
40. Pron G, Belehradek J Jr, Orlowski S, et al. Involvement of membrane bleomycin-binding sites in bleomycin cytotoxicity. *Biochem Pharmacol* 1994;48:301–310.
41. Pron G, Mahrour N, Orlowski S, et al. Internalisation of the bleomycin molecules responsible for bleomycin toxicity: a receptor-mediated endocytosis mechanism. *Biochem Pharmacol* 1999;57:45–56.
42. Poddevin B, Orlowski S, Belehradek J Jr, et al. Very high cytotoxicity of bleomycin introduced into the cytosol of cells in culture. *Biochem Pharmacol* 1991;(42 Suppl):S67–S75.
43. Tounekti O, Pron G, Belehradek J Jr et al. Bleomycin, an apoptosis-mimetic drug that induces two types of cell death depending on the number of molecules internalized. *Cancer Res* 1993;53:5462–5469.
44. Tounekti O, Belehradek J Jr, Mir LM. [Bleomycin is an antineoplastic agent capable of mimicking apoptosis]. *Bull Cancer* 1994;81:1043–1049.
45. Tounekti O, Kenani A, Foray N, et al. The ratio of single- to double-strand DNA breaks and their absolute values determine cell death pathway. *Br J Cancer* 2001;84:1272–1279.
46. Oppenheimer NJ, Rodriguez LO, Hecht SM. Metal binding to modified bleomycins. Zinc and ferrous complexes with an acetylated bleomycin. *Biochemistry* 1980;19:4096–4103.
47. Ehrenfeld GM, Rodriguez LO, Hecht SM, et al. Copper(I)-bleomycin: structurally unique complex that mediates oxidative DNA strand scission. *Biochemistry* 1985;24:81–92.
48. Ehrenfeld GM, Shipley JB, Heimbrook DC, et al. Copper-dependent cleavage of DNA by bleomycin. *Biochemistry* 1987;26:931–942.
49. Levy MJ, Hecht SM. Copper(II) facilitates bleomycin-mediated unwinding of plasmid DNA. *Biochemistry* 1988;27:2647–2650.
50. Petering DH, Mao Q, Li W, et al. Metallobleomycin-DNA interactions: structures and reactions related to bleomycin-induced DNA damage. *Met Ions Biol Syst* 1996;33:619–648.

51. Hoehn ST, Junker HD, Bunt RC, et al. Solution structure of co(iii)-bleomycin–OOH bound to a phosphoglycolate lesion containing oligonucleotide: implications for bleomycin-induced double-strand DNA cleavage. *Biochemistry* 2001;40:5894–5905.
52. Oppenheimer NJ, Chang C, Rodriguez LO, et al. Copper(I) . bleomycin. A structurally unique oxidation–reduction active complex. *J Biol Chem* 1981;256:1514–1517.
53. Sausville EA, Peisach J, Horwitz SB. A role for ferrous ion and oxygen in the degradation of DNA by bleomycin. *Biochem Biophys Res Commun* 1976;73:814–822.
54. Sausville EA, Stein RW, Peisach J, et al. Properties and products of the degradation of DNA by bleomycin and iron(II). *Biochemistry* 1978;17:2746–2754.
55. Sausville EA, Peisach J, Horwitz SB. Effect of chelating agents and metal ions on the degradation of DNA by bleomycin. *Biochemistry* 1978;17:2740–2746.
56. Kane SA, Natrajan A, Hecht SM. On the role of the bithiazole moiety in sequence-selective DNA cleavage by Fe–bleomycin. *J Biol Chem* 1994;269:10,899–10,904.
57. Abraham AT, Zhou X, Hecht SM. DNA cleavage by Fe(II). bleomycin conjugated to a solid support. *J Am Chem Soc* 1999;121:1982–1983.
58. Leitheiser CJ, Rishel MJ, Wu X, et al. Solid-phase synthesis of bleomycin group antibiotics. Elaboration of deglycobleomycin A(5). *Org Lett* 2000;2:3397–3399.
59. Burger RM. Cleavage of nucleic acids by bleomycin. *Chem Rev* 1998;98:1153–1169.
60. Burger RM, Peisach J, Blumberg WE, et al. Iron–bleomycin interactions with oxygen and oxygen analogues. Effects on spectra and drug activity. *J Biol Chem* 1979;254:10,906–10,912.
61. Povirk LF, Wubter W, Kohnlein W, et al. DNA double-strand breaks and alkali-labile bonds produced by bleomycin. *Nucleic Acids Res* 1977;4:3573–3580.
62. Ekimoto H, Kuramochi H, Takahashi K, et al. Kinetics of the reaction of bleomycin–Fe(II)–O_2 complex with DNA. *J Antibiot (Tokyo)* 1980;33:426–434.
63. Burger RM, Berkowitz AR, Peisach J, et al. Origin of malondialdehyde from DNA degraded by Fe(II) × bleomycin. *J Biol Chem* 1980;255:11,832–11,838.
64. Burger RM, Peisach J, Horwitz SB. Mechanism of bleomycin action: in vitro studies. *Life Sci* 1981;28:715–727.
65. Worth L Jr, Frank BL, Christner DF, et al. Isotope effects on the cleavage of DNA by bleomycin: mechanism and modulation. *Biochemistry* 1993;32:2601–2609.
66. Giloni L, Takeshita M, Johnson F, et al. Bleomycin-induced strand-scission of DNA. Mechanism of deoxyribose cleavage. *J Biol Chem* 1981;256:8608–8615.
67. Dedon PC, Plastaras JP, Rouzer CA, et al. Indirect mutagenesis by oxidative DNA damage: formation of the pyrimidopurinone adduct of deoxyguanosine by base propenal. *Proc Natl Acad Sci USA* 1998;95:11,113–11,116.
68. Chaudhary AK, Nokubo M, Reddy GR, et al. Detection of endogenous malondialdehyde-deoxyguanosine adducts in human liver. *Science* 1994;265:1580–1582.
69. Fink SP, Reddy GR, Marnett LJ. Mutagenicity in *Escherichia coli* of the major DNA adduct derived from the endogenous mutagen malondialdehyde. *Proc Natl Acad Sci USA* 1997;94:8652–8657.
70. Mao H, Schnetz-Boutaud NC, Weisenseel JP, et al. Duplex DNA catalyzes the chemical rearrangement of a malondialdehyde deoxyguanosine adduct. *Proc Natl Acad Sci USA* 1999;96:6615–6620.
71. Absalon MJ, Wu W, Kozarich JW, et al. Sequence-specific double-strand cleavage of DNA by Fe–bleomycin. 2. Mechanism and dynamics. *Biochemistry* 1995;34:2076–2086.
72. Dedon PC, Goldberg IH. Free-radical mechanisms involved in the formation of sequence-dependent bistranded DNA lesions by the antitumor antibiotics bleomycin, neocarzinostatin, and calicheamicin. *Chem Res Toxicol* 1992;5:311–332.
73. Steighner RJ, Povirk LF. Bleomycin-induced DNA lesions at mutational hot spots: implications for the mechanism of double-strand cleavage. *Proc Natl Acad Sci USA* 1990;87:8350–8354.

74. Onishi T, Iwata H, Takagi Y. Effects of reducing and oxidizing agents on the action of bleomycin. *J Biochem (Tokyo)* 1975;77:745–752.

75. Burger RM, Peisach J, Horwitz SB. Effects of O2 on the reactions of activated bleomycin. *J Biol Chem* 1982;257:3372–3375.

76. Burger RM, Blanchard JS, Horwitz SB, et al. The redox state of activated bleomycin. *J Biol Chem* 1985;260:15,406–15,409.

77. Templin J, Berry L, Lyman S, et al. Properties of redox-inactivated bleomycins. In vitro DNA damage and inhibition of Ehrlich cell proliferation. *Biochem Pharmacol* 1992;43: 615–623.

78. Absalon MJ, Kozarich JW, Stubbe J. Sequence-specific double-strand cleavage of DNA by Fe–bleomycin. 1. The detection of sequence-specific double-strand breaks using hairpin oligonucleotides. *Biochemistry* 1995;34:2065–2075.

79. Povirk LF. Catalytic release of deoxyribonucleic acid bases by oxidation and reduction of an iron.bleomycin complex. *Biochemistry* 1979;18:3989–3995.

80. Kuroda R, Shinomiya M, Otsuka M. Origin of sequence specific cleavage of DNA by bleomycins. *Nucleic Acids Symp Ser* 1992;27:9–10.

81. Nightingale KP, Fox KR. DNA structure influences sequence specific cleavage by bleomycin. *Nucleic Acids Res* 1993;21:2549–2555.

82. Kuroda R, Satoh H, Shinomiya M, et al. Novel DNA photocleaving agents with high DNA sequence specificity related to the antibiotic bleomycin A2. *Nucleic Acids Res* 1995;23:1524–1530.

83. Kane SA, Hecht SM, Sun JS, et al. Specific cleavage of a DNA triple helix by FeII.bleomycin. *Biochemistry* 1995;34:16,715–16,724.

84. Mascharak PK, Sugiura Y, Kuwahara J, et al. Alteration and activation of sequence-specific cleavage of DNA by bleomycin in the presence of the antitumor drug cis-diamminedichloro-platinum(II). *Proc Natl Acad Sci USA* 1983;80:6795–6798.

85. Hertzberg RP, Caranfa MJ, Hecht SM. DNA methylation diminishes bleomycin-mediated strand scission. *Biochemistry* 1985;24:5286–5289.

86. Hertzberg RP, Caranfa MJ, Hecht SM. Degradation of structurally modified DNAs by bleomycin group antibiotics. *Biochemistry* 1988;27:3164–3174.

87. Bennett RA, Swerdlow PS, Povirk LF. Spontaneous cleavage of bleomycin-induced abasic sites in chromatin and their mutagenicity in mammalian shuttle vectors. *Biochemistry* 1993;32:3188–3195.

88. Dar ME, Jorgensen TJ. Deletions at short direct repeats and base substitutions are characteristic mutations for bleomycin-induced double- and single-strand breaks, respectively, in a human shuttle vector system. *Nucleic Acids Res* 1995;23:3224–3230.

89. Pavon V, Esteve I, Guerrero R, et al. Induced mutagenesis by bleomycin in the purple sulfur bacterium Thiocapsa roseopersicina. *Curr Microbiol* 1995;30:117–120.

90. Steighner RJ, Povirk LF. Effect of in vitro cleavage of apurinic/apyrimidinic sites on bleomycin-induced mutagenesis of repackaged lambda phage. *Mutat Res* 1990;240:93–100.

91. Tates AD, van Dam FJ, Natarajan AT, et al. Frequencies of HPRT mutants and micronuclei in lymphocytes of cancer patients under chemotherapy: a prospective study. *Mutat Res* 1994;307:293–306.

92. Carter BJ, de Vroom E, Long EC, et al. Site-specific cleavage of RNA by Fe(II).bleomycin. *Proc Natl Acad Sci USA* 1990;87:9373–9377.

93. Huttenhofer A, Hudson S, Noller HF, et al. Cleavage of tRNA by Fe(II)–bleomycin. *J Biol Chem* 1992;267:24,471–24,45.

94. Holmes CE, Hecht SM. Fe.bleomycin cleaves a transfer RNA precursor and its "transfer DNA" analog at the same major site. *J Biol Chem* 1993;268:25,909–25,913.

95. Hecht SM. RNA degradation by bleomycin, a naturally occurring bioconjugate. *Bioconjug Chem* 1994;5:513–526.

96. Keck MV, Hecht SM. Sequence-specific hydrolysis of yeast tRNA(Phe) mediated by metal-free bleomycin. *Biochemistry* 1995;34:12,029–12,037.

97. Morgan MA, Hecht SM. Iron(II) bleomycin-mediated degradation of a DNA-RNA heteroduplex. *Biochemistry* 1994;33:10,286–10,293.

98. Holmes CE, Carter BJ, Hecht SM. Characterization of iron (II).bleomycin-mediated RNA strand scission. *Biochemistry* 1993;32:4293–4307.

99. Holmes CE, Duff RJ, van der Marel GA, et al. On the chemistry of RNA degradation by Fe.bleomycin. *Bioorg Med Chem* 1997;5:1235–1248.

100. Moore CW, Del Valle R, McKoy J, et al. Lesions and preferential initial localization of [*S*-methyl-3*H*]bleomycin A2 on *Saccharomyces cerevisiae* cell walls and membranes. *Antimicrob Agents Chemother* 1992;36:2497–2505.

101. Beaudouin R, Lim ST, Steide JA, et al. Bleomycin affects cell wall anchorage of mannoproteins in *Saccharomyces cerevisiae*. *Antimicrob Agents Chemother* 1993;37: 1264–1269.

102. Lim ST, Jue CK, Moore CW, et al. Oxidative cell wall damage mediated by bleomycin–Fe(II) in *Saccharomyces cerevisiae*. *J Bacteriol* 1995;177:3534–3539.

103. Hay J, Shahzeidi S, Laurent G. Mechanisms of bleomycin-induced lung damage. *Arch Toxicol* 1991;65:81–94.

104. Arndt D, Zeisig R, Bechtel D, et al. Liposomal bleomycin: increased therapeutic activity and decreased pulmonary toxicity in mice. *Drug Deliv* 2001;8:1–7.

105. Ramotar D. The apurinic–apyrimidinic endonuclease IV family of DNA repair enzymes. *Biochem Cell Biol* 1997;75:327–336.

106. Genilloud O, Garrido MC, Moreno F. The transposon Tn5 carries a bleomycin-resistance determinant. *Gene* 1984;32:225–233.

107. Collis CM, Hall RM. Identification of a Tn5 determinant conferring resistance to phleomycins, bleomycins, and tallysomycins. *Plasmid* 1985;14:143–151.

108. Blot M, Meyer J, Arber W. Bleomycin-resistance gene derived from the transposon Tn5 confers selective advantage to *Escherichia coli* K-12. *Proc Natl Acad Sci USA* 1991;88: 9112–9116.

109. Semon D, Movva NR, Smith TF, et al. Plasmid-determined bleomycin resistance in *Staphylococcus aureus*. *Plasmid* 1987;17:46–53.

110. Drocourt D, Calmels T, Reynes JP, et al. Cassettes of the *Streptoalloteichus hindustanus* ble gene for transformation of lower and higher eukaryotes to phleomycin resistance. *Nucleic Acids Res* 1990;18:4009.

111. Gennimata D, Davies J, Tsiftsoglou AS. Bleomycin resistance in *Staphylococcus aureus* clinical isolates. *J Antimicrob Chemother* 1996;37:65–75.

112. Gatignol A, Baron M, Tiraby G. Phleomycin resistance encoded by the ble gene from transposon Tn 5 as a dominant selectable marker in *Saccharomyces cerevisiae*. *Mol Gen Genet* 1987;207:342–348.

113. Dumas P, Bergdoll M, Cagnon C, Crystal structure and site-directed mutagenesis of a bleomycin resistance protein and their significance for drug sequestering. *EMBO J* 1994;13:2483–2492.

114. Hayes JD, Wolf CR. Molecular mechanisms of drug resistance. *Biochem J* 1990;272: 281–295.

115. Maruyama M, Kumagai T, Matoba Y, et al. Crystal structures of the transposon Tn5-carried bleomycin resistance determinant uncomplexed and complexed with bleomycin. *J Biol Chem* 2001;276:9992–9999.

116. Blot M, Heitman J, Arber W. Tn5-mediated bleomycin resistance in *Escherichia coli* requires the expression of host genes. *Mol Microbiol* 1993;8:1017–1024.

117. Calcutt MJ, Schmidt FJ. Bleomycin biosynthesis: structure of the resistance genes of the producer organism. *Ann NY Acad Sci* 1994;721:133–137.

118. Sugiyama M, Kumagai T, Matsuo H, et al. Overproduction of the bleomycin-binding proteins from bleomycin-producing *Streptomyces verticillus* and a methicillin-resistant *Staphylococcus aureus* in *Escherichia coli* and their immunological characterisation. *FEBS Lett* 1995;362:80–84.

119. Umezawa H, Hori S, Sawa T, Yoshioka T, et al. A bleomycin-inactivating enzyme in mouse liver. *J Antibiot (Tokyo)* 1974;27:419–424.

120. Sebti SM, Lazo JS. Metabolic inactivation of bleomycin analogs by bleomycin hydrolase. *Pharmacol Ther* 1988;38:321–329.

121. Nishimura C, Suzuki H, Tanaka N, et al. Bleomycin hydrolase is a unique thiol aminopeptidase. *Biochem Biophys Res Commun* 1989;163:788–796.

122. Lazo JS, Humphreys CJ. Lack of metabolism as the biochemical basis of bleomycin-induced pulmonary toxicity. *Proc Natl Acad Sci USA* 1983;80:3064–3068.

123. Ferrando AA, Velasco G, Campo E, et al. Cloning and expression analysis of human bleomycin hydrolase, a cysteine proteinase involved in chemotherapy resistance. *Cancer Res* 1996;56:1746–1750.

124. Jani JP, Mistry JS, Morris G, et al. In vivo sensitization of human lung carcinoma to bleomycin by the cysteine proteinase inhibitor E-64. *Oncol Res* 1992;4:59–63.

125. Enenkel C, Wolf DH. BLH1 codes for a yeast thiol aminopeptidase, the equivalent of mammalian bleomycin hydrolase. *J Biol Chem* 1993;268:7036–7043.

126. Magdolen U, Muller G, Magdolen V, et al. A yeast gene (BLH1) encodes a polypeptide with high homology to vertebrate bleomycin hydrolase, a family member of thiol proteinases. *Biochim Biophys Acta* 1993;1171:299–303.

127. Bromme D, Rossi AB, Smeekens SP, et al. Human bleomycin hydrolase: molecular cloning, sequencing, functional expression, and enzymatic characterization. *Biochemistry* 1996;35:6706–6714.

128. O'Farrell PA, Gonzalez F, Zheng W, et al. Crystal structure of human bleomycin hydrolase, a self-compartmentalizing cysteine protease. *Structure Fold Des* 1999;7:619–627.

129. Pei Z, Calmels TP, Creutz CE, et al. Yeast cysteine proteinase gene ycp1 induces resistance to bleomycin in mammalian cells. *Mol Pharmacol* 1995;48:676–681.

130. Masson JY, Ramotar D. The *Saccharomyces cerevisiae* IMP2 gene encodes a transcriptional activator that mediates protection against DNA damage caused by bleomycin and other oxidants. *Mol Cell Biol* 1996;16:2091–2100.

131. Zheng W, Xu HE, Johnston SA. The cysteine–peptidase bleomycin hydrolase is a member of the galactose regulon in yeast. *J Biol Chem* 1997;272:30,350–30,305.

132. Xu HE, Johnston SA. Yeast bleomycin hydrolase is a DNA-binding cysteine protease. Identification, purification, biochemical characterization. *J Biol Chem* 1994;269:21,177–21,183.

133. Joshua-Tor L, Xu HE, Johnston SA, et al. Crystal structure of a conserved protease that binds DNA: the bleomycin hydrolase, Gal6. *Science* 1995;269:945–950.

134. Zheng W, Johnston SA, Joshua-Tor L. The unusual active site of Gal6/bleomycin hydrolase can act as a carboxypeptidase, aminopeptidase, and peptide ligase. *Cell* 1998;93:103–109.

135. Koldamova RP, Lefterov IM, DiSabella MT, et al. Human bleomycin hydrolase binds ribosomal proteins. *Biochemistry* 1999;38:7111–7117.

136. Schwartz DR, Homanics GE, Hoyt DG, et al. The neutral cysteine protease bleomycin hydrolase is essential for epidermal integrity and bleomycin resistance. *Proc Natl Acad Sci USA* 1999;96:4680–4685.

137. Demple B, Harrison L. Repair of oxidative damage to DNA: enzymology and biology. *Annu Rev Biochem* 1994;63:915–948.

138. Ramotar D, Popoff SC, Demple B. Complementation of DNA repair-deficient *Escherichia coli* by the yeast Apn1 apurinic/apyrimidinic endonuclease gene. *Mol Microbiol* 1991;5:149–155.

139. Demple B, Johnson A, Fung D. Exonuclease III and endonuclease IV remove 3' blocks from DNA synthesis primers in H_2O_2-damaged *Escherichia coli*. *Proc Natl Acad Sci USA* 1986;83:7731–7735.

140. Levin JD, Johnson AW, Demple B. Homogeneous *Escherichia coli* endonuclease IV. Characterization of an enzyme that recognizes oxidative damage in DNA. *J Biol Chem* 1988;263:8066–8071.
141. Levin JD, Demple B. In vitro detection of endonuclease IV-specific DNA damage formed by bleomycin in vivo. *Nucleic Acids Res* 1996;24:885–889.
142. Cunningham RP, Saporito SM, Spitzer SG, et al. Endonuclease IV (nfo) mutant of *Escherichia coli*. *J Bacteriol* 1986;168:1120–1127.
143. Popoff SC, Spira AI, Johnson AW, et al. Yeast structural gene (APN1) for the major apurinic endonuclease: homology to *Escherichia coli* endonuclease IV. *Proc Natl Acad Sci USA* 1990;87:4193–4197.
144. Ramotar D, Popoff SC, Gralla EB, et al. Cellular role of yeast Apn1 apurinic endonuclease/ 3'-diesterase: repair of oxidative and alkylation DNA damage and control of spontaneous mutation. *Mol Cell Biol* 1991;11:4537–4544.
145. Sander M, Ramotar D. Partial purification of Pde1 from *Saccharomyces cerevisiae*: enzymatic redundancy for the repair of 3'-terminal DNA lesions and abasic sites in yeast. *Biochemistry* 1997;36:6100–6106.
146. Bennett RA. The *Saccharomyces cerevisiae* ETH1 gene, an inducible homolog of exonuclease III that provides resistance to DNA-damaging agents and limits spontaneous mutagenesis. *Mol Cell Biol* 1999;19:1800–1809.
147. Johnson RE, Torres-Ramos CA, Izumi T, et al. Identification of APN2, the *Saccharomyces cerevisiae* homolog of the major human AP endonuclease HAP1, and its role in the repair of abasic sites. *Genes Dev* 1998;12:3137–3143.
148. Jilani A, Ramotar D, Slack C, et al. Molecular cloning of the human gene, PNKP, encoding a polynucleotide kinase 3'-phosphatase and evidence for its role in repair of DNA strand breaks caused by oxidative damage. *J Biol Chem* 1999;274:24,176–24,186.
149. Yang X, Tellier P, Masson JY, et al. Characterization of amino acid substitutions that severely alter the DNA repair functions of *Escherichia coli* endonuclease IV. *Biochemistry* 1999;38:3615–3623.
150. Vance JR, Wilson TE. Uncoupling of 3'-phosphatase and 5'-kinase functions in budding yeast. Characterization of *Saccharomyces cerevisiae* DNA 3'-phosphatase (TPP1). *J Biol Chem* 2001;276:15,073–15,081.
151. Abe H, Wada M, Kohno K, et al. Altered drug sensitivities to anticancer agents in radiation-sensitive DNA repair deficient yeast mutants. *Anticancer Res* 1994;14:1807–1810.
152. Keszenman DJ, Salvo VA, Nunes E. Effects of bleomycin on growth kinetics and survival of *Saccharomyces cerevisiae*: a model of repair pathways. *J Bacteriol* 1992;174:3125–3132.
153. Resnick MA, Martin P. The repair of double-strand breaks in the nuclear DNA of *Saccharomyces cerevisiae* and its genetic control. *Mol Gen Genet* 1976;143:119–129.
154. Jentsch S, McGrath JP, Varshavsky A. The yeast DNA repair gene RAD6 encodes a ubiquitin-conjugating enzyme. *Nature* 1987;329:131–134.
155. Madura K, Prakash S, Prakash L. Expression of the *Saccharomyces cerevisiae* DNA repair gene RAD6 that encodes a ubiquitin conjugating enzyme, increases in response to DNA damage and in meiosis but remains constant during the mitotic cell cycle. *Nucleic Acids Res* 1990;18:771–778.
156. He CH, Masson JY, Ramotar D. A *Saccharomyces cerevisiae* phleomycin-sensitive mutant, ph140, is defective in the RAD6 DNA repair gene. *Can J Microbiol* 1996;42:1263–1266.
157. Mages GJ, Feldmann HM, Winnacker EL. Involvement of the *Saccharomyces cerevisiae* HDF1 gene in DNA double-strand break repair and recombination. *J Biol Chem* 1996;271:7910–7915.
158. Feldmann H, Driller L, Meier B, et al. HDF2, the second subunit of the Ku homologue from *Saccharomyces cerevisiae*. *J Biol Chem* 1996;271:27,765–27,769.

159. Downs JA, Lowndes NF, Jackson SP. A role for *Saccharomyces cerevisiae* histone H2A in DNA repair. *Nature* 2000;408:1001–1004.

160. Botstein D, Chervitz SA, Cherry JM. Yeast as a model organism [comment]. *Science* 1997;277:1259–1260.

161. Phizicky EM, Fields S. Protein-protein interactions: methods for detection and analysis. *Microbiol Rev* 1995;59:94–123.

162. Winzeler EA, Davis RW. Functional analysis of the yeast genome. *Curr Opin Genet Dev* 1997;7:771–776.

163. Winzeler EA, Shoemaker DD, Astromoff A, et al. Functional characterization of the *S. cerevisiae* genome by gene deletion and parallel analysis. *Science* 1999;285:901–906.

164. Goldstein AL, McCusker JH, Knop M, et al. Three new dominant drug resistance cassettes for gene disruption in *Saccharomyces cerevisiae*. *Yeast* 1999;15:1541–1553.

165. Lashkari DA, DeRisi JL, McCusker JH, et al. Yeast microarrays for genome wide parallel genetic and gene expression analysis. *Proc Natl Acad Sci USA* 1997;94:13,057–13,062.

166. Mayer G, Launhardt H, Munder T. Application of the green fluorescent protein as a reporter for Ace1-based, two-hybrid studies. *Biotechniques* 1999;27:86–88, 92–94.

167. Pereira RdS. The use of baker's yeast in the generation of asymmetric centers to produce chiral drugs and others compounds. *Crit Rev Biotechnol* 1998;18:25–83.

168. Andrade MA, Sander C, Valencia A. Updated catalogue of homologues to human disease-related proteins in the yeast genome. *FEBS Lett* 1998;426:7–16.

169. Bassett DE Jr, Basrai MA, Connelly C, et al. Exploiting the complete yeast genome sequence. *Curr Opin Genet Dev* 1996;6:763–766.

170. Foury F. Human genetic diseases: a cross-talk between man and yeast. *Gene* 1997;195:1–10.

171. Carney JP. Chromosomal breakage syndromes. *Curr Opin Immunol* 1999;11:443–447.

172. Modrich P, Lahue R. Mismatch repair in replication fidelity, genetic recombination, and cancer biology. *Annu Rev Biochem* 1996;65:101–133.

173. Neff NF, Ellis NA, Ye TZ, et al. The DNA helicase activity of BLM is necessary for the correction of the genomic instability of bloom syndrome cells. *Mol Biol Cell* 1999;10:665–676.

174. Vermeulen W, de Boer J, Citterio E, et al. Mammalian nucleotide excision repair and syndromes. *Biochem Soc Trans* 1997;25:309–315.

175. Moore CW. Further characterizations of bleomycin-sensitive (blm) mutants of *Saccharomyces cerevisiae* with implications for a radiomimetic model. *J Bacteriol* 1991;173:3605–3608.

176. Burns N, Grimwade B, Ross-Macdonald PB, et al. Large-scale analysis of gene expression, protein localization, and gene disruption in *Saccharomyces cerevisiae*. *Genes Dev* 1994;8:1087–1105.

177. He CH, Masson JY, Ramotar D. Functional mitochondria are essential for *Saccharomyces cerevisiae* cellular resistance to bleomycin. *Curr Genet* 1996;30:279–283.

178. He CH, Ramotar D. An allele of the yeast RPB7 gene, encoding an essential subunit of RNA polymerase II, reduces cellular resistance to the antitumor drug bleomycin. *Biochem Cell Biol* 1999;77:375–382.

179. Donnini C, Lodi T, Ferrero I, et al. Allelism of IMP1 and GAL2 genes of *Saccharomyces cerevisiae*. *J Bacteriol* 1992;174:3411–3415.

180. Ramotar D, Masson JY. A *Saccharomyces cerevisiae* mutant defines a new locus essential for resistance to the antitumour drug bleomycin. *Can J Microbiol* 1996;42:835–843.

181. Lodi T, Goffrini P, Ferrero I, et al. IMP2, a gene involved in the expression of glucose-repressible genes in *Saccharomyces cerevisiae*. *Microbiology* 1995;141:2201–2209.

182. Masson JY, Ramotar D. The transcriptional activator Imp2p maintains ion homeostasis in *Saccharomyces cerevisiae*. *Genetics* 1998;149:893–901.

183. Colwill K, Pawson T, Andrews B, et al. The Clk/Sty protein kinase phosphorylates SR splicing factors and regulates their intranuclear distribution. *EMBO J* 1996;15:265–275.

184. Rossi F, Labourier E, Forne T, et al. Specific phosphorylation of SR proteins by mammalian DNA topoisomerase I. *Nature* 1996;381:80–82.
185. Kobe B, Deisenhofer J. The leucine-rich repeat: a versatile binding motif. *Trends Biochem Sci* 1994;19:415–421.
186. Kobe B, Deisenhofer J. A structural basis of the interactions between leucine-rich repeats and protein ligands. *Nature* 1995;374:183–186.
187. Benson JD, Benson M, Howley PM, Struhl K. Association of distinct yeast Not2 functional domains with components of Gcn5 histone acetylase and Ccr4 transcriptional regulatory complexes. *EMBO J* 1998;17:6714–6722.
188. Giaccia AJ, Kastan MB. The complexity of p53 modulation: emerging patterns from divergent signals. *Genes Dev* 1998;12:2973–2983.
189. Prives C, Hall PA. The p53 pathway. *J Pathol* 1999;187:112–126.
190. Polyak K, Xia Y, Zweier JL, et al. A model for p53-induced apoptosis. [see comments]. *Nature* 1997;389:300–305.
191. Alarco AM, Balan I, Talibi D, et al. AP1-mediated multidrug resistance in *Saccharomyces cerevisiae* requires FLR1 encoding a transporter of the major facilitator superfamily. *J Biol Chem* 1997;272:19,304–19,313.
192. Coleman ST, Tseng E, Moye-Rowley WS. *Saccharomyces cerevisiae* basic region-leucine zipper protein regulatory networks converge at the ATR1 structural gene. *J Biol Chem* 1997;272:23,224–23,230.
193. Decottignies A, Lambert L, Catty P, et al. Identification and characterization of SNQ2, a new multidrug ATP binding cassette transporter of the yeast plasma membrane. *J Biol Chem* 1995;270:18,150–18,157.
194. Katzmann DJ, Hallstrom TC, Mahe Y, et al. Multiple Pdr1p/Pdr3p binding sites are essential for normal expression of the ATP binding cassette transporter protein-encoding gene PDR5. *J Biol Chem* 1996;271:23,049–23,054.
195. Katzmann DJ, Epping EA, Moye-Rowley WS. Mutational disruption of plasma membrane trafficking of *Saccharomyces cerevisiae* Yor1p, a homologue of mammalian multidrug resistance protein. *Mol Cell Biol* 1999;19:2998–3009.
196. Oskouian B, Saba JD. YAP1 confers resistance to the fatty acid synthase inhibitor cerulenin through the transporter Flr1p in *Saccharomyces cerevisiae*. *Mol Gen Genet* 1999;261:346–353.
197. Wemmie JA, Moye-Rowley WS. Mutational analysis of the *Saccharomyces cerevisiae* ATP-binding cassette transporter protein Ycf1p. *Mol Microbiol* 1997;25:683–694.

9 Potential Role of PARP Inhibitors in Cancer Treatment and Cell Death

Michèle Rouleau, PhD
and Guy G. Poirier, PhD

CONTENTS

1. THE POLY(ADP-RIBOSE) POLYMERASE FAMILY AND POLY(ADP-RIBOSYLATION)

Poly(ADP-ribose) polymerase-1 (PARP-1) is a multifunctional nuclear enzyme involved in several cellular processes related to the maintenance of genomic integrity (reviewed in refs. *1* and *2*). DNA damage induces a 10- to 500-fold increase in PARP-1 activity, which causes a drastic lowering of the cellular NAD^+ content. PARP-1 binds to DNA strand breaks and hydrolyzes NAD^+ to synthesize poly(ADP-ribose) chains (pADPr) on nuclear protein substrates, including itself. This posttranslational modification has profound effects on the physicochemical and functional properties of proteins because pADPr is negatively charged and can comprise more than 200 ADP-ribose units. The

From: *Cancer Drug Discovery and Development: DNA Repair in Cancer Therapy*
Edited by: L. C. Panasci and M. A. Alaoui-Jamali © Humana Press Inc., Totowa, NJ

automodification of PARP-1 results in its release from DNA strand breaks and alters its catalytic activity. This modification is reversible, through the action of poly(ADP-ribose) glycohydrolase (PARG), which hydrolyzes the polymer into shorter pADPr chains and ADP-ribose units *(3)*.

The involvement of PARP-1 in the maintenance of chromosome integrity is suggested from numerous studies using PARP inhibitors, PARP-1 knockout mice, and PARP-1(–/–) cells (reviewed in refs. *1,2*, and *4*). Although cells devoid of PARP-1 because of display the exact same phenotypes as those devoid of PARP activity due to the presence of inhibitors, both show an increased frequency of sister chromatid exchange, chromosomal fragmentation and fusion, chromosome loss and gain, shortened telomeres, hypersensitivity to γ-irradiation, accumulation of damaged cells in the G_2/M boundary of the cell cycle, and an increased rate of apoptosis. These phenotypes most likely reflect the physiological importance of poly(ADP-ribosyl)ation of protein substrates as well as of PARP-1 itself in several aspects of the DNA-damage response (DDR). Indeed, substrates of poly(ADP-ribosyl)ation reactions are mostly nuclear proteins involved in the metabolism of nucleic acids and in the maintenance of chromatin architecture *(1)*. Also, PARP-1 interacts with several proteins involved in DNA-damage signaling and response pathways, as summarized in Fig. 1. Interactions of PARP-1 with proteins implicated in DNA strand break repair, base excision repair (BER), homologous and nonhomologous recombination, replication, and centrosome duplication have been found (*see* Fig. 1). In particular, the role of PARP-1 in BER, first suggested from the observed inhibition of BER when PARP activity is inhibited *(18,19)* and from interactions of PARP-1 with the known BER proteins XRCC-1) and DNA polymerase β *(6)*, was recently clearly established in the long patch subpathway of BER through a cooperation with flap endonuclease-1 *(7)*. PARP-1 is also implicated in cell death *(20,21)*. In the context of DNA-damaged-induced cell death, it has been suggested to act as a molecular switch from apoptosis to necrosis, depending on the level of DNA stand breaks. Its interactions with known apoptosis proteins p53 and NFκB suggest pathways through which it may function.

PARP-1 is the best characterized of a growing family of enzymes, now including PARP-2 to PARP-7 *(21–28)* (*see* Table 1). These PARPs share a "PARP domain," which includes all of the amino acids necessary for NAD^+ binding and catalysis of poly(ADP-ribosyl)ation, in addition to specific domains that should confer distinct functions to these numerous PARPs. However, studies on these new PARP enzymes suggest that some may also play roles in the maintenance of genomic integrity. As for PARP-1, PARP-2 is activated by DNA strand breaks and is involved in DDR *(22)*. PARP-2 may be responsible for most of the residual PARP activity stimulated by DNA strand breaks in PARP(–/–) cells *(22)*. In addition, it appears that, together, these two enzymes are essential to the mainte-

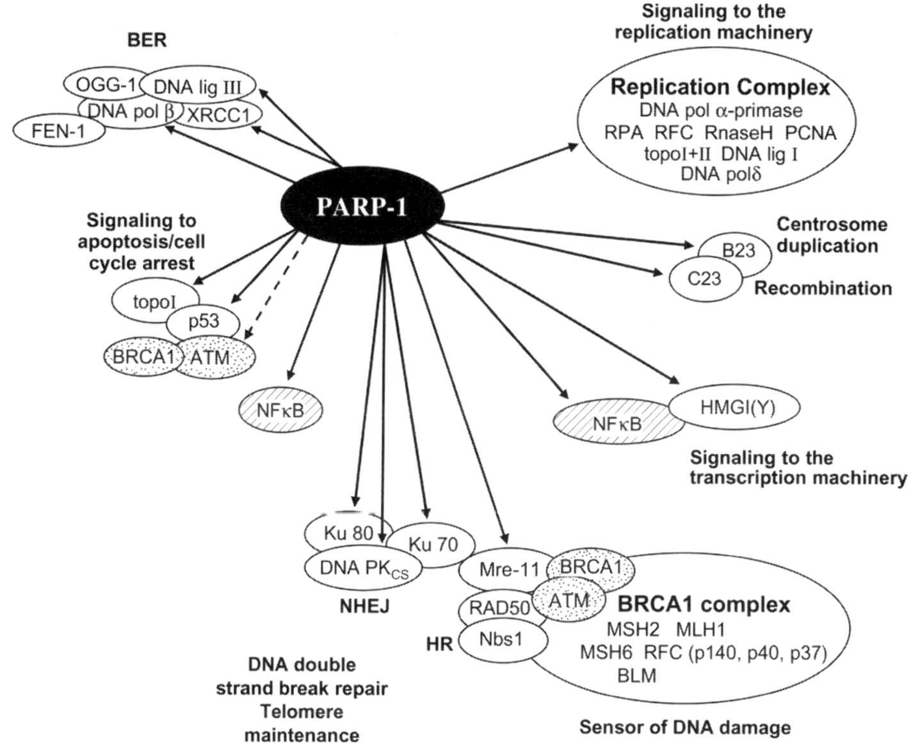

Fig. 1. Summary of physical interactions between PARP-1 and proteins involved in the DDR and cell cycle control. Demonstrated (solid lines) and putative (dashed line) physical interactions between PARP-1 and proteins are indicated by arrows. PARP-1 is required in the long patch subpathway of base excision repair (BER) through cooperation with flap endonuclease-1 (FEN-1) *(5–7)*. PARP-1 also interacts with the DNA-dependent protein kinase complex DNA-PK$_{CS}$/Ku70/Ku80 *(8)* as well as with the Mre-11/Rad50/ Nbs1complex through a Mre-11 interaction recently demonstrated in *C. elegans (9)*. These proteins mediate DNA double-strand-break repair either by nonhomologous end-joining (NHEJ) or by homologous recombination (HR). The breast-cancer-associated gene product BRCA1 *(64)* and ATM (ataxia–telangiectasia mutated) *(11)* may also play a role in double strand break repair. Interactions of PARP-1 with NFκB *(12)*, HMGI(Y) *(13)*, proteins of the replication complex *(14)* and possibly with p53 and ATM suggest that PARP-1 may also play roles in sensing and signaling DNA damage. PARP-1, which localizes at centrosomes *(15)*, also interacts with B23 (nucleophosmin *[16]*), a trigger for centrosome duplication *(17)*.

nance of genomic integrity because the double knockout of PARP-1 and PARP-2 genes is lethal in mice *(19)*. Little is known about PARP-3, but its recent localization to daughter centrioles suggests a role during cellular division *(30)*. Tankyrases (PARP-5 and PARP-6), mostly associated with the Golgi, are also

Table 1
Schematic Representation and Properties of PARP Enzymes

PARP members	Modular organization	MW (kDa)	Activity dependent on DNA damage	Auto- modification	Sensitivity to PARP-1 inhibitors	Intracellular localization	Refs.
PARP-1	N DBD BRCT PD C	113	Yes	Yes	Yes	Nuclear (nucleolus, telomeres) Centrosomes	15,21
sPARP-1	PD	55	No	Yes	Yes	Nuclear	29
PARP-2	DBD PD	62	Yes	Yes	Yes	Nuclear	22
PARP-3	PD	60	?	Yes	?	Centromes	23,30
VPARP (PARP-4)	BRCT PD MVPD	193	No	Yes	Yes	Cytoplasmic (vault particles), and nuclear (mitotic spindles)	24
Tankyrase 1 (PARP-5)	HPS ANKYRIN PD	142	No	Yes	Yes	Cytoplasmic (Golgi, centrosomes) and nuclear (telomeres)	25,31,32
Tankyrase 2 (PARP-6)	ANKYRIN PD	130	?	Yes	?	Cytoplasmic (Golgi, centrosomes) and nuclear (telomeres)	26,27
Ti PARP (PARP-7)	TIL PD	76	?	?	?	?	28

Note: PARP-1, the prototype and best characterized PARP enzyme, comprises a DNA-binding domain (DBD) which includes two zinc fingers and a nuclear localization signal, a central automodification domain that show similarity to the BRCA1 C-terminal domain (BRCT), and a catalytic PARP domain (PD). The BRCT region mediates a number of PARP–protein interactions. sPARP-1 arises from an alternative transcription start site on the PARP-1 gene. Although it comprises only the PD, it is localized in the nucleus and can be automodified. VPARP is a component of vault particles that interacts with the major vault protein through its C-terminal domain (MVPD). Tankyrases include several ankyrin repeats that mediate interactions with the telomere-repeat-binding factor-1 as well as a sterile α-module (hatched boxes). Homopolymeric stretches of histidine, proline, and serine (HPS) are also found at the N-terminal region of tankyrase 1. Ti PARP, whose expression is induced by the halogenated aromatic hydrocarbon TCDD, comprises a sequence specifically expressed in tumor infiltrating lymphocytes (TIL).

found at telomeres and interact with telomere-repeat-binding factor-1, directly involved in the regulation of telomeric length *(25,34)*. Tankyrase-1 is also associated with centrosomes during mitosis *(31)*. A better knowledge of the function of these new PARP enzymes and of the respective stimuli that might control their activity will allow one to determine the contribution of each in the maintenance of genomic integrity. The development of knockout mice for these new PARPs should help to address these issues.

2. DEVELOPMENT OF PARP INHIBITORS

In view of the important role of some PARPs in the maintenance of genomic integrity and of the numerous interactions of PARP-1 with single- and double-strand break sensing enzymes and with apoptosis-related proteins, PARP inhibitors have become an interesting avenue for increasing the cytotoxicity of radiation therapy as well as of DNA-damaging anticancer drugs. The first indication of the potential of PARP-1 inhibitors as coeffectors of cancer therapies came from the elegant work of Durkacz and co-workers, who showed that benzamides increased the toxicity of alkylating agents in L1210 cells, prevented cellular NAD^+ lowering, and caused an increase in the amounts of DNA strand breaks *(35)*. Benzamides are analogs of nicotinamide, the byproduct of NAD^+ hydrolysis by PARPs to polymerize ADP-ribose. However, because millimolar levels of 3-aminobenzamide were necessary for efficient PARP inhibition, it could not be considered clinically useful because of various side effects and its lack of specificity at that concentration. Therefore, subsequent development of PARP inhibitors aimed at an increased potency as well as specificity. A second generation of inhibitors was developed by Suto and co-workers *(36)*. These consisted of dihydroisoquinoline derivatives with particularly high inhibitory potency (micromolar range) when substituted at the 5 position *(36,37)*. The most potent inhibitor of this series was PD128763, which was shown to greatly increase cell killing by X-rays *(38)*. This work also demonstrated that restriction of the carboxamide moiety in the anti conformation improves inhibitor potency, consistent with the observation that this is the biologically active conformation of the carboxamide group of NAD^+ *(39)*. However, while characterizing the role of PARP-1 in BER (using γ-irradiation and alkylating agents), Satoh and co-workers found that benzamides were very weak inhibitors of BER and that 3-aminobenzamide and 6(5*H*)-phenanthridinone could inhibit BER by only 50% *(40)*. This work demonstrated that at physiologically acceptable concentrations of these inhibitors, PARP activity is not completely inhibited and it indicated the need for inhibitors with higher affinity and potency to completely block BER. To efficiently inhibit PARP activity in vivo, it is evaluated that competitive inhibitors must display a K_i in the nanomolar range such that they can compete with high intracellular concentrations of NAD^+ for which PARP-1 has a K_m of about

$50\ \mu M$ *(41)*. The third generation of inhibitors, developed by Agouron Pharmaceuticals in collaboration with the University of Newcastle group, is likely to reach these ranges of potency. These inhibitors include quinazolinones, benzoxazole 4-carboxamide, and benzimidazole 4-carboxamide derivatives *(42–44)*. Nanomolar concentrations of these compounds inhibit PARP-1 and could provide an interesting cytotoxic potential *(45,46)*. These various classes of competitive inhibitors, reviewed recently *(47)*, are summarized in Table 2.

3. PARP INHIBITORS IN CANCER THERAPY

The potential of PARP competitive inhibitors to increase the cytotoxicity of anticancer drugs has been assessed in a variety of cell lines *(45,46,50–54)*. These studies have mostly focused on benzamide, 3-aminobenzamide, PD128763, and the quinazolinone NU1025. Although all of these inhibitors significantly increased the sensitivity of cells to the methylating agent temozolomide, PD128763 and NU1025 were nearly 50-fold more potent than the benzamides at increasing sensitivity to temozolomide, consistent with their higher potency as PARP inhibitors *(50)*. Similarly, tumor cytotoxicity resulting from inhibition of topoisomerase I with topotecan and camptothecin was potentiated by combining these drugs with NU1025 and the benzimidazole-4-carboxamide NU1085 *(46,53)*. In contrast, no potentiation of the topoisomerase II inhibitor etoposide was observed *(53)*. PARP inhibitors were also found to enhance the efficacy of ionizing radiation. The recovery of L1210 cells treated with lethal doses of γ-irradiation was compromised when cells were simultaneously treated with NU1025 *(45)*.

In vivo studies with 3-aminobenzamide, a first-generation inhibitor, was shown to increase the antitumor activity of a number of drugs *(47)*. However, the observed potentiation may have resulted from effects on body temperature and drug clearance, in addition to PARP inhibition. A recent study described the use of a third-generation inhibitor, NU1025, in combination with temozolomide *(55)*. This study demonstrated that an intracerebral injection of NU1025 before the systemic administration of the methylating agent increased its antitumor activity toward lymphomas located in the central nervous system of mice. Noteworthy, an intraperitoneal administration of NU1025 did not improve temozolomide antitumor activity, suggesting that the route of administration and type of formulations used to deliver PARP inhibitors are likely to be crucial to their in vivo potency. Therefore, future development of PARP inhibitors should address these issues, as well as those of specificity toward the now identified numerous PARPs, improved degree of potency, and solubility in aqueous solvents.

4. SPECIFICITY OF INHIBITORS

Because PARPs share a common catalytic domain that binds NAD^+, it is likely that many of the functions attributed to PARP-1 from chemical inhibition studies

Table 2
Classes of Competitive Inhibitors of PARPs

Type		*Structure*	IC_{50}	*Refs.*
Benzamides	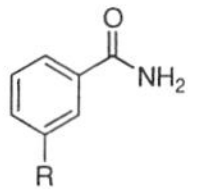	Benzamide: R=H 3-Aminobenzamide: R=NH_2	22 μ*M* 8–33 μ*M*	*37*
Isoquinolines	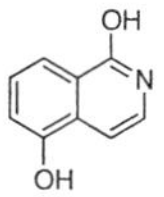	1,5-dihydroxyisoquinoline	0.39 μ*M*	*37*
	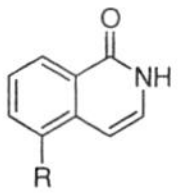	5-aminoisoquinolinone: R=$NH_3^+Cl^-$	0.24 μ*M*	*36*
	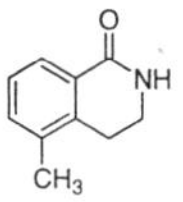	PD128763	0.13–0.4 μ*M*	*36*
	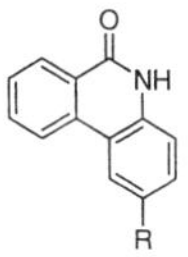	6(5*H*)-phenanthridinone: R=H PJ-34: R=$NHCOCH_2N(CH_3)_2$	0.52 μ*M* 0.02 μ*M*	*37, 48, 49*
	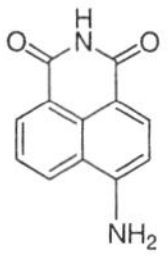	4-amino-1,8-naphthalimide	0.18 μ*M*	*37*
Benzoxazoles	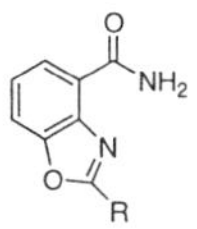	NU1056: R=CH_3 NU1051: R=Phenyl	9.8 μ*M* 2.1 μ*M*	*42*
Benzimidazoles	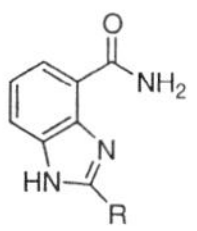	NU1064: R=CH_3 NU1085: R=4'-HOC_6H_4	1 μ*M* 0.08 μ*M*	*44*
Quinazolinones	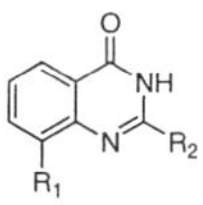	NU1025: R_1=OH, R_2=CH_3 NU1057: R_1=OH, R_2=4'-$NO_2C_6H_4$	0.4 μ*M* 0.2 μ*M*	*43*

reflect the sum of the activities of the PARP enzymes (Table 1 *[56]*). Because PARP-1 automodification should be minimal when cells are treated with chemical inhibitors, it should not be released from DNA strand breaks and may mask breaks from the other proteins involved in BER, a model referred to as the transdominant inhibition of DNA strand break repair *(57)*. On the other hand, cells devoid of PARP-1 (PARP-1 [–/–] cells) do not have the same phenotype as cells devoid of PARP activity (PARP-1 [+/+] cells treated with PARP inhibitors). In the presence of DNA damage, polymer synthesis in PARP-1 (–/–) cells is about 20–30% that of normal (PARP [+/+]) cells, confirming that other PARPs contribute significantly to poly(ADP-ribosylation. It is interesting to note that the transdominant inhibition of strand-break repair by PARP-1 will not occur in PARP-1 (–/–) cells because they lack PARP-1 itself. One of the aims of the future development of competitive PARP inhibitors should be to increase their specificity toward PARPs, which may play roles in cancer development. In addition to PARP-1 and PARP-2, which are involved in DDR, other PARPs could be interesting targets. Particularly, potential applications of VPARP inhibitors to offset multidrug resistance in some tumors has been suggested in view of the upregulation of LRP in tumor cells, a major vault protein involved in the sequestration and exocytosis of drugs *(54)*. Similarly, inhibition of tankyrase-1 and -2 could result in the inhibition of telomere elongation resulting from sustained telomerase activity, a condition correlated with the unlimited proliferation of tumor cells *(54)*. Tankyrases positively regulate telomere length by interacting with telomere-repeat factor 1, a negative regulator of telomere extension by telomerase *(58)*. It has also been suggested that the recently identified Ti PARP (PARP-7) may play a role in tumor progression with the finding of a high degree of similarity in its N-terminal region with a sequence expressed in T cells from a progressing murine mammary carcinoma but not in T cells from regressing tumor *(28)*. Further research may therefore identify Ti PARP as another interesting target for inhibition.

5. STEPS TOWARD THE DEVELOPMENT OF PARP-1-SPECIFIC INHIBITORS

Because only PARP-1 arbors a DNA-binding domain with zinc fingers, noncompetitive inhibition of PARP-1 has been achieved using its DNA-binding domain *(59,60)*. This transdominant inhibition resulted in levels of polymer synthesis of about 20–30% that of untreated cells, similar to what is found in PARP-1 (–/–) cells. This was expected because other PARPs are not inhibited by this noncompetitive inhibitor. More recently, nanomolar concentrations of the 24-kDa N-terminal apoptotic fragment of PARP-1 has been found to inhibit PARP-1 *(61,62)*. Because this very specific PARP-1 inhibition also impairs BER, it could become an avenue of great interest for the improvement of chemotherapy.

A novel and promising strategy for the selective inhibition of the various PARPs may be through "RNA inhibition by feeding" *(63)*. By this approach, the mRNAs encoding the targeted PARPs may be depleted using small antisense RNA. However, because of the very low levels of PARP-1 mRNA and a slow turnover for this protein (Aubin and Poirier, unpublished observations), the success of this approach is uncertain. Indeed, a significant decrease of PARP-1 protein levels required long depletion times using antisense PARP-1 RNA *(64)*.

Because only PARP-1 bears zinc fingers, specific inhibition of PARP-1 has also been obtained with a metal chelation approach developed by Zahradka and Ebisuzaki *(65)*. This method allows one to remove zinc atoms from purified PARP-1 with orthophenantroline. Similarly, Buki and collaborators used 6-amino-1,2-benzopyrone to remove one zinc atom from PARP-1, most probably selectively *(66)*. This caused full inhibition of PARP-1 in vitro and was shown to have some effect on NAD^+ lowering and cell survival in intact cells *(66,67)*.

Recently, a yeast cell survival assay has been established to compare the effect of inhibitors on PARP-1 and PARP-2 *(68)*. These data suggested that some analogs of 6(5*H*)-phenanthridinone differentially inhibits PARP-1 and PARP-2, both in yeast survival assays and with direct enzyme inhibition.

Because PARP enzymes may share a similar structure of their catalytic site, it may be difficult to develop competitive inhibitors with high specificity and selectivity toward the various PARPs. However, it may be possible to exploit their distinct cellular distributions by targeting PARP inhibitors to specific cellular compartments, such as Golgi to inhibit tankyrase-1 and -2 and nucleus for PARP-1 and PARP-2.

6. STRATEGIES FOR THE DEVELOPMENT OF CLINICALLY USEFUL INHIBITORS OF PARPS

The characterization of the active conformation of NAD^+ *(39)*, the development of inhibitors with improved specificity and potency and the analysis of the crystal structure of the catalytic domain of chicken PARP-1 bound to the inhibitor PD128763 *(69)* all have advanced our understanding of the interaction between PARP-1 and the inhibitors and will significantly contribute to the design of new, more potent compounds. The availability of high-potency inhibitors appears crucial for the success of PARP inhibition in vivo, in view of the absence of a direct relationship between inhibition of BER and inhibition of PARP-1 activity. Satoh and co-workers demonstrated that the automodification of 1 PARP-1 molecule with only 25 residues of ADP-ribose was sufficient for a BER event to be completed, suggesting that a nearly complete inhibition of PARPs is necessary for a significant inhibition of BER *(40)*. Moreover, it must be kept in mind that the in vitro activity of inhibitors (i.e., on purified PARP-1) does not necessarily correlate with their in vivo activity *(70)*. A number of factors known to affect PARP-1 activity will have to be considered in the assessment of the potency of new competitive inhibitors, including the fact that they must compete with very high local NAD^+ concentrations, that the K_m of PARP-1 for NAD^+ varies as a function of its automodification status *(71)* and that transcription factors and histones stimulate PARP-1 activity *(1)*. Screening of inhibitors on purified PARPs would benefit from the use of a poly(ADP-ribose) turnover system *(72)*. This system contains poly(ADP-ribose) glycohydrolase (PARG) and may as well include histones or other known coactivators affecting PARP activity and its K_m for NAD^+. The IC_{50} of PARP competitive inhibitors determined in this system should better reflect their potency in vivo. This system could be combined with the recently described scintillation assay that allows for high-throughput screening of PARP inhibitors *(73)*. However, it will remain critical to evaluate the efficiency of these inhibitors in intact cells, where they will encounter physiological distributions and concentrations of NAD^+, PARG activity, and coregulators of PARP activity.

Another important aspect that must be considered in the development of new PARP inhibitors is their solubility in aqueous solvents. Most of the currently available competitive PARP inhibitors are soluble only in organic solvents such as dimethyl sulfoxide and ethanol, which are known to cause side effects irrelevant to the PARP inhibitor used. Moreover, these solvents can also display stimulatory and/or inhibitory effects on PARP-1 *(74)*. To better define the activity of these inhibitors, it will be important to test them in PARP-1 (–/–) versus PARP-1 (+/+) backgrounds and compare with the effect of solvents alone. The need for water-soluble inhibitors is starting to be addressed with the development of water-soluble isoquinoline derivatives (5-aminoquinolinone and PJ-34; *see* Table 2). These should display reduced side effects in comparison to currently available PARP inhibitors. Moreover, because very low concentrations of 5-aminoisoquinolinone and PJ-34 efficiently inhibit PARP-1 *(37,47)*, their specificity should be high because they should not inhibit other NAD^+-binding enzymes such as mono(ADP ribosyl) transferase.

7. CONCLUSIONS

Work conducted in the last two decades indicates that PARP inhibitors can increase the potency of DNA-damaging anticancer agents. The recent development of competitive inhibitors with potency in the nanomolar range and with improved solubility in water suggests that the availability of a pharmacologically acceptable inhibitor is foreseeable. In vivo studies are now required to address the issues of tumor toxicity versus toxicity toward normal cells and tissues, as well as formulations and delivery systems that may enhance tumor targeting. Moreover, the availability of the PARP-1 (–/–) mice and eventually of knockout mice for the other PARPs will help to evaluate inhibitors side effects as well as the importance of the other PARPs in DNA repair and as targets for cancer therapy.

8. ACKNOWLEDGMENTS

The authors would like to thank Dr. N. J. Curtin and Dr. R. J. Griffin for reviewing Table 2, Dr. Donald Poirier for helpful discussions, and Marie-Ève Bonicalzi for critically reviewing the manuscript.

REFERENCES

1. D'Amours D, Desnoyers S, D'Silva I, et al. Poly(ADP-ribosyl)ation reactions in the regulation of nuclear functions. *Biochemical J* 1999;342:249–268.
2. Herceg Z, Wang Z-Q. Functions of poly(ADP-ribose) polymerase (PARP in DNA repair, genomic integrity and cell death. *Mutat Res* 2001;477:97–110.
3. Davidovic L, Vodenicharov M, Affar EB, et al. Importance of poly(ADP-ribose) glycohydrolase in the control of poly(ADP-ribose) metabolism. *Exp Cell Res* 2001;268:7–13.
4. Shall S, de Murcia G. Poly(ADP-ribose) polymerase-1: what have we learned from the deficient mouse model? *Mutat Res* 2000;460:1–15.
5. Masson M, Niedergang C, Schreiber V, et al. XRCC1 is specifically associated with poly(ADP-ribose) polymerase and negatively regulates its activity following DNA damage. *Mol Cell Biol* 1998;18:3563–3571.
6. Dantzer F, de La Rubia G, Ménissier-De Murcia J, et al. Base excision repair is impaired in mammalian cells lacking poly(ADP-ribose) polymerase-1. *Biochemistry* 2000;39:7559–7569.
7. Prasad R, Lavrik OI, Kim S-J, et al. DNA polymerase β-mediated long patch base excision repair. *J Biol Chem* 2001;276:32,411–32,414.
8. Morrison C, Smith GC, Stingl L, et al. Genetic interaction between PARP and DNA-PK in V(D)J recombination and tumorigenesis. *Nature Genet* 1997;17:479–482.
9. Boulton SJ, Gartner A, Reboul J, et al. Combined functional genomic maps of the *C. elegans* DNA damage response. *Science* 2002;295:127–131.
10. Wang Y, Cortez D, Yazdi P, et al. BASC, a super complex of BRCA1-associated proteins involved in the recognition and repair of aberrant DNA structures. *Genes Dev* 2000;14:927–939.
11. Khanna KK, Lavin MF, Jackson SP, et al. ATM, a central controller of cellular responses to DNA damage. *Cell Death Differ* 2001;8:1052–1065.
12. Hassa PO, Hottiger MO. A role of poly(ADP-ribose) polymerase in NF-kappaB transcriptional activation. *Biol Chem* 1999;380:953–959.
13. Ullrich O, Diestel A, Eyüpoglu IY, et al. Regulation of microglial expression of integrins by poly(ADP-ribose) polymerase-1. *Nature Cell Biol* 2001;3:1035–1042.
14. Simbulan-Rosenthal CM, Rosenthal DS, Hilz H, et al. The expression of poly(ADP-ribose) polymerase during differentiation linked DNA replication reveals that it is a component of the multiprotein DNA replication complex. *Biochemistry* 1996;35:11,622–11,633.
15. Kanai M, Uchida M, Hanai S, et al. Poly(ADP-ribose) polymerase localizes to the centrosomes and chromosomes. *Biochem Biophys Res Commun* 2000;278:385–389.
16. Borggrefe T, Wabl M, Akhmedov AT, et al. A B-cell-specific DNA recombination complex. *J Biol Chem* 1998;273:17025–17035.
17. Arlot-Bonnemains Y, Prigent C. A trigger for centrosome duplication. *Science* 2002;295:455–456.
18. Satoh MS, Lindahl T. Role of poly(ADP-ribose) formation in DNA repair. *Nature* 1992;356:356–358.

19. Satoh MS, Poirier GG, Lindahl T. NAD(+)-dependent repair of damaged DNA by human cell extracts. *J Biol Chem* 1993;268:5480–5487.

20. Affar EB, Germain M, Poirier GG. Role of poly(ADP-ribose) polymerase in cell death. In: de Murcia G, Shall S, eds. *From DNA Damage and Stress Signalling to Cell Death,* Oxford University Press, New York, 2000:125–150.

21. Smith S. The world according to PARP. *Trends Biochem Sci* 2001;26:174–179.

22. Amé JC, Rolli V, Schreiber V, et al. PARP-2, a novel mammalian DNA damage-dependent poly(ADP-ribose) polymerase. *J Biol Chem* 1999;274:17860–17868.

23. Johansson M. A human poly(ADP-ribose) polymerase gene family (ADPRTL): cDNA cloning of two novel poly(ADP-ribose) polymerase homologues. *Genomics* 1999;57:442–445.

24. Kickhoefer VA, Siva AC, Kedersha NL, et al. The 193-kd vault protein, VPARP, is a novel poly(ADP-ribose) polymerase. *J Cell Biol* 1999;146:917–928.

25. Smith S, Giriat I, Schmitt A, et al. Tankyrase, a poly(ADP-ribose) polymerase at human telomeres. *Science* 1998;282:1484–1487

26. Kaminker PG, Kim SH, Taylor RD, et al. TANK2, a new TRF1-associated poly(ADP-ribose) polymerase, causes rapid induction of cell death upon overexpression. *J Biol Chem* 2001;276:35,891–35,899.

27. Lyons RJ, Deane R, Lynch DK, et al. Identification of a novel human tankyrase through its interaction with the adaptor protein Grb14. *J Biol Chem* 2001;276:17,172–17,180.

28. Ma Q, Baldwin KT, Renzelli AJ, et al. TCDD-inducible poly(ADP-ribose) polymerase: a novel response to 2,3,7,8-tetracholorodibenzo-*p*-dioxin. *Biochem Biophys Res Commun* 2001;289:499–506.

29. Sallmann FR, Vodenicharov MD, Wang ZQ, et al. Characterization of sPARP-1. An alternative product of PARP-1 gene with poly(ADP-ribose) polymerase independent activity of DNA strand breaks . *J Biol Chem* 2000;275:15,504–15,511.

30. Spenlehauer C, Augustin A, Dumond H, et al. Poly(ADP-ribose) polymerase-3 localized to the daughter centriole is involved in centrosome duplication and cell cycle progression. *Biol Cell* 2001;93:205–209.

31. Smith S, de Lange T. Cell cycle dependent localization of the telomeric PARP, tankyrase, to nuclear pore complexes and centrosomes. *J Cell Sci* 1999;112:3649–3656.

32. Chi NW, Lodish HF. Tankyrase is a golgi-associated mitogen-activated protein kinase substrate that interacts with IRAP in GLUT4 vesicles. *J Biol Chem* 2000;275:38,437–38,444.

33. Schreiber V, Amé J-C, Niedergang C, et al. Characterization of PARP-2, a novel mammalian DNA damage-dependent poly(ADP-ribose) polymerase involved in base excision repair. ADPR 2001: 13th International Symposium on ADP-ribosylation. 200 abstract #5.

34. Sbodio JI, Lodish HF, Chi N-W. Tankyrase-2 oligomerizes with tankyrase-1 and binds to both TRF1 (telomere-repeat-binding factor 1) and IRAP (insuline-responsive aminopeptidase). *Biochem J* 2002;361:451–459.

35. Durkacz BW, Omidiji O, Gray DA, et al. (ADP-ribose)$_n$ participates in DNA excision repair. *Nature* 1980;283:593–596.

36. Suto MJ, Turner WR, Arundel-Suto CM, et al. Dihydroisoquinolines: the design and synthesis of a new series of potent inhibitors of poly(ADP-ribose) polymerase. *Anti-Cancer Drug Des* 1991;7:107–117.

37. Banasik M, Komura H, Shimoyama M, et al. Specific inhibitors of poly(ADP-ribose) synthetase and mono(ADP-ribosyl)transferase. *J Biol Chem* 1992;267:1569–1575.

38. Arundel-Suto, CM, Scavone SV, Turner WR, et al. Effects of PD128763, a new potent inhibitor of poly(ADP-ribose) polymerase, on X-ray induced cellular recovery processes in Chinese Hamster V79 cells. *Radiat Res* 1991;126:367–371.

39. Li H, Goldstein BM. Carboxamide group conformation in the nicotinamide and thiazole-4-carboxamide rings: implications for enzyme binding. *J Med Chem* 1992;35:3560–3567.

40. Satoh MS, Poirier GG, Lindahl T. Dual function for poly(ADP-ribose) synthesis in response to DNA strand breakage. *Biochemistry* 1994;33:7099–7106.
41. Althaus FR, Richter C. *ADP-Ribosylation of Proteins. Enzymology and biological Significance.* Springer-Verlag, Berlin, 1987.
42. Griffin RJ, Pemberton LC, Rhodes D, et al. Novel potent inhibitors of the DNA repair enzyme poly(ADP-ribose) polymerase (PARP). *Anti-Cancer Drug Des* 1995;10:507–514.
43. Griffin RJ, Srinivasan S, Bowman K, et al. Resistance-modifying agents. 5. Synthesis and biological properties of quinazolinone inhibitors of the DNA repair enzyme poly(ADP-ribose) polymerase (PARP). *J Med Chem* 1998;41:5247–5256.
44. White AW, Almassy R, Calvert AH, et al. Resistance-modifying agents. 9. Synthesis and biological properties of benzimidazole inhibitors of the DNA repair enzyme poly(ADP-ribose) polymerase. *J Med Chem* 2000;43:4084–4097.
45. Bowman KJ, White A, Golding BT, et al. Potentiation of anti-cancer agent cytotoxicity by the potent poly(ADP-ribose) polymerase inhibitors NU1025 and NU1064. *Br J Cancer* 1998;78:1269–1277.
46. Delaney CA, Wang LZ, Kyle S, et al. Potentiation of temozolomide and topotecan growth inhibition and cytotoxicity by novel poly(adenosine diphosphoribose) polymerase inhibitors in a panel of human tumor cell lines. *Clin Cancer Res* 2000;6:2860–2867.
47. Curtin NJ, Golding BT, Griffin RJ, et al. New poly(ADP-ribose) polymerase inhibitors for chemo- and radiotherapy of cancer. In: de Murcia G, Shall S, eds. *From DNA Damage and Stress Signalling to Cell Death,* Oxford University Press, New York, 2000:177–206.
48. Li J-H, Serdyuk l, Ferraris DV, et al. Synthesis of substituted 5[*H*]phenanthridin-6-ones as potent poly(ADP-ribose)polymerase-1 (PARP1) inhibitors. *Bioorg Med Chem Let* 2001;11: 1687–1690.
49. Soriano FG, Virag L, Jagtap P, et al. Diabetic endothelial dysfunction: the role of poly(ADP-ribose) polymerase activation. *Nature Med* 2001;7:108–113.
50. Boulton S, Pemberton LC, Porteous JK, et al. Potentiation of temozolomide-induced cytotoxicity: a comparative study of the biological effects of poly(ADP-ribose) polymerase inhibitors. *Br J Cancer* 1995;72:849–856.
51. Liu L, Taverna P, Whitacre CM, et al. Pharmacologic disruption of base excision repair sensitizes mismatch repair-deficient and -proficient colon cancer cells to methylating agents. *Clin Cancer Res* 1999;5:2908–2917.
52. Holl V, Coelho D, Weltin D, et al. Modulation of the antiproliferative activity of anticancer drugs in hematopoietic tumor cell lines by the poly(ADP-ribose) polymerase inhibitor 6(5*H*) phenanthridinone. *Anticancer Res* 2000;20:3233–3241.
53. Bowman KJ, Newell DR, Calvert AH, et al. Differential effects of the poly (ADP-ribose) polymerase (PARP) inhibitor NU1025 on topoisomerase I and II inhibitor cytotoxicity in L1210 cells in vitro. Br J Cancer. 2001;84:106–112.
54. Tentori L, Portarena I, Graziani G. Potential clinical applications of poly(ADP-ribose) polymerase (PARP) inhibitors. *Pharmacol Res* 2002;45:73–85.
55. Tentori L, Leonetti C, Scarsella M, et al. Combined treatment with temozolomide and poly(ADP-ribose) polymerase inhibitor enhances survival of mice bearing hematologic malignancy at the central nervous system site. *Blood* 2002;99:2241–2244.
56. Shah GM, Poirier D, Desnoyers S, et al. Complete inhibition of poly(ADP-ribose) polymerase activity prevents the recovery of C3H10T1/2 cells from oxidative stress. *Biochim Biophys Acta* 1996;1312:1–7.
57. Lindahl T, Satoh MS, Poirier GG, et al. Post-translational modification of poly(ADP-ribose) polymerase induced by DNA strand breaks. *Trends Biochem Sci* 1995;20:405–411.
58. Cook BD, Dynek JN, Chang W, et al. Role for the related poly(ADP-ribose) polymerases tankyrase 1 and 2 at human telomeres. *Mol Cell Biol* 2002;22:332–342.

58. Küpper JH, de Murcia G, Burkle A. Inhibiton of poly(ADP-ribosyl)ation by overexpressing the poly(ADP-ribose) polymerase DNA-binding domain in mammalian cells. *J Biol Chem* 1990;265:18,721–18,724.

60. Molinete M, Vermeulen W, Bürkle A, et al. Overproduction of the poly(ADP-ribose) polymerase DNA-binding domain blocks alkylation-induced DNA repair synthesis in mammalian cells. *EMBO J* 1993;12:2109–2117.

61. D'Amours D, Sallmann FR, Dixit VM, et al. Gain-of-function of poly(ADP-ribose) polymerase-1 upon cleavage by apoptotic proteases: implications for apoptosis. *J Cell Sci* 2001;114:3771–3778.

62. Yung TM, Satoh MS. Functional competition between poly(ADP-ribose) polymerase and its 24-kDa apoptotic fragment in DNA repair and transcription. *J Biol Chem* 2001;276:11,279–11,286.

63. Paddison PJ, Caudy AA, Hannon GJ. Stable suppression of gene expression by RNAi in mammalian cells. *Proc Natl Acad Sci USA* 2002;99:1443–1448.

64. Simbulan-Rosenthal CM, Rosenthal DS, Ding R, et al. Prolongation of the p53 response to DNA strand breaks in cells depleted of PARP by antisense RNA expression. *Biochem Biophys Res Commun* 1998;253:864–868.

65. Zahradka P, Ebisuzaki K. Poly(ADP-ribose) polymerase is a zinc metalloenzyme. *Eur J Biochem* 1984;142:503–509.

66. Buki KG, Bauer PI, Mendeleyev J, et al. Destabilization of Zn2+ coordination in ADP-ribose transferase (polymerizing) by 6-nitroso-1,2-benzopyrone coincidental with inactivation of the polymerase but not the DNA binding function. *FEBS Lett* 1991;290:181–185.

67. Rice WG, Hillyer CD, Harten B, et al. Induction of endonuclease-mediated apoptosis in tumor cells by C-nitroso-substituted ligands of poly(ADP-ribose) polymerase. *Proc Natl Acad Sci USA* 1992;89:7703–7707.

68. Perkins E, Xun D, Nguyen A, et al. Novel inhibitors of poly(ADP-ribose) polymerase/PARP1 and PARP2 identified using a cell-based screen in yeast. *Cancer Res* 2001;61:4175–4183.

69. Ruf A, de Murcia G, Schulz GE. Inhibitor and NAD+ binding to poly(ADP-ribose) polymerase as derived from crystal structures and homology modeling. *Biochemistry* 1998;37:3893–3900.

70. Sestili P, Spadoni G, Balsamini G, et al. Structural requirements for inhibitors of poly(ADP-ribose) polymerase. *J Cancer Res Clin Oncol* 1990;116:615–622.

71. Zahradka P, Ebisuzaki K. A shuttle mechanism for DNA-protein interactions. The regulation of poly(ADP-ribose) polymerase. *Eur J Biochem* 1982;127:579–585.

72. Ménard L, Thibault L, Poirier GG. Reconstitution of an in vitro poly(ADP-ribose) turnover system. *Biochim Biophys Acta* 1990;1049:45–58.

73. Cheung A, Zhang J. A scintillation proximity assay for poly(ADP-ribose) polymerase. *Anal Biochem* 2000;282:24–28.

74. Banasik M, Ueda K. Dual inhibitory effects of dimethyl sulfoxide on poly(ADP-ribose) synthetase. *J Enzyme Inhib* 1999;14:239–250.

10 Relationship Among DNA Repair Genes, Cellular Radiosensitivity, and the Response of Tumors and Normal Tissues to Radiotherapy

David Murray, PhD, and Adrian C. Begg, PhD

CONTENTS

1. INTRODUCTION

Radiotherapy (XRT) is an important treatment modality for many cancers. At the present time, the doses prescribed to tumors are largely based on the clinically determined tolerance of the normal tissues in the radiation field *(1)*. The optimization of XRT treatment plans involves the computation of two factors for a given dose prescription: the tumor control probability (TCP) and the normal tissue complication probability (NTCP). The therapeutic outcome depends on maximizing the TCP for a given clinically acceptable level of NTCP. The 5% level of NTCP is often used in this regard. Local tumor control and, thus, overall patient survival may be increased either by shifting the TCP curve to lower doses or by shifting the NTCP curve to higher doses. One approach to shifting these curves is by appropriate patient selection. This would involve grouping patients

From: *Cancer Drug Discovery and Development: DNA Repair in Cancer Therapy*
Edited by: L. C. Panasci and M. A. Alaoui-Jamali © Humana Press Inc., Totowa, NJ

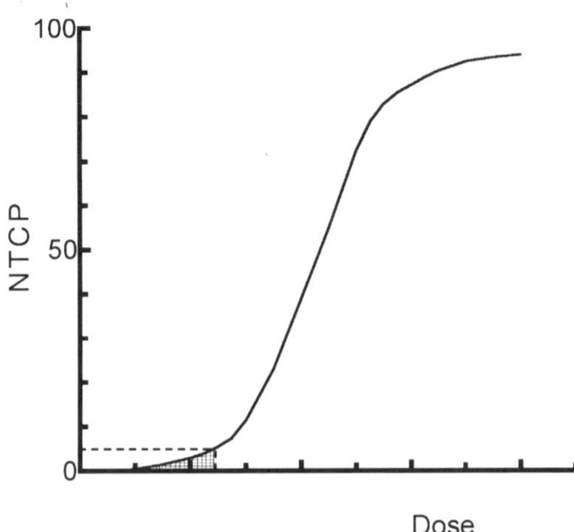

Fig. 1. Hypothetical relationship between NTCP and dose for patients undergoing XRT. The shaded area represents the 5% envelope of NTCP and will probably contain all genetically hypersensitive individuals.

for treatment planning based on pretreatment estimates of their likely normal tissue and/or tumor radiosensitivity. We will illustrate this with respect to NTCPs, although similar reciprocal arguments apply to TCP. Clinically used NTCPs have generally been derived empirically from population averages and do not, therefore, reflect the significant differences among patients in their susceptibility to late radiation sequelae *(2)*. Thus, for a population of nonselected patients receiving XRT for a particular type of tumor, the NTCP curve will be broadened because of interpatient heterogeneity in radiosensitivity. Patients who are at the most radiosensitive end of the spectrum will extend the NTCP curve to lower doses (i.e., broaden it in the critical region for the therapeutic ratio). Indeed, the left-hand 5% envelope of the NTCP curve (*see* Fig. 1) will probably contain all of these hypersensitive individuals. The adjusted NTCP curve for patients who do not exhibit hypersensitivity to XRT (i.e., for the majority of patients) will have a steeper slope and be shifted to the right of that for the nonstratified group, allowing tumor dose escalation to these patients.

Accounting for interpatient heterogeneity in normal tissue tolerance levels or tumor radioresponsiveness (i.e., individualization of XRT treatment plans), is not ordinarily a part of treatment planning. However, it could and should be, if appropriate methodologies were available to discriminate unusually radiosensitive patients or radioresistant tumors prior to treatment. The consequent need for rapid and robust estimates of individual radiosensitivity *(2)* has led to the development of various assays for patient response. The clonogenic cell survival assay has traditionally been the gold standard for evaluating the radiation sensitivity of both normal tissues and tumors. A number of investigators have studied the in vitro response of normal cells obtained from patients as a surrogate for normal tissue radiosensitivity (i.e., for NTCPs of both early- and late-reacting normal tissues) *(2–4)*. A retrospective analysis of a large body of in vitro skin fibroblast clonogenic survival data suggested that the use of predictive assays for normal tissue radiosensitivity might realize a significant therapeutic gain for a subset of patients *(5)*. It should also be noted that there was a significant variation in normal cell radiosensitivity among individuals in these studies, a point to which we will return later. For tumors too, there has been some success in predicting clinical outcome on the basis of in vitro radiosensitivity testing using clonogenic assays (e.g., with cervical carcinomas *[6]* and head and neck cancers *[7]*). Unfortunately, for a number of reasons, such assays are unlikely to be sufficiently rigorous to modify cancer management. This includes the following observations:

1. Their precision and reproducibility is inadequate for routine clinical use *(2,4)*.
2. Because of the requirement for numerous replicate determinations, such assays can take several weeks to generate reliable data.
3. Patient hypersensitivity to XRT is not always associated with cellular hypersensitivity (e.g., refs. *8–10*).

Because of such issues, a number of investigators have turned their attention to the use of molecular predictive techniques involving genetic screening. Although a large number of molecular determinants of radiosensitivity have been identified (it is currently estimated that there are more than 100 "radiosensitivity" genes, e.g., *see* ref. *11*), assays for the various genes/proteins known to be involved in the repair of DNA damage have been at the forefront of this effort. It is these studies that are the focus of the current review. Here, we will examine DNA repair from several perspectives:

1. Their contribution to radioresistance in tumors.
2. Their contribution to normal tissue radiosensitivity.
3. The question of whether defects in repair among the general population might predispose to cancer.

With respect to issue (3), the connection between defects in DNA repair and cancer susceptibility has been known for many years based on the phenotypes of those rare individuals who are afflicted with certain genetically determined clinical syndromes such as xeroderma pigmentosum (XP) *(12)*, ataxia telangiectasia (AT) *(13,14)*, ataxia–telangiectasia-like disorder (ATLD) *(15)*, and Nijmegen breakage syndrome (NBS) *(13,16)* (for a review, see ref. *17*). Even among the broader population, it is apparent based on epidemiological studies using functional repair assays in lymphocytes or other accessible cell types that DNA repair capacity is highly variable among individuals and that low repair capacity is a significant risk factor for the development of some cancers *(18–32)*. More recently, it has been shown that this interindividual variation in DNA repair capacity is genetically determined. Thus, whereas inactivating mutations in DNA repair genes are extremely rare *(33)* and cannot account for the observed variation in DNA repair capability in the general population, a number of common polymorphisms (allelic variants) of genes that encode for DNA repair proteins have now been described that appear to result in diminished (rather than absent) repair function *(29,34–37)*. It has been proposed that these polymorphisms act in concert with environmental carcinogens to increase susceptibility to cancer *(29)*. By extension, they may also be important determinants of the response of cancer patients to anticancer therapeutics.

The suitability of specific repair factors as targets for tumor radiosensitization, especially involving genetic modulation *(38)*, is another important clinical issue but will not be discussed here because of space limitations. It should be noted, however, that a better understanding of how DNA repair pathways are altered in human cancers should ultimately facilitate the rational design of agents that may selectively radiosensitize tumor cells *(39)*.

2. PHENOTYPIC DNA REPAIR

This approach focuses on the physical measurement of different types of ionizing radiation (IR)-induced DNA damage and their rate of removal from the genome, rather than on a particular DNA repair factor or pathway. Such lesions include base damage, apurinic/apyrimidinic (AP) sites, sugar damage, single-strand breaks (SSBs), double-strand breaks (DSBs), DNA interstrand crosslinks (ICLs), and DNA–protein crosslinks (DPCs) (*see* Fig. 2).

2.1. Studies of SSB Rejoining

Several techniques have been applied to the determination of IR-induced SSBs in mammalian cells in the biologically relevant dose range *(40)*. These include alkaline elution *(41)*, alkaline unwinding *(42)*, and the alkaline comet assay, which can detect SSBs in individual cells and thus assess heterogeneity of response *(43–47)*. These various techniques have their advantages and disadvantages.

2.1.1. SSB Rejoining in Normal Cells

Marked interindividual variability in the rate of rejoining of IR-induced SSBs was apparent in lymphocytes from healthy volunteers (e.g., refs. *48* and *49*). A number of studies have reported SSB repair deficits in normal cells derived from cancer patients in comparison to control populations after irradiation in vitro. For example, in a study of squamous cell carcinomas of the head and neck (SCCHN) it was found that oral cavity carcinoma patients (but not laryngeal carcinoma patients) showed slower SSB repair than controls *(50)*. Similarly, the repair of SSBs was slower in lymphocytes from three breast cancer patients than from

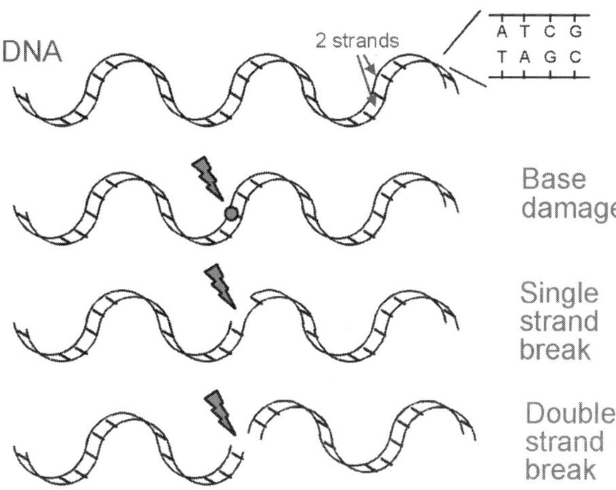

Fig. 2. Some of the classes of IR-induced DNA damage.

three healthy donors *(51)*. As compared to healthy donors, a slower rate of SSB repair and increased residual SSBs was found in lymphocytes from patients who developed thyroid tumors after XRT *(52)*. In contrast, the repair of SSBs in lymphocytes from chronic lymphocytic leukemia (CLL) patients was essentially normal *(53)*, as was SSB rejoining in fibroblasts cultured from individuals with hereditary retinoblastoma *(54)*.

With respect to cellular radiosensitivity, SSB repair capacity was not predictive of IR-induced cytotoxicity in a study of normal and malignant hematopoietic cells and lymphocytes *(53)* or of IR-induced micronuclei (MN) in lymphocytes from cancer patients *(55)*.

A number of investigators have looked for possible relationships between SSB repair status and adverse clinical normal tissue reactions to XRT. For example, Oppitz et al. *(56)* compared clinical reactions with SSB rejoining in skin fibroblasts from 30 patients and cell lines from 4 individuals with AT; fibroblasts from patients with increased early and late side effects exhibited slower SSB rejoining. In a later study, they found that SSB repair in lymphocytes (alkaline comet assay) was better able to discriminate patients with more acute skin reactions than either SSB repair or colony survival in fibroblasts *(57)*. Alapetite et al. *(58)* assessed the response to in vitro irradiation of lymphocytes from breast cancer and Hodgkin's disease patients who developed severe reactions to XRT compared to patients with average reactions and healthy donors. A subgroup of breast cancer overreactors (7/17) displayed increased levels of residual SSBs. Among Hodgkin's disease overreactors, only one patient showed defective repair. Interestingly, the patients with the most severe complications showed impaired SSB rejoining. Thus, impairment in SSB processing may be associated, in specific subgroups of cancer patients, with an increased risk of major XRT-induced complications.

2.1.2. SSB Rejoining in Tumor Cell Lines and Tumors

In general, there is no consistent correlation between the rate or extent of the rejoining of IR-induced SSBs and radiosensitivity for panels of human tumor cell lines, although, naturally, there are some exceptions for individual cell lines. For example, the efficiency of SSB rejoining was not predictive of radioresistance in a panel of 2 normal human cell lines and 12 early-passage human tumor cell lines *(59)*, in a panel of human lung cancer cell lines *(60)*, in 2 squamous carcinoma cell lines and a normal epidermal keratinocyte cell line *(61)*, or among a panel of 11 unrelated mammalian cell lines *(62)*. Although the most radiosensitive of three human ovarian carcinoma cell lines, CH-1, exhibited delayed SSB rejoining compared with the most radioresistant SKOV-3 line, the cell line of intermediate radiosensitivity (A-2780) showed SSB-rejoining kinetics similar to those of the radioresistant SKOV-3 line *(63)*.

2.2. Studies of Base Damage Repair

Methods for detecting specific types of base damage include ^{32}P-post-labeling assays and immunological methods using antibodies to specific structural modifications in irradiated DNA (e.g., refs. *64–66*). Approaches to achieving high sensitivity with immunochemical assays include the sandwich enzyme-linked immunosorbent assay (ELISA) method *(67)* and the coupling of immunochemical recognition with capillary electrophoresis and laser-induced fluorescence detection *(68)*. These assays have been recently reviewed *(69)* but have not yet been widely applied to patient radiosensitivity testing.

2.3. Studies of DSB Rejoining

It is generally believed that, of the various classes of DNA damage, unrepaired or misrepaired DSBs (produced as part of a complex/clustered lesion) are the primary initiators of the sequence of events leading to cytotoxicity following exposure to IR *(70)*. For this reason, considerable effort has been expended on the development of assays for the presence of DSBs in mammalian cells. The most widely used analytical techniques applied to prediction in XRT have been neutral elution *(71)*, pulsed-field gel electrophoresis (PFGE) *(72,73)*, the neutral comet assay *(74)*, and, more recently, constant-field gel electrophoresis (CFGE) *(75)*. Especially in earlier studies, the techniques used were relatively insensitive and required the use of supralethal doses of IR. As noted previously *(76)*, "the significance of investigations of repair mechanisms in populations of dying cells is always moot." Although each of these methods is subject to experimental artifacts, it has been suggested that PFGE is relatively insensitive to chromatin structure effects that plagued neutral elution *(77)*. CFGE is simpler and apparently equally sensitive. The neutral comet assay is particularly useful in that it can provide a means of assessing heterogeneity *(74)*.

2.3.1. DSB Rejoining in Normal Cells

Marked interindividual variability in the repair of IR-induced DSBs was apparent in studies of lymphocytes from 10 healthy volunteers *(48)* and of lymphocytes from both healthy volunteers and cancer patients *(78)*. Several studies have shown decreased repair of DSBs by normal cells (e.g., lymphocytes, fibroblasts) derived from cancer patients when compared to normal donor controls following in vitro irradiation. For example, cancer-associated DSB repair deficiency was reported in fibroblast cell lines derived from two non-Hodgkin's lymphoma patients *(79)* and in lymphocytes from cancer patients *(78)*. Other studies show no such consistent difference. Thus, fibroblasts from three individuals with hereditary retinoblastoma showed normal DSB repair characteristics *(54)*. Similarly, there was no obvious difference in the repair of DSBs in lymphocytes from CLL patients versus normal subjects *(53)*.

Many reports indicate significant correlations between the rate and/or extent of rejoining of IR-induced DSBs and clonogenic survival post-IR, for example, among the following:

1. Eleven fibroblast strains, 9 derived from cervical cancer patients and 2 from radiosensitive individuals *(80)*
2. Three non-transformed human fibroblast cell lines *(81)*
3. Primary human fibroblasts, five normal and two AT *(82)*
4. Fibroblast cell strains derived from seven individuals either with normal radiosensitivity or with genetic abnormalities known to cause altered radiosensitivity *(83)*
5. Fibroblasts from nine pre-XRT cervix cancer patients and two radiosensitive skin fibroblast cell strains *(84)*

Furthermore, among several B-lymphoblast cell lines, those that were deficient in DSB rejoining were invariably radiosensitive, but not all lines that were proficient in DSB rejoining were radioresistant *(85)*. The latter study illustrates an important recurring theme in the literature, namely that repair defects may impart a radiosensitive phenotype but this is not the only possible mechanism. Indeed, not all studies show such a correlation. Thus, in a study of normal hematopoietic cells and lymphocytes from CLL patients and normal donors, IR-induced cytotoxicity did not correlate with DSB repair capacity *(53)*.

A number of studies suggest a relationship between decreased repair of DSBs by normal cells from cancer patients irradiated ex vivo and patient hypersensitivity to severe normal tissue side effects in the clinic:

1. A radiosensitive fibroblast culture established from an acute lymphoblastic leukemia patient who died following XRT *(86)*

2. Lymphocytes of many (but not all) cancer patients who had experienced severe adverse reactions to XRT compared to healthy volunteers *(78)*
3. Lymphocytes from 50 patients with various tumor types, albeit this being a weak correlation *(87)*

In a variation on this theme, a study on 226 breast cancer patients found that the induction of DSBs in in vitro-irradiated lymphocytes correlated with severe XRT-induced skin damage *(88)*. Again, not all studies report such positive correlations. Thus, in a study of 16 pairwise matched head and neck cancer patients assessed at 2–7 yr post-XRT, extreme late reactions were not associated with a difference in DSB repair capacity in fibroblasts *(10)*.

A recent series of studies of breast cancer patients from the Manchester group using PFGE to assess residual DSBs (24 h post-IR) have raised considerable controversy concerning the issue of DSB rejoining as a predictive test for normal tissue reactions. Thus, among a group of 32 patients who were treated uniformly by XRT, there was no correlation between residual DSBs in irradiated keratinocytes and any of the late- or acute-reaction scores *(89)*. In contrast, among 39 patients, there was a significant positive correlation between residual DSBs in fibroblasts and the severity of late normal tissue fibrosis *(90)*, suggesting that fibroblasts, but not keratinocytes, may have a predictive value in this group of patients. However, when this finding was extended using a validation cohort of 50 patients, no significant relationship was found between residual DSBs in fibroblasts and late morbidity/fibrosis for the combined cohort of 88 patients, suggesting that the PFGE fibroblast assay is not a suitable predictor of late IR-induced fibrosis in the breast *(91)*.

Dikomey et al. *(92)* also investigated the relationship among DSBs, cell killing, and fibrosis using 12 confluent skin fibroblast lines derived from breast cancer patients who received postmastectomy XRT. Although the half-times of DSB repair were similar, there were considerable differences in the levels of residual DSBs. The number of residual DSBs was correlated with cell killing (surviving fraction at 3.5 Gy [SF3.5]) after delayed (14 h) but not after immediate plating. There was no significant relationship between residual DSBs and the excess risk of fibrosis determined for the respective patients.

2.3.2. DSB Rejoining in Tumor Cell Lines and Tumors

A number of studies have reported a relationship between intrinsic cellular radiosensitivity (clonogenic survival) and the rate and/or extent of rejoining of IR-induced DSBs. Examples include studies of a panel of 2 normal human cell lines and 12 unrelated early-passage human tumor cell lines *(59)*, 5 human cervical carcinoma cell lines *(93)*, 9 first-passage after-explant SCCHN cell lines *(94)*, 5 mammalian cell lines, including rodent cells and a human melanoma line *(95)*, a panel of 5 SCCHN cell lines, plus a normal fibroblast strain *(96)*, a radiosensitive human renal carcinoma cell line compared with human fibroblasts *(97)*, a small-cell lung carcinoma cell line, U-1285, compared with an undifferentiated large-cell lung carcinoma cell line, U-1810 *(98)*, 3 unrelated human tumor cell lines *(99)*, a panel of 5 human breast cancer cell lines and 1 human bladder cancer cell line *(100)*, 6 human tumor cell lines with SF2 ranging from 0.08 to 0.62 *(101)*, a group of 25 SCCHN and 8 sarcoma cell lines *(102)*, some radioresistant members of a panel of human lung cancer cell lines *(60)*, 9 mammalian (rodent and human) cell lines of broadly differing radiosensitivity *(103)*, a panel of 5 lung carcinoma cell lines *(104)*, some members of a panel of 6 bladder tumor cell lines *(105)*, and 9 cervix carcinoma cell lines, although the correlation was of borderline significance *(106)*. In a panel of seven cervical tumor cell lines, the *fraction* of DSBs remaining at 16 h post-IR (rather than residual DSB levels *per se*) showed a significant correlation with SF2 values *(107)*. Finally, it should be noted that some studies have reported a correlation of DSB *induction* and cell killing post-IR (e.g., refs. *108* and *109*).

On the other hand, several reports indicate no clear relationship between cellular radiosensitivity and the rejoining of IR-induced DSBs. These include studies of two human SCCHN cell lines *(110)*, two subclones of the radioresistant human SCCHN line SQ20B *(111)*, six unrelated human tumor cell lines of differing histology *(112)*, four human melanoma cell lines *(113)*, three unrelated human tumor cell lines *(114)*, three human epithelial cell lines, two squamous carcinoma cell lines, and a nontumorigenic epidermal keratinocyte cell line *(61)*, and clones with differing radiosensitivity derived from the human glioma cell line U-251 MG-Ho *(115)*.

2.4. Summary

There have been several excellent reviews on this subject *(70,116–118)* and the general conclusion is that a single parameter such as DSB rejoining cannot universally predict cellular radiosensitivity. Not surprisingly, then, a single end point cannot always accurately predict clinical outcome (either NTCP or TCP). Whether this lack of general applicability relates to the biological end points themselves or to the assays used is unclear. However, there are several reasons why DSB repair assays could be inadequate in this regard. First, there is the issue of heterogeneity in DSB complexity. The most biologically relevant DSBs are likely to be those that involve clustered damages in which multiple SSBs, damaged bases, or AP sites occur on opposing DNA strands but in close proximity *(119)*, and these lesions are likely to be difficult to repair *(120)*. The types of assay described thus far are unable to discriminate this particularly relevant subgroup of DSBs.

Second, the gene-specific repair or transcription-coupled repair (TCR) of DNA damage may be an important determinant of cellular radioresistance that might not be detected by "global" (i.e., genomewide) repair measurements. Mechanistic details of the transcription-coupled base excision repair (BER) pathway are beginning to emerge *(121–123)*. Although some IR-induced base

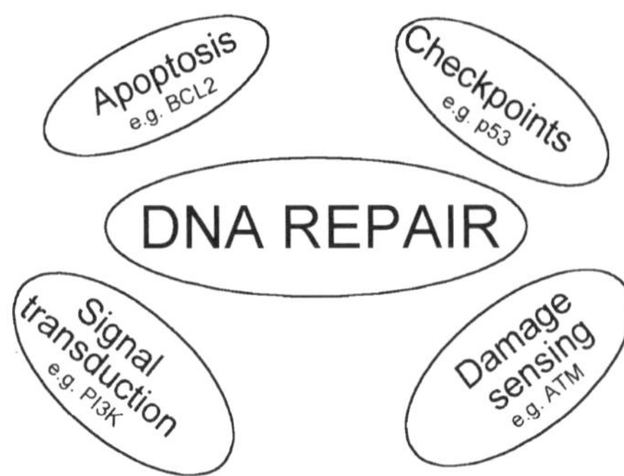

Fig. 3. Cellular responses to IR-induced DNA damage that can influence cell survival postirradiation.

damages are preferentially repaired in the transcribed strand of active genes compared to the overall genome *(124,125)*, this does not appear to be the case for SSBs *(126–128)* or for those IR-induced lesions that block polymerase chain reaction (PCR) *(129)*. There is little data on the preferential repair of DSBs, and what studies there are have, of necessity, used excessively high doses of IR. However, in one study using PFGE, marked intragenomic heterogeneity in DSB repair rates was observed in an adenocarcinoma cell line after 400 Gy *(130)*.

Third, techniques that assess the restoration of DNA integrity as a measure of DSB repair do not discriminate between repaired or misrepaired DNA; that is, they give no insight into whether the cells are rejoining these breaks correctly. Indeed, the fidelity with which human cells rejoin a restriction-enzyme-induced DSB in plasmid DNA has been independently correlated with radiosensitivity in a number of studies (e.g., refs. *131–139*).

Fourth, almost all of the DSB assays used in these studies are relatively insensitive, often requiring very high and, thus, supralethal doses. Their relevance to the response of cells to the lower doses used in the clinic is thus open to question. More sensitive assays (e.g., nuclear focus formation) are now available, but to date their predictive potential has not been tested. Finally, it is clear that pathways other than those for DNA repair can influence cell survival post-IR, such as those for apoptosis, cell cycle checkpoints, and various signal transduction pathways (*see* Fig. 3). These will disturb any one-to-one relationship between DSB induction and/or repair and cell killing.

3. ASSAYS IN WHICH AN EXOGENOUS DNA SEQUENCE IS IRRADIATED AND REPAIRED BY A HOST CELL

As an alternative to direct assays of genomic lesions, irradiated viruses or other DNA probes have been used to characterize the DNA repair proficiency of human cells. Because these probes depend on the host cell for repair factors, the observed repair provides an estimate of the host cells repair capability. The major challenge in applying such assays to the prediction of response to XRT is to irradiate the probe molecules under conditions that will generate a reasonable proportion of the viral population that is being inactivated via a DSB, rather than the much more prevalent base damages and SSBs *(140)*. For example, we (Day and Murray, unpublished data) found that two human malignant glioma cell lines, M059K and M059J, which differ greatly in their radiosensitivity and DSB repair capability *(141)*, showed no difference in their ability to reactivate adenovirus 5 that had been γ-irradiated under several different conditions of scavenging or temperature. Similarly, the ability of several human cell lines of differing radiosensitivity to reactivate γ-irradiated adenovirus 5 showed no obvious relationship to cellular radiosensitivity *(142)*. Other potential limitations of plasmid or viral assays are that the repair of these probes may be different than that of genomic DNA and that the existence of non DNA-repair factors affecting cell survival might also confound such correlations.

Various manifestations of DNA repair/recombination deficiency have been demonstrated in mammalian cell lines using exogenous DNA vectors (e.g., refs. *61, 131, 132,* and *143–145)*. The current status of plasmid assays with regard to the two major DSB repair pathways in mammalian cells (*see* Subheading 4) has been recently reviewed *(39)*.

4. MAMMALIAN DNA REPAIR GENES AND PROTEINS

In contrast to the above sections that focus on DNA damage, this section will focus on studies measuring cellular/tissue levels of DNA repair gene transcripts and/or the levels or activity of repair proteins that mediate the repair of different lesion types. A detailed discussion of the major DNA repair pathways and their key proteins is beyond the scope of this review and we will only briefly describe the most relevant pathways that enable mammalian cells to circumvent the potentially harmful effects of IR-induced DNA lesions (*see* Fig. 4). The successful cloning of the IR-associated DNA repair genes subsequently generated molecular probes that were rapidly applied to the study of laboratory and clinical models of radioresponsiveness. It is these studies that we will describe here.

4.1. Base Excision Repair

Many types of base damage induced by IR are substrates for the BER pathway *(146)*. BER involves the concerted action of a damage-specific DNA glycosylase (which removes the damaged base at the *N*-glycosylic linkage) and an AP endonuclease (which cleaves 5' to the abasic site) or AP lyase (which cleaves 3' to the abasic site). Additional enzymes such as polynucleotide kinase (Pnk) are necessary to restore the appropriate end groups for ligation *(146)*. The enzyme

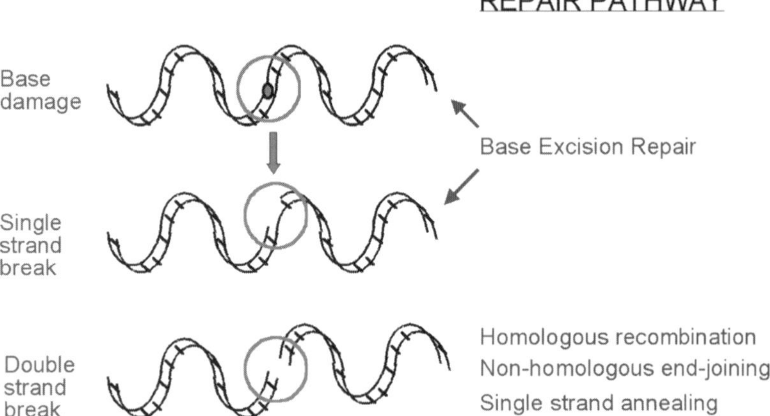

Fig. 4. Major DNA repair pathways involved in the removal of IR-induced DNA lesions in mammalian cells.

poly(ADP-ribose) polymerase (PARP) also plays a critical (although yet to be clearly defined) role in BER.

The repair of AP sites involves one of two BER subpathways. In "short patch" BER the replacement of a single nucleotide is effected by DNA polymerase β (Polβ) and DNA ligase 3 (Lig3) seals the resulting break. In "long patch" BER, 2–10 nucleotides are removed and resynthesized in a process that requires proliferating cell nuclear antigen (PCNA), Polδ/ε and flap endonuclease 1 (Fen1), with Lig1 sealing the resulting break. The long patch subpathway may be responsible for the repair of more complex lesions *(123)*.

Another protein central to BER is Xrcc1, a putative scaffold protein that has no catalytic activity but which, together with Polβ, PARP, and Lig3, coexists in a multiprotein complex involved in the detection of DNA interruptions and in the efficient and coordinated processing of strand breaks *(147,148)*. More recently, Whitehouse and colleagues *(149)* have shown that Xrcc1 also interacts with Pnk, stimulating this enzyme to clean up break termini via its 3' DNA phosphatase and 5' DNA kinase activities. This confirms Xrcc1 as a key player in BER and SSB repair.

The degree of radiosensitization achieved by inactivation of BER is quite modest compared to that observed in cells that are deficient in DSB rejoining. For example, the *XRCC1*-deficient Chinese hamster ovary (CHO) mutant EM9 is only approx twofold radiosensitive compared with wild-type cells *(150)*. Nonetheless, even small differences in radiosensitivity could be clinically important in the course of a standard XRT regimen that is typically given as 30–35 fractions of 2 Gy *(151)*.

4.1.1. Xrcc1

Because *XRCC1* was the first IR-related repair gene to be cloned *(152)*, molecular probes for this gene were quickly applied to clinical material. The human *XRCC1* gene was found to be highly polymorphic in SCCHN cell lines and in other human cell lines, based on Southern blot DNA restriction fragment-length polymorphism (RFLP) analysis *(153)*. Although Northern blotting indicated significant (approx twofold) differences in *XRCC1* mRNA levels among 25 SCCHN cell lines, there was no consistent correlation with intrinsic radiosensitivity or patient response or, indeed, with RFLPs *(153)*. The lack of a correlation between cellular radiosensitivity and *XRCC1* expression is not too surprising considering that (1) the response of cells to IR is modulated by a multitude of gene products, of which the Xrcc1 protein constitutes only one, (2) the repair of DSBs is generally acknowledged to be the most important determinant of radiosensitivity, whereas Xrcc1 is predominantly involved in the repair of SSBs, (3) inactivation of the *XRCC1* gene only results in approx twofold increase in radiosensitivity, whereas a defective DSB repair gene can impart much greater differences in radiosensitivity, and (4) the Xrcc1 protein may not be rate limiting for BER and, consequently, small differences in its expression may have little impact on the overall efficiency of repair. Indeed, in the Dunphy study *(153)*, no tumor examined had *XRCC1* mRNA levels below the baseline established using normal fibroblasts, placenta, and keratinocytes.

4.1.2. Hap1

Hap1 (Ape1/Ref1) is the major AP endonuclease in human cells. Surprisingly, blastocysts from homozygous null *APE⁻* embryos were not radiosensitive *(154)*. Nonetheless, a correlation was found between Hap1 protein levels and radiosensitivity in primary cervical carcinoma biopsy tissue *(155)*, suggesting that Hap1 in this setting has a critical rate-limiting function. In contrast, Hap1 levels in 11 human tumor and fibroblast cell lines varied approx fivefold but bore no relationship to radiosensitivity, suggesting that established cell lines might not mimic the clinical situation *(156)*. A later study of the expression of Hap1 protein in 88 samples of early-stage invasive cervical cancer showed no relationship to patient survival *(157)*.

Bobola et al. *(158)* reported that Hap1 activity in 84 human adult gliomas exhibited an approx 550-fold range. The mean for high-grade gliomas was 3.5-fold higher than for low-grade tumors. In 58 of these cases, it was possible to analyze Hap1 activity in histologically normal brain adjacent to tumor tissue: the mean activity was 7.3-fold higher in gliomas than in normal brain. Increased tumor activity was observed in 93% of the tumor/normal pairs, indicating that elevation of Hap1 activity is characteristic of human gliomagenesis. Similar differences in Hap1 protein levels were observed by Western blotting. On this basis, it was suggested that increased Hap1 activity accompanying gliomagenesis might result in resistance to XRT, although this was not studied directly. Hap1 protein levels were also markedly elevated in prostate cancers *(159)*. The level of Hap1 staining increased from low in benign hypertrophy to intense in prostatic intraepithelial neoplasia and prostate cancers, suggesting that Hap1 may play a role in the early development of prostate tumors. Hap1 protein levels were also elevated in some germ cell tumors from patients with testicular cancer of various histologies, including seminomas, yolk sac tumors, and malignant teratomas *(160)*. The authors suggested that the elevated expression of Hap1 observed in human testicular cancer might be related to its relative resistance to therapy.

4.1.3. Ogg1

One of the most prevalent forms of base damage caused by IR is 8-oxoguanine. Mouse knockouts for Ogg1, the DNA glycosylase-AP lyase that removes 8-oxoguanine from DNA, show clear manifestations of 8-oxoguanine repair deficiency and yet the associated phenotype is very mild, possibly because of the presence of backup systems for 8-oxoguanine repair *(161,162)*. Indeed, there is considerable degeneracy among the glycosylases with respect to the lesions they can process. Human cells also possess a second glycosylase, Ogg2, that can process 8-oxopurines *(122)*. This redundancy may make it difficult to utilize these factors either for predictive purposes or as targets for manipulation in XRT, and little has been done with these genes/enzymes in the clinical context.

4.1.4. Polb and Other BER Factors

Embryonic fibroblasts from mice with a homozygous deletion of the *POLβ* gene are not radiosensitive *(163–165)*. These cells probably circumvent the defect by switching from Polβ-dependent mechanisms to long patch BER that utilizes other polymerases. Possibly for this reason, there has been little activity in this area from the perspective of radiosensitivity testing. Studies of other BER factors such as Fen1 and Pnk in the context of clinical radiosensitivity prediction have not been reported at this time.

4.2. Nonhomologous End-Joining

Mammalian cells have evolved efficient mechanisms to detect and repair DSBs *(166,167)*. IR-induced DSBs are repaired by at least two pathways. The first of these is nonhomologous end-joining (NHEJ), which appears to dominate in mammalian cells, especially in the G_1/G_0 phase *(167,168)*. NHEJ proteins catalyze the direct rejoining of broken DNA ends in a process that requires no or as little as one basepair of homology *(169)*. Ku70 (Xrcc6) is believed to initiate this pathway by binding to a DSB, followed by the sequential binding of Ku86 (Xrcc5, sometimes referred to as Ku80) and the catalytic subunit of the DNA-dependent protein kinase, DNA-PK$_{cs}$ (Xrcc7), to generate the active DNA-PK complex. Ku70 also interacts with the Mre11 protein, an interaction that may serve to recruit the Mre11/Rad50/Nbs1 (MRN) complex, which is involved in microhomology searching *(170)*, into the NHEJ pathway *(171)*. Other proteins involved in NHEJ include Xrcc4 and Lig4, which associate strongly with each other *(168)* and which mediate tail removal and gap filling/ligation.

In general, mutation of any of the NHEJ genes in mammalian cells leads to DSB repair deficiency and marked radiosensitivity. This has been clearly shown for Xrcc4 *(172,173)*, Ku70 *(174,175)*, Ku86 *(168)*, and DNA-PK$_{cs}$ *(141,168)*. The components of this pathway therefore represent an attractive target for modulation in XRT. With respect to the MRN complex, the *MRE11* gene is mutated in a human syndrome similar to AT that has been termed "ataxia–telangiectasia-like disorder" or ATLD *(15)*. Similarly, the *NBS1* gene is mutated in a human radiosensitivity syndrome, Nijmegen breakage syndrome (NBS), that exhibits both similarities and (clinical) differences to AT *(13,16)*. Fibroblasts from these patients are radiosensitive and defective in mounting cellular responses to DNA damage. Disruption of *RAD50* in embryonic stem cells resulted in blastocysts that were acutely radiosensitive *(176)*.

A number of investigators have measured levels of some or all of the three components of the DNA-PK holoenzyme (either mRNA or protein levels) or overall DNA-PK activity in tumor cell lines and clinical material. Several studies have shown altered levels or activity of Ku in cancer cells. An altered regulation of Ku70/Ku86 DNA-binding activity in human breast and bladder tumors was suggested to be related to tumor progression *(177)*. The levels of Ku70 and Ku86 protein were reduced in colon adenomas and carcinomas compared with normal human colon tissue, suggesting that the loss of this activity may be important in the development of colon tumors *(178)*. DNA-PK$_{cs}$ expression was unaltered in that study. Decreased Ku70/Ku86 expression was also reported in malignant melanomas of the oral cavity, and it was again suggested that this might influence the progression of these tumors *(179)*. In contrast, increased Ku DNA-binding activity and Ku protein levels were found in fresh tumor tissue for each member of a group of patients with transitional cell carcinoma of the bladder when compared to normal bladder tissue *(180)*. An overactive NHEJ system, specifically aberrant Ku70/Ku86 activity, has also been associated with chromosomal instability in myeloid leukemias *(181)*. Thus, several manifestations of the disregulation of DNA-PK activity appear to occur in different tumor types. Not surprisingly, other studies show no alteration in these activities. For example, in an investigation of 134 specimens, there was no difference in the expression of the DNA-PK$_{cs}$, Ku70, Ku86, Xrcc4, and Nbs1 proteins between cancerous tissues and the corresponding normal tissues *(182)*.

Some studies have suggested that levels of DNA-PK components can be predictive of cellular radiosensitivity. For example, in a comparison of two human SCCHN cell lines, the radioresistant UM-SCC-1 line that was more proficient in rejoining IR-induced DSBs had 1.6-fold greater DNA-PK activity than the radiosensitive UM-SCC-14A line *(183)*. The constitutive level of DNA ligase activity was, however, similar in the two cell lines. A correlation between radiosensitivity and DNA-PK content/activity was also found in a panel of five lung carcinoma cell lines that again paralleled the cells' ability to rejoin IR-induced DSBs *(104)*. Among a panel of 53 cervical carcinoma tumors, there was a borderline significant correlation between SF2 and pretreatment Ku70 expression, but no such correlation was apparent for Ku86 *(184)*. Other studies have reported contrary findings. Thus, no differences in nuclear staining intensity for the DNA-PK$_{cs}$, Ku70, Ku86, Xrcc4, and Nbs1 proteins were apparent among 134 specimens from various normal and tumor tissues with differing radiosensitivity *(182)*. Similarly, in a panel of nine human malignant glioma cell lines, DNA-PK activity was present in all cell extracts, as were the DNA-PK$_{cs}$, Ku70, and Ku86 proteins, but there was no correlation between DNA-PK activity and inherent radiosensitivity *(185)*. Likewise, among biopsies from patients with previously untreated SCCHNs, no relationship was apparent between SF2 and the expression of either the DNA-PK$_{cs}$, Ku70, or Ku86 proteins *(186)*.

With respect to the predictive power of DNA-PK components for tumor radiosensitivity, several studies have generated positive findings. For example, in pre-XRT biopsy samples from patients with resectable advanced rectal carcinoma, the expression of Ku proteins was inversely correlated with tumor radiosensitivity and with disease-free survival *(187)*. Similarly, in the above-mentioned panel of 53 cervical carcinomas treated with XRT, those with a low number of Ku70 or Ku86 positive cells were more radiosensitive, and there was improved survival among patients whose tumors had low Ku70 expression *(184)*. Among biopsy specimens of 44 patients with oropharyngeal carcinoma and 32 patients with hypopharyngeal carcinoma who had been treated with XRT, the response to therapy was superior for primary tumors that stained weakly for DNA-PK$_{cs}$ protein *(188)*. In contrast, there was no correlation between the DNA-PK activity in a panel of nine glioma cell lines and the tumor response of the donor patients to XRT *(185)*.

Noguchi and colleagues *(189)* examined 67 patients with progressive esophageal cancer treated with chemotherapy–XRT and observed a significant correlation between DNA-PK$_{cs}$ levels and the effect of therapy. Surprisingly, however, the high-DNA-PK$_{cs}$-expressing group showed a greater response than the low-expression group. A caveat here is that the assay of response was a short-term assay (regression) and not a long-term one (e.g., local control), which would have been a better measure of treatment sensitivity.

With respect to DNA-PK components and normal tissue radiosensitivity, the expression of the *KU70* and *KU86* genes as well as DNA-PK activity were studied in 11 normal human fibroblast lines *(190)*. Differences in cell survival were previously shown to correlate with the level of non-repaired DSBs *(92)*. The *KU70* and *KU86* mRNA levels were similar for the 11 cell lines investigated, as was the DNA-PK activity. Despite the correlation between radiosensitivity and DSB repair capacity, there was no correlation between DSB repair capacity and DNA-PK activity in these fibroblast lines. In 10 skin fibroblast cell lines established from cancer patients with different normal tissue reactions to XRT and in control cells, the levels of DNA-PK$_{cs}$, Ku70, Ku86, Xrcc4, and Lig4 were similar *(191)*. The in vitro activities of DNA-PK and Xrcc4/Lig4 showed some differences, but these did not correlate with either the normal tissue response of the patient in vivo or with cellular radiosensitivity in vitro. Among 16 pairwise matched head and neck cancer patients 2–7 yr after curative XRT, patients with extreme late reactions showed no evidence for an *ATM, NBS1, MRE11,* or *RAD50* mutation in their dermal fibroblasts *(10)*.

4.3. The Rad51 Homology-Directed Repair Pathway

The second mammalian pathway for DSB repair involves Rad51-dependent homology-directed repair (HDR). In contrast to NHEJ, HDR depends on extensive regions of homology and provides a mechanism for the error-free repair of DSBs in S and G$_2$, presumably because this mechanism ideally involves homologous recombination between sister chromatids. Important proteins include Rad51, which is a DNA-dependent ATPase that catalyzes strand exchange between homologous DNA molecules in the presence of replication protein A (RPA) *(192,193)*. Mammalian cells also contain at least five Rad51 paralogs— Rad51B, Rad51C, Rad51D, Xrcc2, and Xrcc3—that are involved in HDR. The Rad52 protein interacts with Rad51 to promote homologous pairing and strand exchange by Rad51 *(193,194)*. The MRN complex discussed earlier in the context of NHEJ is also involved in HDR processes, possibly involving the resection of 5' DNA ends. Another HDR factor, Rad54, belongs to the SNF2/SWI2 family of DNA-dependent ATPases and may function with Rad51 in chromatin remodeling *(195,196)*.

As with many HDR factors, homozygous disruption of *RAD51* in the mouse resulted in embryonic lethality *(197,198)*. Knockout mutants of many HDR genes have been generated in chicken DT40 cells. Disruption of *RAD51* in DT40 cells resulted in lethality *(199)*. Disruption of all five Rad51 paralogs in the DT40 background resulted in mutants that are viable and defective in HDR, exhibit mild radiosensitivity, and show reduced Rad51 nuclear focus formation post-IR *(200,201)*. Similarly, rodent mutant cells that are deficient in Xrcc2 (*irs1*) *(131)* and Xrcc3 (*irs1SF*) *(202)* are twofold to threefold sensitive to IR and are defective in the HDR of DSBs *(203,204)*. Xrcc2-deficient embryonic cells also show increased radiosensitivity *(205)*. Surprisingly, disruption of *RAD52* in mouse or chicken DT40 cells causes little radiosensitization *(206,207)*. The *RAD52⁻* DT40 mutants also show normal Rad51 nuclear focus formation post-IR. Thus, there may be some functional redundancy for Rad52 activity in mammalian cells. Mouse embryonic stem cells with homozygous disruption of another HDR gene,

RAD54, are markedly radiosensitive *(208)*. Similar results were obtained using chicken DT40 mutants *(209)*.

There have been surprisingly few studies to date describing the expression of HDR genes or proteins in cancer cell lines or clinical material. Yanagisawa and colleagues *(210)* compared *XRCC1, XRCC3, KU86, RAD51,* and *RAD52* mRNA levels in the radioresistant N10 and parental KB human carcinoma cell lines and observed higher basal expression of *XRCC1, XRCC3,* and *RAD51* mRNA in the radioresistant line. Bishay et al. *(211)* analyzed the basal expression of several DNA repair genes (including *LIG1, LIG3, KU86, DNA-PKcs, PARP, RAD51,* and *RAD52*) in blood samples obtained from 32 healthy male donors irradiated ex vivo. The variability of mRNA levels was of similar magnitude to that for spontaneous or IR-induced MN frequency, and *RAD51* gene expression was negatively correlated with IR-induced MN frequency. With respect to normal cells, the levels of Rad51 protein differed little among 10 skin fibroblast cell lines established from cancer patients with different normal tissue reactions to XRT and in control cells and could not, therefore, explain the associated differences in either the normal tissue response of the patients in vivo or in cellular radiosensitivity in vitro *(191)*.

The Brca1 and Brca2 tumor suppressor proteins appear to be important for HDR, but the mechanism of their involvement is unclear *(212)*. Both proteins have been identified as components of the Rad51 multiprotein complex, and they appear to participate in a common HDR pathway for DSBs and ICLs *(213)*. However, the role of Brca1 and Brca2 in the rejoining of IR-induced DSBs remains uncertain in the face of recent conflicting studies. The role of alterations in these proteins on DNA repair activity and cancer therapeutics has been discussed in this volume by Hakem and colleagues and also elsewhere *(214)* and will not be further considered here.

4.4. IR-Induced DSB Repair Nuclear Foci

Following irradiation, multiple distinct nuclear multiprotein foci accumulate at sites of DNA damage and are believed to be essential for the rejoining of DSBs. There has been considerable recent interest in studying these foci from the perspective of predicting response to XRT. In part, this is because, like assays for DSB rejoining, these nuclear foci integrate the functional impact of a number of important repair proteins. Foci involving Rad51 may represent nuclear domains for HDR *(215,216)*. An early step in nuclear focus formation is the DSB-activated phosphorylation of the histone H2A family protein, H2AX. Phosphorylated H2AX (γ-H2AX) forms foci at DSBs within minutes of irradiation *(217)*. Additional proteins such as Brca1 are then recruited to these γ-H2AX-tagged sites, and at still later times, the Rad50 and Rad51 proteins become colocalized with these foci *(217)*. The Xrcc2 and Xrcc3 proteins are also components of the Rad51 foci-forming complex *(203,204)*. RPA also participates in Rad51 nuclear focus formation in irradiated human cells *(218,219)*. It has been suggested that RPA foci observed several hours after irradiation represent irreparable lesions and, as such, might be particularly useful in identifying radiosensitive cells *(219)*.

The MRN complex also forms nuclear foci at sites of IR-induced DSBs that are distinct from those involving the Rad51 HDR complex *(220,221)*. The MRN complex is preformed and associates with DSBs early after irradiation, remaining associated until DSB repair is almost complete; however, this does not happen in mutant cells such as NBS fibroblasts *(16,222)* or in the absence of Ku70 *(171)*. Mre11 foci *(223)* appear to represent a later critical binding event in the processing of DSBs.

The relationship between IR-induced γ-H2AX foci and DSBs appears to be reasonably good *(224,225)*, and foci can readily be measured at clinically relevant doses. This opens up the possibility of using this response to assess DSBs quantitatively in XRT patients as a radiosensitivity predictor. However, because neither DSB induction nor repair appear to be particularly robust predictors of radiosensitivity based on the use of other assays, this approach may not produce a clinically useful predictor. Moreover, cells would have to be taken from the patient and irradiated in vitro, with all the attendant problems and potential artifacts associated with such explant methods.

4.5. Nucleotide Excision Repair

Nucleotide excision repair (NER) is the major pathway for repairing bulky DNA adducts and intrastrand crosslinks, as typified by ultraviolet (UV) light *(146)*. Although IR does induce several UV-like lesions as well as DNA crosslinks, NER is not ordinarily associated with the cellular response to IR. It may, however, play a role in the repair of some types of IR-induced DNA lesion in hypoxic cells *(226)*. The rate-limiting incision step of NER is mediated by a complex of 10 proteins that includes Xpa and RPA (involved in damage recognition), the Ercc1-Xpf endonuclease (incises 5' to the lesion), Xpd and Xpb (DNA helicases associated with the TFIIH basal transcription complex), Xpg (the 3' endonuclease), Csb (involved in coupling repair to transcription), and Xpc-HR23 and Xpe (involved in global genomic NER, but not in TCR). Of the NER–incision complex proteins, Ercc1 and Xpf are unusual insofar as they have an additional role in the repair of DNA crosslinks, a process that may require some type of HDR event *(227–229)*.

Some human XPG strains derived from patients who also exhibit severe symptoms of Cockayne's syndrome display a modest degree of hypersensitivity to IR *(230,231)*. This effect appears to relate to the fact that the Xpg NER protein also functions as an accessory factor in BER, such that these cells are defective in the global and (especially) transcription-coupled BER of oxidative base damage such as thymine glycol *(232–234)*. Other than these XPG strains, NER-deficient mammalian cells do not display overt radiosensitivity under aerated conditions.

However, some NER-deficient and crosslink-repair-deficient rodent mutant cell lines such as UV20 (*ERCC1⁻*) and UV41 (*XPF⁻*) are markedly radiosensitive in the absence of molecular oxygen *(226)*, apparently because these cells are unable to repair some type of DNA crosslink induced by IR under hypoxic conditions *(235)*. Targeting the *ERCC1* and *XPF* genes or gene products (whose inactivation selectively sensitizes hypoxic cells), whether by chemical or biological means, may therefore provide a useful approach to circumventing the effects of hypoxia in limiting the radiocurability of some human tumors.

4.6. Measurement of Gene Induction

The studies referred to above generally relate to the constitutive pretreatment (basal) expression of DNA repair genes/proteins. A number of studies indicate that the expression of some repair factors can be induced in various cell types following modest doses of IR in vitro *(68,210,236–243)*. The consequences of such events remain to be established. Although inducible responses may play some role in the IR response of tumor cells, it has been noted that for SCCHN cell lines, there is no evidence for an involvement of these responses either in inherent differences in tumor cell radiosensitivity or in the success or failure of XRT *(244)*. This is supported by recent elegant studies using yeast gene deletion libraries that show no relationship between the genes induced or suppressed by IR and those necessary for survival *(11)*. Given the level of conservation of many of the genes and pathways, it is likely that this could also be the case in humans.

5. POLYMORPHISMS

5.1. Polymorphisms in DNA Repair Genes

DNA repair genes play a critical role in protecting the genome from the harmful effects of cancer-causing agents. An inability to appropriately process or repair DNA damage caused by such agents can lead to genetic instability and to increased risk of cancer development. In fact, deficiencies in DNA-damage sensing and repair pathways appear to be fundamental to the etiology of most cancers *(245)*. The relationship between reduced repair and cancer susceptibility is most apparent in repair-deficiency syndromes such as XP in which the inactivation of DNA repair genes is associated with a high risk of cancer (e.g., ref. *246*; *see also* Chapter 13 in this volume). However, these individuals are quite rare. Associations between decreased repair activity and increased cancer risk among the general population initially became apparent from studies using the types of assays already discussed in this chapter, namely phenotypic repair of DNA damage induced by some external agent, the use of plasmid or viral reactivation assays, or the measurement of levels and splicing patterns of repair-gene mRNAs and repair proteins themselves. Many individuals have now been found to exhibit a decreased repair phenotype (typically 60–75% of normal) that is associated with an increased cancer risk from various environmental carcinogens, including IR *(37)*. It has also been shown that a number of genetic polymorphisms (variant alleles with a frequency of approx 5% or greater) occur in various repair genes, including *XRCC1, ERCC1, XPD, XPF,* and *XRCC3 (34)*. More recently, Mohrenweiser et al. *(37)* identified 127 amino acid substitution variants in 37 repair genes. More than half of these were missense events (i.e., they involved the exchange of a dissimilar amino acid) at evolutionarily conserved sites. Another five resulted in proteins with altered termination of translation and one amino acid insertion variant was detected. The polymorphism frequency averaged 0.047, with individual frequencies ranging from <0.01 to 0.43, such that these events have the potential to be population risk factors for cancer because they affect a large number of people. These polymorphisms are often accompanied by modest reductions in repair capability, and their impact on cancer risk is only beginning to be apparent *(36,247)*.

In addition to contributing to increased susceptibility to cancer and cancer progression, this genetic variability in DNA repair in the general population may contribute to sensitivity to IR and, thus, to tumor response and to the extreme reactions to XRT that are observed in some cancer patients. It is important to recognize that germline polymorphisms have the potential to affect both the tumor and normal tissue responses to XRT, whereas acquired tumor-specific defects should only affect the tumor response. Following this reasoning, lymphocytes would provide an easily accessible tissue for studying polymorphisms relevant to normal tissue reactions, although they are likely to be less reliable/ relevant for predicting tumor response. Repair polymorphisms might also have a significant impact on patient response to chemotherapy, but this is beyond the scope of the current chapter. It should also be noted that the complexity of DNA repair pathways and their associated enzymes makes it likely that these phenotypes will be polygenic (i.e., they will represent the aggregate effect of several alleles).

5.2. DNA Repair-Gene Polymorphisms As Cancer Risk Factors

As noted earlier, polymorphisms in repair genes, including DSB repair genes, are believed to result in increased genetic instability and to confer an increased cancer risk. Here, we will describe just a few examples in which such polymorphisms have been associated with various types of cancer.

5.2.1. XRCC1

The *XRCC1* gene has been found to contain several common coding polymorphisms (allelic variants) that might impact on the DNA repair activity of the encoded protein and, thus, on cancer susceptibility. Shen et al. *(34)* identified amino acid substitutions in conserved regions of *XRCC1*, including an arginine to tryptophan change at codon 194 (nucleotide substitution of 26304 C→T) in exon 6 ([Arg194Trp]), and an arginine to glutamine change at codon 399 (nucleotide substitution of 28152 G→A) in exon 10 ([Arg399Gln]). Another missense polymorphism occurs at [Arg280His]. The most widely studied *XRCC1* polymorphism is [Arg399Gln]. The *XRCC1* codon 399 is located within the central Brca1 C-terminus (BRCT) domain that is conserved in several proteins involved in DNA repair, cell cycle control, and recombination and is believed to be a protein–protein interface that interacts with PARP *(248(250)*. The codon 194 polymorphism is in the linker region of the N-terminal domain that is involved in binding to DNA single-nucleotide gaps and gapped DNA–Polβ complexes *(251)*.

The realization that some of these polymorphisms can alter the activity of the repair protein initially emerged from studies of biomarkers of DNA damage. In particular, the *XRCC1* [Arg399Gln] polymorphism is associated with increased levels of sister chromatid exchanges (SCEs) *(252,253)*, DNA adducts *(247,252,254)* and mutations *(254)*, as well as with a prolonged cell cycle arrest post-IR *(255)*. Additionally, hamster cell lines with amino acid substitutions within the BRCT domain display reduced repair of SSBs and hypersensitivity to IR *(256)*. The *XRCC1* [Arg280His] polymorphism has also been found to predispose to bleomycin-induced chromatid aberrations *(257)*.

As shown in Table 1, several epidemiological studies have shown a positive association between the variant *XRCC1* 399-Gln allele and increased cancer risk in various selected study populations, either with or without a known history of carcinogen exposure *(258–263)*. In marked contrast, some studies report that the *XRCC1* 399-Gln allele is associated with a *decreased* risk of some cancers (i.e., that persons with the common 399-Arg/Arg genotype have increased risk *[264–271]*). Perhaps not too surprisingly, other studies have shown no cancer risk associated with the [Arg399Gln] polymorphism *(262,270,272–277)*. Similarly, the variant *XRCC1* 194-Trp allele has been associated with an increased risk of developing some cancers *(263,267,274,275)*, whereas the combined 194-Arg and 399-Gln genotypes imparted a significantly increased risk for gastric cardia cancer *(270)*. Some studies have reported that the common 194-Arg/Arg genotype is associated with an increased cancer risk *(258,271,276,278)*, whereas other studies suggest that the [Arg194Trp] polymorphism is not a risk factor for some cancers *(261,269,270,272,276,277)*. A third *XRCC1* polymorphism, [Arg280His], was associated with an increased risk of cancers in some study populations *(271,276)* but not in others *(277)*.

These combined studies suggest a role for the Xrcc1 protein in the repair of DNA damage in human cells and tissues. They also indicate that the 399-Gln allele of *XRCC1* and, to a lesser extent, the 194-Trp allele, have important phenotypes from the perspective of cancer risk. However, this relationship is complex, and for different patient populations, tumor sites and/or environmental factors, it can apparently take very different forms. This variability may partly reflect the fact that the frequency of the variant allele can vary greatly in different study populations, and particularly in those with different ethnic composition, especially considering that studies to date have generally been quite small. As noted previously *(261)*, differences among epidemiological studies might also arise if the biological impact of an *XRCC1* polymorphism depends on context. Indeed, the effects of any genetic variant is likely to depend on other genetic factors, such as the activity of competing biochemical pathways operating in the particular tissue, in addition to unknown environmental factors that interact with the variant allele *(279)*. It has also been noted that unmeasured genetic factors (e.g., other alleles in linkage disequilibrium with the *XRCC1* allele) could contribute to such differences *(261)*. Furthermore, cancer risk may depend on selective pressures exerted on the cell. For example, if the variant protein has reduced repair activity, then the cells may be more susceptible to undergoing apoptosis, thereby decreasing the associated cancer risk *(266)*. In this event, cancer risk would be very dependent on the activity of apoptotic mechanisms.

5.2.2. OTHER BER GENES

Although its functional significance is unclear, the [Ser326Cys] missense polymorphism in exon 7 of the *OGG1* gene has been suggested, based on in vitro data, to reduce the activity of the enzyme *(280,281)*. In contrast, DNA repair activity of *OGG1* in human lymphocytes did not appear to be dependent on [Ser326Cys] status *(282)*.

The *OGG1* [Ser326Cys] variant has been associated with an increased risk of various types of cancer. For example, Japanese males with the 326-Cys/Cys genotype were at an increased risk of developing tobacco-related squamous cell carcinomas and nonadenocarcinomas of the lung, although the risk for other histological subtypes of lung cancer or overall risk was not significant *(283)*. Similarly, a limited association was found between the *OGG1* [Ser326Cys] polymorphism and risk of lung adenocarcinoma among Japanese *(284)*. The [Ser326Cys] polymorphism (and especially the homozygous 326-Cys genotype) was also an important risk factor for smoking- and alcohol-related orolaryngeal cancers *(285)*, lung cancers *(286)*, and squamous cell carcinomas of the esophagus among a Chinese population, although there was no significant interaction between smoking and the Cys/Cys genotype in the latter study *(287)*. The frequencies of two *OGG1* variants, [Ser326Cys] and 11657 (A→G), were significantly associated with prostate cancer risk *(288)*. In contrast, there was no clear association between the [Ser326Cys] polymorphism and gastric cancer among Japanese and non-Japanese Brazilians *(289)*, breast cancer *(290)*, or lung cancer in a Caucasian population *(291)*.

Seven amino acid substitution variants were identified in the repair domain of Hap1 *(292)*. Three of these, [Leu104Arg], [Glu126Asp] and [Arg237Ala],

Table 1
Cancer Risk Associated with Common Missense *XRCC1* Polymorphisms

Gene [polymorphism]	Tumor site[a]	Association	High-risk allele	Ref.
XRCC1 [Arg399Gln]	Head and neck	Yes, among current smokers	399-Gln	258
	Adenocarcinoma of the lung	Yes	399-Gln	259
	SCC of the lung	Yes, in persons with low cigarette use	399-Gln	260
	Colorectal	Yes	399-Gln	263
	Breast	Yes, in African-Americans, but not whites	399-Gln	261
	Bladder	Yes	399-Arg	264
	Head and neck	Yes	399-Arg	265
	Non-melanoma skin	Yes, for both basal cell and SCCs	399-Arg	266
	Esophageal SCC	Yes, for alcohol-associated cancers	399-Arg	268
	Gastric cardia	Yes	399-Arg	270
	SCCHN	Yes; potential interaction with tobacco use	399-Arg	267
	Therapy-related AML	Yes	399-Arg	269
	Prostate	Yes	399-Arg	271
	Esophagus	No		274
	Lung	No		275
	Pancreas	No, except for a slight interaction of 399-Gln with smoking		262
	Gastric	Slight (nonsignificant)		270
		Yes, especially gastric cardia	194-Arg + 399-Gln	
	Lung	No		276
	Lung (non-small-cell)	No		277

234

	Cancer site	Association	Allele	Ref.
	Bladder	No		273
	Malignant melanoma	No		272
XRCC1 [Arg194Trp]	Colorectal	Yes	194-Trp	263
	Esophagus	Yes for esophageal SCC	194-Trp	274
	Lung	Yes (borderline)		
		Greater-than-additive risk for *XRCC1* 194-Trp plus *XPD* 751-Lys	194-Trp	275
	Gastric cancer	Slight (nonsignificant)		
		Yes, especially for gastric cardia cancer	Combined 194-Arg/399-Gln	270
	SCCHN	Slight	194-Trp	267
	Lung	Yes, among African-American and Caucasian heavy smokers in Los Angeles County	194-Arg	278
	Head and neck	Yes	194-Arg	258
	Prostate	Yes	194-Arg	271
	Lung	Yes, among alcohol drinkers	194-Arg	276
	Lung	No		276
	Therapy-related AML	No		269
	Lung (non-small-cell)	No		277
	Malignant melanoma	No		272
	Breast	No		261
XRCC1 [Arg280His]	Lung	Yes, in Chinese tin miners	280-His	276
	Prostate	Yes (slight)	280-His	271
	Lung (non-small-cell)	No		277

[a] SCC: squamous cell carcinoma; AML: acute myeloblastic leukemia.

resulted in approx 50% reductions in repair activity. A fourth variant, [Asp283Gly], is similar to a previously characterized mutant [Asp283Ala] and exhibits approx 10% repair capacity. The most common substitution, [Asp148Glu], had no impact on endonuclease and DNA-binding activities nor did a [Gly306Ala] substitution. A [Gly241Arg] variant showed slightly enhanced endonuclease activity relative to wild type. Pieretti et al. *(293)* examined normal and tumor cells from women with ovarian or endometrial cancer and found two somatic missense polymorphisms in *HAP1* ([Gln51His] and [Asp148Glu]) and two missense ([Ala3Pro] and [Ser326Cys]) and one intronic polymorphism in *OGG1*. Two substituting somatic mutations in *HAP1*, [Pro112Leu] and [Arg237Cys], and one missense mutation, [Trp188X], that would probably result in a truncated protein, were identified in endometrial tumors, but no mutations were identified in *HAP1* in the ovarian tumors or in *OGG1* at either tumor site. These polymorphisms are obvious candidates for epidemiological studies of cancer susceptibility and sensitivity to IR and XRT.

5.2.3. XRCC3 and Other HDR Genes

The Xrcc3 protein is required for the assembly and stabilization of Rad51 *(294)*. An *XRCC3* polymorphism in position 18067 of exon 7 (C→T variant) results in the substitution of threonine with methionine ([Thr241Met]), which may significantly alter the characteristics of the protein *(272)*. The *XRCC3* 241-Met allele has been associated with elevated DNA adduct levels in lymphocytes *(247)*. The [Thr241Met] polymorphism was associated with increased levels of bulky DNA adducts and risk of developing bladder cancers, especially in nonsmokers *(273)*. Similarly, the 241-Met genotype was associated with the development of malignant melanoma *(272)*.

Kuschel et al. *(295)* studied several polymorphisms in HDR genes (*NBS1*, *RAD52, RAD51, XRCC2,* and *XRCC3*) in breast cancer patients. Genotype frequencies differed between cases and controls for two polymorphisms in *XRCC3*: [Thr241Met] and the rare noncoding IVS5 (A→G) substitution at nucleotide 17893. Homozygous carriers of 241-Met had an increased risk, whereas the IVS5 allele was associated with a decreased risk. The rare *XRCC2* polymorphism [Arg188His] was associated with a marginally significant increase in risk. In another study of HDR genes in 127 breast cancer patients, loss of heterozygosity was found in the *RAD51* region in 32% of tumors, in the *RAD52* region in 16%, in the *RAD54* region in 20%, and in the *BRCA1* and *BRCA2* regions in 49% and 44%, respectively *(296)*.

In contrast, the *XRCC3* [Thr241Met] polymorphism was not associated with risk of developing AML *(269)*, lung cancers *(277,278)*, or SCCHN, although it may play a role in the etiology of a subset of these tumors *(297)*. Similarly, but in contrast to the above-mentioned study by Matullo and colleagues *(273)*, Stern and colleagues *(298)* found little evidence of a positive association between the

241-Met allele and bladder cancer. Among heavy smokers, individuals with the 241-Met allele had a slightly increased risk of bladder cancers. There was some evidence for a gene–gene interaction between the *XRCC1* codon 194 and *XRCC3* codon 241 polymorphisms.

Considering that the *XRCC3* [Thr241Met] polymorphism has been associated with an increased risk of some cancers, it was surprisingly reported that the [Thr241Met] variant Xrcc3 protein is functionally proficient for HDR, suggesting that the increased cancer risk associated with this variant is not the result of an HDR defect *(299)*.

5.2.4. NHEJ Genes

Kuschel et al. *(295)* studied several polymorphisms in NHEJ genes (*KU70*, *KU86*, and *LIG4*) in breast cancer patients. A conservative polymorphism in *LIG4* (T→C at nucleotide 1977) was associated with a decrease in breast cancer risk. No significant association was found for the other polymorphisms.

5.2.5. NER Genes

Because of the presumed role of the encoded enzymes in removing bulky adducts induced by various environmental carcinogens, NER genes (and especially *XPD*) have been extensively studied as cancer-susceptibility factors. However, they are not likely to be especially important from the perspective of responses to XRT, with the possible exception of *XPG* (and potentially *XPF* and *ERCC1* for hypoxic tumors), as described in Subheading 4.5, so they will only be minimally described here.

A number of *XPG* polymorphisms have been described *(300)*. The *XPG* [Asp1104His] (exon 15 G→C) polymorphism was a marginal risk factor for breast cancers *(301)*. Although loss of heterozygosity at the *XPG* locus is common in prostate cancer patients, the lack of *XPG* mutations suggests that *XPG* is not the target gene *(302)*. With respect to *XPF* and *ERCC1* polymorphisms, a weak association was found between the development of melanoma and the homozygous TT genotype in position 2063 of the 5'UTR of the *XPF* gene, as well as with the exon 11 TT genotype in position 30028 of the *XPF* gene *(272)*. Both of these polymorphisms are conservative. There seemed to be a strong additive effect of the *XRCC3* 241-Met genotype and the *XPF* exon 11 or 5'UTR T alleles on the development of melanoma. No associations with melanoma were apparent for two conservative *ERCC1* polymorphisms in that study. The missense *XPF* [Pro379Ser] polymorphism was not associated with an elevated risk of non-small-cell lung cancer *(277)*. *ERCC1* polymorphisms (specifically, the A→C polymorphism at nucleotide 8092, which may alter mRNA stability) have recently been reported to be slight risk factors for SCCHN *(303)* and adult gliomas *(304)*.

5.3. Polymorphisms As Predictive Factors for Responses to XRT and As Risk Factors for IR-Induced Cancers

It is apparent that germline polymorphisms or other alterations in DNA repair genes generates an altered repair phenotype in a significant percentage of the population and that these events are more common in cancer patients than in the normal population. This finding has the potential to impact on the response of both normal tissues and tumors to XRT. Similar alterations arising during tumorigenesis presumably also have the potential to influence tumor responses to therapy. Given the number of polymorphisms that have now been identified in this context, we anticipate that there will be a surge of information with respect to XRT responses in the near future, and a few such studies have begun to appear in the literature. For example, rare polymorphic microsatellites in the *XRCC1*, *XRCC3*, and *KU86* genes were analyzed for possible linkage to cancer and clinical radiosensitivity by comparing volunteers with no cancer history, cancer patients who were hypersensitive to XRT, and cancer patients with normal reactions to XRT *(305)*. There was a significant association between these rare *XRCC1* and *XRCC3* polymorphisms and both cancer status and clinical radiosensitivity.

Hu and colleagues *(255)* examined whether the *XRCC1* [Arg194Trp] and [Arg399Gln] polymorphisms might contribute to radiosensitivity in lymphocytes collected from 135 cancer-free women, as indicated by the duration of the IR-activated G_2 arrest. There was a significant association between the 399-Gln genotype and prolonged G_2 arrest. In this same study population, the *HAP1* [Asp148Glu] polymorphism was also associated with cellular radiosensitivity, with the 148-Glu/Glu genotype having a prolonged G_2 arrest *(255)*. Thus, missense variants of both *XRCC1* and *HAP1* have the potential to contribute to patient hypersensitivity to XRT. When screening for individual radiosensitive patients in the clinic (as opposed to studying broad populations), it will probably be important to simultaneously test for alterations in a number of key radiosensitivity genes. This scenario was illustrated in a recent study of four profoundly radiosensitive XRT cancer patients *(306)*. In two of these patients, hypersensitivity to XRT was attributed to defects in the *ATM* gene. In another patient, a defect in *LIG4* was the causative factor. A novel and incompletely defined molecular defect was implicated in a fourth patient who presented with symptoms of XP. A similar scenario was apparent in a study assessing whether mutations in five genes involved in the cellular response to DSBs—*ATM*, *NBS1*, *MRE11*, *RAD50*, and *LIG4*—might underlie the severe late toxicity observed in some patients after XRT for head and neck cancer, compared to those experiencing mild or little late toxicity *(10)*. Other radiosensitivity end points examined were fibroblast clonogenic survival, induction and repair of DSBs, and chromosomal aberrations. This analysis was small and subject to selection bias; however, of these factors, the only parameter that correlated with clinical toxicity was the induction of chromosomal aberrations after 6 Gy in vitro *(10)*. Unfortunately, this study did not have sufficient power to address the question of whether studying a larger, inclusive panel of polymorphisms in radiosensitivity genes in a larger cohort of patients would have predictive capability for late normal tissue injury. Another important caveat with respect to normal tissue damage after XRT is the likelihood that there are genes and pathways playing crucial roles in the pathogenesis of tissue damage that are unrelated to DNA damage and its repair (e.g., those involved in inflammation, the transforming growth factor [TGF]β pathway, etc.) With sufficient knowledge of these pathways, polymorphisms in these genes could be as valuable for predicting XRT-induced morbidity as those in DNA repair genes.

It is possible that cancer-associated NER defects, although not directly affecting radiosensitivity, might underlie genetic instability that could contribute to tumor responses to XRT. It should also be noted that in one microarray study of gene activation by IR, four repair genes were induced, but, surprisingly, these were all in the NER pathway (*P48, XPC, GADD45,* and *PCNA*) and were not associated with DSB repair *(242)*.

Finally, many of the case-control studies described in Section 5.2 considered cancer risk in the context of gene–environment interactions. Environmental exposure to IR is obviously an important such factor. Indeed, exposure to IR has been linked to various cancers, including leukemias and cancers of the breast and lung *(307)*. The *XRCC1* 399-Arg/Arg genotype was, in fact, associated with an increased breast cancer risk from occupational exposure to IR *(261)*. Further evidence that inherited hypersensitivity to IR may contribute to breast carcinogenesis comes from a recent study of the relationship between missense variant alleles of *XRCC1, XRCC3,* and *HAP1*, cell cycle delay following IR, and breast cancer risk *(308)*. As had been reported previously *(309)*, there was a significant relationship between cell cycle delay and risk of breast cancer. Furthermore, in the control group, a prolonged G_2 arrest was significantly associated with the number of *HAP1* 148-Glu and *XRCC1* 399-Gln variant alleles. In contrast, there was no significant gene–environment interaction between the *XRCC1* genotype and a history of IR exposure among nonmelanoma skin cancer patients *(266)*.

6. CONCLUSIONS

Despite major advances in our knowledge of the genetic factors that control cellular radiosensitivity, it is still difficult to predict the clinically observed heterogeneity of response of patients (both normal tissues and tumors) to XRT. It is probably fair to say that no single DNA repair-related parameter will be able to reliably predict cellular radiosensitivity. Clearly, there are a number of reasons why a particular cell type might be radiosensitive. In fact, more than 100 genes have now been identified that would have to be taken into account in a rigorous screening of radiosensitivity parameters. Cellular radiosensitivity is determined by multiple integrated events, including the activation of cell cycle checkpoints and apoptosis and their integration with DNA repair, involving various signal transduction pathways *(118,245) (see* Fig. 3). An excellent overview of the many proteins involved in the regulation of the DNA damage response, cell-cycle progression, signal transduction, and apoptosis can be found elsewhere *(310,311).* Alterations in some of these factors would lead to a differential tolerance to various types of DNA lesions among different cell types without the necessity for overt differences in damage reparability. In others cases, there may be a direct connection. For example, exposure to IR was found to cause the mitogen-activated protein kinase (MAPK)-dependent induction of both the *ERCC1*/Ercc1 and *XRCC1*/Xrcc1 genes/proteins in human prostate cancer cells *(312).* Furthermore, inhibiting MAPK diminished the ability of these cells to remove IR-induced AP sites and resulted in increased levels of MN post-IR, suggesting that MAPK signaling plays an important role in the regulation of repair factors following IR exposure. Also, overexpressing the 24 kDa fibroblast growth factor (FGF)-2 isoform in two cell lines, HeLa 3A and CAPAN A3, was associated with radioresistance and with a twofold increase in the activity of DNA-PK that was, in turn, associated with an overexpression of DNA-PK$_{cs}$ but without modification of Ku expression or activity *(313).* This effect was the result of the activation of the *DNA-PKcs* gene by FGF-2, and in HeLa 3A cells, it was accompanied by a faster repair of DSBs. Another important signaling response to DNA damage involves the p53 tumor suppressor protein, which is also important for the repair of IR-induced DNA damage. However, a discussion of these relationships is beyond the scope of the current chapter and the reader should consult one of the recent reviews on this subject (e.g., ref. *214*) as well as Chapter 5 in this volume.

Until quite recently, the molecular mechanisms of the response of human cells to IR were generally studied one gene at a time (i.e., using a candidate gene approach). This was very time-consuming and depended on having advance knowledge of genes that may be involved in the phenotype being studied or on finding an association by chance. For example, the studies described in Subheading 4.2.1. relating the expression of NHEJ factors to radiosensitivity were potentially compromised by our incomplete knowledge of the genes/proteins involved in that pathway *(314),* such that limiting screening to known repair factors might not always give correlations with radiosensitivity. However, the advent of cDNA microarrays *(315,316)* and, more recently, protein chip technologies *(317–319)* has enabled the rapid screening of thousands of genes and proteins without the prerequisite for knowledge of their activity. The expression profile of genes believed to be involved in radiosensitivity has recently been characterized by microarrays in cervical cancer cell lines *(320)* and in head and neck cancer *(321).* As our knowledge of the critical radiosensitivity genes develops, it should be possible to use array methodologies to define an optimal subset of genes involved in the response to XRT. Such an approach using partial gene subsets has already been applied in some cases (e.g., ref. *322*). Future studies will require that all of the genes with roles in relevant pathways be screened for variation so that the genotyping studies will emulate the repair capacity phenotyping studies in monitoring the function of entire pathways *(323)*.

Finally, the impact of repair gene and other radiosensitivity gene polymorphisms on cell killing and on the prediction of normal tissue and tumor radiosensitivity is clearly an area in which more information is needed.

ACKNOWLEDGMENTS

DM is grateful for support by Operating Grant MOP43986 from the Canadian Institutes of Health Research and by Operating Grant RI-202 from the Alberta Cancer Board Research Initiative Program.

REFERENCES

1. Emami B, Lyman J, Brown A, et al. Tolerance of normal tissue to therapeutic irradiation. *Int J Radiat Oncol Biol Phys* 1991;15:109–122.
2. Peters LJ. Radiation therapy tolerance limits. For one or for all?—Janeway Lecture. *Cancer* 1996;77:2379–2385.
3. West CM. Invited review: intrinsic radiosensitivity as a predictor of patient response to radiotherapy. *Br J Radiol* 1995;68:827–837.
4. Budach W, Classen J, Belka C, et al. Clinical impact of predictive assays for acute and late radiation morbidity. *Strahlenther Onkol* 1998;174 Suppl. 3:20–24.
5. Tucker SL, Geara FB, Peters LJ, et al. How much could the radiotherapy dose be altered for individual patients based on a predictive assay of normal-tissue radiosensitivity? *Radiother Oncol* 1996;38:103–113.
6. West CM, Davidson SE, Roberts SA, et al. The independence of intrinsic radiosensitivity as a prognostic factor for patient response to radiotherapy of carcinoma of the cervix. *Br J Cancer* 1997;76:1184–1190.
7. Girinsky T, Lubin R, Pignon JP, et al. Predictive value of in vitro radiosensitivity parameters in head and neck cancers and cervical carcinomas: preliminary correlations with local control and overall survival. *Int J Radiat Oncol Biol Phys* 1993;25:3–7.
8. Marcou Y, D'Andrea A, Jeggo P, et al. Normal cellular radiosensitivity in an adult Fanconi anaemia patient with marked clinical radiosensitivity. *Radiother Oncol* 2001;60:75–79.
9. Severin DM, Leong T, Cassidy B, et al. Novel DNA sequence variants in the hHR21 DNA repair gene in radiosensitive cancer patients. *Int J Radiat Oncol Biol Phys* 2001;50:1323–1331.
10. Borgmann K, Roper B, El-Awady R, et al. Indicators of late normal tissue response after radiotherapy for head and neck cancer: fibroblasts, lymphocytes, genetics, DNA repair, and chromosome aberrations. *Radiother Oncol* 2002;64:141–152.
11. Birrell GW, Brown JA, Wu HI, et al. Transcriptional response of *Saccharomyces cerevisiae* to DNA-damaging agents does not identify the genes that protect against these agents. *Proc Natl Acad Sci USA* 2002;99:8778–8783.
12. Friedberg EC. How nucleotide excision repair protects against cancer. *Nat Rev Cancer* 2001;1:22–33.

13. Shiloh Y. Ataxia-telangiectasia and the Nijmegen breakage syndrome: related disorders but genes apart. *Annu Rev Genet* 1997;31:635–662.

14. Meyn MS. Ataxia-telangiectasia, cancer and the pathobiology of the ATM gene. *Clin Genet* 1999;55:289–304.

15. Stewart GS, Maser RS, Stankovic T, et al. The DNA double-strand break repair gene hMRE11 is mutated in individuals with an ataxia-telangiectasia-like disorder. *Cell* 1999;99:577–587.

16. Petrini JH. The mammalian Mre11–Rad50-nbs1 protein complex: integration of functions in the cellular DNA-damage response. *Am J Hum Genet* 1999;64:1264–1269.

17. Thompson LH, Schild D. Recombinational DNA repair and human disease. *Mutat Res* 2002;509:49–78.

18. Setlow R. Variations in DNA repair among humans. In: Harris C, Autrup H, eds. *Human Carcinogenesis.* Academic Press, New York, 1983;231–254.

19. Kovacs E, Almendral A. Reduced repair synthesis in healthy women having first degree relatives with breast cancer. *Eur J Cancer Clin Oncol* 1987;51:1051–1057.

20. Knight R, Parshad R, Price F, et al. X-ray induced chromatid damage in relation to DNA repair and cancer incidence in family members. *Int J Cancer* 1993;54:589–593.

21. Wei Q, Matanoski GM, Farmer ER, et al. DNA repair and susceptibility to basal cell carcinoma: a case-control study. *Am J Epidemiol* 1994;140:598–607.

22. Scott D, Spreadborough A, Levine E, et al. Genetic predisposition in breast cancer. *Lancet* 1994;344:1444.

23. Helzsouer KJ, Harris EL, Parshad R, Fogel S, Bigbee W, Sanford K. Familial clustering of breast cancer: possible interactions between DNA repair exposure on the development of breast cancer. *Int J Cancer* 1995;64:14–17.

24. Parshad R, Price F, Bohr V, et al. Deficient DNA repair capacity, a predisposing factor in breast cancer. *Br J Cancer* 1996;74:1–5.

25. Helzsouer K, Harris E, Parshad R, et al. DNA repair proficiency: potential susceptibility factor for breast cancer. *J Natl Cancer Inst* 1996;88:754–755.

26. Wei Q, Cheng L, Hong WK, et al. Reduced DNA repair capacity in lung cancer patients. *Cancer Res* 1996;56:4103–4107.

27. Wu X, Gu J, Hsu T, et al. Differences in sensitivity to benzopyrene-diol-epoxide. (BPDE) as a marker of lung cancer susceptibility. *Proc Am Assoc Cancer Res* 1997;38:618.

28. Reed E. Ovarian cancer: Molecular abnormalities. In: Bertino JR, ed. *Encyclopedia of Cancer, vol. II.* Academic Press, San Diego, 1997;1192–1200.

29. Mohrenweiser H, Jones I. Variation in DNA repair is a factor in cancer susceptibility: a paradigm for the promises and perils of individual and population risk estimation? *Mutat Res* 1998;400:15–24.

30. Cheng L, Eicher SA, Guo Z, et al. Reduced DNA repair capacity in head and neck cancer patients. *Cancer Epidemiol Biomarkers Prev* 1998;7:465–468.

31. Berwick M, Vineis P. Markers of DNA repair and susceptibility to cancer in humans: an epidemiologic review. *J Natl Cancer Inst* 2000;92:874–897.

32. Spitz MR, Wu X, Wang Y, et al. Modulation of nucleotide excision repair capacity by XPD polymorphisms in lung cancer patients. *Cancer Res* 2001;61:1354–1357

33. Yu Z, Chen J, Ford B, et al. Human DNA repair systems: an overview. *Environ Mol Mutagen* 1999;33:3–20.

34. Shen R, Jones I, Mohrenweiser H. Nonconservative amino acid substitution variants exist at polymorphic frequency in DNA repair genes in healthy humans. *Cancer Res* 1998;58:604–608.

35. Ford BN, Ruttan CC, Kyle VL, et al. Identification of single nucleotide polymorphisms in human DNA repair genes. *Carcinogenesis* 2000;21:1977–1981.

36. Goode EL, Ulrich CM, Potter JD. Polymorphisms in DNA repair genes and associations with cancer risk. *Cancer Epidemiol Biomarkers Prev* 2002;11:1513–1530.

37. Mohrenweiser HW, Xi T, Vazquez-Matias J, et al. Identification of 127 amino acid substitution variants in screening 37 DNA repair genes in humans. *Cancer Epidemiol Biomarkers Prev* 2002;11:1054–1064.

38. Begg AC, Vens C. Genetic manipulation of radiosensitivity. *Int J Radiat Oncol Biol Phys* 2001;49:367–371.

39. Willers H, Xia F, Powell SN. Recombinational DNA repair in cancer and normal cells: the challenge of functional analysis. *J Biomed Biotechnol* 2002;2:86–93.

40. Ahnstrom G. Techniques to measure DNA single-strand breaks in cells: a review. *Int J Radiat Biol* 1988;54:695–707.

41. Kohn KW. Principles and practice of DNA filter elution. *Pharmacol Ther* 1991;49:55–77.

42. Elmendorff-Dreikorn K, Chauvin C, Slor H, et al. Assessment of DNA damage and repair in human peripheral blood mononuclear cells using a novel DNA unwinding technique. *Cell Mol Biol (Noisy-le-grand)* 1999;45:211–218.

43. Olive PL, Banath JP, Durand RE. Heterogeneity in radiation-induced DNA damage and repair in tumor and normal cells measured using the "comet" assay. *Radiat Res* 1990;122:86–94.

44. Tice RR, Andrews PW, Singh NP. The single cell gel assay: a sensitive technique for evaluating intercellular differences in DNA damage and repair. *Basic Life Sci* 1990;53:291–301.

45. Singh NP, McCoy MT, Tice RR, et al. A simple technique for quantitation of low levels of DNA damage in individual cells. *Exp Cell Res* 1988;175:184–191.

46. Singh NP, Tice RR, Stephens RE, et al. A microgel electrophoresis technique for the direct quantitation of DNA damage and repair in individual fibroblasts cultured on microscope slides. *Mutat Res* 1991;252:289–296.

47. Olive PL. DNA damage and repair in individual cells: applications of the comet assay in radiobiology. *Int J Radiat Biol* 1999;75:395–405.

48. Banath JP, Fushiki M, Olive PL. Rejoining of DNA single- and double-strand breaks in human white blood cells exposed to ionizing radiation. *Int J Radiat Biol* 1998;73:649–660.

49. Palyvoda O, Mukalov I, Polanska J, et al. Radiation-induced DNA damage and its repair in lymphocytes of patients with head and neck cancer and healthy donors. *Anticancer Res* 2002;22:1721–1725.

50. Cloos J, Steen I, Timmerman AJ, et al. DNA-damage processing in blood lymphocytes of head-and-neck-squamous-cell-carcinoma patients is dependent on tumor site. *Int J Cancer* 1996;68:26–29.

51. Nascimento PA, da Silva MA, Oliveira EM, et al. Evaluation of radioinduced damage and repair capacity in blood lymphocytes of breast cancer patients. *Braz J Med Biol Res* 2001;34:165–176.

52. Leprat F, Alapetite C, Rosselli F, et al. Impaired DNA repair as assessed by the "comet" assay in patients with thyroid tumors after a history of radiation therapy: a preliminary study. *Int J Radiat Oncol Biol Phys* 1998;40:1019–1026.

53. Myllyperkio MH, Koski TR, Vilpo LM, et al. Gamma-irradiation-induced DNA single- and double-strand breaks and their repair in chronic lymphocytic leukemia cells of variable radiosensitivity. *Hematol Cell Ther* 1999;41:95–103.

54. Woods WG, Lopez M, Kalvonjian SL. Normal repair of γ radiation-induced single-strand and double-strand DNA breaks in retinoblastoma fibroblasts. *Biochim Biophys Acta* 1982;698:40–48.

55. Malcolmson AM, Davies G, Hanson JA, et al. Determination of radiation-induced damage in lymphocytes using the micronucleus and microgel electrophoresis 'Comet' assays. *Eur J Cancer* 1995;31A:2320–2323.

56. Oppitz U, Denzinger S, Nachtrab U, et al. Radiation-induced comet-formation in human skin fibroblasts from radiotherapy patients with different normal tissue reactions. *Strahlenther Onkol* 1999;175:341–346.

57. Oppitz U, Schulte S, Stopper H, et al. In vitro radiosensitivity measured in lymphocytes and fibroblasts by colony formation and comet assay: comparison with clinical acute reactions to radiotherapy in breast cancer patients. *Int J Radiat Biol* 2002;78:611–616.

58. Alapetite C, Thirion P, de la Rochefordiere A, et al. Analysis by alkaline comet assay of cancer patients with severe reactions to radiotherapy: defective rejoining of radioinduced DNA strand breaks in lymphocytes of breast cancer patients. *Int J Cancer* 1999;83:83–90.

59. Schwartz JL, Rotmensch J, Giovanazzi S, et al. Faster repair of DNA double-strand breaks in radioresistant human tumor cells. *Int J Radiat Oncol Biol Phys* 1988;15:907–912.

60. Bergqvist M, Brattstrom D, Stalberg M, et al. Evaluation of radiation-induced DNA damage and DNA repair in human lung cancer cell lines with different radiosensitivity using alkaline and neutral single cell gel electrophoresis. *Cancer Lett* 1998;133:9–18.

61. Daza P, Schubler H, McMillan TJ, et al. Radiosensitivity and double-strand break rejoining in tumorigenic and non-tumorigenic human epithelial cell lines. *Int J Radiat Biol* 1997;72: 91–100.

62. Roos WP, Binder A, Bohm L. Determination of the initial DNA damage and residual DNA damage remaining after 12 hours of repair in eleven cell lines at low doses of irradiation. *Int J Radiat Biol* 2000;76:1493–1500.

63. Bacova G, Hunakova LE, Chorvath M, et al. Radiation-induced DNA damage and repair evaluated with 'comet assay' in human ovarian carcinoma cell lines with different radiosensitivities. *Neoplasma* 2000;47:367–374.

64. Cadet J, Weinfeld M. Detecting DNA damage. *Anal Chem* 1993;65:675A-682A.

65. Chen SK, Tsai MH, Hwang JJ, et al. Determination of 8-oxoguanine in individual cell nucleus of γ-irradiated mammalian cells. *Radiat Res* 2001;155:832–836.

66. van der Schans GP, Timmerman AJ, Bruijnzeel PL. Detection of single-strand breaks and base damage in DNA of human white blood cells as a tool for biological dosimetry of exposure to ionizing radiation. *Mil Med* 2002;167(2 Suppl):5–7.

67. Timmerman AJ, Mars-Groenendijk RH, van der Schans GP, et al. A modified immunochemical assay for the fast detection of DNA damage in human white blood cells. *Mutat Res* 1995;334:347–356.

68. Le XC, Xing JZ, Lee J, et al. Inducible repair of thymine glycol detected by an ultrasensitive assay for DNA damage. *Science* 1998;280:1066–1069.

69. Weinfeld M, Xing JZ, Lee J, et al. Immunofluorescence detection of radiation-induced DNA base damage. *Mil Med* 2002;167(2 Suppl):2–4.

70. Olive PL. The role of DNA single- and double-strand breaks in cell killing by ionizing radiation. *Radiat Res* 1998;150(5 Suppl):S42–S51.

71. Bradley MO, Kohn KW. X-ray induced DNA double strand break production and repair in mammalian cells as measured by neutral filter elution. *Nucleic Acids Res* 1979;7:793–804.

72. Blöcher D, Einspenner M, Zajackowski J. CHEF electrophoresis, a sensitive technique for the determination of DNA double-strand breaks. *Int J Radiat Biol* 1989;56:437–448.

73. Elia MC, De Luca JG, Bradley MO. Significance and measurement of DNA double strand breaks in mammalian cells. *Pharmacol Ther* 1991;51:291–327.

74. Olive PL, Wlodek D, Banath JP. DNA double-strand breaks measured in individual cells subjected to gel electrophoresis. *Cancer Res* 1991;51:4671–4676.

75. Wlodek D, Banath J, Olive PL. Comparison between pulsed-field and constant-field gel electrophoresis for measurement of DNA double-strand breaks in irradiated Chinese hamster ovary cells. *Int J Radiat Biol* 1991;60:779–790.

76. Lett JT. Cellular radiation biology in consolidation and transition. *Br J Cancer* 1987;8(Suppl): 145–152.

77. Olive PL. DNA organization affects cellular radiosensitivity and detection of initial DNA strand breaks. *Int J Radiat Biol* 1992;62:389–396.

78. Muller WU, Bauch T, Stuben G, et al. Radiation sensitivity of lymphocytes from healthy individuals and cancer patients as measured by the comet assay. *Radiat Environ Biophys* 2001;40:83–89.

79. Hannan MA, Blocher D, Sigut D, et al. DNA double strand breaks in fibroblast cell lines, from non-Hodgkin's lymphoma patients, showing increased sensitivity to chronic γ irradiation. *Cancer Lett* 1991;57:137–143.

80. Eastham AM, Marples B, Kiltie AE, et al. Fibroblast radiosensitivity measured using the comet DNA-damage assay correlates with clonogenic survival parameters. *Br J Cancer* 1999;79:1366–1371.

81. Badie C, Iliakis G, Foray N, et al. Induction and rejoining of DNA double-strand breaks and interphase chromosome breaks after exposure to X rays in one normal and two hypersensitive human fibroblast cell lines. *Radiat Res* 1995;144:26–35.

82. Wurm R, Burnet NG, Duggal N, et al. Cellular radiosensitivity and DNA damage in primary human fibroblasts. *Int J Radiat Oncol Biol Phys* 1994;30:625–633.

83. Zhou PK, Sproston AR, Marples B, et al. The radiosensitivity of human fibroblast cell lines correlates with residual levels of DNA double-strand breaks. *Radiother Oncol* 1998;47:271–276.

84. Kiltie AE, Orton CJ, Ryan AJ, et al. A correlation between residual DNA double-strand breaks and clonogenic measurements of radiosensitivity in fibroblasts from preradiotherapy cervix cancer patients. *Int J Radiat Oncol Biol Phys* 1997;39:1137–1144.

85. Evans HH, Ricanati M, Horng MF, et al. DNA double-strand break rejoining deficiency in TK6 and other human B-lymphoblast cell lines. *Radiat Res* 1993;134:307–315.

86. Badie C, Iliakis G, Foray N, et al. Defective repair of DNA double-strand breaks and chromosome damage in fibroblasts from a radiosensitive leukemia patient. *Cancer Res* 1995;55:1232–1234.

87. Muller WU, Bauch T, Streffer C, et al. Does radiotherapy affect the outcome of the comet assay? *Br J Radiol* 2002;75:608–614.

88. Ruiz de Almodóvar JM, Guirado D, Núñez MI, et al. Individualisation of radiotherapy in breast cancer patients: possible usefulness of a DNA damage assay to measure normal cell radiosensitivity. *Radiother Oncol* 2002;62:327–333.

89. Kiltie AE, Barber JB, Swindell R, et al. Lack of correlation between residual radiation-induced DNA damage, in keratinocytes assayed directly from skin, and late radiotherapy reactions in breast cancer patients. *Int J Radiat Oncol Biol Phys* 1999;43:481–487.

90. Kiltie AE, Ryan AJ, Swindell R, et al. A correlation between residual radiation-induced DNA double-strand breaks in cultured fibroblasts and late radiotherapy reactions in breast cancer patients. *Radiother Oncol* 1999;51:55–65.

91. Dickson J, Magee B, Stewart A, et al. Relationship between residual radiation-induced DNA double-strand breaks in cultured fibroblasts and late radiation reactions: a comparison of training and validation cohorts of breast cancer patients. *Radiother Oncol* 2002;62:321–326.

92. Dikomey E, Brammer I, Johansen J, et al. Relationship between DNA double-strand breaks, cell killing, and fibrosis studied in confluent skin fibroblasts derived from breast cancer patients. *Int J Radiat Oncol Biol Phys* 2000;46:481–490.

93. Kelland LR, Edwards SM, Steel GG. Induction and rejoining of DNA double-strand breaks in human cervix carcinoma cell lines of differing radiosensitivity. *Radiat Res* 1988;116:526–538.

94. Schwartz JL, Mustafi R, Beckett MA, et al. Prediction of the radiation sensitivity of human squamous cell carcinoma cells using DNA filter elution. *Radiat Res* 1990;123:1–6.

95. Eguchi-Kasai K, Kosaka T, Sato K, et al. Reparability of DNA double-strand breaks and radiation sensitivity in five mammalian cell lines. *Int J Radiat Biol* 1991;59:97–104.

96. Giaccia AJ, Schwartz JL, Shieh J, et al. The use of asymmetric-field inversion electrophoresis to predict tumor cell radiosensitivity. *Radiother Oncol* 1992;24:231–238.

97. Wei K, Wandl E, Karcher KH. X-ray induced DNA double-strand breakage and rejoining in a radiosensitive human renal carcinoma cell line estimated by CHEF electrophoresis. *Strahlenther Onkol* 1993;169:740–744.

98. Cedervall B, Sirzea F, Brodin O, et al. Less initial rejoining of X-ray-induced DNA double-strand breaks in cells of a small cell (U-1285) compared to a large cell (U-1810) lung carcinoma cell line. *Radiat Res* 1994;139:34–39.

99. Muller WU, Bauch T, Streffer C, et al. Comet assay studies of radiation-induced DNA damage and repair in various tumour cell lines. *Int J Radiat Biol* 1994;65:315–319.

100. Nunez MI, Villalobos M, Olea N, et al. Radiation-induced DNA double-strand break rejoining in human tumour cells. *Br J Cancer* 1995;71:311–316.

101. Whitaker SJ, Ung YC, McMillan TJ. DNA double-strand break induction and rejoining as determinants of human tumour cell radiosensitivity. A pulsed-field gel electrophoresis study. *Int J Radiat Biol* 1995;67:7–18.

102. Schwartz JL, Mustafi R, Beckett MA, et al. DNA double-strand break rejoining rates, inherent radiation sensitivity and human tumour response to radiotherapy. *Br J Cancer* 1996;74:37–42.

103. Dikomey E, Dahm-Daphi J, Brammer I, et al. Correlation between cellular radiosensitivity and non-repaired double-strand breaks studied in nine mammalian cell lines. *Int J Radiat Biol* 1998;73:269–278.

104. Sirzen F, Nilsson A, Zhivotovsky B, et al. DNA-dependent protein kinase content and activity in lung carcinoma cell lines: correlation with intrinsic radiosensitivity. *Eur J Cancer* 1999;35:111–116.

105. Price ME, McKelvey-Martin VJ, Robson T, et al. Induction and rejoining of DNA double-strand breaks in bladder tumor cells. *Radiat Res* 2000;153:788–794.

106. Eastham AM, Atkinson J, West CM. Relationships between clonogenic cell survival, DNA damage and chromosomal radiosensitivity in nine human cervix carcinoma cell lines. *Int J Radiat Biol* 2001;77:295–302.

107. Marples B, Longhurst D, Eastham AM, et al. The ratio of initial/residual DNA damage predicts intrinsic radiosensitivity in seven cervix carcinoma cell lines. *Br J Cancer* 1998;77:1108–1114.

108. Radford IR. Evidence for a general relationship between the induced level of DNA double-strand breakage and cell-killing after X-irradiation of mammalian cells. *Int J Radiat Biol* 1986;49:611–620.

109. Ruiz de Almodovar JM, Nunez MI, McMillan TJ, et al. Initial radiation-induced DNA damage in human tumour cell lines: a correlation with intrinsic cellular radiosensitivity. *Br J Cancer* 1994;69:457–462.

110. Smeets MFMA, Mooren EHM, Begg AC. Radiation-induced DNA damage and repair in radiosensitive and radioresistant human tumour cells measured by field inversion gel electrophoresis. *Int J Radiat Biol* 1993;63:703–713.

111. Smeets MF, Mooren EH, Abdel-Wahab AH, et al. Differential repair of radiation-induced DNA damage in cells of human squamous cell carcinoma and the effect of caffeine and cysteamine on induction and repair of DNA double-strand breaks. *Radiat Res* 1994;140:153–160.

112. Olive PL, Banath JP, MacPhail HS. Lack of a correlation between radiosensitivity and DNA double-strand break induction or rejoining in six human tumor cell lines. *Cancer Res* 1994;54:3939–3946.

113. McKay MJ, Kefford RF. The spectrum of in vitro radiosensitivity in four human melanoma cell lines is not accounted for by differential induction or rejoining of DNA double strand breaks. *Int J Radiat Oncol Biol Phys* 1995;31:345–352.

114. Woudstra EC, Brunsting JF, Roesink JM, et al. Radiation induced DNA damage and damage repair in three human tumour cell lines. *Mutat Res* 1996;362:51–59.

115. Wang J, Hu L, Gupta N, et al. Induction and characterization of human glioma clones with different radiosensitivities. *Neoplasia* 1999;1:138–144.

116. Nunez MI, McMillan TJ, Valenzuela MT, et al. Relationship between DNA damage, rejoining and cell killing by radiation in mammalian cells. *Radiother Oncol* 1996;39:155–165.

117. Foray N, Arlett CF, Malaise EP. Radiation-induced DNA double-strand breaks and the radiosensitivity of human cells: a closer look. *Biochimie* 1997;79:567–575.

118. McMillan TJ, Tobi S, Mateos S, et al. The use of DNA double-strand break quantification in radiotherapy. *Int J Radiat Oncol Biol Phys* 2001;49:373–377.

119. Ward JF. The complexity of DNA damage: relevance to biological consequences. *Int J Radiat Biol* 1994;66:427–432.

120. Sutherland BM, Bennett PV, Sutherland JC, et al. Clustered DNA damages induced by x rays in human cells. *Radiat Res* 2002;157:611–616.

121. Le Page F, Klungland A, Barnes DE, et al. Transcription coupled repair of 8-oxoguanine in murine cells: the ogg1 protein is required for repair in nontranscribed sequences but not in transcribed sequences. *Proc Natl Acad Sci USA* 2000;97: 8397–8402.

122. Boiteux S, Le Page F. Repair of 8-oxoguanine and Ogg1-incised apurinic sites in a CHO cell line. *Progr Nucleic Acid Res Mol Biol* 2001;68:95–105.

123. Lindahl T. Keynote: past, present, and future aspects of base excision repair. *Progr Nucleic Acid Res Mol Biol* 2001;68:xvii–xxx.

124. Leadon SA, Cooper PK. Preferential repair of ionizing radiation-induced damage in the transcribed strand of an active human gene is defective in Cockayne syndrome. *Proc Natl Acad Sci USA* 1993;90:10499–10503.

125. Avrutskaya AV, Leadon SA. Measurement of oxidative DNA damage and repair in specific DNA sequences. *Methods* 2000;22:127–134.

126. Bunch RT, Gewirtz DA, Povirk LF. Ionizing radiation-induced DNA strand breakage and rejoining in specific genomic regions as determined by an alkaline unwinding/Southern blotting method. *Int J Radiat Biol* 1995;68:553–562.

127. Ljungman M. Repair of radiation-induced DNA strand breaks does not occur preferentially in transcriptionally active DNA. *Radiat Res* 1999;152:444–449.

128. May A, Bohr VA. Gene-specific repair of γ-ray-induced DNA strand breaks in colon cancer cells: no coupling to transcription and no removal from the mitochondrial genome. *Biochem Biophys Res Commun* 2000;269:433–437.

129. Ploskonosova II, Baranov VI, Gaziev AI. PCR assay of DNA damage and repair at the gene level in brain and spleen of γ-irradiated young and old rats. *Mutat Res* 1999;434:109–117.

130. Sak A, Stuschke M. Repair of ionizing radiation induced DNA double-strand breaks (dsb) at the c-myc locus in comparison to the overall genome. *Int J Radiat Biol* 1998;73:35–43.

131. Debenham PG, Jones NJ, Webb MB. Vector-mediated DNA double-strand break repair analysis in normal, and radiation-sensitive, Chinese hamster V79 cells. *Mutat Res* 1988;199:1–9.

132. Thacker J. The use of integrating DNA vectors to analyse the molecular defects in ionising radiation-sensitive mutants of mammalian cells including ataxia telangiectasia. *Mutat Res* 1989;220:187–204.

133. Powell S, McMillan TJ. Clonal variation of DNA repair in a human glioma cell line. *Radiother Oncol* 1991;21:225–232.

134. Powell SN, Whitaker SJ, Edwards SM, et al. A DNA repair defect in a radiation-sensitive clone of a human bladder carcinoma cell line. *Br J Cancer* 1992;65:798–802.

135. Ganesh A, North P, Thacker J. Repair and misrepair of site-specific DNA double-strand breaks by human cell extracts. *Mutat Res* 1993;299:251–259.

136. Powell SN, McMillan TJ. The repair fidelity of restriction enzyme-induced double strand breaks in plasmid DNA correlates with radioresistance in human tumor cell lines. *Int J Radiat Oncol Biol Phys* 1994;29:1035–1040.

137. Britten RA, Liu D, Kuny S, et al. Differential level of DSB repair fidelity effected by nuclear protein extracts derived from radiosensitive and radioresistant human tumour cells. *Br J Cancer* 1997;76:1440–1447.

138. Powell SN, Mills J, McMillan TJ. Radiosensitive human tumour cell lines show misrepair of DNA termini. *Br J Radiol* 1998;71:1178–1184.

139. Macann AM, Britten RA, Poppema S, et al. DNA double-strand break rejoining in human follicular lymphoma and glioblastoma tumor cells. *Oncol Rep* 2000;7:299–303.

140. Bennett CB, Rainbow AJ. DNA damage and biological expression of adenovirus: a comparison of liquid versus frozen conditions of exposure to γ rays. *Radiat Res* 1989;120:102–112.

141. Lees-Miller SP, Godbout R, Chan DW, et al. Absence of p350 subunit of DNA-activated protein kinase from a radiosensitive human cell line. *Science* 1995;267:1183–1185.

142. Eady JJ, Peacock JH, McMillan TJ. Host cell reactivation of γ-irradiated adenovirus 5 in human cell lines of varying radiosensitivity. *Br J Cancer* 1992;66:113–118.

143. Krisch RE, Tan PM, Flick MB. A sensitive SV40 viral probe assay for DNA strand breaks and their biological repair in higher cells: techniques and preliminary results. *Radiat Res* 1985;101:356–372.

144. Smith-Ravin J, Jeggo PA. Use of damaged plasmid to study DNA repair in X-ray sensitive (xrs) strains of Chinese hamster ovary (CHO) cells. *Int J Radiat Biol* 1989;56:951–961.

145. Sikpi MO, Liu X, Lurie AG, et al. Alteration of irradiated shuttle vector processing by exposure of human lymphoblast host cells to single or split γ-ray doses. *Int J Radiat Biol* 1994;65:157–164.

146. Hoeijmakers JHJ. Genome maintenance mechanisms for preventing cancer. *Nature* 2001;411:366–374.

147. Masson M, Niedergang C, Schreiber V, et al. XRCC1 is specifically associated with poly(ADP-ribose) polymerase and negatively regulates its activity following DNA damage. *Mol Cell Biol* 1998;18:3563–3571.

148. Tomkinson AE, Chen L, Dong Z, et al. Completion of base excision repair by mammalian DNA ligases. *Prog Nucleic Acid Res Mol Biol* 2001;68:151–164.

149. Whitehouse CJ, Taylor RM, Thistlethwaite A, et al. XRCC1 stimulates human polynucleotide kinase activity at damaged DNA termini and accelerates DNA single-strand break repair. *Cell* 2001;104:107–117.

150. Thompson LH, Brookman KW, Dillehay LE, et al. A CHO-cell strain having hypersensitivity to mutagens, a defect in DNA strand-break repair, and an extraordinary baseline frequency of sister chromatid exchange. *Mutat Res* 1982;95:427–440.

151. Wong CS, Hill RP. Experimental radiotherapy. In: Tannock IF, Hill RP, eds. *The Basic Science of Oncology, 3rd ed.* McGraw-Hill,New York, 1998:322–349.

152. Thompson LH, Brookman KW, Jones NJ, et al. Molecular cloning of the human XRCC1 gene, which corrects defective DNA strand break repair and sister chromatid exchange. *Mol Cell Biol* 1990;10:6160–6171.

153. Dunphy EJ, Beckett MA, Thompson LH, et al. Expression of the polymorphic human DNA repair gene XRCC1 does not correlate with radiosensitivity in the cells of human head and neck tumor cell lines. *Radiat Res* 1992;130:166–170.

154. Ludwig DL, MacInnes MA, Takiguchi Y, et al. A murine AP-endonuclease gene-targeted deficiency with post-implantation embryonic progression and ionizing radiation sensitivity. *Mutat Res* 1998;409:17–29.

155. Herring CJ, West CM, Wilks DP, et al. Levels of the DNA repair enzyme human apurinic/apyrimidinic endonuclease (APE1, APEX, Ref-1) are associated with the intrinsic radiosensitivity of cervical cancers. *Br J Cancer* 1998;78:1128–1133.

156. Herring CJ, Deans B, Elder RH, et al. Expression levels of the DNA repair enzyme HAP1 do not correlate with the radiosensitivities of human or HAP1-transfected rat cell lines. *Br J Cancer* 1999;80:940–945.

157. Schindl M, Oberhuber G, Pichlbauer EG, et al. DNA repair–redox enzyme apurinic endonuclease in cervical cancer: evaluation of redox control of HIF-1α and prognostic significance. *Int J Oncol* 2001;19:799–802.

158. Bobola MS, Blank A, Berger MS, et al. Apurinic/apyrimidinic endonuclease activity is elevated in human adult gliomas. *Clin Cancer Res* 2001;7:3510–3518.

159. Kelley MR, Cheng L, Foster R, et al. Elevated and altered expression of the multifunctional DNA base excision repair and redox enzyme Ape1/ref-1 in prostate cancer. *Clin Cancer Res* 2001;7:824–830.

160. Robertson KA, Bullock HA, Xu Y, et al. Altered expression of Ape1/ref-1 in germ cell tumors and overexpression in NT2 cells confers resistance to bleomycin and radiation. *Cancer Res* 2001;61:2220–2225.

161. Nishimura S. Mammalian Ogg1/Mmh gene plays a major role in repair of the 8-hydroxyguanine lesion in DNA. *Prog Nucleic Acid Res Mol Biol* 2001;68:107–123.

162. Epe B. Role of endogenous oxidative DNA damage in carcinogenesis: what can we learn from repair-deficient mice? *Biol Chem* 2002;383:467–475.

163. Sobol RW, Horton JK, Kühn R, et al. Requirement of mammalian DNA polymerase-β in base-excision repair. *Nature* 1996;379:183–186.

164. Miura M, Watanabe H, Okochi K, et al. Biological response to ionizing radiation in mouse embryo fibroblasts with a targeted disruption of the DNA polymerase beta gene. *Radiat Res* 2000;153:773–780.

165. Raaphorst GP, Ng CE, Yang DP. Comparison of response to radiation, hyperthermia and cisplatin in parental and polymerase beta knockout cells. *Int J Hyperthermia* 2002;18:33–39.

166. Norbury CJ, Hickson ID. Cellular responses to DNA damage. *Annu Rev Pharmacol Toxicol* 2001;41:367–401.

167. Jackson SP. Sensing and repairing DNA double-strand breaks. *Carcinogenesis* 2002;23:687–696.

168. Jeggo PA. Identification of genes involved in repair of DNA double-strand breaks in mammalian cells. *Radiat Res* 1998;150(5 Suppl):S80–S91.

169. Liang F, Han M, Romanienko PJ, et al. Homology-directed repair is a major double-strand break repair pathway in mammalian cells. *Proc Natl Acad Sci USA* 1998;95:5172–5177.

170. Paull TT, Gellert M. A mechanistic basis for Mre11-directed DNA joining at microhomologies. *Proc Natl Acad Sci USA* 2000;97:6409–6414.

171. Goedecke W, Eijpe M, Offenberg HH, et al. Mre11 and Ku70 interact in somatic cells, but are differentially expressed in early meiosis. *Nature Genet* 1999;23:194–198.

172. Gao Y, Sun Y, Frank KM, et al. A critical role for DNA end-joining proteins in both lymphogenesis and neurogenesis. *Cell* 1998;95:891–902.

173. Pierce AJ, Hu P, Han M, et al. DNA end-binding protein modulates homologous repair of double-strand breaks in mammalian cells. *Genes Dev* 2001;15:3237–3242.

174. Gu Y, Seidl KJ, Rathbun GA, et al. Growth retardation and leaky SCID phenotype of Ku70-deficient mice. *Immunity* 1997;7:653–665.

175. Gao Y, Chaudhuri J, Zhu C, et al. A targeted DNA-PKcs-null mutation reveals DNA-PK-independent functions for KU in V(D)J recombination. *Immunity* 1998;9:367–376.

176. Luo G, Yao MS, Bender CF, et al. Disruption of mRad50 causes embryonic stem cell lethality, abnormal embryonic development, and sensitivity to ionizing radiation. *Proc Natl Acad Sci USA* 1999;96:7376–7381.

177. Pucci S, Mazzarelli P, Rabitti C, et al. Tumor specific modulation of KU70/80 DNA binding activity in breast and bladder human tumor biopsies. *Oncogene* 2001;20:739–747.

178. Rigas B, Borgo S, Elhosseiny A, et al. Decreased expression of DNA-dependent protein kinase, a DNA repair protein, during human colon carcinogenesis. *Cancer Res* 2001;61:8381–8384.

179. Korabiowska M, Tscherny M, Grohmann U, et al. Decreased expression of Ku70/Ku80 proteins in malignant melanomas of the oral cavity. *Anticancer Res* 2002;22:193–196.

180. Stronati L, Gensabella G, Lamberti C, et al. Expression and DNA binding activity of the Ku heterodimer in bladder carcinoma. *Cancer* 2001;92:2484–2492.

181. Gaymes TJ, Mufti GJ, Rassool FV. Myeloid leukemias have increased activity of the nonhomologous end-joining pathway and concomitant DNA misrepair that is dependent on the Ku70/86 heterodimer. *Cancer Res* 2002;62:2791–2797.

182. Sakata K, Matsumoto Y, Tauchi H, et al. Expression of genes involved in repair of DNA double-strand breaks in normal and tumor tissues. *Int J Radiat Oncol Biol Phys* 2001;49:161–167.

183. Polischouk AG, Cedervall B, Ljungquist S, et al. DNA double-strand break repair, DNA-PK, and DNA ligases in two human squamous carcinoma cell lines with different radiosensitivity. *Int J Radiat Oncol Biol Phys* 1999;43:191–198.

184. Wilson CR, Davidson SE, Margison GP, et al. Expression of Ku70 correlates with survival in carcinoma of the cervix. *Br J Cancer* 2000;83:1702–1706.

185. Allalunis-Turner MJ, Lintott LG, Barron GM, et al. Lack of correlation between DNA-dependent protein kinase activity and tumor cell radiosensitivity. *Cancer Res* 1995;55:5200–5202.

186. Bjork-Eriksson T, West C, Nilsson A, et al. The immunohistochemical expression of DNA-PKCS and Ku (p70/p80) in head and neck cancers: relationships with radiosensitivity. *Int J Radiat Oncol Biol Phys* 1999;45:1005–1010.

187. Komuro Y, Watanabe T, Hosoi Y, et al. The expression pattern of Ku correlates with tumor radiosensitivity and disease free survival in patients with rectal carcinoma. *Cancer* 2002;95:1199–1205.

188. Sakata K, Matsumoto Y, Satoh M, et al. Clinical studies of immunohistochemical staining of DNA-dependent protein kinase in oropharyngeal and hypopharyngeal carcinomas. *Radiat Med* 2001;19:93–97.

189. Noguchi T, Shibata T, Fumoto S, et al. DNA-PKcs expression in esophageal cancer as a predictor for chemoradiation therapeutic sensitivity. *Ann Surg Oncol* 2002;9:1017–1022.

190. Kasten U, Plottner N, Johansen J, Overgaard J, Dikomey E. Ku70/80 gene expression and DNA-dependent protein kinase (DNA-PK) activity do not correlate with double-strand break (dsb) repair capacity and cellular radiosensitivity in normal human fibroblasts. *Br J Cancer* 1999;79:1037–1041.

191. Carlomagno F, Burnet NG, Turesson I, et al. Comparison of DNA repair protein expression and activities between human fibroblast cell lines with different radiosensitivities. *Int J Cancer* 2000;85:845–849.

192. Baumann P, Benson FE, West SC. Human Rad51 protein promotes ATP-dependent homologous pairing and strand transfer reactions in vitro. *Cell* 1996;87:757–766.

193. Benson FE, Baumann P, West SC. Synergistic actions of Rad51 and Rad52 in recombination and DNA repair. *Nature* 1998;391:401–404.

194. McIlwraith MJ, Van Dyck E, Masson JY, et al. Reconstitution of the strand invasion step of double-strand break repair using human Rad51 Rad52 and RPA proteins. *J Mol Biol* 2000;304:151–164.

195. Kingston RE, Bunker CA, Imbalzano AN. Repression and activation by multiprotein complexes that alter chromatin structure. *Genes Dev* 1996;10:905–920.

196. Swagemakers SM, Essers J, de Wit J, et al. The human RAD54 recombinational DNA repair protein is a double-stranded DNA-dependent ATPase. *J Biol Chem* 1998;273:28,292–28,297.

197. Tsuzuki T, Fujii Y, Sakumi K, et al. Targeted disruption of the Rad51 gene leads to lethality in embryonic mice. *Proc Natl Acad Sci USA* 1996;93:6236–6240.

198. Lim DS, Hasty P. A mutation in mouse rad51 results in an early embryonic lethal that is suppressed by a mutation in p53. *Mol Cell Biol* 1996;16:7133–7143.

199. Sonoda E, Sasaki MS, Buerstedde JM, et al. Rad51-deficient vertebrate cells accumulate chromosomal breaks prior to cell death. *EMBO J* 1998;17:598–608.

200. Takata M, Sasaki MS, Sonoda E, et al. The Rad51 paralog Rad51B promotes homologous recombinational repair. *Mol Cell Biol* 2000;20:6476–6482.

201. Takata M, Sasaki MS, Tachiiri S, et al. Chromosome instability and defective recombinational repair in knockout mutants of the five Rad51 paralogs. *Mol Cell Biol* 2001;21: 2858–2866.
202. Fuller LF, Painter RB. A Chinese hamster ovary cell line hypersensitive to ionizing radiation and deficient in repair replication. *Mutat Res* 1988;193:109–121.
203. Johnson RD, Liu N, Jasin M. Mammalian XRCC2 promotes the repair of DNA double-strand breaks by homologous recombination. *Nature* 1999;401:397–399.
204. Pierce AJ, Johnson RD, Thompson LH, et al. XRCC3 promotes homology-directed repair of DNA damage in mammalian cells. *Genes Dev* 1999;13:2633–2638.
205. Deans B, Griffin CS, Maconochie M, et al. Xrcc2 is required for genetic stability, embryonic neurogenesis and viability in mice. *EMBO J* 2000;19:6675–6685.
206. Yamaguchi-Iwai Y, Sonoda E, Buerstedde JM, et al. Homologous recombination, but not DNA repair, is reduced in vertebrate cells deficient in RAD52. *Mol Cell Biol* 1998;18:6430–6435.
207. Rijkers T, Van Den Ouweland J, Morolli B, et al. Targeted inactivation of mouse RAD52 reduces homologous recombination but not resistance to ionizing radiation. *Mol Cell Biol* 1998;18:6423–6429.
208. Essers J, Hendriks RW, Swagemakers SM, et al. Disruption of mouse RAD54 reduces ionizing radiation resistance and homologous recombination. *Cell* 1997;89:195–204.
209. Bezzubova O, Silbergleit A, Yamaguchi-Iwai Y, et al. Reduced X-ray resistance and homologous recombination frequencies in a RAD54–/– mutant of the chicken DT40 cell line. *Cell* 1997;89:185–193.
210. Yanagisawa T, Urade M, Yamamoto Y, et al. Increased expression of human DNA repair genes, XRCC1, XRCC3 and RAD51, in radioresistant human KB carcinoma cell line N10. *Oral Oncol* 1998;34:524–528.
211. Bishay K, Ory K, Olivier MF, et al. DNA damage-related RNA expression to assess individual sensitivity to ionizing radiation. *Carcinogenesis* 2001;22:1179–1183.
212. Thompson LH, Schild D. Homologous recombinational repair of DNA ensures mammalian chromosome stability. *Mutat Res* 2001;477:131–153.
213. Chen Y, Lee WH, Chew HK. Emerging roles of BRCA1 in transcriptional regulation and DNA repair. *J Cell Physiol* 1999;181:385–392.
214. Xia F, Powell SN. The molecular basis of radiosensitivity and chemosensitivity in the treatment of breast cancer. *Semin Radiat Oncol* 2002;12:296–304.
215. Raderschall E, Golub EI, Haaf T. Nuclear foci of mammalian recombination proteins are located at single-stranded DNA regions formed after DNA damage. *Proc Natl Acad Sci USA* 1999;96:1921–1926.
216. Raderschall E, Bazarov A, Cao J, et al. Formation of higher-order nuclear Rad51 structures is functionally linked to p21 expression and protection from DNA damage-induced apoptosis. *J Cell Sci* 2002;115:153–164.
217. Paull TT, Rogakou EP, Yamazaki V, et al. A critical role for histone H2AX in recruitment of repair factors to nuclear foci after DNA damage. *Curr Biol* 2000;10:886–895.
218. Golub EI, Gupta RC, Haaf T, et al. Interaction of human rad51 recombination protein with single-stranded DNA binding protein, RPA. *Nucleic Acids Res* 1998;26:5388–5393.
219. MacPhail SH, Olive PL. RPA foci are associated with cell death after irradiation. *Radiat Res* 2001;155:672–679.
220. Maser RS, Monsen KJ, Nelms BE, et al. hMre11 and hRad50 nuclear foci are induced during the normal cellular response to DNA double-strand breaks. *Mol Cell Biol* 1998;17:6087–6096.
221. Nelms BE, Maser RS, MacKay JF, et al. In situ visualization of DNA double-strand break repair in human fibroblasts. *Science* 1998;280:590–592.
222. Carney JP, Maser RS, Olivares H, et al. The hMre11/hRad50 protein complex and Nijmegen breakage syndrome: linkage of double-strand break repair to the cellular DNA damage response. *Cell* 1998;93:477–486.

223. Mirzoeva OK, Petrini JH. DNA damage-dependent nuclear dynamics of the Mre11 complex. *Mol Cell Biol* 2001;21:281–288.
224. Rogakou EP, Boon C, Redon C, et al. Megabase chromatin domains involved in DNA double-strand breaks in vivo. *J Cell Biol* 1999;146:905–916.
225. Sedelnikova OA, Rogakou EP, Panyutin IG, et al. Quantitative detection of (125)IdU-induced DNA double-strand breaks with gamma-H2AX antibody. *Radiat Res* 2002;158:486–492.
226. Murray D, Macann A, Hanson J, et al. ERCC1/ERCC4 5'-endonuclease activity as a determinant of hypoxic cell radiosensitivity. *Int J Radiat Biol* 1996;69:319–327.
227. Thompson LH. Evidence that mammalian cells possess homologous recombinational repair pathways. *Mutat Res* 1996;363:77–88.
228. Andersson BS, Sadeghi T, Siciliano MJ, et al. Nucleotide excision repair genes as determinants of cellular sensitivity to cyclophosphamide analogs. *Cancer Chemother Pharmacol* 1996;38:406–416.
229. Legerski RJ, Richie C. Mechanisms of repair of interstrand crosslinks in DNA. In: Andersson BS, Murray D, eds. *Clinically Relevant Resistance in Cancer Chemotherapy*. Kluwer Academic, Norwell, MA, 2002:109–128.
230. Arlett CF, Harcourt SA, Lehmann AR, et al. Studies on a new case of xeroderma pigmentosum (XP3BR) from complementation group G with cellular sensitivity to ionizing radiation. *Carcinogenesis* 1980;1:745–751.
231. Cooper PK, Leadon SA. Defective repair of ionizing radiation damage in Cockayne's syndrome and xeroderma pigmentosum group G. *Ann NY Acad Sci* 1994;726:330–332.
232. Cooper PK, Nouspikel T, Clarkson SG, et al. Defective transcription-coupled repair of oxidative base damage in Cockayne syndrome patients from XP group G. *Science* 1997;275:990–993.
233. Klungland A, Hoss M, Gunz D, et al. Base excision repair of oxidative DNA damage activated by XPG protein. *Mol Cell* 1999;3:33–42.
234. Bessho T. Nucleotide excision repair 3' endonuclease XPG stimulates the activity of base excision repair enzyme thymine glycol DNA glycosylase. *Nucleic Acids Res* 1999;27:979–983.
235. Murray D, Rosenberg E. The importance of the ERCC1/ERCC4[XPF] complex for hypoxic-cell radioresistance does not appear to derive from its participation in the nucleotide excision repair pathway. *Mutat Res* 1996;364:217–226.
236. Bases R, Franklin WA, Moy T, et al. Enhanced excision repair activity in mammalian cells after ionizing radiation. *Int J Radiat Biol* 1992;62:427–441.
237. Ikushima T, Aritomi H, Morisita J. Radioadaptive response: efficient repair of radiation-induced DNA damage in adapted cells. *Mutat Res* 1996;358:193–198.
238. Hermeking H, Lengauer C, Polyak K, et al. 14-3-3 Sigma is a p53-regulated inhibitor of G2/M progression. *Mol Cell* 1997;1:3–11.
239. Marples B, Joiner MC. Modification of survival by DNA repair modifiers: a probable explanation for the phenomenon of increased radioresistance. *Int J Radiat Biol* 2000;76:305–312.
240. Amundson SA, Do KT, Shahab S, et al. Identification of potential mRNA biomarkers in peripheral blood lymphocytes for human exposure to ionizing radiation. *Radiat Res* 2000;154:342–346.
241. Joiner MC, Marples B, Lambin P, et al. Low-dose hypersensitivity: current status and possible mechanisms. *Int J Radiat Oncol Biol Phys* 2001;49:379–389.
242. Tusher VG, Tibshirani R, Chu G. Significance analysis of microarrays applied to the ionizing radiation response. *Proc Natl Acad Sci USA* 2001;98:5116–5121.
243. Leskov KS, Criswell T, Antonio S, et al. When X-ray-inducible proteins meet DNA double strand break repair. *Semin Radiat Oncol* 2001;11:352–372.
244. Schwartz JL. The role of constitutive and inducible processes in the response of human squamous cell carcinoma cell lines to ionizing radiation. *Radiat Res* 1994;138(1 Suppl):S37–39.

245. Khanna KK, Jackson SP. DNA double-strand breaks: signaling, repair and the cancer connection. *Nature Genet* 2001;27:247–254.

246. van Steeg H, Kraemer KH. Xeroderma pigmentosum and the role of UV-induced DNA damage in skin cancer. *Mol Med Today* 1999;5:86–94.

247. Matullo G, Palli D, Peluso M, et al. XRCC1, XRCC3, XPD gene polymorphisms, smoking and (32)P-DNA adducts in a sample of healthy subjects. *Carcinogenesis* 2001;22:1437–1445.

248. Bork P, Hofmann K, Bucher P, et al. A superfamily of conserved domains in DNA damage-responsive cell cycle checkpoint proteins. *FASEB J* 1997;11:68–76.

249. Masson M, Niedergang C, Schreiber V, et al. XRCC1 is specifically associated with poly(ADP-ribose) polymerase and negatively regulates its activity following DNA damage. *Mol Cell Biol* 1998;18:3563–3571.

250. Zhang X, Morera S, Bates PA, et al. Structure of an XRCC1 BRCT domain: a new protein-protein interaction module. *EMBO J* 1998;17:6404–6411.

251. Marintchev A, Mullen MA, Maciejewski MW, et al. Solution structure of the single-strand break repair protein XRCC1 N-terminal domain. *Nat Struct Biol* 1999;6:884–893.

252. Duell EJ, Wiencke JK, Cheng TJ, et al. Polymorphisms in the DNA repair genes XRCC1 and ERCC2 and biomarkers of DNA damage in human blood mononuclear cells. *Carcinogenesis* 2000;21:965–971.

253. Abdel-Rahman S, El-Zein RA. The 399Gln polymorphism in the DNA repair gene XRCC1 modulates the genotoxic response induced in human lymphocytes by the tobacco-specific nitrosamine NNK. *Cancer Lett* 2000;159:63–71.

254. Lunn R, Langlois R, Hsieh L, et al. XRCC1 polymorphisms: effects on aflatoxin B1-DNA adducts and glycophorin A variant frequency. *Cancer Res* 1999;59:2557–2561.

255. Hu JJ, Smith TR, Miller MS, et al. Amino acid substitution variants of APE1 and XRCC1 genes associated with ionizing radiation sensitivity. *Carcinogenesis* 2001;22:917–922.

256. Shen MR, Zdzienicka MZ, Mohrenweiser H, et al. Mutations in hamster single-strand break repair gene XRCC1 causing defective DNA repair. *Nucleic Acids Res* 1998;26:1032–1037.

257. Tuimala J, Szekely G, Gundy S, et al. Genetic polymorphisms of DNA repair and xenobiotic-metabolizing enzymes: role in mutagen sensitivity. *Carcinogenesis* 2002;23:1003–1008.

258. Sturgis EM, Castillo EJ, Li L, et al. Polymorphisms of DNA repair gene XRCC1 in squamous cell carcinoma of the head and neck. *Carcinogenesis (London)* 1999;20:2125–2129.

259. Divine KK, Gilliland FD, Crowell RE, et al. The XRCC1 399 glutamine allele is a risk factor for adenocarcinoma of the lung. *Mutat Res* 2001;461:273–278.

260. Park JY, Lee SY, Jeon HS, et al. Polymorphism of the DNA repair gene XRCC1 and risk of primary lung cancer. *Cancer Epidemiol Biomarkers Prev* 2002;11:23–27.

261. Duell EJ, Millikan RC, Pittman GS, et al. Polymorphisms in the DNA repair gene XRCC1 and breast cancer. *Cancer Epidemiol Biomarkers Prev* 2001;10:217–222.

262. Duell EJ, Holly EA, Bracci PM, et al. A population-based study of the Arg399Gln polymorphism in X-ray repair cross- complementing group 1 (XRCC1) and risk of pancreatic adenocarcinoma. *Cancer Res* 2002;62:4630–4636.

263. Abdel-Rahman SZ, Soliman AS, Bondy ML, et al. Inheritance of the 194Trp and the 399Gln variant alleles of the DNA repair gene XRCC1 are associated with increased risk of early-onset colorectal carcinoma in Egypt. *Cancer Lett* 2000;159:79–86.

264. Stern MC, Umbach DM, van Gils CH, et al. DNA repair gene XRCC1 polymorphisms, smoking and bladder cancer risk. *Cancer Epidemiol Biomark Prev* 2001;10:125–132.

265. Watson MA, Olshan AF, Weissler MC, et al. XRCC1 polymorphisms (Arg194Trp and Arg399Gln) and risk for head and neck cancer. *Proc Am Assoc Cancer Res* 2000;41:557 (abstract 3551).

266. Nelson HH, Kelsey KT, Mott LA, et al. The XRCC1 Arg399Gln polymorphism, sunburn, and non-melanoma skin cancer: evidence of gene-environment interaction. *Cancer Res* 2002;62:152–155.

267. Olshan AF, Watson MA, Weissler MC, et al. XRCC1 polymorphisms and head and neck cancer. *Cancer Lett* 2002;178:181–186.

268. Lee JM, Lee YC, Yang SY, et al. Genetic polymorphisms of XRCC1 and risk of the esophageal cancer. *Int J Cancer* 2001;95:240–246.

269. Seedhouse C, Bainton R, Lewis M, et al. The genotype distribution of the XRCC1 gene indicates a role for base excision repair in the development of therapy-related acute myeloblastic leukemia. *Blood* 2002;100:3761–3766.

270. Shen H, Xu Y, Qian Y, et al. Polymorphisms of the DNA repair gene XRCC1 and risk of gastric cancer in a Chinese population. *Int J Cancer* 2000;88:601–606.

271. van Gils CH, Bostick RM, Stern MC, et al. Differences in base excision repair capacity may modulate the effect of dietary antioxidant intake on prostate cancer risk: an example of polymorphisms in the XRCC1 gene. *Cancer Epidemiol Biomarkers Prev* 2002;11:1279–1284.

272. Winsey SL, Haldar NA, Marsh HP, et al. A variant within the DNA repair gene XRCC3 is associated with the development of melanoma skin cancer. *Cancer Res* 2000;60:5612–5616.

273. Matullo G, Guarrera S, Carturan S, et al. DNA repair gene polymorphisms, bulky DNA adducts in white blood cells and bladder cancer in a case-control study. *Int J Cancer* 2001;92:562–567.

274. Xing D, Qi J, Miao X, et al. Polymorphisms of DNA repair genes XRCC1 and XPD and their associations with risk of esophageal squamous cell carcinoma in a Chinese population. *Int J Cancer* 2002;100:600–605.

275. Chen S, Tang D, Xue K, et al. DNA repair gene XRCC1 and XPD polymorphisms and risk of lung cancer in a Chinese population. *Carcinogenesis* 2002;23:1321–1325.

276. Ratnasinghe D, Yao SX, Tangrea JA, et al. Polymorphisms of the DNA repair gene XRCC1 and lung cancer risk. *Cancer Epidemiol Biomarkers Prev* 2001;10:119–123.

277. Butkiewicz D, Rusin M, Enewold L, et al. Genetic polymorphisms in DNA repair genes and risk of lung cancer. *Carcinogenesis* 2001;22:593–597.

278. David-Beabes GL, London SJ. Genetic polymorphism of XRCC1 and lung cancer among African-Americans and Caucasians. *Lung Cancer* 2001;34:333–339.

279. Weiss K, Terwilliger H. How many diseases does it take to map a gene with SNPs? *Nature Genet* 2000;26:151–157.

280. Kohno T, Shinmura K, Tosaka M, et al. Genetic polymorphisms and alternative splicing of the hOGG1 gene, that is involved in the repair of 8-hydroxyguanine in damaged DNA. *Oncogene* 1998;16:3219–3225.

281. Dherin C, Radicella JP, Dizdaroglu M, et al. Excision of oxidatively damaged DNA bases by the human alpha-hOgg1 protein and the polymorphic alpha-hOgg1(Ser326Cys) protein which is frequently found in human populations. *Nucleic Acids Res* 1999;27:4001–4007.

282. Janssen K, Schlink K, Gotte W, et al. DNA repair activity of 8-oxoguanine DNA glycosylase 1 (OGG1) in human lymphocytes is not dependent on genetic polymorphism Ser326/Cys326. *Mutat Res* 2001;486:207–216.

283. Sugimura H, Kohno T, Wakai K, et al. hOGG1 Ser326Cys polymorphism and lung cancer susceptibility. *Cancer Epidemiol Biomarkers Prev* 1999;8:669–674.

284. Ito H, Hamajima N, Takezaki T, et al. A limited association of OGG1 Ser326Cys polymorphism for adenocarcinoma of the lung. *J Epidemiol* 2002;12:258–265.

285. Elahi A, Zheng Z, Park J, et al. The human OGG1 DNA repair enzyme and its association with orolaryngeal cancer risk. *Carcinogenesis* 2002;23:1229–1234.

286. Le Marchand L, Donlon T, Lum-Jones A, et al. Association of the hOGG1 Ser326Cys polymorphism with lung cancer risk. *Cancer Epidemiol Biomarkers Prev* 2002;11:409–412.

287. Xing DY, Tan W, Song N, et al. Ser326Cys polymorphism in hOGG1 gene and risk of esophageal cancer in a Chinese population. *Int J Cancer* 2001;95:140–143.

288. Xu J, Zheng SL, Turner A, et al. Associations between hOGG1 sequence variants and prostate cancer susceptibility. *Cancer Res* 2002;62:2253–2257.

289. Hanaoka T, Sugimura H, Nagura K, et al. hOGG1 exon7 polymorphism and gastric cancer in case-control studies of Japanese Brazilians and non-Japanese Brazilians. *Cancer Lett* 2001;170:53–61.
290. Vogel U, Nexo BA, Olsen A, et al. No association between OGG1 Ser326Cys polymorphism and breast cancer risk. *Cancer Epidemiol Biomarkers Prev* 2003;12:170–171.
291. Wikman H, Risch A, Klimek F, et al. hOGG1 polymorphism and loss of heterozygosity (LOH): significance for lung cancer susceptibility in a caucasian population. *Int J Cancer* 2000;88:932–937.
292. Hadi MZ, Coleman MA, Fidelis K, et al. Functional characterization of Ape1 variants identified in the human population. *Nucleic Acids Res* 2000;28:3871–3879.
293. Pieretti M, Khattar NH, Smith SA. Common polymorphisms and somatic mutations in human base excision repair genes in ovarian and endometrial cancers. *Mutat Res* 2001;432:53–59.
294. Bishop D, Ear U, Bhattacharyya A, et al. XRCC3 is required for assembly of Rad51 complexes in vivo. *J Biol Chem* 1998;273:21482–21488.
295. Kuschel B, Auranen A, McBride S, et al. Variants in DNA double-strand break repair genes and breast cancer susceptibility. *Hum Mol Genet* 2002;11:1399–1407.
296. Gonzalez R, Silva JM, Dominguez G, et al. Detection of loss of heterozygosity at RAD51, RAD52, RAD54 and BRCA1 and BRCA2 loci in breast cancer: pathological correlations. *Br J Cancer* 1999;81:503–509.
297. Shen H, Sturgis EM, Dahlstrom KR, et al. A variant of the DNA repair gene XRCC3 and risk of squamous cell carcinoma of the head and neck: a case-control analysis. *Int J Cancer* 2002;99:869–872.
298. Stern MC, Umbach DM, Lunn RM, et al. DNA repair gene XRCC3 codon 241 polymorphism, its interaction with smoking and XRCC1 polymorphisms, and bladder cancer risk. *Cancer Epidemiol Biomarkers Prev* 2002;11:939–943.
299. Araujo FD, Pierce AJ, Stark JM, et al. Variant XRCC3 implicated in cancer is functional in homology-directed repair of double-strand breaks. *Oncogene* 2002;21:4176–4180.
300. Emmert S, Schneider TD, Khan SG, et al. The human XPG gene: gene architecture, alternative splicing and single nucleotide polymorphisms. *Nucleic Acids Res* 2001;29:1443–1452.
301. Kumar R, Hoglund L, Zhao C, et al. Single nucleotide polymorphisms in the XPG gene: determination of role in DNA repair and breast cancer risk. *Int J Cancer* 2003;103:671–675.
302. Hyytinen ER, Frierson HF Jr, Sipe TW, et al. Loss of heterozygosity and lack of mutations of the XPG/ERCC5 DNA repair gene at 13q33 in prostate cancer. *Prostate* 1999;41:190–195.
303. Sturgis EM, Dahlstrom KR, Spitz MR, et al. DNA repair gene ERCC1 and ERCC2/XPD polymorphisms and risk of squamous cell carcinoma of the head and neck. *Arch Otolaryngol Head Neck Surg* 2002;128:1084–1088.
304. Chen P, Wiencke J, Aldape K, et al. Association of an ERCC1 polymorphism with adult-onset glioma. *Cancer Epidemiol Biomarkers Prev* 2000;9:843–847.
305. Price EA, Bourne SL, Radbourne R, et al. Rare microsatellite polymorphisms in the DNA repair genes XRCC1, XRCC3 and XRCC5 associated with cancer in patients of varying radiosensitivity. *Somat Cell Mol Genet* 1997;23:237–247.
306. Rogers PB, Plowman PN, Harris SJ, et al. Four radiation hypersensitivity cases and their implications for clinical radiotherapy. *Radiother Oncol* 2000;57:143–154.
307. Ron E. Ionizing radiation and cancer risk: evidence from epidemiology. *Radiat Res* 1998;150(5 Suppl):S30–41.
308. Hu JJ, Smith TR, Miller MS, et al. Genetic regulation of ionizing radiation sensitivity and breast cancer risk. *Environ Mol Mutagen* 2002;39:208–215.
309. Lavin MF, Bennett I, Ramsay J, et al. Identification of a potentially radiosensitive subgroup among patients with breast cancer. *J Natl Cancer Inst* 1994;86:1627–1634.

310. Rosen EM, Fan S, Rockwell S, et al. The molecular and cellular basis of radiosensitivity: implications for understanding how normal tissues and tumors respond to therapeutic radiation. *Cancer Invest* 1999;17:56–72.

311. Rosen EM, Fan S, Goldberg ID, et al. Biological basis of radiation sensitivity. Part 2: Cellular and molecular determinants of radiosensitivity. *Oncology (Huntingt)* 2000;14:741–757.

312. Yacoub A, Park JS, Qiao L, et al. MAPK dependence of DNA damage repair: ionizing radiation and the induction of expression of the DNA repair genes XRCC1 and ERCC1 in DU145 human prostate carcinoma cells in a MEK1/2 dependent fashion. *Int J Radiat Biol* 2001;77:1067–1078.

313. Ader I, Muller C, Bonnet J, et al. The radioprotective effect of the 24 kDa FGF-2 isoform in HeLa cells is related to an increased expression and activity of the DNA dependent protein kinase (DNA-PK) catalytic subunit. *Oncogene* 2002;21:6471–6479.

314. Dai Y, Kysela B, Hanakahi LA, et al. Nonhomologous end joining and V(D)J recombination require an additional factor. *Proc Natl Acad Sci USA* 2003;100:2462–2467.

315. Duggan DJ, Bittner M, Chen Y, et al. Expression profiling using cDNA microarrays. *Nature Genet* 1999;21:10–14.

316. Hegde P, Qi R, Abernathy K, et al. A concise guide to cDNA microarray analysis. *Biotechniques* 2000;29:548–562.

317. MacBeath G. Protein microarrays and proteomics. *Nature Genet* 2002;(32 Suppl):526–532.

318. Ng JH, Ilag LL. Biomedical applications of protein chips. *J Cell Mol Med* 2002;6:329–340.

319. Jain KK. Recent advances in oncoproteomics. *Curr Opin Mol Ther* 2002;4:203–209.

320. Achary MP, Jaggernauth W, Gross E, et al. Cell lines from the same cervical carcinoma but with different radiosensitivities exhibit different cDNA microarray patterns of gene expression. *Cytogenet Cell Genet* 2000;91:39–44.

321. Hanna E, Shrieve DC, Ratanatharathorn V, et al. A novel alternative approach for prediction of radiation response of squamous cell carcinoma of head and neck. *Cancer Res* 2001;61:2376–2380.

322. Bishay K, Ory K, Lebeau J, et al. DNA damage-related gene expression as biomarkers to assess cellular response after γ irradiation of a human lymphoblastoid cell line. *Oncogene* 2000;19:916–923.

323. Mohrenweiser HW, Jones IM. Uncertainty of response to ionizing radiation due to genotype: potential role for variation in DNA repair genes. *Radiat Res* 2000;154:722–723.

11 Strand-Break Repair and Radiation Resistance

George Shenouda, MBB Ch, PhD, FRCP

CONTENTS

1. INTRODUCTION

Radiotherapy is at the dawn of a new era, in which molecular radiation biology is coming into use for clinical decision-making. During the last decade of the 20th century, new research and technical developments were achieved. A new three-dimensional (3D) treatment optimization technology became available. In addition new biological principles are better understood for application in the area of clinical radiation oncology.

It is conceivable that the two areas of recent advances are important in order to improve the radiation therapy outcomes such as local tumor control and quality of life. Better technology alone, allowing a better 3D dose delivery, will have a minimal impact on outcome, if basic fractionation and radiobiological information are lacking.

The most common error leading to local tumor failure is geographic miss. It is defined as the inability to appropriately identify the target volume and/or to achieve an acceptable cancericidal dose-homogeneity within the target volume.

From: *Cancer Drug Discovery and Development: DNA Repair in Cancer Therapy*
Edited by: L. C. Panasci and M. A. Alaoui-Jamali © Humana Press Inc., Totowa, NJ

These factors are not discussed in this chapter. The main discussion will focus on radiation resistance as defined by the inability to achieve local control resulting from intrinsic radiobiological factors, assuming no geographic miss or setup errors in treatment delivery.

2. MOLECULAR BIOLOGY OF RADIATION SENSITIVITY

A large number of genetic factors influence the radiation sensitivity of neoplastic and normal tissues. For the neoplastic cells, the state of proto-oncogenes expression is the first and the most important factor. Proto-oncogenes generally encode for proteins responsible for the signal transduction cascade of growth factors, which, under normal conditions, will stimulate cell division. Functionally, there are four main groups of proto-oncogenes, categorized according to their influence on normal cell proliferation. The following are examples of such proto-oncogenes: (1) autocrine growth factors (hst, sis), (2) growth factor receptors (erb, fms), (3) signal transduction factors (ras, src), and (4) nuclear transcription factors (myc, fos, jun).

Several mechanisms influence the genome to activate such proto-oncogenes. These mechanisms include structural alteration, amplification or loss of control mechanism by insertional mutagenesis, transduction, and translocation. Most of these mechanisms are triggered by external factors, including chemical agents, radiation, and viruses.

A second important group of genes that influence cellular radiosensitivity is the tumor suppressor genes. The classical example is the retinoblastoma (RB) gene. The Rb gene has been shown to be affected in retinoblastoma, breast, lung carcinoma, and osteosarcoma. It has an important function in the regulation of cell cycle progression *(1)*.

One of the most commonly mutated or deleted genes in human cancer, p53, also belongs to the tumor suppressor group. The p53 gene product is a DNA-binding transcription factor, which can induce apoptosis and cell cycle arrest in the G_1-phase of the cell cycle. It influences the decision of the cell either to rest and repair the induced DNA damage or to eliminate itself by programmed cell death because of a too severe level of damage *(2)*. Because of its central role in handling DNA damage, p53 is mutated in many cancers, including glioblastoma multiforme, colorectal, breast, brain, and lung carcinomas. If p53 is mutated in a cell, the risk that this cell line may develop neoplasia is immediately increased because such cells are likely to inappropriately repair DNA damage to its genome. On the other hand, radiation therapy may be the ideal way to treat a tumor with a p53 mutation, because the tumor cells will handle poorly the DNA damage inflicted by the ionizing radiation. The situation, however, is not always as simple because p53 mutations may also decrease the ability of the cell to undergo apoptosis. This may lead to increased cell survival, even though the cells are damaged and not repaired with a high degree of fidelity.

In addition to the oncogenes and tumor suppressor genes that are largely responsible for tumor induction, a large number of other genes may also promote tumor development and alter radiation sensitivity of mammalian cells. At least four gene families can be identified: (1) cell cycle control genes (cycline A-E, CDC2, GADD45), (2) DNA repair genes (XPA-G, DNAPK, RAD1), (3) DNA processing and topology genes (TOPO1), and (4) detoxification and stress response genes (GSH, HSP). It is clear that if some of these genes have an impaired function, the processing of normal and damaged DNA will be affected. This may, in turn, promote tumor development and alter radiation sensitivity. It is likely that the status of many of the previous mentioned genes, with regard to polymorphism, amplification and transcription factors, and mutations, may be combined as a useful molecular predictor of radiation sensitivity both for tumors and for normal tissues as well.

3. REVIEW OF REPAIR KINETICS FOR IONIZING RADIATION-INDUCED DNA DAMAGE

Ionizing radiation interacts with matter by depositing energy in the target structure within about 10^{-19}–10^{-14} s. The energy deposition event is followed by radiochemical processes leading to altered target molecules (DNA), which are the substrates for subsequent enzymatic repair reactions, taking place in the time range of seconds to days. These physical, chemical, and biological processes together determine, finally, the effect of radiation on the irradiated unicellular or multicellular organisms. A variety of lesions can be detected in the DNA of cells exposed to ionizing radiation. Such DNA damage includes DNA–protein crosslinks (DPCs), base alterations and base detachments, sugar alterations, bulky lesions, and DNA single- and double-strand breaks.

3.1. Induction of DNA–Protein Crosslinks in Mammalian Cells by Ionizing Radiation

DNA–protein crosslinks are formed by covalent linkage between DNA and proteins of the nuclear matrix. Primarily, those DNA regions containing actively transcribing and presumably also replicating sequences are involved in the linkage to protein. Irradiation of cells such as the Chinese hamster V79 lung fibroblasts will induce about 160 DPCs/cell/Gy *(3)*. DPCs are induced linearly with radiation dose and the yield of DPCs in the irradiated Chinese hamster V79 cells is reduced in the presence of reducing agents such as cysteamine *(4)*. Mammalian cells are capable of repairing radiation-induced DPCs. The DPCs are repaired with biphasic kinetics during the postirradiation period. Human fibroblasts irradiated with 50 Gy exhibit a rapid component of repair of DPCs with a half-life of about 2 h and another slow component with a half-life of 12–13 h *(5)*. In actively transcribing DNA sequences, DPCs are not only preferentially induced, but also the removal of DPCs in these regions is faster as compared to nontranscribing DNA regions *(3)*. The mechanism of repair of radiation-induced DPCs seems to require an intact nuclear membrane, as indicated by the absence of DPC repair in metaphase cells, lacking an intact nuclear membrane *(3)*. In general, DPC seem to play a minor role in radiation-induced cell killing. The radiation-induced DPCs may be relevant for cell aging because there is a parallel in radiation and age-induced changes of the properties of chromatin in mammalian cells *(6)*.

3.2. Induction of Base Damage in Mammalian Cells by Ionizing Radiation

Damage to the four DNA bases is a major type of lesions caused by ionizing radiation in mammalian cells *(7)*. It is estimated that after a given dose of ionizing radiation, the total number of damaged bases in human lung fibroblasts may be twice as large as the number of DNA single-strand breaks *(8)*. Most of the ionizing-radiation-induced base damage was studied in consideration to thymine. There is a linear induction of thymine damage in human lung fibroblasts, EMT6 mouse sarcoma cells, and HeLa cells irradiated with γ-rays up to 30 Gy. The frequency of induction of thymine damage depends on the chromatin structure. Thymine damage is found to be about three times more frequent in replicating chromatin, and a similar high yield is also reported for actively transcribing chromatin in HeLa cells *(9)*. The repair of radiation-induced thymine damage is usually completed within 1 h after exposure to radiation. The thymine damage is removed by endonuclease activity *(10)*. The chromatin structure of cells is important, not only for the frequency of induction of thymine damage but also for the rate of repair. In active chromatin of HeLa cells, the rate of removal of thymine damage is higher than in inactive chromatin *(9)*. Thymine damage does not play a significant role in radiation-induced cell killing. However, the thymine damage induced by radiation plays a central role in the induction of point mutation *(11)*.

3.3. Induction of DNA Single-Strand Breaks by Ionizing Radiation in Mammalian Cells

Induction of single-strand breaks (SSBs) has been studied for a large variety of mammalian cells and is found to increase in proportion to the radiation dose *(12–15)*. The same induction frequency for SSBs is observed in radiation-resistant Chinese hamster and in radiation-sensitive mouse cells *(4,16)*. Likewise, no difference in SSB induction is detected in normal human fibroblasts and in radiation-sensitive cells from retinoblastoma patients *(17)*. Hypoxia decreases the yield of radiation-induced SSBs relative to oxic conditions *(18,19)*. The decrease is by a factor of 3 for Chinese hamster V79 fibroblasts *(20)*. Exogenously supplied thiol compounds reduce the yield of SSB in human fibroblasts and in mouse and Chinese hamster cells irradiated under a hypoxic condition *(4,21)*. SSBs are preferably induced by radiation in transcriptionally active DNA sequences *(22)*. This is explained by the better accessibility of OH radicals to DNA during transcription. SSBs are rejoined exponentially with time (first-order reaction) during postirradiation incubation of cells at 37°C. Up to three components of SSB repair kinetics can be distinguished. The first component comprises 70–90% of all SSBs, which are rejoined independently of dose with a half-life of 2–10 min. The second, slower component represents a dose-dependent rejoining of SSBs with half-life of about 10 min. The third component is attributed to the rejoining of DNA double-strand breaks with a half-life $\geq$ 1 h. SSBs are rejoined with the same rate in both radiation-resistant and radiation-sensitive human cell lines *(17,23,24)*. Radiation-induced SSBs are rejoined even under extreme hypoxic conditions in Chinese hamster V79 cells, but the amount of SSBs left unrejoined after an incubation period of 1 h is 20% in cells under extreme hypoxia and only 10% under oxic conditions *(20)*. Rejoining of SSBs within 1 h after irradiation of human fibroblasts under oxic condition is only 70% in glutathione-depleted cells compared to glutathione-proficient cells, where rejoining of SSBs radiation-induced under hypoxia is about 100% *(25,26)*. Although it is not easy to draw a general conclusion, most studies published to date show that the initial radiation-induced DNA damage is the same in radioresistant as in radiosensitive cell lines. Considering the large number of events arising after the DNA-damage induction in a cell, it is not surprising that damage induction is not always different between pairs of radiosensitive and radioresistant cells. Differences in radiosensitivity may arise from differences in the molecular events taking place during the biochemical cascade triggered by exposure to ionizing radiation. In addition, the spatial molecular configuration of the DNA may explain differences in radiosensitivity by allowing different degrees of access to damaged DNA by repair enzymes.

3.4. Induction of DNA Double-Strand Breaks by Ionizing Radiation

After irradiation with supralethal doses of ionizing radiation, a linear relationship between double-strand breaks (DSBs) and dose has been observed for various types of mammalian cells *(27–29)*. The yield of radiation-induced DSBs is lower by a factor of 25 than the yield of SSBs based on 1000 SSBs/mammalian cell/Gy vs 40 DSBs/cell/Gy *(27,30)*. In oxic cells, the yield of radiation-induced DSBs is about threefold higher than in anoxic cells *(18,31)*. No difference is observed between radiation-sensitive and radiation-resistant cell lines for the induction of DSBs *(32,33)*.

The DSBs are rejoined during postradiation incubation of various types of cells. The rejoining of radiation-induced DSBs is exponential with time (first-order reaction) and is independent of dose. A great variation in the half-time constants is observed for the DSB rejoining postirradiation. The half-life for rejoining of DSBs ranges from 2–6 h.

The kinetics of DSB rejoining is independent of the cell cycle phase during which the cells were irradiated. The same rate of DSB rejoining is observed for radiation-resistant and radiation-sensitive cell lines *(29,33,34)*. The mechanism of rejoining of radiation-induced DSBs seem to involve a recombinational process *(35)* and simple ligation *(36)*.

There is strong evidence that radiation-induced DSBs lead to cell killing. Radiation-sensitive human fibroblasts, from ataxia–telangiectasia patients, show a higher level of unrejoined chromosome breaks compared to normal cells after exposure to radiation. Yeast mutants, completely deficient in DSB rejoining, are killed by about one DSB per cell. They do not repair potentially lethal damage and sublethal damage. There is no evidence for the involvement of DSB in the formation of point mutations.

4. MOLECULAR DETERMINANTS OF RADIATION SENSITIVITY IN MAMMALIAN CELLS

4.1. Damage Induction

There are now a number of reports that suggest that cells can differ in the amount of DNA damage detected immediately after irradiation in the absence of DNA repair. The level of initially radiation-induced DNA damage is directly related to the radiation sensitivity in different cell lines *(37)*. The measurement of the induced DNA damage using neutral filter elution techniques could distinguish between radiation-resistant and radiation-sensitive cell lines *(38,39)*. Using this technique, a difference between reduced sensitivity of a mutant cell line EM9 was detected as compared to its parental cell line *(40)*. Another technique of pulsed-field gel electrophoresis was used to measure the relationship between ionizing-radiation-induced DNA damage and radiation sensitivity. In addition, it has been found that some chemotherapy resistant cell lines, that show a cross-resistance to ionizing radiation have decreased level of DNA-induced damage *(41)*.

Although the physical damage of ionizing radiation is identical in all cells, the subsequent chemical reactions will either increase or decrease the eventual biological effects of irradiation. Oxygen is a very strong modifier of the amount of radiation-induced damage. The level of thiol compounds has also modifying effect on radiation sensitivity. The effect of thiols can be large when the cells are irradiated under hypoxic condition.

4.2. Oncogene-Mediated Radiation Resistance

It has been observed that overexpression of certain oncogenes, in addition to increasing malignant potential, causes an increase in radiation resistance in naturally occurring tumors *(42,43)*. Several studies have demonstrated a link between the expression of certain oncogenes and cellular radiation resistance in vitro. Oncogene-mediated resistance has been demonstrated following transfection with the ras oncogene into murine hematopoietic cells or NIH 3T3 cells *(44,45)*. Furthermore, a synergistic increase in the level of radiation resistance and changes in the transformed phenotype associated with increased tumorigenicity have been observed with cotransfection of ras and myc oncogenes in primary rat embryo cells *(46,47)*.

Although oncogene-mediated radiation resistance is well documented, little is known about the mechanisms underlying the phenomenon. Oncogene-mediated radiation resistance resulting from H-ras and v-myc transfection did not correlate with induction or repair of DNA DSBs measured by pulsed-field gel electrophoresis *(48)*.

Recent evidence is accumulating and showing that nuclear matrix-mediated organization of DNA is altered with increasing tumorigenicity. The nuclear matrix, which originally was operationally defined as the nuclear structure remaining after detergent, high salt, and DNaseI extraction, is the structure that provides higher-order DNA organization within the nucleus. The organization of DNA by the nuclear matrix is believed to be pivotal to the perpetuation of specific transcriptional patterns in one cell's progeny, thus maintaining cell and tissue lineage identity. One aspect of spatial organization of DNA is the arrangement of nuclear DNA organized into loop domains, which are anchored to the nuclear matrix. DNA in loop domains forms negatively supercoiled structures, which lead to periodic associations of the chromatin structure with the underlying matrix through nucleoprotein interactions called matrix-associated regions. The matrix-associated regions are believed to allow the maintenance of contiguous DNA regions with different superhelical densities. Differences in superhelical densities have been implicated in the spatial control of DNA replication, control of gene transcription, and DNA-damage repair *(49,50)*. Changes in the DNA matrix association could alter the completeness of the isolation between contiguous loop domains and affect loop stability in the presence of DNA damage leading to inefficient repair.

4.3. Potential Role of Poly(ADP-Ribose) Polymerase in Radiation-Induced Apoptosis

Oxidative stress to cells can trigger DNA strand damage, which then will activate the nuclear enzyme poly(ADP-ribose) polymerase (PARP). Rapid activation of the enzyme depletes the intracellular concentration of its substrate, NAD^+, thus slowing the rate of glycolysis, electron transport, and subsequent ATP formation.

The surveillance and repair of the genome are vital to the survival of a cell when exposed to potentially genotoxic insult such as ionizing radiation. Following

DNA-induced damage, the nuclear enzyme PARP binds tightly to single-strand nicks and double-strand breaks. Using NAD^+ as its substrate, it catalyzes the extensive polymerization of ADP-ribose using glutamine residues on itself or on other nuclear proteins *(51,52)*. PARP activation mediates cell death in ischemia–reperfusion injury after cerebral ischemia *(53,54)*, inflammatory injury, and reactive oxygen species-induced injury *(55)*. The molecular mechanisms of PARP-induced cell death are unknown but is proposed to be related to the cellular depletion of NAD^+ and ATP.

Early studies with primary fibroblasts derived from ataxia–telangiectasia (AT) patients uncovered a severe sensitivity to ionizing radiation *(56,57)*. It was subsequently discovered that AT cells contain a high number of unrepaired DNA breaks after being exposed to ionizing radiation *(58,59)*. It was observed that the persistent presence of unrepaired DNA damage in AT cells resulted in the chronic steady-state activation of PARP. The intracellular pools of NAD^+ are therefore severely depleted. Marecki and McCord *(60)* showed that when they treated AT cells in culture with PARP inhibitors such as 3-aminobenzamide or 1,5 dihydroisoquinoline, the cellular growth rate of AT cells reached those observed in normal fibroblast cultures. The improvement of cellular growth and NAD^+ levels in AT cells with PARP inhibition suggest that the cellular metabolic status of AT cells is compromised and that the inhibition of PARP may relieve some of the drain on cellular pyridine nucleotides and ATP. This suggests that inhibitors of PARP may play a significant role in future research to modulate the response of cell responses to ionizing radiation.

The role of apoptosis-inducing-factor (AIF) in PARP-mediated cell death was studied by Woon et al *(61)*. AIF is a mitochondrial flavoprotein that is released in response to death stimuli in mammalian cells. AIF is translocated into the nucleus after genotoxic agent treatment of cells. PARP activation, induced by DNA damage, leads to a decrease in cellular content of NAD^+. This is perceived by the mitochondria, and AIF is translocated from the cytosol to the nucleus, where it initiates nuclear condensation. Once the nucleus condenses, the cell is doomed to die. Cytosolic AIF acts on the mitochondria and initiates the release of caspase activators, which will facilitate the dissolution of the cell.

Future research is awaited to investigate the role of PARP inhibitors in the modulation of mammalian cells to ionizing radiation. It should be important to evaluate the effect of modulation of PARP activity in neoplastic as well as in the normal tissue counterpart where radio-sensitizing, or radio-protecting properties can be invested to improve the therapeutic ratio in radiotherapy.

4.4. Bcl-2 Control of Apoptosis Propensity in Mammalian Cells

Bcl-2 is the first of an expanding family of proteins including Bax, Bcl-x, Mcl-1 and ced-9. Abnormal expression of bcl-2 has been observed in many neoplastic processes, including lymphoma, prostate, breast, and tumors of neural origin

(62–65). In cells that have abundant expression, the protein appears to be localized in the cell membranes—mainly the mitochondrial and nuclear membranes—and endoplasmic reticulum. There is evidence to suggest that bcl-2 regulates the transport of calcium in the cell. Because calcium is known to be important in apoptosis, it seems reasonable to speculate that bcl-2 blocks apoptosis by limiting the availability of calcium required for the apoptotic process.

An alternative pathway proposes that bcl-2 protein is involved in the antioxidant pathway of the cell *(66)*. Cells transfected with bcl-2 expression vector had twice the intracellular level of the antioxidant glutathione (GSH) as the nontransfected counterpart. This is an important pathway and the importance of GSH has been demonstrated in relation to apoptosis. When GSH levels in cells are elevated, apoptosis can be blocked *(67)* and when the levels are lowered by agents, such as buthionine sulfoximine (BSO), apoptosis is enhanced *(68)*.

Story et al. in 1994 *(69)* established a cell culture system for exploring the biochemical and molecular basis for radiation resistance resulting from the block of the apoptosis pathway. This model system consists of two cell lines derived from a murine lymphoma; one line is sensitive to radiation and has apoptosis propensity, and the other cell line is radioresistant. It was demonstrated that the absence of an apoptotic response in the resistant line is caused by deregulated expression of bcl-2. Cells expressing a high level of bcl-2 have also a high level of GSH content compared to the non-bcl-2-expressing counterpart. The radioresistant cell lines could be sensitized to the effect of radiation by inhibiting their ability to sensitize GSH after triggering apoptosis with radiation. These results are consistent with the role of GSH in bcl-2-expressing cells in blocking the signals that mediate apoptosis and not by modifying the initial damage to the cell induced by ionizing radiation.

5. HOW DO WE COMBAT HYPOXIA?

The realization that hypoxia is a common characteristic of human tumors that adversely affects patient prognosis suggests that targeting hypoxia will be an effective means of improving treatment. Scientists and clinicians are using two fundamentally different approaches to tackle the problem of hypoxia. The first approach is to improve or restore normal tumor oxygenation, and the second is to exploit the unique property of tumor hypoxia for targeting treatment to the tumor. The success of these two approaches will ultimately depend on the relative importance of hypoxia in radioresistance.

5.1. Improve Oxygenation

Attempts to increase the oxygen supply to hypoxic potentially viable tumor cells has been a major goal of experimental and clinical research for the last 40 yr. Various strategies have been considered, including hyperbaric or increased oxygen breathing, the administration of hypoxic cell sensitisers, and, more recently, erythropoietin to improve the hemoglobin level and to avoid repeated transfusions. Although most of the early attempts to overcome hypoxia have yielded mixed results, in head and neck malignancy the large meta-analysis of these trials has shown that oxygen modification results in a significant improvement in local control and disease-specific survival *(70,71)*.

5.1.1. ERYTHROPOIETIN

Erythropoietin (EPO) is a glycoprotein hormone produced by the kidney in response to tissue hypoxia that stimulates red blood cell production in the bone marrow. Currently, there is active interest in using recombinant human EPO in patients with low hemoglobin levels in order to improve tumor oxygenation. The hypothesis is that some hypoxic tumors may result from low hemoglobin levels in anemic patients. Hemoglobin concentration has been shown to be an important prognostic factor for the outcome of various cancer types treated with radiotherapy. Most of the clinical studies published have shown better tumor control in patients with higher hemoglobin levels than in patients with hemoglobin levels in the lower part of/or below the normal range. This has been demonstrated in the uterine cervix *(72,73)*, in head and neck cancer *(74,75)*, in bronchogenic carcinoma *(76,77)*, in bladder carcinoma *(78,79)*, and in prostate cancer *(80)*.

At the present time, several phase III trials are being conducted to test the hypothesis that an increase of hemoglobin with EPO during radiation treatment will improve the outcome mainly in carcinoma of the uterine cervix and in head and neck malignancy.

5.1.2. ARCON ACCELERATED RADIOTHERAPY COMBINED WITH CARBOGEN AND NICOTINAMIDE

The ARCON protocol is currently being evaluated in the clinic. Carbogen (95% O_2 + 5% CO_2) is used to reduce diffusion-limited or chronic hypoxia, and nicotinamide is used to reduce acute hypoxia resulting from temporary vascular shutdown. The use of these agents simultaneously has indeed been shown to increase the radiation-damaging effect in a variety of rodent tumor models *(81,82)*.

Promising results have been obtained in several nonrandomized clinical studies using this combination in conjunction with accelerated radiotherapy. To limit clonogenic repopulation during therapy, the overall duration of the radiotherapy course is reduced, generally by delivering multiple fractions per day. This accelerated radiotherapy is combined with inhalation of carbogen. Three clinical studies have been done to test the enhancing effect of these three components of ARCON, individually and in combination, in several experimentally induced tumor and normal tissues. Phase I and II clinical trials have shown the feasibility and tolerability of ARCON and have produced promising results in terms of tumor control. In particular, in cancers of the head and neck and bladder, the local tumor control rates are higher than in other studies, and phase III trials for these tumor types are underway *(83,84)*.

5.1.3. RADIOSENSITIZERS

This approach is based on the concept that these compounds could mimic the effects of oxygen at the time of radiation delivery, thereby increasing DNA damage and restoring radiosensitivity. However, most of the compounds developed could not be administered to patients at effective concentrations with acceptable toxicity. Nimorazole has been widely used in a large double-blind randomized phase III trial in Denmark. It was reported to significantly improve the effect of radiotherapy of supraglottic and pharyngeal tumors, and the toxicity of the drug was mild *(85)*. This study was a multicenter, randomized, and balanced double-blind trial with the objective of assessing the efficacy and tolerance of nimorazole given as a hypoxic radiosensitizer in conjunction with primary radiotherapy of invasive carcinoma of the supraglottic larynx and pharynx. Four hundred twenty-two patients (414 eligible) with pharynx and supraglottic larynx carcinoma were double-blind randomized to receive the hypoxic cell radiosensitizer nimorazole or placebo in association with conventional primary radiotherapy (62–68 Gy, 2 Gy per fraction, five fractions per week). The median observation time was 112 mo. Cox multivariate regression analysis showed the most important prognostic parameters for loco-regional control to be positive neck nodes (relative risk 1.84 [1.38–2.45], T3–T4 tumor (relative risk 1.65 [1.25–2.17]) and nimorazole (relative risk 0.69 [0.52–0.90]). The same parameters were also significantly related to the probability of dying from cancer. The compliance to radiotherapy was good and 98% of the patients received the planned dose. Late radiation-related morbidity was observed in 10% of the patients, irrespective of nimorazole treatment. Drug-related side effects were minor and tolerable, with transient nausea and vomiting being the most frequent complications.

5.2. Bioreductive Drugs

Bioreductive drugs are compounds that are reduced by biological enzymes to their toxic, active metabolites. They are designed such that this metabolism occurs only or preferentially in the absence of oxygen. The use of these drugs in combination with traditional therapies has the potential to greatly improve treatment outcome by increasing cytotoxicity to the hypoxic fraction. Tirapazamine (TPZ) is the leading compound in this class of agents and has shown promising results in a number of clinical trials when used in combination with cisplatin and/or radiotherapy *(86–88)*. The mechanism of this preferential toxicity is mediated by an enzymatically catalyzed one-electron reduction of TPZ. This yields a highly reactive radical capable of causing cell death by producing various types of DNA damage. In the presence of oxygen, the TPZ radical is rapidly oxidized back to the nontoxic parental compound, thus minimizing toxicity to well-oxygenated tissues.

Preclinical in vitro testing has shown that TPZ has a synergistic effect on cell killing when given prior to cisplatin. The mechanism of this synergism has yet to be elucidated, but has been postulated to involve the inhibition of cisplatin-induced DNA crosslink repair *(89,90)*.

Current clinical phase III studies are underway to investigate the role of concomitant radiation, cisplatin, and tirapazamine versus concomitant radiation and cisplatin in patients with advanced head and neck cancer. The result of these studies are awaited and will help to better define the role of bioreductive drugs in the combined modality treatment of human malignancies where hypoxic cell are present and play an important role in radiation resistance.

6. CONCLUSION

This review summarizes our current knowledge of the mechanisms involved in radiation-induced DNA damage in mammalian cells. It is a rapidly advancing area of radiation research but is still incomplete. Future research is needed to complete the gaps in our current knowledge of the mechanism controlling radiation response. The development of future strategies to manipulate radiation sensitivity in addition to other therapeutic modalities, such as combined chemo-radiation and/or radiosensitizers, will depend on a better understanding of the complex and multiple mechanisms controlling cell response to ionizing radiation.

ACKNOWLEDGMENT

The author would like to thank Mrs. Giulia Dagostino for her expert help in the preparation of this manuscript.

REFERENCES

1. Morgenbesser SD, Williams BO, Jacks T, et al. P53-depdendent apoptosis produced by Rb-deficiency in the developing mouse lens. *Nature* 1994;871:72–74.
2. Lane D. P53, guardian of the genome. *Nature* 1992;858:15–16.
3. Oleinick NL, Chiu S, Friedman LR, et al. DNA–protein crosslinks: new insights into their formation and repair in irradiated mammalian cells. In: Simic MG, Grossman L, Upton AC, eds. *Mechanism of DNA Damage and Repair,* Plenum, New York, 1986:181–192.
4. Radford IR. Effect of radiomodifying agents on the ratios of X-ray-induced lesions in cellular DNA: use in lethal lesions determination. *Int J Radiat Biol* 1986;49:621–637.
5. Fornace AJ, Little JB. DNA crosslinking induced by X-rays and chemical agents. *Biochim Biophys Acta* 1977;477:343–355.
6. Cutler RG. Crosslinkage hypothesis of aging: DNA adducts in chromatin as a primary aging process. In: Smith KD, ed. *Aging, Carcinogenesis, and Radiation Biology.* Plenum, New York, 1976;443–492.
7. Cerutti PA. Base damage induced by ionizing radiation. In: Wang SY, ed. *Photochemistry and Photobiology of Nucleic Acids.* Wang. Academic, New York, 1976;vol II:375–401.
8. Cerutti PA. Effects of ionizing radiation on mammalian cells. *Naturwissenschaften* 1974;61:51–59.

9. Patil MS, Locher SE, Hariharan PV. Radiation induced thymine base damage and its excision repair in active and inactive chromatin of HeLa cells. *Int J Radiat Biol* 1985;48:691–700.

10. Brent TP. Purification and characterization of human endonucleases specific for damaged DNA. Analysis of lesions induced by UV or X-radiation. *Biochim Biophys Acta* 1976;454: 172–183.

11. Grosovsky AJ, De Boer JG, Drobetsky EA, et al. DNA sequence analysis of ionizing radiation induced mutation in mammalian cells. Proc. *8(th) Int Congr Radiat Res* 1987; abstract 217.

12. Ahnström, G Edvarsson KA. Radiation-induced single-strand breaks in DNA determined by rate of alkaline strand separation and hydroxylapatite chromatography: an alternative to velocity sedimentation. *Int J Radiat Biol* 1974;26:493–497.

13. Coquerelle T, Bopp A, Kessler B, et al. Strand breaks and 5'-end groups in DNA of irradiated thymocytes. *Int J Radiat Biol* 1973;24:397–404.

14. Kohn KW, Erickson, LC, Ewing RAG, et al. Fractionation of DNA from mammalian cells by alkaline elution. *Biochemistry* 1976;15:4629–4637.

15. Roots R, Yang, TC, Craise I, et al. Impaired repair capacity of DNA breaks induced in mammalian cellular DNA by accelerated heavy ions. *Radiat Res* 1979;78:38–49.

16. Sakai K, Okada S. Radiation-induced DNA damage and cellular lethality in cultured mammalian cells. *Radiat Res* 1984;98:479–490.

17. Woods WG, Lopez, M, Kalvonjian M. Normal repair of gamma-radiation induced single- and double-strand DNA breaks in retinoblastoma fibroblasts. *Biochim Biophys Acta* 1982;698:40–48.

18. Lennartz M, Coquerelle T, Bopp A, et al. Oxygen effect on strand breaks and specific endgroups in DNA of irradiated thymocytes. *Int J Radiat Biol* 1975;27: 577–587.

19. Palcic B, Skarsgard LD. The effects of oxygen on DNA single-strand breaks produced by ionizing radiation in mammalian cells. *Int J Radiat Biol* 1972;21:417–433.

20. Koch CJ, Painter RB. The effects of extreme hypoxia on the repair DNA single-strand breaks in mammalian cells. *Radiat Res* 1975;64:256–269.

21. Van der Schans GP, Centen HB, Lohman PHM. The induction of gamma endonuclease-susceptible sites by gamma rays in CHO cells and their cellular repair are not affect by the presence of thiol compounds during irradiation. *Mutat Res* 1979;59:119–122.

22. Chiu S, Oleinick NL, Friedman LR, et al. Hypersensitivity of DNA in transcriptionally active chromatin to ionizing radiation. *Biochim Biophys Acta* 1982;699:15–21.

23. Hariharan PV, Eleczko S, Smith BP, et al. Normal rejoining of DNA strand breaks in ataxia telangiectasia fibroblast lines after low X-ray exposure. *Radiat Res* 1981;86:589–597.

24. Paterson MC, Smith BP, Lohman PHM, et al. Defective excision repair of gamma-ray damaged DNA in human (ataxia telangiectasia) fibroblasts. *Nature* 1976;260:444–446.

25. Edgren M, Revesz L, Larsson A. Induction and repair of single-strand DNA breaks after X-irradiation of human fibroblasts deficient in glutathione. *Int J Radiat Biol* 1981;40:355–363.

26. Revesz L, Edgren M. Glutathione dependent yield and repair of single-strand breaks in irradiated cells. *Br J Cancer Res* 1984;49(Suppl VI):55–60.

27. Blöcher D. DNA double-strand breaks in Ehrlich ascites tumour cells at low doses of X-rays. Determination of induced breaks by centrifugation at reduced speed. *Int J Radiat Biol* 1982;42:317–328.

28. Corry PM, Cole A. Radiation-induced double-strand scissions of the DNA of mammalian metaphase cells. *Radiat Res* 1968;36:528–543.

29. Lehman AR, Stevens S. The production and repair of double-strand breaks in cells from normal humans and from patients with Ataxia telangiectasia. *Biochim Biophys Acta* 1977;474:49–60.

30. Elkind MM. DNA repair and cell repair, are they related? *Int J Radiat Oncol Biol Phys* 1979;5:1089–1094.

31. Fankenberg-Schwager M, Frankenberg D, et al. Repair of DNA double-strand breaks in irradiated yeast cells under nongrowth conditions. *Radiat Res* 1980;82:498–510.

32. Van der Schans GP, Centen HB, Lohman PHM. DNA lesions induced by ionizing radiation. In: Natarajan AT, Obe G, Altmann H, eds. *Progress in Mutation Research*. Elsevier, Amsterdam, 1982;vol 4, 285–299.
33. Van der Schans GP, Paterson MC, Cross WG. DNA strand breaks and rejoining in cultured human fibroblasts exposed to fast neutron or gamma rays. *Int J Radiat Biol* 1983;44:75–85.
34. Coquerelle T, Weibezahn KF. Rejoining of DNA double-strand breaks in human fibroblasts and its impairment in one ataxia telangiectasia and two Fanconi strains. *J Supramol Struct Cell Biochem* 1981;17:369–376.
35. Resnick MA. The repair of double strand breaks in DNA: a model involving recombination. *J Theor Biol* 1976;59:97–106.
36. Weibezahn KF, Coquerelle T. Radiation-induced DNA double-strand breaks are rejoined by ligation and recombination process. *Nucleic Acids Res* 1981;9:3139–3150.
37. Radford IR. The level of induced DNA double-strand breakage correlates with cell killing after X-irradiation. *Int J Radiat Biol* 1990;48:45–54.
38. Schwartz JL, Rotmensch J, Giovanazzi SM, et al. Faster repair of DNA double-strand breaks in radioresistant human tumour cells. *Intl J Radiat Oncol Biol Phys* 1988;15:907–912.
39. Kelland LR, Edwards SM, Steel G. Induction and rejoining of DNA double-strand breaks in human cervix carcinoma cell lines of differing radiosensitivity. *Radiat Res* 1988;116:526–538.
40. Green A, Pager A, Stout PM, et al. Relationships between DNA damage and the survival of radiosensitive mutant Chinese hamster cell lines exposed to gamma-radiation. Part I: intrinsic radiosensitivity. *Intl J Radiat Biol* 1992;61:465–472.
41. Alaoui-Jamali MA, Batist G, Lehnert S. Radiation-induced damage to DNA in drug and radiation-resistant sublines of a human breast cancer cell line. *Radiat Res* 1992;129:37–42.
42. Chang EH, Pirollo KF, Zou ZQ, et al. Oncogenes in radioresistant, noncancerous skin fibroblasts from a cancer-prone family. *Science* 1987;234:1036–1039.
43. Kasid UN, Pfeifer A, Weichselbaum RR, et al. The ras oncogene is associated with a radiation-resistant human laryngeal cancer. *Science* 1987;234:1039–1041.
44. FitzGerald TJ, Daugherty C, Kase K, et al. Activated human N-ras oncogene enhances X-irradiation repair of mammalian cells in vitro less effectively at low dose rate. *Am J Cln Oncol* 1985;8:517–522.
45. Sklar MD. The ras oncogenes increase the intrinsic resistance of NIH 3T3 cells to ionizing radiation. *Science* 1988;239:645–647.
46. Ling CC, Endlich B. Radioresistance induced by oncogenic transformation. *Radiat Res* 1989;120:267–279.
47. McKenna WG, Weiss MC, Bakanauskas VJ, et al. The role of the H-ras oncogene in radiation resistance and metastasis. *Int J Radiat Oncol Biol Phys* 1990;18:849–859.
48. Iliakis G, Metzger L, Muschel RJ, et al. Induction and repair of DNA double strand breaks in radiation-resistant cells obtained by transformation of primary rate embryo cells with oncogenes H-ras and v-myc. *Cancer Res* 1990;50:6575–6579.
49. Gordon DJ, Milner AW, Beaney RP, et al. The increase in radioresistance of V79 cells cultured as spheroids is correlated to changes in nuclear morphology. *Radiat Res* 1990;121:174–179.
50. Kapiszewska M, Wright WD, Lange CS, et al. DNA supercoiling changes in nucleoids from irradiated L5178Y-S and -R cells. *Radiat Res* 1989;119:569–575.
51. Burzio LO, Riquelme PT, Koide SS. ADP ribosylation of rate liver nucleosomal core histones. *J Biol Chem* 1979;254:3029–3037.
52. Riquelme PT, Burzio LO, Koide SS. Poly(ADP-ribose) synthetase activity in rat testis mitochondria. *J Biol Chem* 1979;254:3018–3028.
53. Endres M, Wang ZQ, Namura S, et al. Ischemic brain injury is mediated by the activation of poly(ADP-ribose) polymerase. *J Cereb Blood Flow Metab* 1997;17 :1143.
54. Eliasson MJ, Sampei K, Mandir AS, et al. Poly(ADP-ribose) polymerase gene disruption renders mice resistant to cerebral ischemia. *Nature Med* 1997;3:1089.

55. Szabo C, Dawson VL. Role of poly(ADP-ribose) synthetase in inflammation and ischaemia-reperfusion. *Trends Pharmacol Sci* 1998;19:287.
56. Paterson MC, Smith BP, Lohman PH, et al. Defective excision repair of gamma-ray-damaged DNA in human (ataxia telangiectasia) fibroblasts. *Nature* 1976;260:444–447.
57. Lavin MF, Davidson M. Repair of strand breaks in superhelical DNA of ataxia telangiectasia lymphoblastoid cells. *Cell J Sci* 1981;48:383–391.
58. Cole J, Arlett CF, Green MH, et al. Comparative human cellular radiosensitivity: II. The survival following gamma-irradiation of unstimulated (G0) T-lymphocytes, T-lymphocyte lines, lymphoblastoid cell lines and fibroblasts from normal donors, from ataxia–telangiectasia patients and from ataxia–telangiectasia heterozygotes. *Int J Radiat Biol* 1988;54:929–943.
59. Cornforth MN, Bedford JS. On the nature of a defect in cells from individuals with ataxia-telangiectasia. *Science* 1985;227:1589–1591.
60. Marecki JC, McCord JM. The inhibition of poly (ADP-ribose) polymerase enhances growth rates of ataxia telangiectasia cells. *Arch Biochem Biophys* 2002;402:227–234.
61. Yu SW, Wang H, Poitras MF, et al. Mediation of poly(ADP-ribose) polymerase-1-dependent cell death by apoptosis-inducing factor. *Science* 2002;297(5579):259–263.
62. Tsujimoto Y, Cossman E, Jaffe E, et al. Involvement of the bcl-2 gene in human follicular lymphoma. *Science* 1985;228:1440–1443.
63. McDonnell T, Troncosco P, Brisbay S, et al. Expression of the protooncogene bcl-2 in the prostate and its association with emergence of androgen-independent prostate cancer. *Cancer Res* 1992;52:6940–6944.
64. Bhargava V, Kell D, Van de Rijn M, et al. Bcl-2 immunoreactivity in breast carcinoma correlates with hormone receptor positivity. *Am J Pathol* 1994;145:535–540.
65. Reed J, Meister L, Tanaka S, et al. Differential expression of bcl-2 protooncogene in neuroblastoma and other human tumor cell lines of neural origin. *Cancer Res* 1991;51:6529–6538.
66. Hockenberry D, Oltvai Z, Yin XM, et al. Bcl-2 functions in an antioxident pathway to prevent apoptosis. *Cell* 1993;75:241–251.
67. Sandstrom P, Mannic M, Buttke T. Inhibition of activation-induced death in T cell hybridomas by thiol antioxidants: oxidative stress as a mediator of apoptosis. *J Leukocyte Biol* 1994;55:221–226.
68. Kane D, Saralin T, Auton S, et al. Bcl-2 inhibition of neural cell death: decreased generation of reactive oxygen species. *Science* 1993;262:1274–1276.
69. Story M, Vochringer D, Malone C, et al. Radiation-induced apoptosis in sensitive and resistant cells isolated from a mouse lymphoma. *Int J Radiat Biol* 1994;66:659–669.
70. Overgaard J, Horsman MR. Modification of hypoxia-induced radioresistance in tumors by the use of oxygen and sensitizers. *Semin Radiat Oncol* 1996;6:10–21.
71. Saunders M, Dische S. Clinical results of hypoxic cell radiosensitization from hyperbaric oxygen to accelerated radiotherapy, carbogen and nicotinamide. *Br J Cancer* 1996;27(Suppl): S271–S278.
72. Overgaard J. Sensitization of hypoxic tumor cells-clinical experience. *Int J Radiat Biol* 1989;56:801–811.
73. Werner-Wasik M, Schmid CH, Bornstein L, et al. Prognostic factors for local and distant recurrence in stage I and II cervical carcinoma. *Int J Radiat Oncol Biol Phys* 1995;32:1309–1317.
74. Fein DA, Le WR, Henlon AL, et al. Pretreatment hemoglobin level influences local control and survival of T1–T2 squamous cell carcinomas of the glottic larynx. *J Clin Oncol* 1995;13:2077–2083.
75. Dubray B, Mosseri V, Brunin F, et al. Anemia is associated with lower local–regional control and survival after radiation therapy for head and neck cancer: a prospective study. *Radiology* 1996;201:553–558.
76. Dische S, Warburton MF, Saunders MI. Radiation myelitis and survival in the radiotherapy of lung cancer. *Int J Radiat Oncol Biol Phys* 1988;15:75–81.

77. Sasai K, Ono K, Hiraoka M, et al. The effect of arterial oxygen content on the results of radiation therapy for epidermoid bronchogenic carcinoma. *Int J Radiat Oncol Biol Phys* 1989;16:1477–1481.
78. Cole CJ, Pollack A, Zagars GK, et al. Local control of muscle-invasive bladder cancer: preoperative radiotherapy and cystectomy versus cystectomy alone. *Int J Radiat Oncol Biol Phys* 1955;32:331–340.
79. Greven KM, Solin LJ, Hanks GE. Prognostic factors in patients with bladder carcinoma treated with definitive irradiation. *Cancer* 1990;65:908–912.
80. Dunphy EP, Peterson IA, Cox RS, et al. The influence of initial hemoglobin and blood pressure levels on results of radiation therapy for carcinoma of the prostate. *Int J Radiat Oncol Biol Phys* 1989;16:1173–1178.
81. Siemann DW, Horsman MR, Chaplin DJ. The radiation response of KHT sarcomas following nicotinamide treatment and carbogen breathing. *Radiother Oncol* 1994;31: 117–122.
82. Fenton BM, Lord EM, Paoni SF. Enhancement of tumor perfusion and oxygenation by carbogen and nicotinamide during single- and multifraction irradiation. *Radiat Res* 2000;153:75–83.
83. Kaanders JH, Pop LA, Marres HA, et al. Accelerated radiotherapy with carbogen and nicotinamide (ARCON) for laryngeal cancer. *Radiother Oncol* 1998;48:115–122.
84. Hoskin PJ, Saunders MI, Dische S. Hypoxic radiosensitizers in radical radiotherapy for patients with bladder carcinoma: hyperbaric oxygen, misonidazole, and accelerated radiotherapy, carbogen, and nicotinamide. *Cancer* 1999;86:1322–1328.
85. Overgaard J, Hansen HS, Overgaard M, et al. A randomized double-blind phase III study of nimorazole as a hypoxic radiosensitizer of primary radiotherapy in supraglottic larynx and pharynx carcinoma. Results of the Danish Head and Neck Cancer Study (DAHANCA) Protocol 5–85. *Radiother Oncol* 1998;46:135–146.
86. Von Pawel J, von Rosemeling R, Gatzemeier U, et al. Tirapazamine plus cisplatin versus cisplatin in advanced non-small-cell lung cancer: a report of the international CATAPULT I study group. Cisplatin and tirapazamine in subjects with advanced previously untreated non-small-cell lung tumors. *J Clin Oncol* 2000;18:1351–1359.
87. Rischin D, Peters L, Hicks R, et al. Phase I trial of concurrent tirapazamine, cisplatin, and radiotherapy in patients with advanced head and neck cancer. *J Clin Oncol* 2001;19:535–542.
88. Craighead PS, Pearcey R, Stuart G. A phase I/II evaluation of tirapazamine administered intravenously concurrent with cisplatin and radiotherapy in women with locally advanced cervical cancer. *Int J Radiat Oncol Biol Phys* 2000;48:791–795.
89. Kovacs MS, Hocking DJ, Evans JW, et al. Cisplatin anti-tumor potentiation by tirapazamine results from a hypoxia-dependent cellular sensitization to cisplatin. *Br J Cancer* 1999;80: 1245–1251.
90. Goldberg Z, Evans J, Birrell G, et al. An investigation of the molecular basis for the synergistic interaction of tirapazamine and cisplatin. *Int J Radiat Oncol Phys* 2001;49:175–182.

12 V(D)J Recombination and DNA Double-Strand-Break Repair

From Immune Deficiency to Tumorigenesis

*Despina Moshous, MD, PhD
and Jean-Pierre de Villartay, PhD*

CONTENTS

1. INTRODUCTION

The vertebrate immune system encounters an almost limitless number and variety of foreign antigens. Innate immunity provides the first line of defense against infection. Adaptive immune responses also play a crucial role in defense against pathogens and depend on the generation of a vast repertoire of soluble and membrane-bound antigen receptors expressed on B- and T-lymphocytes respectively. The variable regions of the Immunoglobulin (Ig) and the T-cell receptor (TCR) are encoded by a modest number of gene segments. Immunological diversity is achieved through a series of programmed DNA rearrangements, termed V(D)J recombination, the only known form of site-specific DNA rearrangement in vertebrates. This assembly process is highly conserved throughout evolution and present in most of the vertebrates analyzed so far.

From: *Cancer Drug Discovery and Development: DNA Repair in Cancer Therapy*
Edited by: L. C. Panasci and M. A. Alaoui-Jamali © Humana Press Inc., Totowa, NJ

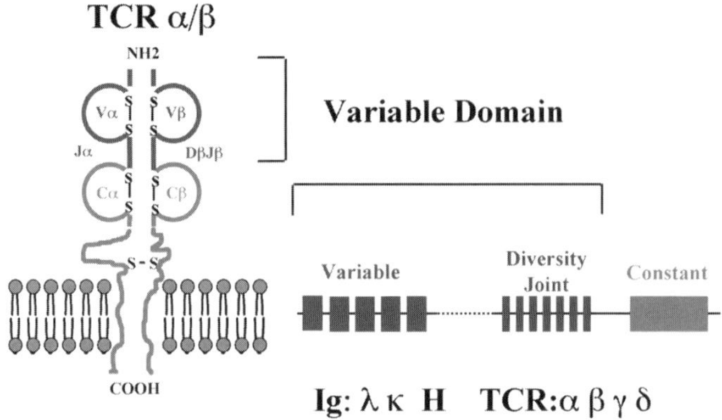

Fig. 1: The variable antigen-binding domain of the TCR and Ig are composed of V (variable), D (diversity), and J (joining) segments.

V(D)J recombination allows the somatic DNA rearrangement of germline V (variable), D (diversity), and J (joining) coding segments to form a contiguous exon for the variable antigen-binding domain of the receptor (*see* Fig. 1) *(1)*. Sequence diversity is in part combinatorial because the one V, one J, and, in some cases, one D segment, which will form the functional V(D)J unit, are randomly selected (*see* Fig. 2). Another source of diversity is of junctional nature, originating from imprecise joining of the V, D, and J gene segments. Additional mechanisms contribute to an increased diversification of the coding joints. Random integration of nontemplated nucleotides to coding ends (N diversity) is mediated through the lymphoid-specific enzyme terminal deoxynucleotidyl transferase (TdT) *(2)*. The resulting primary mRNA transcript is fused to several constant (C) region exons giving rise to the transcription of one of the antigen-receptor chains that will be processed, glycosylated, and ultimately exported to the cell surface or secreted.

2. STRUCTURE AND EXPRESSION OF IMMUNOGLOBULIN AND TCR GENES

During lymphocyte development, seven complex gene loci can undergo V(D)J recombination; the Ig heavy, κ and λ light chains of the Immunoglobulins, and the α, β, γ, and δ loci of the T-cell receptor *(1)*. These gene loci contain arrays of tandemly repeated V, D, and J gene segments, which form clusters of up to several hundred V(D)J segments. In humans, the DNA segments for the Ig heavy chain are located on chromosome 14; those for the κ and λ light chain are located on chromosomes 2 and 22, respectively. The genes coding for the α, β, and γ loci of the TCR are found on chromosome 14, 7q, and 7p, respectively, whereas the

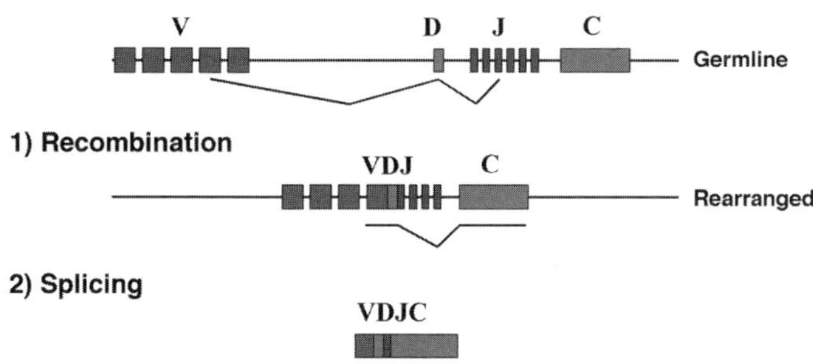

Fig. 2: Rearrangement and expression of Ig and TCR genes.

δ locus is encoded by elements situated between the J and Vα elements on chromosome 14. The site specificity of the reaction is ensured through the presence of conserved recombination signal sequences (RSSs) at the border of all coding regions. These RSSs consist of a highly conserved palindromic heptamer, which is separated from an A/T-rich nonamer by a spacer sequence of either 12 or 23 basepairs (bp). Rearrangement occurs exclusively between gene segments, which are flanked by RSSs of different spacer lengths, a restriction known as the "12/23 rule" *(3)* so that only coding sequences of different types can be joint.

Although based on a common lymphoid recombinase machinery and highly conserved RSS, V(D)J recombination is remarkably regulated and both lineage-specific and developmental-stage-specific. More precisely, Ig genes are rearranged in B-cell progenitors and TCR genes in T-cell precursors. Furthermore, because of allelic exclusion, a single lymphocyte and its clonal progeny will almost always express only one functional Ig or TCR molecule despite the theoretical possibility of encoding a second receptor by the other allele. milar to the proposal of Alt and colleagues in the "accessibility hypothesis" in 1985 *(4)*, the packaging of the immunoglobulin and TCR loci into chromatin seems to be different in B- and T-cells and the activity of the locus may be regulated through conformational changes of the chromatin structure (chromatin remodeling) *(5)*. Regulated accessibility guarantees that V(D)J recombination progresses in a developmentally ordered way during lymphopoiesis.

2.1. The V(D)J Recombinase Machinery

V(D)J recombination can be roughly divided into three parts (*see* Fig. 3). The initialization of this process is performed by the recombination-activating genes RAG1 and RAG2, whose restricted expression renders this phase specific to immature lymphocytes. As a result of the development of cell-free reaction systems, this phase has been investigated extensively. The RAG proteins guide recombination events to RSSs and introduce DNA double-strand breaks (DSBs)

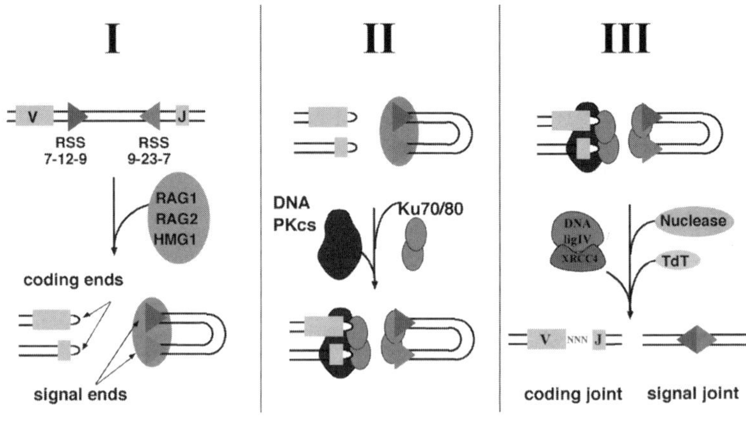

Fig. 3: The V(D)J recombination.

(6) precisely between the RSS and the adjacent coding segment, producing two types of terminus: coding ends and signal ends *(7–9)*. The recombination-activating RAG1 and RAG2 proteins are necessary and sufficient to initiate V(D)J recombination on an accessible antigen-receptor locus *(6)*, although high-mobility group (HMG) proteins regulate their action. The lymphoid-specific proteins RAG1 and RAG2 have originally been isolated on the basis of their ability to mediate V(D)J recombination on artificial substrates in nonlymphoid cells such as NIH 3T3 fibroblasts *(10,11)*. Located on human chromosome 11p13 *(12)*, the *RAG1* gene codes for 1043 amino acids and the *RAG2* gene coder for a protein of 527 amino acids. The two protein sequences are not related to each other. The amino acid sequence and the overall genomic organization of the RAG factors are highly conserved throughout evolution *(13)*, whereas no homologs have been found in lower organisms. Interestingly, the genes display an unique organization: Separated by 15–18 kb in humans, they face each other in a transcriptional inverse orientation. Their coding regions and most of the 3' untranslated sequences are organized in a single exon *(11,14,15)*. This particularity implies that RAG1 and RAG2 together could form a transposase *(16)* and supports the theory that RAG1 and RAG2 were originally components of a transposable element *(16)*. A "synaptic complex" consisting of RAG1, RAG2, a 12-RSS, and a 23-RSS initiates the V(D)J recombination reaction. A DNA DSB is introduced precisely at the boundary of the heptamer and the adjacent coding segment. First, a single-stranded nick is generated at each RSS in one DNA strand, the resulting 3'OH then performs a nucleophilic attack on the phosphodiester bond on the opposite strand. This transesterification reaction leads to the creation of a covalently

sealed hairpin, referred to as "coding end", and a blunt-ended 5'phosphorylated RSS, referred to as "signal end" (reviewed in refs. *8* and *17*). The cleavage reaction is facilitated by the presence of the HMG proteins. The RAG proteins remain bound to the generated coding and signal ends in a "RAG postcleavage complex," which possibly allows one to stabilize and direct the adequate joining of the cleaved ends in the following steps.

2.2. The Link Between V(D)J Recombination and DNA Repair

The naturally occurring murine *scid* (severe combined immunodeficiency) mutation blocks the formation of coding joints *(18)* and is characterized by accumulation of covalently sealed coding ends *(19)*, which are normal intermediates in the V(D)J recombination process. Further investigations revealed that cells from *scid* mice also exhibit an increased sensitivity to ionizing radiation as a result of a general defect in DNA DSB repair *(20–22)*. These findings provided the first direct link between V(D)J joining and DNA DSB repair. Additional evidence for the functional overlap between the general DNA repair machinery of the cell and the V(D)J recombination process was provided by the characterization of other mutants that display defects in both V(D)J recombination and DSB repair *(23)*. A large panel of x-ray-sensitive Chinese hamster ovary (CHO) cell lines was tested for their ability to perform V(D)J recombination after the introduction of the RAG1 and RAG2 genes. These cell lines displayed not only impaired repair of radiation-induced DSB, but also showed defects in V(D)J recombination. They comprise four "x-ray cross-complementation groups" (termed XRCC4 through XRCC7). Except for murine *scid* (XRCC7), all mutants are defective in both signal and coding joint formation *(23,24)*.

The XRCC5, XRCC6, and XRCC7 factors define the DNA-PK/Ku70/Ku86 complex. The Ku heterodimer is composed of 70-kDa and 86-kDa subunits and is able to bind to altered DNA structures such as broken ends, single-stranded gaps, and hairpins. Ku possesses helicase activity also and functions as the DNA-binding component within the DNA-dependent protein kinase (DNA-PK) complex *(25)*. XRCC5 mutants are characterized by an absence of the DNA-end-binding property *(26,27)* of the heterodimer Ku. As the XRCC5 mutants do not generate a functional 86-kDa Ku subunit, they lack DNA-PK activity *(28)*. The functional defect can be restored by transfecting the mutant cells with Ku86 cDNA (28–30). The phenotype of the XRCC6 mutant cell line or *sxi-1* comprises defects in DSB repair, DNA end binding, and V(D)J recombination and can be complemented upon transfection with Ku70 cDNA (31–33). The group of the XRCC7 mutants, which includes the mouse *scid* mutation, lacks DNA-PK activity *(34–36)*. The murine *scid* condition is caused by an intragenic mutation at the 3' end of the gene encoding the catalytic subunit of DNA-dependent protein kinase (DNA-PKcs) *(37,38)*. The mutation leads to the expression of a truncated and, therefore, catalytically inactive DNA-PKcs protein. The complete DNA-PKcs knockout has provided evidence that in mice, DNA-PK activity is required for coding, but not for signal joint formation *(39,40)*. In addition to the murine *scid* condition, an equine form of DNA-PKcs deficiency in Arabian SCID foals displaying autosomal recessive severe combined immunodeficiency and radiosensitiviy has been identified *(41)*, as well as a canine form in Jack Russell terriers *(42)*. Whereas in *scid* mice DNA-PKcs deficiency seems to generate an incomplete block in V(D)J recombination with "leaky" coding joint formation and only a modest defect in signal end ligation, DNA-PKcs deficiency in horses profoundly blocks both coding and signal end-joining. In the canine condition, both coding and signal end-joining are impaired, but the defect is intermediate between that observed in *scid* mice and SCID foals. DNA-PKcs deficiencies in humans have not been identified, with the exception of a human glioma cell line (MO59J) displaying increased radiosensitivity and a defect in the repair of DNA DSBs *(43)*.

The defect in XRCC4, a 38-kDa nuclear phosphoprotein, was identified in the X-ray-sensitive CHO cell line XR-1 via a complementation cloning method *(44)*. It has been shown that XRCC4 interacts with *(45,46)* and stabilizes the DNA ligase IV *(47)*. The XRCC4/DNA ligase IV complex plays a fundamental role in DNA nonhomologous end-joining (NHEJ). A region in the carboxy-terminal tail of DNA ligase IV, located between the two BRCA1 C terminus (BRCT) domains, is necessary and sufficient to confer binding to XRCC4 *(48)*. Recently, the crystal structure of human XRCC4 bound to a polypeptide that corresponds to the DNA ligase IV sequence has been reported *(49)*. In the complex, a single ligase chain binds asymmetrically to an XRCC4 dimer *(50)*.

3. V(D)J RECOMBINATION REPRESENTS A CRITICAL CHECKPOINT OF THE DEVELOPMENT OF THE IMMUNE SYSTEM

3.1. Animal Models

The crucial role of the RAG1 and RAG2 proteins with regard to the lymphocyte development in vivo was first demonstrated by the observation of a complete block in both T- and B-cell maturation in mice with a targeted disruption of the RAG1 *(51)*, or RAG2 gene *(52)*. Targeted disruptions of all the known components of the NHEJ factors involved in V(D)J recombination have also been realized. The corresponding deficient mice models are all characterized by severe combined immunodeficiency resulting from the inability to properly perform V(D)J recombination and from increased radiosensitivity. In addition, they show quite different phenotypes with regard to cellular proliferation and neuronal development (for a review, *see* ref. *53*). Targeted disruption of Ku70 *(33,54,55)*, Ku86 *(56,57)* or DNA-PKcs *(39,40,58)* result in viable mice with severe combined immunodeficiency. The DNA-PKcs-deficient mice show no growth defect *(39,40)*, whereas the inactivation of the gene coding for Ku70 or

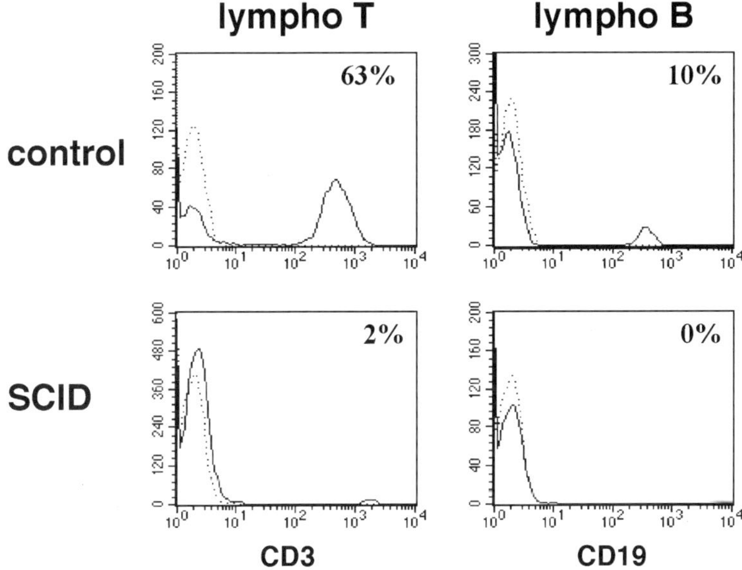

Fig. 4: Lack of mature B- and T-cells in human SCIDs.

Ku86 results in a cellular growth defect *(56)* and premature senescence *(59,60)*. In contrast, mice lacking XRCC4 *(61)* or DNA ligase IV *(62,63)* die at a late stage of embryogenesis because of extensive apoptotic death of newly generated postmitotic neurons. Gene-targeted mutation of XRCC4 in primary murine cells causes growth defects, premature senescence, ionized radiation (IR) sensitivity, and inability to support V(D)J recombination *(61,64)*. As expected, V(D)J recombination in fibroblasts is impaired in all of these murine models.

3.2. The Human T–B–SCID Conditions

Human severe combined immunodeficiency (SCID), characterized by a profound blockage in the T-cell differentiation, represents one of the most serious forms within the group of primary immunodeficiencies. About 20% of human SCID display a complete absence of both T- and B-lymphocytes while natural killer (NK) cells are present and functional (*see* Fig. 4). In the absence of treatment, this condition is lethal within the first year of life because of multiple and extended infections. The only curative therapy available consists of allogenic bone marrow transplantation. The lack of both T- and B-lymphocytes suggested an early arrest in the common lymphoid development (*see* Fig. 5). An important mechanism shared by both T- and B-cell maturation comprises the V(D)J recombination, which is a crucial prerequisite for the acquisition of the antigen receptor on the cell surface. Thus, defects in the V(D)J joining process were an attractive hypothesis to explain the T–B–SCID phenotype. Not surprisingly, when thinking

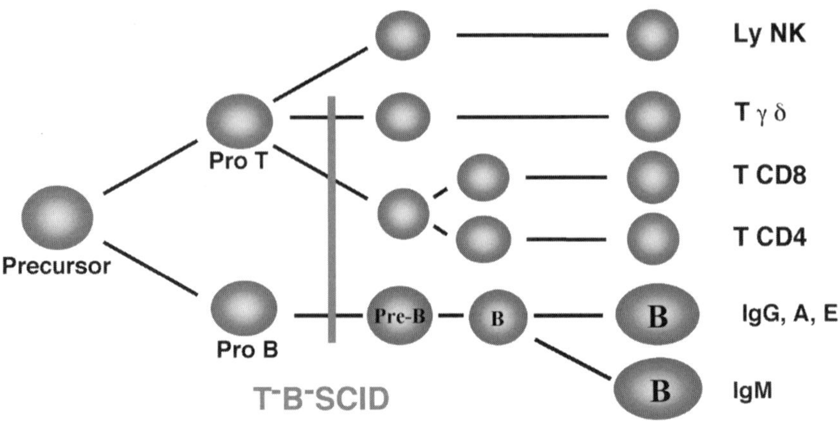

Fig. 5: Early arrest of T- and B-cell maturation in human SCIDs.

of the multitude of factors involved in V(D)J recombination, the causes for human T–B–SCID are not homogenous. For one subset of patients, mutations in the *RAG1* or *RAG2* gene were identified *(65–68)*. A particular condition of T–B–SCID is the Omenn syndrome. These patients, who lack peripheral B cells, are characterized by the presence of a large number of activated T-lymphocytes, with very restricted heterogeneity, that infiltrate tissues. The observation that T–B–SCID and Omenn syndrome can occur in siblings of the same family *(69)* indicated that the two disorders could be based on the same molecular defect. Indeed, mutations in the *Rag* genes have been identified in Omenn patients *(66,70)*, and it is now well established that a residual V(D)J recombinase activity in these patients leads to the generation of an oligoclonal T-cell repertoire in the thymus with further restriction in the periphery *(71)*. Interestingly, identical mutations can cause either T–B–SCID or Omenn syndrome. This may be explained by the fact that a leaky mutation in one of the *Rag* genes with a residual recombinase activity is not sufficient *per se* to generate the Omenn phenotype *(68)*. An attractive hypothesis for the missing link may be the necessity of a second factor (e.g., an endogenous or exogenous antigen) that will drive the few generated T-cell clones to noncontrolled activation and expansion as a result of the absence of a normal balanced T-cell repertoire that is able to exert inhibitory retrocontrol.

Finally, T–B–SCID patients in whom the *RAG1* and *RAG2* genes are found normal show the additional feature of an increased sensitivity to ionizing irradiation and thus constitute a distinctive subgroup of "radiosensitive" SCID patients, hereafter referred to as "RS-SCID" (*see* Fig. 6). Increased radiosensitivity in bone marrow cells as well as in skin fibroblasts pointed to a defect in the general DNA repair/V(D)J recombination machinery of the cell *(72)*. This condition is

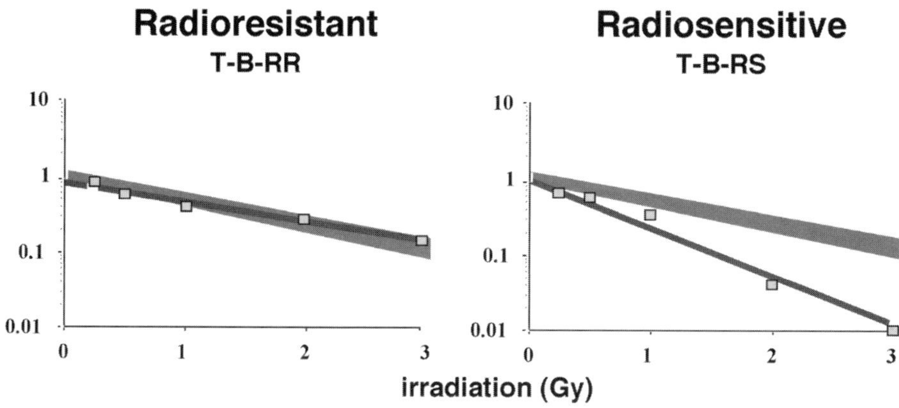

Fig. 6: Two types of human T–SCID.

reminiscent of the defect specific to the *scid* mouse in which a lack of DNA-PKcs alters the final phase of the V(D)J recombination process. A genomewide search for linkage in RS-SCID families localized the gene to a 6.5 cM region on the short arm of human chromosome 10 delimited by the polymorphic markers D10S1664 and D10S674 *(73)*, a region to which is also mapped the locus for the so-called A-SCID, a particular form of SCID that occurs at a high frequency in Athabascan-speaking Native-Americans *(74)*. A systematical survey of the sequence data released by the Sanger Center for 24 bacterial artificial chromosomes (BACs) covering the RS-SCID region led to the identification of a novel DNA repair/ V(D)J recombination protein that was named Artemis *(75)*. Mutations in the *Artemis* gene were found in all RS-SCID patients and a wild-type form of Artemis complemented the V(D)J recombination defect in RS-SCID cells.

4. FUNCTION OF ARTEMIS IN V(D)J RECOMBINATION

Artemis is a protein of 78 KDa encoded by 692 amino acids. The entire cDNA sequence comprises 2354 bp. To date, no global homologs of Artemis were described in other species besides the initial finding of a region showing 20% identity with the DNA repair protein SNM1 in mouse and PSO2 in yeast. Cloning of the murine *Artemis* cDNA sequence revealed 78% protein identity, suggesting a high degree of conservation among higher eukaryotes. The genomic organization of the Artemis gene shows the presence of 14 transcribed exons with sizes between 52 and 1160 bp. Artemis is ubiquitously, but weakly, expressed, as expected for a protein belonging to the general DNA repair machinery. The low expression could be a common feature in the SNM1 protein family, as the basal expression of mSNM1 protein was shown to be also quite low *(76)*. Interestingly, no increased expression is observed in the sites of V(D)J recombination (thymus

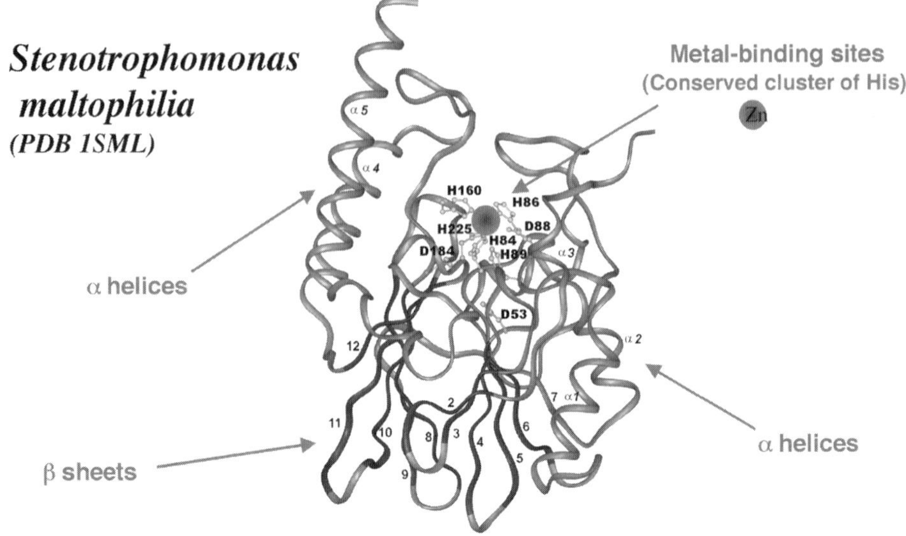

Fig. 7: The metallo-β-lactamase fold; example of *Stenotrophomonas maltophilia.*

or the bone marrow) or meiotic recombination (ovary, testis). A detailed sequence analysis of the Artemis protein nonetheless indicated that it belongs to the metallo-β-lactamase superfamily *(75)*. Metallo-β-lactamases are enzymes first described in bacteria where they are responsible for cleavage of the β-lactam ring of certain antibiotics. The metallo-β-lactamase fold consists of a four-layer β-sandwich with two mixed β-sheets flanked by α-helices, with the metal-binding sites located at one edge of the β-sandwich *(77)* (*see* Fig. 7). Five sequence motifs consisting mostly in histidine and aspartic acids constitute critical residues of the catalytic site *(78,79)*. Motifs 1 to 4 are clearly conserved in Artemis, and motif 5 was not easily recognizable. However, the sequence downstream of motif 4 presented significant homology with a wide variety of proteins, including the DNA repair factors SNM1 and PSO2 as well as the RNA cleavage and polyadenylation-specific factor (CPSF). Based on these homologies, we named this new domain β-CASP. The β-CASP domain, which is observed in all living organisms, is always associated with the metallo-β-lactamase domain. Several highly conserved residues in the β-CASP domain are likely to play a key role in the structure and/or function of this family within the metallo-β-lactamase super-family. This suggested that the metallo-β-lactamase fold may be functional and may, thus, confer a catalytic acitivity to Artemis.

Several distinct enzymatic activities constitute the V(D)J recombinase. The RAG1 and RAG2 proteins initiate the reaction and the five known factor of the NHEJ are responsible for signaling and repairing the DNA damage. Up to now, there is one important activity that has remained somehow elusive: the hairpin-opening activity. Indeed, the RAG1/2-generated DSB is left as a hairpin-sealed coding end that obviously needs to be processed before it can be repaired. Several candidates have been suggested for this activity. The RAD50/MRE11/NBS1 complex, which is known to participate in NHEJ *(80,81)*, was found at the site of the rearranging TCR genes, arguing for its possible involvement in the DNA repair phase of the V(D)J recombination *(82)*. Moreover, MRE11 was found to have hairpin-opening activity in vitro, although only in the nonphysiologic Mn^{2+} conditions. The finding of normal coding joint formation in Nijmegen breakage syndrome (NBS) patients does not support strongly this candidate. The RAG1/2 complex has also been proposed to carry the hairpin-opening activity. Indeed, this complex remains on the broken ends following the initial DNA DSB and displays hairpin-opening activity in vitro *(83–86)*. One major drawback of this hypothesis is that hairpins at coding ends accumulate in the murine scid where RAG1/2 proteins are present and normal. Although DNA-PKcs may be argued to be necessary to activate RAG1/2 for this activity, there are no experimental data to support this hypothesis. We proposed that Artemis could be the factor responsible for hairpin opening for several reasons *(75)*. First, it is unlikely that Artemis participates directly in the DNA DSB rejoining phase, as one would expect a much stronger clinical phenotype if this was the case. After all, DNA–LigIV and XRCC4 knock-out (KO) mice are embryonic lethal and NBS patients have evident extraimmunological manifestations ("birdlike" face, growth retardation, cancer predisposition) as is the case for defects in the other players of the NHEJ (review in ref. *87*). Moreover, the rejoining of linearized DNA constructs introduced in RS-SCID fibroblasts and the V(D)J signal joint formation are normal in Artemis-deficient patients (*88*, and our unpublished observations). In fact, human RS-SCID patients and *scid* mice are the only two situations where a V(D)J recombination associated DNA repair defect affects only the formation of the V(D)J coding joints. In all of the other situations analyzed in the context of a defective V(D)J recombination, the signal joints are absent. Given the accumulation of hairpin structures in scid mice and the likely hydrolase activity of Artemis through its metallo-β-lactamase domain, it was tempting to propose that Artemis is indeed the hairpin opener. It is exactly what Lieber and colleagues demonstrated recently in a very elegant paper *(89)*. Artemis forms a complex with and is phosphorylated by DNA-PKcs, which activates the endonuclease activity of Artemis, leading to opening of RAG1/2-generated hairpin structures.

5. CAN WE CONSIDER ARTEMIS AS A "CARETAKER"?

DNA is constantly exposed to DNA damage. Among the different types of DNA lesion, DNA DSB represent the most deleterious harm to the cell because one single DSB may already lead to cellular death. Thus, it is not surprising that during evolution, sophisticated mechanisms have emerged to guarantee efficient repair of DNA DSBs that may be induced accidentally or as part of cellular mechanisms such as meiosis or V(D)J recombination. In mammalian cells, two major pathways of DNA DSB repair exist: the homologous recombination (HR) and the nonhomologous end-joining (NHEJ) *(90)*. These repair mechanisms display a remarkable conservation between the different species, emphasizing their biological importance. Whereas HR leads, in general, to a precise repair as a result of the use of the identical sister chromatid as a template, NHEJ may result in the addition or deletion of nucleotides. Both pathways restore the continuity of the broken chromosome and therefore preserve the integrity of the chromosome by preventing chromosomal translocation or deletion. Because defective response to DNA damage is a common feature of cancer cells, it has been proposed that tumorigenesis may be linked to defects in the DNA DSB repair (for recent reviews, *see* refs. *53*, *91*, and *92*). Neoplastic transformation seems to require consecutive defects in both of the two major groups of protein belonging to the tumor suppressor genes, the "gatekeepers" and the "caretakers" *(93)*. Gatekeepers control cellular proliferation and cell death. They regulate the cell cycle progression and, upon introduction of DNA damage into the cell, they induce apoptosis or cell cycle arrest to allow appropriate DNA repair by the different repair systems. A well-known member of this group is p53. Caretakers belong to the DNA repair machinery itself. Thus, they are directly involved in DNA repair and their inactivation leads to genomic instability. Furthermore, the invalidation of both surveillance systems, the gatekeepers and caretakers, increases considerably the susceptibility to tumorigenesis. We will concentrate in this review on the role of the NHEJ factors as "genomic guardians" and will not emphasize the probable link of deficiencies in HR to cancerogenesis (for further information, *see* ref. *91*).

5.1. Illegitimate V(D)J Recombination Leads to Lymphoid Malignancies

In the physiological situation, the V(D)J recombination process concerns exclusively the V, D, and J gene segments of the immunoglobulin and TCR gene loci. In vertebrates, millions of lymphoid precursors undergo V(D)J rearrangements each day. The fact that the RSS is the unique element *in cis* that is necessary and sufficient to initiate the site-specific reconnaissance by the V(D)J recombinase *(94)* represents a considerable danger with regard to the stability of the genome. RSS-like sequences might be an erroneous aim for the V(D)J recombinase machinery, leading to illegitimate recombination events followed by genomic deletions and inversions or chromosomal translocations, with the eventual possibility of a transposase activity of the RAG1 and RAG2 factors, which has been shown in vitro *(16,95)*. If these "pseudo-RSS" or "cryptic RSS," the number of which is estimated at about 10 millions in the whole genome, are located near proto-oncogenes, an illegitimate recombination event may result in the inapropriate activation of these oncogenes. The restricted expression of the

RAG genes to immature lymphoid cell lineages directs these oncogenic mechanisms toward the tissues of the immune system, explaining why specific chromosomal translocations between gene segments of the TCR/Ig locus and cellular proto-oncogenes are a characteristic feature of lymphoid neoplasms. In the absence of functional systems, the definition of translocations caused by illegitimate recombination requires the presence of several criteria *(96)*: the involvement of a TCR/Ig gene segment in the translocation, the identification of a pseudo-RSS in the vicinity of the germline sequence of the proto-oncogene, the presence of recurrent breaks at the border of the RSS in the immune locus and at the border of the pseudo-RSS in the nonimmune locus, the addition and deletion of nucleotides at one breakpoint, and, finally, the generation of a signal joint at the other breakpoint. The in vivo evaluation of these criteria is not always possible. Despite multiple in vivo and in vitro studies, no precise rules have been established that allow one to predict the intrinsic oncogenic potential of a given RSS/pseudo-RSS, besides the obligatory presence of a consensus sequence 5'-CAC-3' in the heptamer *(97)*. The formal confirmation that a pseudo-RSS is indeed involved in the process of illegitimate V(D)J recombination requires functional assays. In acute lymphoblastic T leukemia (T-ALL), the translocations of *LMO2*, *TAL1*, and *TAL2* are compatible with an illegitimate recombination event between a TCR locus and a locus of a proto-oncogene containing a fortuitous but functional RSS *(96)*. The fragile site with regard to *LMO2* could function like a RSS-12 with an efficacity to induce V(D)J recombination, which is only slightly reduced compared to a consensus RSS-12 *(98)*. In the case of non-Hodgkin's B-cell lymphomas (B-NHL), the translocations concern the genes *BCL1* and *BCL2*. Only the breakpoints at the IgH locus seem to be induced by an illegitimate V(D)J recombination process, whereas the DNA DSB on the side of the proto-oncogene may be initiated first by other mechanisms and may then invade the synaptic V(D)J complex *(96)*. The transforming events leading to the development of B-cell lymphomas occur mainly at two distinct developmental stages: during the V(D)J recombination in B-cell precursors in the bone marrow and—in particular—during the maturation of B-cells in the germinal center following T-cell-dependent immune reponses *(99)*. The difference between the translocations in T- versus B-lymphocytes might be explained by the fact that the accessibility during the V(D)J recombination process seems to be more restricted with regard to the Ig loci in B-cells than with regard to the TCR loci in T-cells *(100,101)*, facilitating a "complete" illegitimate V(D)J recombination process in T-cells, whereas nonimmune loci in B-cells are less accessible to the V(D)J recombinase, thus preventing, to some extent, "complete" illegitimate V(D)J recombination in B-cells. Because of the existence of RAG1- and RAG2-mediated transposition in vitro, it is not excluded that this mechanism might play a role in vivo with regard to the generation of chromosomal translocations. Recent results emphasize this possibility *(102)*.

5.2. NHEJ Factors of the V(D)J Recombination Are "Caretakers"

Defects of NHEJ components with alteration of the V(D)J recombination process predispose the deficient mice in a selective way to develop lymphoid neoplasia. This link has been initially established by the obervation that *scid* mice with a deficiency in DNA-PKcs display the delayed apparition of thymic lymphomas with a moderate incidence of about 15% *(18)*. These lymphomas reveal translocations in the majority of the cases. To prevent illegimitate rearrangement events, the V(D)J recombination process is highly regulated and submitted to tight surveillance. Despite the fact that the "caretaker" ATM is not required for the V(D)J recombination process (as illustrated by the normal formation of coding joint and signal joint formation in ATM-deficient patients with ataxia–telangiectasia), it seems to be important for the surveillance of the process, especially with regard to efficient prevention of aberrant V(D)J recombination events. Recently, ATM has been shown to colocalize precisely at the DNA double-strand breakpoints during V(D)J recombination with regard to the Igk locus in a pre-B-cell line and the TCRα locus in primary murine thymocytes *(103)*. When DNA repair during V(D)J recombination fails, ATM may thus carry out its "caretaker" function by preventing oncogenic translocations that may lead to the development of neoplasms.

The targeted disruption of the NHEJ genes has contributed considerably to the understanding of the function and the biological importance of the NHEJ. As describe earlier, for each of the well-known NHEJ factors (with the exception of the recently identified Artemis gene), deficient mice models have been developed. These murine models share common features such as the severe combined immunodeficiency and the increased radiosensitivity, which are found in all cases, but they also display quite important phentotypical differences (for a review, *see* ref. *53*). Haploinsufficiency of Ku86 and DNA ligase IV leads to increased chromosomal instability in primary murine fibroblasts *(104)*. The generation of soft tissue sarcomas has recently been reported in DNA ligase IV haploinsufficiency *(105)*. DNA ligase IV –/– or Ku70 –/– murine embryonic fibroblasts (MEFs) reveal a genomic instability with chromosomal abnormalities *(53)*. The complete inactivation of the gene coding for Ku70 or Ku86 leads to an increase in chromosomal aberrations *(104,106,107)* and a multiplication of telomeric fusions *(108–110)*. The expression of hypomorphic variant forms of Ku86 provokes multiple myeloma in humans *(111)*. The first pieces of evidence for a link between defective NHEJ and tumorigenesis was provided by the observation that Ku70 –/– mice develop T-cell lymphomas *(55)* and, that as mentioned earlier, DNA-PKcs deficiency predisposes the *scid* mice to the apparition of thymic lymphoblastic lymphomas *(58)*. Additional evidence for this correlation was supplied through the analysis of crosses of NHEJ-deficient mice with p53 or ATM mutant mice (for review, *see* refs. *53* and *92*). The p53 –/– background rescues the embryonic lethality of both XRCC4 –/– and DNA ligase IV –/– mice. Whereas p53 –/– mice develop thymice lymphomas at approx 5 mo, double-mutant mice develop early-onset pro-B-cell lymphomas. This has been shown for XRCC4 –/–/p53 –/– mice *(112)*, which develop disseminated B-cell lymphoma at about 6 wk of age, displaying characteristic t(12;15) translocations involving the IgH locus and c-*myc*. DNA ligase IV –/–/p53 –/– mice *(113)*, Ku80 –/–/ p53 –/– mice *(106,114)* and DNA-PKcs $^{scid/scid}$/p53 –/– mice *(115)* also present B-cell lymphomas. In contrast to the thymic lymphomas arising spontaneously in *scid* mice, the B-cell lymphomas in the case of NHEJ deficiency on a p53-negative background can be suppressed by inactivation of the Rag genes, suggesting that the mechanism of tumorigenesis in the case of NHEJ deficiency may, indeed, be initiated by inappropriate repair of DNA DSB generated by the V(D)J recombinase *(115)*. In the same way as for p53 deficiency, the inactivation of ATM rescues the embryonic lethality of the DNA ligase IV-deficient mice *(116)*. These results illustrate the crucial role of these factors in maintaining genomic stability and also provide the evidence that the developmental defect in neurogenesis may, indeed, be caused by an excessive apoptotic death of the neurons via the ATM/p53 pathway. Increased tumor development has also be observed in Ku80 –/– mice on a p53 –/+ background *(114)*. These mice not only acquire lymphomas but also other tumors like sarcomas, and the inactivation of Ku80 seems to accelerate this process, which can already be observed in the p53 –/+ mice *(91)*. Taken altogether, all of these murine models recapitulate the hypothesis of Kinzler and Vogelstein for tumorigenesis, where p53 –/– provides the gatekeeper-defective background. These results therefore strongly support the notion that the five known factors of the NHEJ (DNA-PKcs, Ku70, Ku86, XRRC4, and DNA ligase IV) are indeed guardians of the genome and that they have to be considered as genomic caretakers.

5.3. Hypomorphic Mutations of Artemis Lead to B-Cell Lymphomas

In addition to RS-SCID, several other diseases are characterized by an immunodeficiency associated with increased sensitivity to ionizing radiation (IR) such as AT and NBS *(117)*. In our survey of such conditions, we came across four patients who presented a combined immune deficiency characterized by a severe B- and T-lymphocytopenia and a profound hypogammaglobulinemia (our unpublished observations). Mutations leading to truncation of Artemis in the C-terminus domain were responsible for these conditions. The low T- and B-cell count, the partial V(D)J recombination activity in fibroblast of some of these patients, and the incomplete capacity of these truncated Artemis forms to complement the IR sensitivity of an Artemis fully deficient cell line attested for the hypomorphic characteristic of these new mutations. Hypomorphic mutations in NHEJ factors have already been described in the case of DNA–ligase IV *(118)*, for which complete loss-of-function mutations are not expected as DNA–ligIV KO proved to be embryonic lethal in mice *(61,62)*. Interestingly, two of these patients developed disseminated B-cell lymphomas.

The parallel between the NHEJ-defective animal models and the situation of the patients harboring hypomorphic mutation of Artemis strongly suggests that Artemis may as well be considered as a "caretaker" factor. The development of an appropriate model of Artemis deficiency should help accredit this hypothesis.

REFERENCES

1. Tonegawa S. Somatic generation of antibody diversity. *Nature* 1983;302:575–581.
2. Gilfillan S, Dierich A, Lemeur M, et al. Mice lacking TdT: mature animals with an immature lymphocyte repertoire. *Science* 1993;261:1175–1178.
3. Hiom K, Gellert M. Assembly of a 12/23 paired signal complex: a critical control point in V(D)J recombination. *Mol Cell* 1998;1:1011–1019.
4. Yancopoulos GD, Alt FW. Developmentally controlled and tissue-specific expression of unrearranged VH gene segments. *Cell* 1985;40:271–281.
5. Roth DB, Roth SY. Unequal access: regulating V(D)J recombination through chromatin remodeling. *Cell* 2000;103:699–702.
6. McBlane JF, van Gent DC, Ramsden DA, et al. Cleavage at a V(D)J recombination signal requires only RAG1 and RAG2 proteins and occurs in two steps. *Cell* 1995;83:387–395.
7. Bogue M, Roth DB. Mechanism of V(D)J Recombination. [review]. *Curr Opin Immunol* 1996;8:175–180.
8. Fugmann SD, Lee AI, Shockett PE, et al. The RAG proteins and V(D)J recombination: complexes, ends, and transposition. *Annu Rev Immunol* 2000;18:495–527.
9. Critchlow SE, Jackson SP. DNA end-joining: from yeast to man. *Trends Biochem Sci* 1998;23:394–398.
10. Schatz DG, Oettinger MA, Baltimore D. The V(D)J recombination activating gene, RAG-1. *Cell* 1989;59:1035–1048.
11. Oettinger MA, Schatz DG, Gorka C, et al. RAG-1 and RAG-2, adjacent genes that synergistically activate V(D)J recombination. *Science* 1990;248:1517–1523.
12. Oettinger MA, Stanger B, Schatz DG, et al. The recombination activating genes, RAG 1 and RAG 2, are on chromosome 11p in humans and chromosome 2p in mice. *Immunogenetics* 1992;35:97–101.
13. Bernstein RM, Schluter SF, Bernstein H, et al. Primordial emergence of the recombination activating gene 1 (RAG1): sequence of the complete shark gene indicates homology to microbial integrases. *Proc Natl Acad Sci USA* 1996;93:9454–9459.
14. Zarrin AA, Fong I, Malkin L, et al. Cloning and characterization of the human recombination activating gene 1 (RAG1) and RAG2 promoter regions. *J Immunol* 1997;159:4382–4394.
15. Ichihara Y, Hirai M, Kurosawa Y. Sequence and chromosome assignment to 11p13-p12 of human RAG genes. *Immunol Lett* 1992;33:277–284.
16. Agrawal A, Eastman QM, Schatz DG. Transposition mediated by RAG1 and RAG2 and its implications for the evolution of the immune system. *Nature* 1998;394:744–751.
17. Agrawal A, Schatz DG. RAG1 and RAG2 form a stable postcleavage synaptic complex with DNA containing signal ends in V(D)J recombination. *Cell* 1997;89:43–53.
18. Bosma MA, Custer RP, Bosma MJ. A severe combined immunodeficiency mutation in the mouse. *Nature* 1983;301:527–530.
19. Roth DB, Menetski JP, Nakajima PB, et al. V(D)J recombination: broken DNA molecules with covalently sealed (hairpin) coding ends in scid mouse thymocytes. *Cell* 1992;70:983–991.

20. Biedermann KA, Sun J, Giaccia AJ, et al. *scid* mutation in mice confers hypersensitivity to ioniting radiation and a deficiency in DNA double-strand break repair. *Proc Natl Acad Sci USA* 1991;88:1394–1397.
21. Fullop GM, Phillips RA. The scid mutation in mice causes a general defect in DNA repair. *Nature* 1990;346:479–482.
22. Hendrickson EA, Qin XQ, Bump EA, et al. A link between double-strand break-related repair and V(D)J recombination: the scid mutation. *Proc Natl Acad Sci USA* 1991;88:4061–4065.
23. Pergola F, Zdzienicka MZ, Lieber MR. V(D)J recombination in mammalian cell mutants defective in DNA double-strand break repair. *Mol Cell Biol* 1993;13:3464–3471.
24. Taccioli GE, Rathbun G, Oltz E, et al. Impairment of V(D)J recombination in double-strand break repair mutants. *Science* 1993;260:207–210.
25. Finnie NJ, Gottlieb TM, Blunt T, et al. DNA-dependent protein kinase activity is absent in xrs-6 cells: implications for site-specific recombination and DNA double-strand break repair. *PNAS* 1995;92:320–324.
26. Getts RC, Stamato TD. Absence of a Ku-like DNA end binding activity in the xrs double-strand DNA repair-deficient mutant. *J Biol Chem* 1994;269:15981–15984.
27. Rathmell WK, Chu G. A DNA end-binding factor involved in double-strand break repair and V(D)J recombination. *Mol Cell Biol* 1994;14:4741–4748.
28. Taccioli GE, Gottlieb TM, Blunt T, et al. Ku80: product of the XRCC5 gene and its role in DNA repair and V(D)J recombination. *Science* 1994;265:1442–1445.
29. Boubnov NV, Hall KT, Wills Z, et al. Complementation of the ionizing radiation sensitivity, DNA end binding, and V(D)J recombination defects of double-strand break repair mutants by the p86 Ku autoantigen. *Proc Natl Acad Sci USA* 1995;92:890–894.
30. Smider V, Rathmell WK, Lieber MR, et al. Restoration of X-ray resistance and V(D)J recombination in mutant cells by Ku cDNA. Science 1994;S183:288–291.
31. Weaver DT. What to do at an end: DNA double-strand-break repair. *Trends Genet* 1995;11:388–392.
32. Lee SE, Pulaski CR, He DM, et al. Isolation of mammalian cell mutants that are X-ray sensitive, impaired in DNA double-strand break repair and defective for V(D)J recombination. *Mutat Res* 1995;336:279–291.
33. Gu Y, Jin S, Gao Y, et al. Ku70-deficient embryonic stem cells have increased ionizing radiosensitivity, defective DNA end-binding activity, and inability to support V(D)J recombination. *Proc Natl Acad Sci USA* 1997;94:8076–8081.
34. Kirchgessner CU, Patil CK, Evans JW, et al. DNA-dependent kinase (p350) as a candidate gene for the murine SCID defect. *Science* 1995;267:1178–1183.
35. Blunt T, Finnie NJ, Taccioli GE, et al. Defective DNA-dependent protein kinase activity is linked to V(D)J recombination and DNA repair defects associated with the murine scid mutation. *Cell* 1995;80:813–823.
36. Boubnov NV, Weaver DT. *SCID* cells are deficient in Ku and replication protein A phosphorylation by the DNA-dependent protein kinase. *Mol Cell Biol* 1995;15:5700–5706.
37. Blunt T, Gell D, Fox M, et al. Identification of a nonsense mutation in the carboxyl-terminal region of DNA-dependent protein kinase catalytic subunit in the scid mouse. *Proc Natl Acad Sci USA* 1996;93:10,285–10,290.
38. Danska JS, Holland DP, Mariathasan S, et al. Biochemical and genetic defects in the DNA-dependent protein kinase in murine scid lymphocytes. *Molecular & Cellular Biology* 1996;16:5507–5517.
39. Taccioli GE, Amatucci AG, Beamish HJ, et al. Targeted disruption of the catalytic subunit of the DNA-PK gene in mice confers severe combined immunodeficiency and radiosensitivity. *Immunity* 1998;9:355–366.
40. Gao Y, Chaudhuri J, Zhu C, et al. A targeted DNA-PKcs-null mutation reveals DNA-PK-independent functions for KU in V(D)J recombination. *Immunity* 1998;9:367–376.

41. Wiler R, Leber R, Moore BB, et al. Equine severe combined immunodeficiency—a defect in V(D)J recombination and DNA-dependent protein kinase activity. *Proc Natl Acad Sci USA* 1995;92:11,485–11,489.

42. Meek K, Kienker L, Dallas C, et al. SCID in Jack Russell terriers: a new animal model of DNA-PKcs deficiency. *J Immunol* 2001;167:2142–2150.

43. Allalunis-Turner MJ, Zia PK, Barron GM, et al. Radiation-induced DNA damage and repair in cells of a radiosensitive human malignant glioma cell line. *Radiat Res* 1995;144:288–293.

44. Li Z, Otevrel T, Gao Y, et al. The XRCC4 gene encodes a novel protein involved in DNA double-strand break repair and V(D)J recombination. *Cell* 1995;83:1079_1089.

45. Critchlow SE, Bowater RP, Jackson SP. Mammalian DNA double-strand break repair protein XRCC4 interacts with DNA ligase IV. *Curr Biol* 1997;7:588–598.

46. Grawunder U, Zimmer D, Kulesza P, et al. Requirement for an interaction of XRCC4 with DNA ligase IV for wild-type V(D)J recombination and DNA double-strand break repair in vivo. *J Biol Chem* 1998;273:24,708–24,714.

47. Bryans M, Valenzano MC, Stamato TD. Absence of DNA ligase IV protein in XR-1 cells: evidence for stabilization by XRCC4. *Mutat Res* 1999;433:53–58.

48. Grawunder U, Zimmer D, Leiber MR. DNA ligase IV binds to XRCC4 via a motif located between rather than within its BRCT domains. *Curr Biol* 1998;8:873–876.

49. Junop MS, Modesti M, Guarne A, et al. Crystal structure of the Xrcc4 DNA repair protein and implications for end joining. *EMBO J* 2000;19:5962–5970.

50. Sibanda BL, Critchlow SE, Begun J, et al. Crystal structure of an Xrcc4–DNA ligase IV complex. *Nat Struct Biol* 2001;8:1015–1019.

51. Mombarts P. RAG-1-deficient mice have no mature B and T lymphocytes. *Cell* 1992;68:869–877.

52. Shinkai Y, Rathbun G, Lam KP, et al. RAG-2 deficient mice lack mature lymphocytes owing to inability to initiate V(D)J arrangement. *Cell* 1992;68:855–867.

53. Ferguson DO, Sekiguchi JM, Chang S, et al. The nonhomologous end-joining pathway of DNA repair is required for genomic stability and the suppression of translocations. *Proc Natl Acad Sci USA* 2000;97:6630–6633.

54. Ouyang H, Nussenzweig A, Kurimasa A, et al. Ku70 is required for DNA repair but not for T cell antigen receptor gene recombination in vivo. *J Exp Med* 1997;186:921–929.

55. Li GC, Ouyang H, Li X, et al. Ku70: a candidate tumor suppressor gene for murine T cell lymphoma. *Mol Cell* 1998;2:1–8.

56. Nussenzweig A, Chen CH, Soares VD, et al. Requirement for Ku80 in growth and immunoglobulin V(D)J recombination. *Nature* 1996;382:551–555.

57. Zhu CM, Bogue MA, Lim DS, et al. Ku86-deficient mice exhibit severe combined immunodeficiency and defective processing of V(D)J recombination rntermediates. *Cell* 1996;86:379–389.

58. Jhappan C, Morse HC 3rd, Fleischmann RD, et al. DNA-PKcs: a T-cell tumour suppressor encoded at the mouse scid locus. *Nature Genet* 1997;17:483–486.

59. Gu Y, Seidl KJ, Rathbun GA, Zhu C, et al. Growth retardation and leaky SCID phenotype of Ku70-deficient mice. *Immunity* 1997;7:653–665.

60. Vogel H, Lim DS, Karsenty G, et al. Deletion of Ku86 causes early onset of senescence in mice. *Proc Natl Acad Sci USA* 1999;96:10770–10775.

61. Gao Y, Sun Y, Frank KM, et al. A critical role for DNA end-joining proteins in both lymphogenesis and neurogenesis. *Cell* 1998;95:891–902.

62. Frank KM, Sekiguchi JM, Seidl KJ, et al. Late embryonic lethality and impaired V(D)J recombination in mice lacking DNA ligase IV. *Nature* 1998;396:173–177.

63. Barnes DE, Stamp G, Rosewell I, et al. Targeted disruption of the gene encoding DNA ligase IV leads to lethality in embryonic mice. *Curr Biol* 1998;31:1395–1398.

64. Lieber MR. The biochemistry and biological significance of nonhomologous DNA end joining: an essential repair process in multicellular eukaryotes. *Genes Cells* 1999;4:77–85.

65. Schwarz K, Gauss GH, Ludwig L, et al. Rag mutations in human B cell-negative scid. *Science* 1996;274:97–99.
66. Villa A, Santagata S, Bozzi F, et al. Partial V(D)J recombination activity leads to Omenn syndrome. *Cell* 1998;93:885–896.
67. Corneo B, Moshous D, Callebaut I, et al. Three-dimensional clustering of human RAG2 gene mutations in severe combined immune deficiency. *J Biol Chem* 2000;275:12,672–12,675.
68. Corneo B, Moshous D, Gungor T, et al. Identical mutations in RAG1 or RAG2 genes leading to defective V(D)J recombinase activity can cause either T-B-severe combined immune deficiency or Omenn syndrome. *Blood* 2001;97:2772–2776.
69. de Saint-Basile G, Le Deist F, de Villartay JP, et al. Restricted heterogeneity of T lymphocytes in combined immunodeficiency with hypereosinophilia (Omenn's syndrome). *J Clin Invest* 1991;87:1352–1359.
70. Villa A, Santagata S, Bozzi F, et al. Omenn syndrome: a disorder of Rag1 and Rag2 genes. *J Clin Immunol* 1999;19:87–97.
71. Signorini S, Imberti L, Pirovano S, et al. Intrathymic restriction and peripheral expansion of the T-Cell repertoire in Omenn syndrome. *Blood* 1999;94:3468–3478.
72. Cavazzana-Calvo M, Le Deist F, De Saint Basile G, et al. Increased radiosensitivity of granulocyte macrophage colony-forming units and skin fibroblasts in human autosomal recessive severe combined immunodeficiency. *J Clin Invest* 1993;91:1214–1218.
73. Moshous D, Li L, Chasseval R, et al. A new gene involved in DNA double-strand break repair and V(D)J recombination is located on human chromosome 10p. *Hum Mol Genet* 2000;9:583–588.
74. Li L, Drayna D, Hu D, et al. The gene for severe combined immunodeficiency disease in Athabascan- speaking Native Americans is located on chromosome 10p. *Am J Hum Genet* 1998;62:136–144.
75. Moshous D, Callebaut I, de Chasseval R, et al. Artemis, a novel DNA double-strand break repair/V(D)J recombination protein, is mutated in human severe combined immune deficiency. *Cell* 2001;105:177–186.
76. Dronkert ML, de Wit J, Boeve M, et al. Disruption of mouse SNM1 causes increased sensitivity to the DNA interstrand cross-linking agent mitomycin C. *Mol Cell Biol* 2000;20:4553–4561.
77. Wang Z, Fast W, Valentine AM, et al. Metallo-beta-lactamase: structure and mechanism. *Curr Opin Chem Biol* 1999;3:614–622.
78. Aravind L. An evolutionary classification of the metallo-β-lactamase fold. *In Silico Biology* 1999;2L69–91.
79. Daiyasu H, Osaka K, Ishino Y, et al. Expansion of the zinc metallo-hydrolase family of the beta-lactamase fold. *FEBS Lett* 2001;503:1–6.
80. Varon R, Vissinga C, Platzer M, et al. Nibrin, a novel DNA double-strand break repair protein, is mutated in Nijmegen breakage syndrome. *Cell* 1998;93:467–476.
81. Carney JP, Maser RS, Olivares H, et al. The hMre11/hRad50 protein complex and Nijmegen breakage syndrome: linkage of double-strand break repair to the cellular DNA damage response. *Cell* 1998;93:477–486.
82. Chen HT, Bhandoola A, Difilippantonio MJ, et al. Response to RAG-mediated V(D)J cleavage by NBS1 and gamma-H2AX. *Science* 2000;290:1962–1965.
83. Besmer E, Mansilla-Soto J, Cassard S, et al. Hairpin coding end opening is mediated by RAG1 and RAG2 proteins. *Mol Cell* 1998;2:817–828.
84. Shockett PE, Schatz DG. DNA hairpin opening mediated by the RAG1 and RAG2 proteins. *Mol Cell Biol* 1999;19:4159–4166.
85. Qiu JX, Kale SB, Yarnell Schultz H, et al. Separation-of-function mutants reveal critical roles for RAG2 in both the cleavage and joining steps of V(D)J recombination. *Mol Cell* 2001;7:77–87.

86. Yarnell Schultz H, Landree MA, Qiu JX, et al. Joining-deficient RAG1 mutants block V(D)J recombination in vivo and hairpin opening in vitro. *Mol Cell* 2001;7:65–75.

87. Roth DB, Gellert M. New guardians of the genome. *Nature* 2000;404:823–825.

88. Nicolas N, Moshous D, Cavazzana-Calvo M, et al. A human severe combined immunodeficiency (SCID) condition with increased sensitivity to ionizing radiations and impaired V(D)J rearrangements defines a new DNA recombination/repair deficiency. *J Exp Med* 1998;188:627–634.

89. Ma Y, Pannicke U, Schwarz K, et al. Hairpin opening and overhang processing by an artemis/DNA-dependent protein kinase complex in nonhomologous end joining and V(D)J recombination. *cell* 2002;Published online March 1, 2002.

90. Haber JE. Partners and pathways repairing a double-strand break. *Trends Genet* 2000;16: 259–264.

91. Pierce AJ, Stark JM, Araujo FD, et al. Double-strand breaks and tumorigenesis. *Trends Cell Biol* 2001;11:S52–59.

92. Ferguson DO, Alt FW. DNA double strand break repair and chromosomal translocation: lessons from animal models. *Oncogene* 2001;20:5572–5579.

93. Kinzler KW, Vogelstein B. Cancer-susceptibility genes, gatekeepers and caretakers. *Nature* 1997;386:761–763.

94. Grawunder U, West RB, Lieber MR. Antigen receptor gene rearrangement. *Curr Opin Immunol* 1998;10:172–180.

95. Hiom K, Melek M, Gellert M. DNA transposition by the RAG1 and RAG2 proteins: a possible source of oncogenic translocations. *Cell* 1998;94:463–470.

96. Marculescu R, Le T, Simon P, et al. V(D)J-mediated translocations in lymphoid neoplasms: a functional assessment of genomic instability by cryptic sites. *J Exp Med* 2002;195:85–98.

97. Lewis SM, Agard E, Suh S, et al. Cryptic signals and the fidelity of V(D)J joining. *Mol Cell Biol* 1997;17:3125–3136.

98. Raghavan SC, Kirsch IR, Lieber MR. Analysis of the V(D)J recombination efficiency at lymphoid chromosomal translocation breakpoints. *J Biol Chem* 2001;276:29126–29133.

99. Küppers R, Klein U, Hansmann M-L, et al. Cellular origin of human B-cell lymphomas. *N Engl J Med* 1999;341:1520–1529.

100. Bories JC, Demengeot J, Davidson L, et al. Gene-targeted deletion and replacement mutations of the T-cell receptor beta-chain enhancer: the role of enhancer elements in controlling V(D)J recombination accessibility. *Proc Natl Acad Sci USA* 1996;93:7871–7876.

101. Ferrier P, Krippl B, Blackwell TK, et al. Separate elements control DJ and VDJ rearrangement in a transgenic recombination substrate. *EMBO J* 1990;9:117–125.

102. Lee GS, Neiditch MB, Sinden RR, et al. Targeted Transposition by the V(D)J recombinase. *Mol Cell Biol* 2002;22:2068–2077.

103. Perkins EJ, Nair A, Cowley DO, et al. Sensing of intermediates in V(D)J recombination by ATM. *Genes Dev* 2002;16:159–164.

104. Karanjawala ZE, Grawunder U, Hsieh CL, et al. The nonhomologous DNA end joining pathway is important for chromosome stability in primary fibroblasts. *Curr Biol* 1999;9:1501–1504.

105. Sharpless NE, Ferguson DO, O'Hagan RC, et al. Impaired nonhomologous end-joining provokes soft tissue sarcomas harboring chromosomal translocations, amplifications, and deletions. *Mol Cell* 2001;8:1187–1196.

106. Difilippantonio MJ, Zhu J, Chen HT, et al. DNA repair protein Ku80 suppresses chromosomal aberrations and malignant transformation. *Nature* 2000;404:510–514.

107. d'Adda di Fagagna F, Hande MP, Tong WM, et al. Effects of DNA nonhomologous end-joining factors on telomere length and chromosomal stability in mammalian cells. *Curr Biol* 2001;11:1192–1196.

108. Bailey SM, Meyne J, Chen DJ, et al. DNA double-strand break repair proteins are required to cap the ends of mammalian chromosomes. *Proc Natl Acad Sci USA* 1999;96:14,899–14,904.

109. Hsu HL, Gilley D, Blackburn EH, et al. Ku is associated with the telomere in mammals. *Proc Natl Acad Sci USA* 1999;96:12454–12458.

110. Samper E, Goytisolo FA, Slijepcevic P, et al. Mammalian Ku86 protein prevents telomeric fusions independently of the length of TTAGGG repeats and the G-strand overhang. *EMBO Rep* 2000;1:244–252.

111. Tai YT, Teoh G, Lin B, et al. Ku86 variant expression and function in multiple myeloma cells is associated with increased sensitivity to DNA damage. *J Immunol* 2000;165:6347–6355.

112. Gao Y, Ferguson DO, Xie W, et al. Interplay of p53 and DNA-repair protein XRCC4 in tumorigenesis, genomic stability and development. [in process citation]. *Nature* 2000;404: 897–900.

113. Frank KM, Sharpless NE, Gao Y, et al. DNA ligase IV deficiency in mice leads to defective neurogenesis and embryonic lethality via the p53 pathway. *Mol Cell* 2000;5:993–1002.

114. Lim DS, Vogel H, Willerford DM, et al. Analysis of Ku80-mutant mice and cells with deficient level of p53. *Mol Cell Biol* 2000;20:3772–3780.

115. Vanasse GJ, Halbrook J, Thomas S, et al. Genetic pathway to recurrent chromosome translocations in murine lymphoma involves V(D)J recombinase. *J Clin Invest* 1999;103: 1669–1675.

116. Sekiguchi J, Ferguson DO, Chen HT, et al. Genetic interactions between ATM and the nonhomologous end-joining factors in genomic stability and development. *Proc Natl Acad Sci USA* 2001;98:3243–3248.

117. Shiloh Y. Ataxia–telangiectasia and the Nijmegen breakage syndrome: related disorders but genes apart. *Annu Rev Genet* 1997;31:635–662.

118. O'Driscoll M, Cerosaletti KM, Girard PM, et al. DNA ligase IV mutations identified in patients exhibiting developmental delay and immunodeficiency. *Mol Cell* 2001;8:1175–1185.

13 Inherited Disorders of Genomic Instability and Cancer Susceptibility

Gary M. Kupfer, MD

CONTENTS

1. INTRODUCTION

Clues to carcinogenesis have been elucidated by knowledge of basic cancer genes. Knowledge of the genes implicated in this process has come from the study of relatively obscure genetic diseases, which include cancer predisposition. One hallmark of these diseases is genomic instability. Indeed, it is the characteristic of genomic instability that has given rise to the notion that such instability is a forerunner of cancer and, in fact, is a basic mechanism of cancer. Even more interestingly, the underlying phenotypes of some of these diseases involve hypersensitivity to DNA damage, which implies that the molecules under normal circumstances in their wild-type state could prove to be targets of modern molecular chemotherapy. Ironically, the very molecular basis of this hypersensitivity can, in the end, be a cause of acquisition of clinical resistance.

Furthermore, the increase in mutation acquisition is the basis of genomic instability, whereby cells attain what is termed "the mutator phenotype." From an evolutionary perspective, whereas the loss of most genetic material may, in fact, be deleterious, the pressure to select for a favorable change is, indeed, the forerunner of clonal expansion.

Finally, the gain of mutations may indeed make the acquisition of resistance an even more likely event. Given that most chemotherapy targets DNA

From: *Cancer Drug Discovery and Development: DNA Repair in Cancer Therapy*
Edited by: L. C. Panasci and M. A. Alaoui-Jamali © Humana Press Inc., Totowa, NJ

Table 1
Disorders of Gross Genomic Instability

Disorder	Inheritance	Gene product
Li–Fraumeni syndrome	AD	p53
Mismatch repair deficiency		
HNPCC	AD	hMSH2, hMLH1, hPMS2, hMSH6
Turcot's		hMLH1, hPMS2
Muir–Torre		hMSH2, hMLH1
Familial breast cancer	AD	BRCA1, BRCA2
Fanconi anemia	AR	FANCA, -B, -C, -D1 (=BRCA2), -D2, -E, -F, -G
Ataxia–telangiectasia	AR	ATM
Helicase disorders		
Bloom's	AR	BLM
Werner's	AR	WRN
Rothmun–Thomsun	AR	RECQL4
Nijmegen breakage syndrome	AR	nibrin
AT variant	AR	MRE11

chemically, the loss of the ability to detect damage and arrest in the cell cycle or repair or activate apoptosis can lead to cell tolerance of DNA damage.

The hypersensitivity phenotype *per se* is reminiscent of more than the syndromes discussed in this chapter. The collection of disorders that include xeroderma pigmentosum involve disorders that display hypersensitivity but that do not necessarily exhibit gross genomic instability. For this discussion, we restrict ourselves to gross genomic instability as a phenomenon that may at times cause growth disadvantage in the "normal cell" but that generates genetic loss leading to growth advantage, enabling the malignant cell to make the transition from hypersensitive to resistant cell (Table 1).

2. DOMINANT SYNDROMES OF GENOMIC INSTABILITY

2.1. Li–Fraumeni Syndrome

Some of the first well-described "cancer families" became part of the pedigree that helped identify *p53* as the aberrant gene product that explains the Li–Fraumeni syndrome (LFS) *(1)*. Inheritance of the *p53* gene predisposes these families to rhabdomyosarcoma, soft tissue sarcoma, osteosarcoma, early breast cancer, brain tumors, adrenal cortex tumors, and acute leukemia. The clinical criteria used to define those likely affected originally were a sarcoma of bone or soft tissue in an under 45-yr-old coupled with similarly affected relatives. Affected people in LFS families experience a 50% risk of cancer by age 30.

p53 is the most commonly mutated gene in human cancer. Because of the cancers implicated by *p53* mutations fall in the LFS spectrum, it was a natural extension to search for *p53* mutations in affected families. Most cases of LFS are indeed caused by heritable germline mutations in *p53* producing an aberrant protein with either no or abnormal function. A majority are missense mutations, suggesting that at least some mutants act in a dominant-negative fashion *(2–4)*.

2.1.1. p53 Function

p53 is known to have many functions, more fully reviewed elsewhere, involving transcriptional activation and repression, modulation of DNA repair, cell cycle, and apoptosis. p53 is a phosphoprotein targeted by ATM and chk2, among others. p53 acts via an oligomerization domain into a tetramer, a function that has implications for effects by mutations. p53 is also modified via acetylation by coactivators, which enhances its role as a transcription factor. It has as its target of activation the cyclin-dependent kinase inhibitor p21, among many other genes. In addition, p53 is linked to DNA-damage checkpoints at G_1–S, G_2, and the mitotic spindle, presumably through its induction of p21 *(5–8)*.

p53 is directly modulated by DNA damage through interaction with ATM, WRN, and chk2 *(9,10)*. In addition, p53 becomes uncoupled from mdm2, which is responsible for targeting p53 for ubiquitin-linked proteolysis *(11–13)*. The end result is generally either arrest at G_1–S or induction of apoptosis *(14)*.

DNA damage itself appears to be the signal for G_1–S arrest, and p53 is stabilized and acts as a transcriptional activator for p21, which may also play a role in G_2/M arrest as well *(8,15)*. Other p53-inducible genes, including gadd45, mdm2, and bax, play roles in apoptosis and cell cycle arrest *(16)*.

2.1.2. p53 and DNA Repair

p53 has been implicated directly in DNA repair in addition to the indirect role by virtue of its involvement in cell cycle arrest and triggering of apoptosis. For example, p53 has been shown to be coupled with nucleotide excision repair and is phosphorylated by cyclin-activating kinase, a component of two human transcription factor IIH (TFIIH). The interaction of p53 with damaged DNA may be regulated in a dynamic fashion with replication protein A (RPA) *(17)*. Knockout of p53 function by E6 also results in ultraviolet (UV) sensitivity *(18)*.

Mutant p53 is implicated in a majority of all human cancers *(19–21)*. These mutants take the form of knockout mutations as well as those in which p53 is stabilized and/or able to form nonfunctional oligomers.

2.1.3. LFS Cells

Cells derived from LFS patients exhibit chromosomal aberrations in response to DNA damage. For the most part, LFS cells display radiosensitivity, although some reports have detailed radioresistance *(2,22)*. Certainly, by the time p53 mutant cells are analyzed in the setting of tumors, significant radioresistance is seen. Although a transient arrest in G_1–S can be seen, LFS cells fail to permanently arrest in G_1–S as in normal cells, which is rescued by provision of wild-type p53 *(23)*. G_2 arrest is also abrogated by mutant p53, which allows the propagation of genetic damage to be passed onto daughter cells *(24)*.

2.1.4. CLINICAL IMPLICATIONS OF P53

Mutation or loss of p53 is the most important means of resistance in human oncology and is clearly associated with poor clinical outcome *(25–26)*. p53 appears to be an inhibitor of transcription of the *mdr* gene, and thus loss of p53 leads to mdr expression *(27)*. Also, p53 gain-of-function mutants can activate mdr in combination with ets *(28)*. This is exemplified by the fact that pediatric tumors are generally more chemoresponsive and exhibit fewer p53 mutations than adult tumors *(25,29,30)*.

Modern molecular techniques are under investigation to both restore wild-type p53 or to chemosensitize cells lacking wild-type p53. Adenoviral transduction of p53 into null cells restores apoptosis and chemosensitivity *(31)*. In addition, drug development is underway to turn on p53 target genes and inhibit p53 mutant cells *(32)*. However, chemotherapy is still a necessity to effect cell killing; thus, restoring chemosensitivity does not help reduce side effects suffered by normal tissue. Chemical targeting of wild-type p53 to inhibit its normal action is one recent proposed strategy to spare normal tissue from chemotherapy-induced toxicity *(33)*.

2.2. Breast Cancer Genes (BRCA1 and BRCA2)

Familial breast cancer accounts for at least 10% of breast cancer cases. Because a lifetime risk of breast cancer in women increases to 25% in extreme old age, analysis of potential breast cancer genes considers the rarer occurrence of breast cancer in younger women. *BRCA1* and *BRCA2* germline mutations follow the two-hit hypothesis model in which susceptible women contain one mutant allele but whose tumor has the other normal allele deleted *(34–36)*.

The recent characterization of BRCA1 binding with Fanconi anemia proteins and of BRCA2 as the Fanconi group D1 protein (*see* Subheading 3.1.) firmly establishes familial breast cancer as a genomic instability syndrome. Cells derived from knockout mice show propensity to the development of chromosomal breakage with radials, deletions, and translocations. These findings are recapitulated in tumors mutant for BRCA1 or BRCA2 *(37)*.

2.2.1. BRCA1 AND BRCA2 PROTEINS

BRCA1 and BRCA2 are large, ubiquitously nuclear proteins, which are maximally expressed during S–G_2. BRCA1 is an 1863-amino-acid protein and BRCA2 is a 3418-amino-acid protein. Both are nuclear proteins with nuclear localization signals and contain domains involving protein–protein interactions and which contain transactivation activity. Additionally, motifs occur that are consistent with other DNA-binding and cell cycle proteins, although BRCA proteins contain no obvious functional motifs, no resemblance to other proteins, or have no lower eukaryotic homologs. Localization is aberrantly cytoplasmic in tumor samples, implying the lack of ability to act normally in the nucleus *(38–41)*.

However, BRCA proteins do exhibit several important actions that give some indication as to function. BRCA1, phosphorylated by atm, atr, and chk2, activates chk1, the kinase that responds to DNA damage at a G_2–M checkpoint *(13, 42, 43)*. Further, BRCA1 appears to have chromatin remodeling activity and has associations with several histone-modifying enzymes *(44,45)*. Several mutations associated with cancers are at these contact points, suggesting the importance of this function in carcinogenesis.

Most intriguingly, BRCA1 has been linked to an associated supercomplex, dubbed BASC *(45,46)*. Within this protein complex mismatch repair (MMR) proteins, replication factors, Nijmegen repair protein, atm, RAD51, and helicases have been detected. Furthermore, FANCD2, a recently cloned Fanconi anemia (FA) complementation group gene, was found to bind to BRCA1 *(47)*. All of these proteins involved different aspects of DNA repair, as these proteins are involved in nonhomologous end-joining, homologous recombination, and mismatch repair. BRCA1 binds directly to DNA without apparent specificity and facilitates homology-driven repair *(48)*. Restoration of mutations of BRCA1 results in correction of DNA crosslinker and radiation sensitivity and chromosomal stability. BRCA1 and BRCA2 stably interact in somatic and germline cells as well *(38)*. In addition, mutations in the *BRCA2* gene have been shown to account for the FA-D1 complementation group of FA *(49)*.

2.2.2. Clinical Implications of BRCA1 and BRCA2

Cells that contain *BRCA1* mutations can be sensitive to DNA-damaging agents *(50,51)*. Experimentally, provision of the complementing cDNA will correct this phenotype. However, in a similar fashion to p53, an accumulation of genetic loss over time either induced by cell stress from chemotherapy or from normal cellular activity theoretically leads to resistance to DNA-damaging agents. Indeed, BRCA1 and BRCA2 appear linked to p53, as the phenotype in knockout mice is predicated on p53 inactivation *(52)*. Thus the very nature of the molecular defect itself engenders gain of function through genetic loss via evolutionary pressure.

2.3. Mismatch Repair

2.3.1. Hereditary Nonpolyposis Colorectal Cancer

Hereditary nonpolyposis Colorectal Cancer (HNPCC) patients exhibit a familial association of colorectal or endometrial cancers, along with associated gastric, ovarian, renal, brain, biliary, and skin cancers. Colon cancers in this population demonstrate proximal colon location, multifocality, better survival, and poor mucinous differentiation *(53–55)*.

The inevitable process of accumulation of mispairing in DNA replication occurs because of the innate error rate. Especially, certain stretches of DNA are particularly prone to errors, such as tandem repeats, which are otherwise known as microsatellite regions, thus leading to microsatellite instability. These regions can accumulate mutations by way of slippage during replication, leading to nucleotide mismatches. In addition, MMR has been implicated in the regulation of recombination intermediates *(53–57)*.

The MMR proteins MSH2 and MSH6 appear to recognize a mismatch followed by binding of MLH1 and PMS1 *(58)*. This complex then recruits the machinery for the downstream correction of the mismatch. Nicking, excision, synthesis, and ligation then ensue. Defects in MMR lead to microsatellite instability and hypermutability, probably as a direct result of replication slippage and lack of surveillance *(59,60)*.

The prototype disease attributable to MMR defects is the HNPCC syndrome. HNPCC syndrome is an autosomal dominant disorder. It accounts for up 5% of all colon cancers and is associated with other cancers, including endometrial, ovarian, stomach, pancreatic, and genitourinary cancers. These data imply a carrier frequency as high as 0.5%. Early onset is the norm, with cancers arising by age 50. About 30–40% of HNPCC syndrome family members will develop other cancers *(55,61)*.

The first clues for the proteins involved in the biochemistry of MMR came from the observation that the predominant mutations were simple repeats, resembling microsatellites *(62,63)*. The MMR genes were first isolated in yeast and then humans when it became apparent that similar microsatellite mutants existed in the yeast mutants as well. The MMR genes *MSH2*, *MLH1*, *MSH6*, *PMS2*, and *PMS1* are responsible for HNPCC. The protein products of these genes participate in a protein complex, which serves as a recognition factor for mispairing and which stimulates downstream effects of repair *(64,65)*. *MSH2* and *MLH1* account for most cases *(56,60,66–69)*. HNPCC evolves in a classic two-hit process whereby the first hit is a germline mutation followed by the second arising from a somatic lesion. This differs from the sporadic inactivation of MMR genes in tumors in which methylation of promoters appears to be the predominant mode of inactivation *(70–72)*.

As described, microsatellite instability appears to be a primary feature of cells defective in MMR *(73)*. Cells containing germline defects in MMR derived from patients display at least two orders of magnitude increase at specific loci. Human knockout cell lines undergo frame shifts of up to 1000-fold increased frequency, which is corrected by insertion of the respective wild-type MMR gene *(74,75)*. Mouse knockout experiments reveal mutated microsatellite loci in all the MMR gene mutants *(76,77)*. Tumors from HNPCC patients contain more than 90% incidence of mutations at microsatellite markers, and these markers contain mutations at 15% of sporadic tumors. This most likely results from the lack of ability to detect microsatellite formation after replication-induced slippage. Genes containing an inordinate amount of repeats may be affected selectively *(78,79)*.

Interestingly, MMR genes appear to engender specific loci mutations. For example, p53 in colon cancer often acquires multiple mutations in the form of base substitutions and frame shifts. Other affected genetic loci include APC and k-ras *(80)*. Teleologically, these mutations make sense because they affect genes that, when knocked out or activated, lead to growth advantage by decreased cell death or increased cell proliferation.

2.3.2. Muir–Torre Syndrome

Muir–Torre Syndrome (MTS) is a subset of HNPCC that reportedly encompasses mutations in *hMSH2* or *hMLH1*. Patients go on to develop sebaceous gland tumors but appear to have an overall good prognosis *(81–84)*.

2.3.3. Turcot's Syndrome

This is another subset in HNPCC that has overlap with familial adenomatous polyposis (FAP), another familial colon cancer syndrome with mutations at the APC locus. Mutations in Turcot's have been found indeed at APC as well as *MLH* and *PMS2*. Turcot's patients have been reported to develop brain tumors, including medulloblastoma and glioma *(85–89)*.

2.3.4. Clinical Implications of HNPCC

Given the 5% incidence of HNPCC in all colon cancers, HNPCC represents an important factor in the assessment of outcome for patients. In a disease with a lifetime risk of up to 80% of developing colon cancer, it is critical to decide if premalignant interventions should be made. An intervention such as partial colectomy would have to be justified in a careful analysis of the natural history of HNPCC *(71,90)*.

In addition, HNPCC tumors have been especially difficult to treat because of the inherent resistance displayed by cells defective for MMR. Paradoxically, the deficiency in repair ability results in resistance to certain chemotherapeutic agent damage). Indeed, alkylating agents that methylate guanine (O^6) depend on MMR for toxicity. Cells mutated for MMR genes have sensitivity to drugs such as cisplatin restored by provision of the normal cDNAs *(92,93)*.

The resulting death appears to arise by several potential mechanisms. First, attempted repair of the O^6-guanine results in maintenance of a mismatch in the complementary strand. Replication forks are thus impeded and a DNA strand break ensues. MMR has also been postulated to play a role in other repair systems, such as nucleotide excision repair, especially in transcription-linked repair, in the role of sensing. MMR proteins may play a role in resolution of recombination intermediates and the repair of double-strand breaks. MMR also may interact with cell cycle checkpoint machinery, especially at the G_2/M checkpoint. Arrest at G_2/M inducible by alkylating agents or by radiation fails to occur in MMR-deficient cells, but MMR-deficient cells are uniquely resistant to alkylating agents, topoisomerase poisons, and other minor groove-binding agents *(94–96)*. This implies that the fixation of DNA damage that leads to apoptosis fails to occur because of the lack of the checkpoint. Ionizing radiation, which can also lead to a G_2/M checkpoint, does not cause differential cytotoxicity in MMR-deficient cells, suggesting that the checkpoint induced is qualitatively different. Alternatively, the ability to engage a G_2 checkpoint that results in cell death may be mediated by MMR proteins. G_2 arrest leading to apoptosis, typical after DNA damage with alkylators, fails to occur in resistant cells. When overexpressed, the MMR proteins lead to the induction of apoptosis.

Resistance to cisplatin has also been described in MMR-deficient cells *(94–96)*. This implies that a sensor function of the MMR proteins may work because of the relative torsion that alkylation and DNA crosslinking provide. Consistent with this is the fact that MMR-deficient cells exhibit no difference in ionizing radiation sensitivity. Contrastingly, other crosslinkers result in hypersensitivity in MMR-deficient cells. That this checkpoint engagement may lead not only to repair but also to death is a theme reiterated in the p53-deficiency phenotype. In any case, the lack of MMR proteins may paradoxically lead to engagement of other repair systems, as evidenced by reports of reactivation of a reporter crosslinked with cisplatin *(98)*.

3. AUTOSOMAL RECESSIVE SYNDROMES

3.1. Fanconi Anemia

Fanconi anemia (FA) is a rare disorder with a frequency of 1 in 300,000, with features of congenital skeletal abnormalities, bone marrow failure, and propensity to malignancy. FA has traditionally been a classic disease marked by the description of patients, but in 2004, we could define it more on a molecular and biochemical basis, as seven genes have been cloned which when mutant lead to disease *(99–102)*.

The clinical features of FA are numerous and heterogeneous, making the diagnosis a challenge *(100,101,103,104)*. Classically, FA patients exhibit bony abnormalities, growth retardation, and café au lait spots. In addition, gonadal failure can occur that gives some implication to the idea that FA genes are involved in development and gamete production. Perhaps most striking, probably at least a third of all FA patients exhibit none of the classic features of FA but look completely normal. This implies the possibility that more patients with FA exist and that the genes affected may consequently play a larger role in cancer *(105–107)*.

Fanconi anemia patients develop predominately myeloid malignancies, although numerous other cancers arise, including squamous cell carcinomas of the reproductive tract, skin, and gastrointestinal system. Liver tumors have been described as well, although these have been associated with use of androgens for treatment of bone marrow failure, as has been generally noted in other bone marrow failure syndromes *(100,101,103,104,108)*.

3.1.1. BIOLOGY OF FA

The biology of FA is inherent in the description of how cells respond to DNA crosslinking agents. The first clue to FA biology came from the observation that patients treated for leukemia suffer inordinate degrees of toxicity from treatment with chemotherapy. This toxicity is recapitulated on a cellular level because cells derived from patients are hypersensitive as well. This hypersensitivity has been captured in a cell-based assay used as a diagnostic tool *(109–111)*. Hypersensitivity of FA cells appears to be specifically restricted to bifunctional alkylating agents, as matched, isogenic pairs of mutant and complemented cells have been used to demonstrate. Drugs that crosslink DNA include cisplatin, cyclophosphamide, diepoxybutane, mitomycin C (MMC), and nitrogen mustards. These agents, although chemically diverse, are unique in their ability to form adducts with two adjacent guanine residues on DNA, leading to both intrastrand and interstrand crosslinks. Although it is unclear which type of lesion is the lethal one, what is clear is that monofunctional agents do not confer hypersensitivity in FA cells *(112–114)*.

Biochemical data in the past concerning FA have been extremely lacking. However, with the cloning of several genes, the field has exploded. First, the FA complex has been described in which FANCA, FANCC, FANCG, and FANCF have been found to bind together *(115–120)*. In fact, in all of the complementation groups except FA-D2, the complex fails to form, indicating that it is important in FA pathogenesis. Immunoprecipitation–immunoblotting has confirmed most of the interactions. Recent two-hybrid data have also confirmed these data.

Most recently, the *FANCD2* gene has been cloned, and this development has been the most exciting to date *(121)*. Uniquely, the FANCD2 mutant cells contain an intact FA nuclear complex, but the FANCD2 is not a binding member of the complex *(47,122)*. However, the FANCD2 protein is nonetheless dependent on the complex, as it becomes modified by a monoubiquitin upon DNA damage and the during S-phase only in the presence of the wild-type FA complex. Furthermore, the FANCD2 protein binds and colocalizes with BRCA1 and is phosphorylated by ATM kinase. These findings are especially interesting in light of the fact that *BRCA1* associates in a supercomplex with numerous DNA repair proteins *(47,102,121,123)*. The report that BRCA2 is the FA-D1 complementation group gene places FA squarely in the DNA-damage response pathway *(102)*.

3.1.2. CLINICAL IMPLICATIONS OF FA

The hypersensitivity of FA cells was first evident by the observation of Fanconi anemia patients exhibiting extreme hypersensitivity to DNA crosslinking agents. As a result, children undergoing chemotherapy suffered inordinate amounts of toxicity. Currently, patients with FA are initially provided supportive therapy with blood banking, hematopoietic growth factors, and androgens. Close monitoring is performed via bone marrow aspiration in order to detect changes indicative of myelodysplasia. Numerous investigators now promote the early performance of bone marrow transplantation if a related, matched donor is available. Once frank myelodysplasia has occurred, ultimate cure with the use of transplant is reduced. Transplant in FA patients has its own set of issues, as FA patients exhibit a high degree of graft failure and toxicity *(124)*. Because of the toxicity, providing enough conditioning has been a real issue, and graft failure has been a significant one. Use of agents such as ATG, fludaribine, and reduced doses of conditioning have lowered the percentage of graft failures as well as procedure-related toxicity in recent years *(93,125,126)*. Graft-vs-host disease has also been ameliorated by use of cord blood transplantation. In vitro fertilization has been used in conjunction with selection for human leukocyte antigen (HLA) match and selection against affected FA embryos *(127)*.

FA has also been proposed as the prototypical disease for gene therapy. Efforts toward this end are underway, using viral vectors to infect CD34+ cells *(128–130)*. FA exhibits somatic reversion, whereby recombination events favoring correction of the FA phenotype can occur, indicating that FA cells can be prone to survival advantage to help with the acquisition of a corrective gene *(131–135)*.

3.2. Ataxia–Telangiectasia

Ataxia–telangiectasia (AT) is an autosomal recessive disorder affecting upward of 1 per 40,000–100,000. Its name describes the disease: a syndrome of neurologic sequelae and blood vessel malformations. Patients eventually develop marked and progressive ataxia, dysarthria, and choreoathetosis. Immunodeficiency is also part of the syndrome, with deficiencies reported in immunoglobulin subclasses and T-cell function. Patients develop sinus and respiratory infections as a result, which are exacerbated by the inactivity caused by decreasing muscle strength resulting from diminished reflexes. Most dramatically, patients with AT contend with a 30–40% lifetime risk of cancer, most commonly lymphoid malignancy. Heterozygosity may be an issue, as some conflicting research have hinted at an increased risk of cancer in carriers of AT *(136–139)*.

Chromosomes from AT patients show increased breakage, restricted to specific ones in lymphoblasts but random in fibroblasts. Similar to FA, the hallmark of AT appears to be a relationship to ionizing radiation sensitivity and some degree of chromosomal instability *(140–143)*. Genomic instability is characteristic of AT cells, and spontaneous chromosomal breakage occurs in cells derived from AT patients. However, AT cells maintain normal ability to undergo DNA repair, indeed displaying radioresistant DNA synthesis, implying that a signaling pathway may be aberrant. Also, some investigators have noted a qualitative defect in the fidelity of the repair process, suggesting that a subset of the damage may be critical *(145,146)*.

Focus, however, has remained on checkpoint interaction, as biochemical analysis of AT cells show that p53-related response to DNA damage is impaired, placing the AT defect upstream of p53 *(146,147)*, although p53–/– cell lines are not radiation sensitive.

3.2.1. The AT Gene Product

As in FA, AT was thought initially to composed of multiple complementation groups, but subsequent cloning of the atm kinase revealed that all groups contained mutations in atm. Positional cloning efforts yielded the cloning of the *ATM* (ataxia telangiectasia mutated) gene, located at chromosome 11q22–23, which encodes a 350-kDa protein *(140,148,149)*. The ATM protein is homologous to a phosphatidylinositol kinase motif, which is found in numerous other DNA damage proteins, such as DNA-PK (protein kinase) *(150–153)*.

As a kinase, ATM has been found to target several downstream effector proteins, which places it in the vanguard of the DNA-damage response, including p53, c-abl, chk2, and RPA *(154)*. Even more striking is the association of the ATM protein in the BRCA1 supercomplex, again placing a striking genetic-disease-associated protein in the hub of the cell response to DNA damage *(45,46)*. Most recently, the FANCD2 protein, part of the FA epistasis pathway, has also been shown to be phosphorylated by atm, further linking two distinct cancer susceptibility syndromes *(123)*.

In addition, the ATM protein functions in both cytoplasm and nucleus. In the nucleus, ATM appears to function as part of DNA-damage surveillance machinery, which activates checkpoints via p53. In the cytoplasm, ATM activates NFKβ in response to reactive intermediates created by radiation and other sources of double-strand breakage, many of which are very potent in formation of free oxygen radicals. ATM also appears to phosphorylate IKβ, the target of nuclear factor κ-β (NFκβ) *(155)*.

Mouse knockouts have recapitulated the human clinical findings *(156,157)*. Interestingly, AT knockouts have decreased fertility, as in FA, and this mirrors the ovarian hypoplasia and sperm production in humans. The atm protein has been linked to synaptonemal complexes on meiotic chromosomes, making a potential role in homologous recombination even stronger *(158)*.

AT cells lack the G_1–S checkpoint, indicating that ATM signals through p53. Induction of p53, p21, and GADD45 in response to ionizing radiation is abnormal

(159). However, the paradox that AT cells are hypersensitive and p53 knockout cells are resistant remains. Recent evidence has firmly placed ATM in the classification of a DNA binding protein *(160)*.

3.2.2. CLINICAL IMPLICATIONS OF AT

AT patients who receive ionizing therapy and other chemotherapy double-strand-break inducers generally undergo a greater degree of clinical toxicity, but successful therapy can be designed *(142,161–163)*. AT heterozygotes have been shown to display intermediate sensitivity *(164)*. Because of the notion that AT heterozygotes have increased cancer risk *(165,166)*, the association between increased toxicity and heterozygosity has been postulated. AT has also been implicated as a target therapeutically. Wortmannin, an ATM inhibitor, sensitizes proliferating cells to radiation by inhibiting ATM kinase activity *(167,168)*.

3.3. Helicase Disorders

In the realm of resolution of DNA damage is the inevitable disentanglement of DNA strands to allow access of DNA repair enzymes. DNA helicases are energy-requiring enzymes involved in a myriad of processes, including replication, transcription, repair, and recombination. In *Escherichia coli*, the RecQ family of helicases is involved in recombination. Mutants display illegitimate recombination. Five human RecQ homologs have been discovered, and three of them are involved in autosomal recessive disorders characterized by premature aging, genomic instability, and cancer predisposition *(169–171)*.

3.3.1. BLOOM'S SYNDROME

Bloom's syndrome (BS) is a rare autosomal recessive syndrome of growth retardation, telangiectasia manifest by facial erythema, immunodeficiency, and skull abnormalities. BS patients also are predisposed to cancer, as they develop mostly leukemias and lymphomas in about half of the patients. Miscellaneous solid tumors account for the remainder and show up in older BS patients. Interestingly, BS patients also are relatively infertile and have reduced germ cell production. BS patients are immunodeficient and prone to respiratory infections *(169,170,172)*.

As is the case of FA, BS patients are unique in their hypersensitivity to a wide array of DNA-damaging agents. Spontaneous chromosomal breakage is also a feature of BS, exhibiting similar forms to that seen in FA. This phenomenon has been mostly demonstrated by increased rates of quadriradial formation and sister chromatid exchange, which is diagnostic of BS. However, this phenomenon may not be associated with a hypersensitivity phenotype, as BS cells are not universally hypersensitive to various forms of DNA damage. However, many BS patients come to attention by virtue of excessive toxicity to chemotherapy during cancer therapy. Many BS patients show characteristic facial erythema during the first summer of life, and cells from patients may be hypersensitive to ionizing radiation and ultraviolet light. In spite of the hypersensitivity that has been demonstrated, in vitro DNA repair defects have not been shown *(170,172–174)*.

3.3.1.1. BLM Gene Product. The gene for BS was cloned in the classic positional cloning approach after its locus was confined by microcell-mediated chromosome transfer to approximate it to chromosome 15 *(175–177)*. The *BLM* gene encodes a 160-kDa protein that resembles a DEAD box helicase that has homologs in yeast (sgs1) and in other DNA repair disorders, such as xeroderma pigmentosum (XPB, XPD, and ERCC6).

Curiously, the BLM, as the FANCD2 and ATM proteins, interacts with the BRCA1 supercomplex, suggesting that the BLM protein is part of the master regulatory or sensor response to DNA damage *(45,46)*. However, the BLM protein has been thought to act in a p53- and ATM-independent fashion at the G_2 phase of the cell cycle *(178)*. In response to DNA damage, BLM enters discrete foci in promyelocytic leukemia bodies along with RAD51 and RPA. *BLM* also has been associated with a function in nonhomologous end-joining, as evidenced by its importance with proper localization of RAD50, MRE11, and NBS1 *(179,180)*.

As a helicase. the BLM protein resembles the RecQ family of *E. coli* and contains seven helicase motifs *(172)*. As a RecQ helicase, BLM has perhaps greater logic in proposed mechanisms of action in regulation of recombination, but its chief role in genomic stability maintenance remains unclear. It may have a normal role in unwinding and separating DNA at replication, mitosis, and during homologous recombination or to allow relaxation of DNA locally to facilitate the access of DNA repair machinery *(181)*. BLM has been shown to unwind many different DNA structures, which is stimulated by its interaction with RPA *(182)*. The RPA/ MSH4/blm/topoisomerase complex may act as a negative regulator of recombination preventing supranormal recombinatory events *(183)*.

BLM may couple replication to repair as well. Interactions with topoisomerase III results in the ability to unwind double-stranded DNA, suggestive of a role in the resolution of replication-fork intermediates *(184)*. Also, depletion of BLM from Xenopus results in replication inhibition.

3.3.1.2. Clinical Complications of BS. Often the only way BS patients are diagnosed is by recognition of a patient with inordinate amounts of toxicity after the onset of cancer. Dose reduction is often necessary to allow for the possibility of adequate treatment *(185–187)*. Propagation of resistance, such as to 6TG, can also be evident in BS cells *(188)*. Unusual tumors in the pediatric setting can also be suggestive of BS *(189)*.

3.3.2. Werner's Syndrome

Werner's syndrome (WS) is a rare autosomal recessive disorder with a combination of an aging phenotype and cancer susceptibility *(171,190)*. WS cells display a mutator phenotype with chromosome deletions and rearrangements. WS patients fall within two subcategories, one of which mimics aging and the other not. The age-mimicking WS exhibits premature graying and hair thinning, decreased skin elasticity, cataracts, osteoporosis, atherosclerosis, and non-insulin-dependent diabetes mellitus. The nonaging features can include loss of subcutaneous fat and skin ulcers. Often, WS patients are not evident until adolescence, when the growth spurt fails to occur and WS patients are also cancer-prone, with the predominant cancer being sarcoma *(191)*.

Although cells from WS patients display genomic instability and poor growth capacity, they do not exhibit outright hypersensitivity to DNA damage. WS cell lines show shortened life-spans and grow more slowly. Although cells are specifically hypersensitive to certain agents such as nitroquinolones, they are not universally so to a wide array of others, such as radiation *(192,193)*. A mutator phenotype is evident by generation of deletion mutants and resistance in the presence of genotoxic agents *(194)*.

3.3.2.1. WRN Gene Product. *WRN* was first positionally cloned after localization to chromosome 8p12-11. WRN is a 1432-amino-acid protein with similarity to RecQ DNA helicases, including BLM *(195–197)*. Mutations are generally truncations as a result of nonsense, insertion, deletion, and frame-shift mutants, resulting in erroneously localized proteins or in helicase domain deletions *(198,199)*. Generally, these mutants result in mislocalized, helicase+ proteins *(192,200)*. WRN appears to localize to nucleoli preferentially as well as non-promyelocytic leukemia (PML) body nuclear foci *(201,202)*.

WRN plays a direct role in p53 biology by direct interaction with p53, and WS cells display diminished p53-dependent activity, such as apoptosis and transcriptional activity *(203–205)*. Evidence exists that the p53–WRN complex acts at replication forks to coordinate the cell cycle checkpoints in which p53 clearly is involved as well as in DNA repair and replication *(206,207)*. WRN has been purported to have proofreading exonuclease activity as well *(196,208,209)*. The DNA-end binding proteins Ku70 and Ku86 also interact with WRN *(210)*. More dramatic is the interaction with DNA polymerase δ in the absence of proliferating cell nuclear antigen (PCNA), which indicates special DNA synthesis at sites of damage *(204,211–213)*. Nuclear foci clearly form in response to DNA damage *(207)*.

3.3.3. Rothmund–Thomson Syndrome

Rothmund–Thomson syndrome (RTS) is characterized by skin and skeletal abnormalities, cataracts, premature aging, and cancer susceptibility, especially osteosarcoma and myelodysplasia *(169,214)*. As one of the premature aging syndromes, the family of *E. coli* RECQ helicases has been examined for mutations. *RECQL4* helicase was cloned and mutations have been found to account for RTS, which segregate among affected probands and involve the helicase domain *(215)*.

Rothmund–Thomson syndrome patients exhibit photosensitivity with poikiloderma along with early graying and hair loss. Osteogenic sarcoma patients with any of the RTS-associated findings should be considered to have RTS unless proven otherwise *(169,214)*.

3.4. Nijmegen Breakage Syndrome

Nijmegen breakage syndrome (NBS) is another in the set of rare radiation-hypersensitivity syndromes. NBS is an autosomal recessive disorder with a range of phenotype that overlaps with AT *(169,216,217)*. Indeed, NBS has been termed a variant of AT. Clinical features include growth retardation, craniofacial abnormalities, skin findings such as vitiligo and café au lait spots, skeletal defects, and propensity to infection. Radiation hypersensitivity is a hallmark of the disease, along with a predisposition to cancer, most notably lymphomas. Immunodeficiency is a major aspect in NBS patients, with consequent infections.

Chromosome rearrangements are found in NBS cells, specifically in a subset of chromosomes in lymphoid cells at sites of immunoglobulin and T-cell receptors *(218)*. Hypersensitivity of patients to ionizing radiation is recapitulated in cells derived from patients, who also display radioresistant DNA synthesis that occurs upon failure to arrest at G_1–S *(137,219)*. As is the case for so many other genomic instability disorders, cells from NBS patients display poor growth as well as radiation hypersensitivity. NBS cells resemble ATM cells in the failure to arrest appropriately at G_1–S in response to radiation concomitantly with absence of p53 induction.

The *NBS* gene encodes a 95-kDa protein that binds with MRE11 and RAD50 to form a nuclease-containing protein complex that appears to be involved in homologous and nonhomologous recombination. Following a similar theme, both NBS, RAD50, and MRE11 can be found in the BRCA1 supercomplex *(45,46)*, and a mutation in mre11 can mimic the AT phenotype *(220)*. NBS, RAD50, and MRE11 also form foci as part of the radiation response and serves to maintain genomic stability at telomeres *(221,222)*.

4. IMPLICATIONS FOR THERAPY AND CONCLUSIONS

Genetic disorders of genetic instability have shed enormous light upon cancer biology in general and upon therapy-related issues in particular. For example, much of the biology of p53 and its implications for cancer and drug resistance comes from the study of the Li–Fraumeni syndrome.

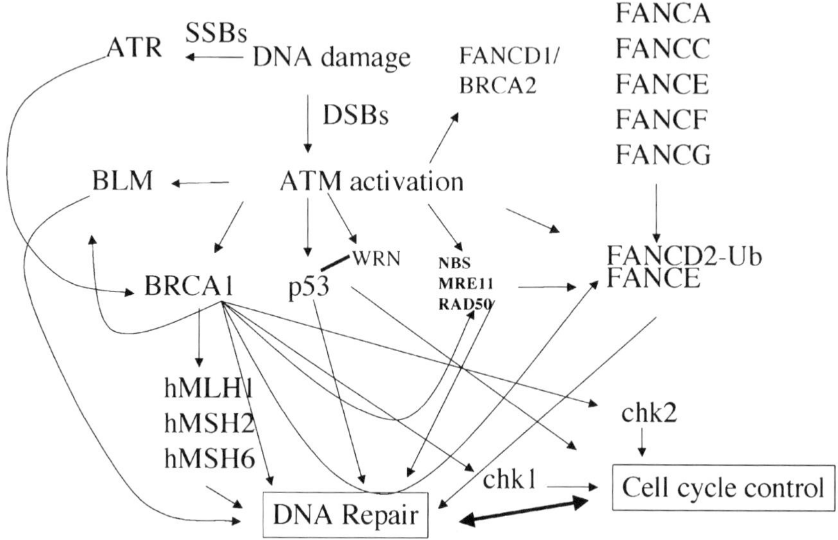

Fig. 1. An incomplete rendering of the multitude of interactions between DNA damage response pathways.

Paradoxically, however, the biology of cells derived from patients with an autosomal dominant LFS, for example, may indeed be different in that it initially has a DNA-damage hypersensitivity phenotype. It is this hypersensitivity phenotype that in fact delivers evolutionary pressure whereby the cells undergo genetic loss that results in a growth advantage. This advantage translates into cancer. It is this evolutionary pressure that also results in the acquisition of chemotherapeutic resistance, which is the ultimate end point of genetic loss.

Interestingly, emerging evidence suggests that many of these disorders are linked functionally and physically in complex pathways. The BRCA1-associated supercomplex, for example, contains mismatch repair proteins, ATM, NBS complex, and BLM *(46)*. Also, BRCA1 has been reported to bind to FANCD2 as well as BRCA2, and ATM phosphorylates BRCA1, NBS, FANCD2, and p53 *(38,47,123,223–225)*. Thus, all of the disease entities presented in this review may have at least some indirect interaction. The associated gene products may represent an extremely complex and regulated way for the cell to respond to DNA damage. This pathway and interconnectedness is illustrated in Fig. 1.

Patients with the autosomal recessive disorders elaborated upon in this chapter are distinguished in their exhibition of excessive DNA damage and cytotoxicity in response to certain chemotherapeutic agents. Part and parcel of this phenomenon is the fact that these patients require chemotherapy or bone marrow transplantation for treatment of the malignancy to which they are prone or require bone marrow transplantation for treatment of the underlying disorder. These patients require extraordinary care to avoid life-threatening side effects from the administration of cytotoxic agents.

REFERENCES

1. Malkin D. The Li–Fraumeni syndrome. In: Vogelstein B, Kinzler KW, eds. *The Genetic Basis of Human Cancer.* McGraw-Hill, New York, 1998.
2. Boyle JM, Spreadborough A, Greaves MJ, Birch JM, Varley JM, Scott D. The relationship between radiation-induced G(1)arrest and chromosome aberrations in Li–Fraumeni fibroblasts with or without germline TP53 mutations. *Br J Cancer* 2001;85:293–296.
3. Varley JM, McGown G, Thorncroft M, et al. Significance of intron 6 sequence variations in the TP53 gene in Li–Fraumeni syndrome. *Cancer Genet Cytogenet* 2001;129:85–87.
4. Chompret A. The Li–Fraumeni syndrome. *Biochimie* 2002;84:75–82.
5. Jin S, Levine AJ. The p53 functional circuit. *J Cell Sci* 2001;114:4139–4140.
6. Khosravi R, Maya R, Gottlieb T, et al. Rapid ATM-dependent phosphorylation of MDM2 precedes p53 accumulation in response to DNA damage. *Proc Natl Acad Sci USA* 1999;96:14,973–14,937.
7. Oren M. Regulation of the p53 tumor suppressor protein. *J Biol Chem* 1999;274: 36,031–36,034.
8. Smith PJ, Jones CJ. p53 and the integrated response to DNA damage. In: Smith PJ, Jones CJ, eds. *DNA Recombination and Repair.* Oxford University Press, New York, 1999:202–231.
9. Hasegawa D, Kojima S, Tatsumi E, et al. Elevation of the serum Fas ligand in patients with hemophagocytic syndrome and Diamond-Blackfan anemia. *Blood* 1998;91:2793–2799.
10. Nakamura Y. ATM: the p53 booster. *Nature Med* 1998;4:1231–1232.
11. Blander G, Zalle N, Leal JF, et al. The Werner syndrome protein contributes to induction of p53 by DNA damage. *FASEB J* 2000;14:2138–2140.
12. Lev Bar-Or R, Maya R, Segel LA, et al. Generation of oscillations by the p53–Mdm2 feedback loop: a theoretical and experimental study. *Proc Natl Acad Sci USA* 2000;97: 11,250–11,255.
13. Maya R, Balass M, Kim ST, et al. ATM-dependent phosphorylation of Mdm2 on serine 395: role in p53 activation by DNA damage. *Genes Dev* 2001;15:1067–1077.
14. de Rozieres S, Maya R, Oren M, et al. The loss of mdm2 induces p53-mediated apoptosis. *Oncogene* 2000;19:1691–1697.
15. Winters ZE, Ongkeko WM, Harris AL, Norbury CJ. p53 regulates Cdc2 independently of inhibitory phosphorylation to reinforce radiation-induced G2 arrest in human cells. *Oncogene* 1998;17:673–684.
16. Schwartz D, Almog N, Peled A, et al. Role of wild type p53 in the G2 phase: regulation of the gamma-irradiation-induced delay and DNA repair. *Oncogene* 1997;15:2597–2607.
17. Abramova NA, Russell J, Botchan M, et al. Interaction between replication protein A and p53 is disrupted after UV damage in a DNA repair-dependent manner. *Proc Natl Acad Sci USA* 1997;94:7186–7191.
18. Ko LJ, Shieh SY, Chen X, et al. p53 is phosphorylated by CDK7–cyclin H in a p36MAT1-dependent manner. *Mol Cell Biol* 1997;17:7220–7229.
19. Liu MC, Gelmann EP. P53 gene mutations: case study of a clinical marker for solid tumors. *Semin Oncol* 2002;29:246–257.
20. Shono T, Tofilon PJ, Schaefer TS, et al. Apoptosis induced by adenovirus-mediated p53 gene transfer in human glioma correlates with site-specific phosphorylation. *Cancer Res* 2002;62:1069–1076.

21. DiGiammarino EL, Lee AS, Cadwell C, et al. A novel mechanism of tumorigenesis involving pH-dependent destabilization of a mutant p53 tetramer. *Nature Struct Biol* 2002; 9:12–16.

22. Mirzayans R, Aubin RA, Bosnich W, et al. Abnormal pattern of post-gamma-ray DNA replication in radioresistant fibroblast strains from affected members of a cancer-prone family with Li–Fraumeni syndrome. *Br J Cancer* 1995;71:1221–1230.

23. Delia D, Goi K, Mizutani S, et al. Dissociation between cell cycle arrest and apoptosis can occur in Li–Fraumeni cells heterozygous for p53 gene mutations. *Oncogene* 1997;14: 2137–2147.

24. Bunz F, Hwang PM, Torrance C, et al. Disruption of p53 in human cancer cells alters the responses to therapeutic agents. *J Clin Invest* 1999;104:263–269.

25. Ding HF, Fisher DE. Mechanisms of p53-mediated apoptosis. *Crit Rev Oncol* 1998;9:83–98.

26. Ruley HE. p53 and response to chemotherapy and radiotherapy. Important Adv Oncol 1996:37–56.

27. Zhan M, Yu D, Lang A, et al. Wild type p53 sensitizes soft tissue sarcoma cells to doxorubicin by down-regulating multidrug resistance-1 expression. *Cancer* 2001;92:1556–1566.

28. Sampath J, Sun D, Kidd VJ, et al. Mutant p53 cooperates with ETS and selectively up-regulates human MDR1 not MRP1. *J Biol Chem* 2001;276:39,359–39,367.

29. Naf D, Kupfer GM, Suliman A, et al. Functional activity of the Fanconi anemia protein FAA requires FAC binding and nuclear localization. *Mol Cell Biol* 1998;18:5952–5960.

30. Wallace-Brodeur RR, Lowe SW. Clinical implications of p53 mutations. *Cell Mol Life Sci* 1999;55:64–75.

31. Swisher SG, Roth JA, Nemunaitis J, et al. Adenovirus-mediated p53 gene transfer in advanced non-small-cell lung cancer. *J Natl Cancer Inst* 1999;91:763–771.

32. Foster BA, Coffey HA, Morin MJ, et al. Pharmacological rescue of mutant p53 conformation and function. *Science* 1999;286:2507–2510.

33. Komarov PG, Komarova EA, Kondratov RV, et al. A chemical inhibitor of p53 that protects mice from the side effects of cancer therapy. *Science* 1999;285:1733–1737.

34. Williams WR, Anderson DE. Genetic epidemiology of breast cancer: segregation analysis of 200 Danish pedigrees. *Genet Epidemiol* 1984;1:7–20.

35. Williams WR, Anderson DE. Genetic epidemiology of breast cancer: further clarification and a response to King and Elston. *Genet Epidemiol* 1985;2:170–176.

36. Tavtigian SV, Simard J, Rommens J, et al. The complete BRCA2 gene and mutations in chromosome 13q-linked kindreds. *Nature Genet* 1996;12:333–337.

37. Moynahan ME, Cui TY, Jasin M. Homology-directed dna repair, mitomycin-C resistance, and chromosome stability is restored with correction of a Brca1 mutation. *Cancer Res* 2001;61:4842–4850.

38. Chen J, Silver DP, Walpita D, et al. Stable interaction between the products of the BRCA1 and BRCA2 tumor suppressor genes in mitotic and meiotic cells. *Mol Cell* 1998;2:317–328.

39. Couch F, Weber B. Breast cancer. In: Vogelstein B, Kinzler KW, eds. *The Genetic Basis of Human Cancer*. McGraw-Hill, New York, 1998.

40. Thakur S, Zhang HB, Peng Y, et al. Localization of BRCA1 and a splice variant identifies the nuclear localization signal. *Mol Cell Biol* 1997;17:444–452.

41. Chen Y, Chen CF, Riley DJ, et al. Aberrant subcellular localization of BRCA1 in breast cancer. *Science* 1995;270:789–791.

42. Xu B, Kim S, Kastan MB. Involvement of Brca1 in S-phase and G(2)-phase checkpoints after ionizing irradiation. *Mol Cell Biol* 2001;21:3445–3450.

43. Shiloh Y. ATM and ATR: networking cellular responses to DNA damage. *Curr Opin Genet Dev* 2001;11:71–77.

44. Muchardt C, Yaniv M. When the SWI/SNF complex remodels . . . the cell cycle. *Oncogene* 2001;20:3067–3075.

45. Bochar DA, Wang L, Beniya H, et al. BRCA1 is associated with a human SWI/SNF-related complex: linking chromatin remodeling to breast cancer. *Cell* 2000;102:257–65.

46. Wang Y, Cortez D, Yazdi P, et al. BASC, a super complex of BRCA1-associated proteins involved in the recognition and repair of aberrant DNA structures. *Genes Dev* 2000;14: 927–939.

47. Garcia-Higuera I, Taniguchi T, Ganesan S, et al. Interaction of the Fanconi anemia proteins and BRCA1 in a common pathway. *Mol Cell* 2001;7:249–262.

48. Venkitaraman AR. Cancer susceptibility and the functions of BRCA1 and BRCA2. *Cell* 2002;108:171–182.

49. Howlett NG, Taniguchi T, Olson S, et al. Biallelic inactivation of BRCA2 in Fanconi anemia. *Science* 2002;297:606–609.

50. Bhattacharyya A, Ear US, Koller BH, et al. The breast cancer susceptibility gene BRCA1 is required for subnuclear assembly of Rad51 and survival following treatment with the DNA cross-linking agent cisplatin. *J Biol Chem* 2000;275:23,899–23,903.

51. Coleman CN. Molecular biology in radiation oncology. Radiation oncology perspective of BRCA1 and BRCA2. *Acta Oncol* 1999;38(Suppl 13):55–59.

52. Greenblatt MS, Chappuis PO, Bond JP, et al. TP53 mutations in breast cancer associated with BRCA1 or BRCA2 germ-line mutations: distinctive spectrum and structural distribution. *Cancer Res* 2001;61:4092–4097.

53. Bocker T, Ruschoff J, Fishel R. Molecular diagnostics of cancer predisposition: hereditary non-polyposis colorectal carcinoma and mismatch repair defects. *Biochim Biophys Acta* 1999;1423:O1–O10.

54. Kolodner RD. Guarding against mutation. *Nature* 2000;407:687–689.

55. Peel DJ, Ziogas A, Fox EA, et al. Characterization of hereditary nonpolyposis colorectal cancer families from a population-based series of cases. *J Natl Cancer Inst* 2000;92: 1517–1522.

56. Muller A, Fishel R. Mismatch repair and the hereditary non-polyposis colorectal cancer syndrome (HNPCC). *Cancer Invest* 2002;20:102–109.

57. Flores-Rozas H, Clark D, Kolodner RD. Proliferating cell nuclear antigen and Msh2p– Msh6p interact to form an active mispair recognition complex. *Nature Genet* 2000;26: 375–378.

58. Hall MC, Wang H, Eric DA, et al. High affinity cooperative DNA binding by the yeast Mlh1-Pms1 heterodimer. *J Mol Biol* 2001;312:637–647.

59. Marti TM, Kunz C, Fleck O. DNA mismatch repair and mutation avoidance pathways. *J Cell Physiol* 2002;191:28–41.

60. Kolodner RD, Marsischky GT. Eukaryotic DNA mismatch repair. *Curr Opin Genet Dev* 1999;9:89–96.

61. Rovella V, Carrara S, Crucitti SC, et al. Familial microsatellite-stable nonpolyposis colorectal cancer: incidence and characteristics in a clinic-based population. *Ann Oncol* 2001;12: 813–818.

62. Boland RC. Hereditary nonpolyposis colorectal cancer. In: Vogelstein B, Kinzler KW, eds. *The Genetic Basis of Human Cancer.* McGraw-Hill, New York, 1998.

63. Thibodeau SN, Bren G, Schaid D. Microsatellite instability in cancer of the proximal colon. *Science* 1993;260:816–819.

64. Drummond JT, Li GM, Longley MJ, Modrich P. Isolation of an hMSH2-p160 heterodimer that restores DNA mismatch repair to tumor cells. *Science* 1995;268:1909–1912.

65. Prolla TA, Baker S, Liskay RM. Genetics of DNA Mismatch repair, microsatellite instability, and cancer. In: Nickoloff JA, Hoekstra MF, eds. *DNA Damage and Repair.* Humana, Totowa, NJ, 1998.

66. Boyer JC, Umar A, Risinger JI, et al. Microsatellite instability, mismatch repair deficiency, and genetic defects in human cancer cell lines. *Cancer Res* 1995;55:6063–6070.

67. Fishel R, Kolodner RD. Identification of mismatch repair genes and their role in the development of cancer. *Curr Opin Genet Dev* 1995;5:382–395.
68. Kolodner RD. Mismatch repair: mechanisms and relationship to cancer susceptibility. *Trends Biochem Sci* 1995;20:397–401.
69. Jacob S, Praz F. DNA mismatch repair defects: role in colorectal carcinogenesis. *Biochimie* 2002;84:27–47.
70. Esteller M, Levine R, Baylin SB, et al. MLH1 promoter hypermethylation is associated with the microsatellite instability phenotype in sporadic endometrial carcinomas. *Oncogene* 1998;17:2413–2417.
71. Menigatti M, Di Gregorio C, Borghi F, et al. Methylation pattern of different regions of the MLH1 promoter and silencing of gene expression in hereditary and sporadic colorectal cancer. *Genes Chromosomes Cancer* 2001;31:357–361.
72. Pedroni M, Sala E, Scarselli A, et al. Microsatellite instability and mismatch-repair protein expression in hereditary and sporadic colorectal carcinogenesis. *Cancer Res* 2001;61:896–899.
73. Strand M, Prolla TA, Liskay RM, et al. Destabilization of tracts of simple repetitive DNA in yeast by mutations affecting DNA mismatch repair. *Nature* 1993;365:274–276.
74. Koi M, Umar A, Chauhan DP, et al. Human chromosome 3 corrects mismatch repair deficiency and microsatellite instability and reduces *N*-methyl-*N'*-nitro-*N*-nitrosoguanidine tolerance in colon tumor cells with homozygous hMLH1 mutation. *Cancer Res* 1994;54:4308–4312.
75. Glaab WE, Tindall KR. Mutation rate at the hprt locus in human cancer cell lines with specific mismatch repair-gene defects. *Carcinogenesis* 1997;18:1–8.
76. Narayanan L, Fritzell JA, Baker SM, et al. Elevated levels of mutation in multiple tissues of mice deficient in the DNA mismatch repair gene Pms2. *Proc Natl Acad Sci USA* 1997;94:3122–3127.
77. Andrew SE, McKinnon M, Cheng BS, et al. Tissues of MSH2-deficient mice demonstrate hypermutability on exposure to a DNA methylating agent. *Proc Natl Acad Sci USA* 1998;95:1126–1130.
78. Olschwang S. Germline mutation and genome instability. *Eur J Cancer Prev* 1999;8(Suppl 1):S33–S37.
79. Parniewski P, Jaworski A, Wells RD, et al. Length of CTG.CAG repeats determines the influence of mismatch repair on genetic instability. *J Mol Biol* 2000;299:865–874.
80. Suzuki M, Ohwada M, Saga Y, et al. Are DNA mismatch repair deficiencies responsible for accumulation of genetic alterations in epithelial ovarian cancers? *Cancer Genet Cytogenet* 2001;124:152–158.
81. Liang SB, Furihata M, Takeuchi T, et al. Reduced human mismatch repair protein expression in the development of precancerous skin lesions to squamous cell carcinoma. *Virchows Arch* 2001;439:622–627.
82. Kruse R, Rutten A, Lamberti C, et al. Muir–Torre phenotype has a frequency of DNA mismatch-repair-gene mutations similar to that in hereditary nonpolyposis colorectal cancer families defined by the Amsterdam criteria. *Am J Hum Genet* 1998;63:63–70.
83. Mathiak M, Rutten A, Mangold E, et al. Loss of DNA mismatch repair proteins in skin tumors from patients with Muir–Torre syndrome and MSH2 or MLH1 germline mutations: establishment of immunohistochemical analysis as a screening test. *Am J Surg Pathol* 2002;26:338–343.
84. Lynch HT, Leibowitz R, Smyrk T, et al. Colorectal cancer and the Muir–Torre syndrome in a Gypsy family: a review. *Am J Gastroenterol* 1999;94:575–580.
85. Miyaki M, Nishio J, Konishi M, et al. Drastic genetic instability of tumors and normal tissues in Turcot syndrome. *Oncogene* 1997;15:2877–2881.
86. Foulkes WD. A tale of four syndromes: familial adenomatous polyposis, Gardner syndrome, attenuated APC and Turcot syndrome. *Q J Med* 1995;88:853–863.

87. Mori T, Nagase H, Horii A, et al. Germ-line and somatic mutations of the APC gene in patients with Turcot syndrome and analysis of APC mutations in brain tumors. *Genes Chromosomes Cancer* 1994;9:168–172.

88. Luk GD. Diagnosis and therapy of hereditary polyposis syndromes. *Gastroenterologist* 1995;3:153–167.

89. Malkhosyan S, Rampino N, Yamamoto H, et al. Frameshift mutator mutations. *Nature* 1996;382:499–500.

90. Church JM. Prophylactic colectomy in patients with hereditary nonpolyposis colorectal cancer. *Ann Med* 1996;28:479–482.

91. Fedier A, Schwarz VA, Walt H, et al. Resistance to topoisomerase poisons due to loss of DNA mismatch repair. *Int J Cancer* 2001;93:571–576.

92. Karran P. Mechanisms of tolerance to DNA damaging therapeutic drugs. *Carcinogenesis* 2001;22:1931–1937.

93. Buermeyer AB, Wilson-Van Patten C, Baker SM, et al. The human MLH1 cDNA complements DNA mismatch repair defects in Mlh1-deficient mouse embryonic fibroblasts. *Cancer Res* 1999;59:538–541.

94. Lin X, Howell SB. Effect of loss of DNA mismatch repair on development of topotecan-, gemcitabine-, and paclitaxel-resistant variants after exposure to cisplatin. *Mol Pharmacol* 1999;56:390–395.

95. Lin X, Kim HK, Howell SB. The role of DNA mismatch repair in cisplatin mutagenicity. *J Inorg Biochem* 1999;77:89–93.

96. Colella G, Marchini S, D'Incalci M, et al. Mismatch repair deficiency is associated with resistance to DNA minor groove alkylating agents. *Br J Cancer* 1999;80:338–343.

97. Hawn MT, Umar A, Carethers JM, et al. Evidence for a connection between the mismatch repair system and the G2 cell cycle checkpoint. *Cancer Res* 1995;55:3721–3725.

98. Cenni B, Kim HK, Bubley GJ, et al. Loss of DNA mismatch repair facilitates reactivation of a reporter plasmid damaged by cisplatin. *Br J Cancer* 1999;80:699–704.

99. Kupfer GM, Naf D, D'Andrea AD. Molecular biology of Fanconi anemia. *Hematol Oncol Clin North Am* 1997;11:1045–1060.

100. D'Andrea AD, Grompe M. Molecular biology of Fanconi anemia genes: implications for diagnosis and therapy. *Blood* 1997;5:1725–1726.

101. Alter BP. Modern review of congenital hypoplastic anemia. *J Pediatr Hematol Oncol* 2001;23:383–284.

102. Howlett NG, Taniguchi T, Olson S, et al. Biallelic inactivation of BRCA2 in Fanconi anemia. *Science* 2002;5581:606–609.

103. Liu J, Buchwald M, Walsh C, et al. Fanconi anemia and novel strategies for therapy. *Blood* 1994;84:3995.

104. Auerbach AD, Weiner MA, Warburton D, et al. Acute myeloid leukemia as the first hematologic manifestation of Fanconi anemia. *Am J Hematol* 1982;12:289–300.

105. Giampietro PF, Verlander PC, Davis JG, et al. Diagnosis of Fanconi anemia in patients without congenital malformations: an international Fanconi Anemia Registry Study. *Am J Med Genet* 1997;68:58–61.

106. Gillio AP, Verlander PC, Batish SD, et al. Phenotypic consequences of mutations in the Fanconi anemia FAC gene: an International Fanconi Anemia Registry study. *Blood* 1997;90:105–110.

107. Dokal I. The genetics of Fanconi's anaemia. *Baillieres Best Pract Res Clin Haematol* 2000;13:407–425.

108. Auerbach AD, Allen RG. Leukemia and preleukemia in Fanconi anemia patients. A review of the literature and report of the International Fanconi Anemia Registry. *Cancer Genet Cytogenet* 1991;51:1–12.

109. Giampietro PF, Adler-Brecher B, Verlander PC, et al. The need for more accurate and timely diagnosis in Fanconi anemia: a report from the International Fanconi Anemia Registry. *Pediatrics* 1993;91:1116–1120.
110. MacMillan ML, Auerbach AD, Davies SM, et al. Haematopoietic cell transplantation in patients with Fanconi anaemia using alternate donors: results of a total body irradiation dose escalation trial. *Br J Haematol* 2000;109:121–129.
111. Jakobs PM, Sahaayaruban P, Saito H, et al. Immortalization of four new Fanconi anemia fibroblast cell lines by an improved procedure. *Somat Cell Mol Genet* 1996;22:151–157.
112. Kupfer GM, D'Andrea AD. The effect of the Fanconi Anemia polypeptide, FAC, upon p53 induction and G2 checkpoint regulation. *Blood* 1996;88:1019–1025.
113. Yamashita T, Wu N, Kupfer G, et al. Clinical variability of Fanconi anemia (type C) results from expression of an amino terminal truncated Fanconi anemia complementation group C polypeptide with partial activity. *Blood* 1996;87:4424–4432.
114. Marathi UK, Howell SR, Ashmun RA, et al. The Fanconi anemia complementation group C protein corrects DNA interstrand cross-link-specific apoptosis in HSC536N cells. *Blood* 1996;88:2298–2305.
115. Kupfer GM, Naf D, Suliman A, et al. The Fanconi anaemia proteins, FAA and FAC, interact to form a nuclear complex. *Nature Genet* 1997;17:487–490.
116. de Winter JP, Waisfisz Q, Rooimans MA, et al. The Fanconi anaemia group G gene FANCG is identical with XRCC9. *Nature Genet* 1998;20:281–283.
117. de Winter JP, Leveille F, van Berkel CG, et al. Isolation of a cDNA representing the Fanconi anemia complementation group E gene. *Am J Hum Genet* 2000;67:1306–1308.
118. de Winter JP, Rooimans MA, van Der Weel L, et al. The Fanconi anaemia gene FANCF encodes a novel protein with homology to ROM. *Nature Genet* 2000;24:15–16.
119. de Winter JP, van Der Weel L, de Groot J, et al. The Fanconi anemia protein FANCF forms a nuclear complex with FANCA, FANCC and FANCG. [in process citation]. *Hum Mol Genet* 2000;9:2665–2674.
120. Huber PA, Medhurst AL, Youssoufian H, ç Investigation of Fanconi anemia protein interactions by yeast two-hybrid analysis. *Biochem Biophys Res Commun* 2000;268:73–77.
121. Timmers C, Taniguchi T, Hejna J, et al. Positional cloning of a novel Fanconi anemia gene, FANCD2. *Mol Cell* 2001;7:241–248.
122. Garcia-Higuera I, Kuang Y, Naf D, Wasik J, D'Andrea AD. Fanconi anemia proteins FANCA, FANCC, and FANCG/XRCC9 interact in a functional nuclear complex. *Mol Cell Biol* 1999;19:4866–4873.
123. Taniguchi T, Garcia-Higuera I, Xu B, et al. Convergence of the Fanconi anemia and ataxia telangiectasia signaling pathways. *Cell* 2002;109:459–472.
124. Gluckman E, Devergie A, Schaison G, et al. Bone marrow transplantation in Fanconi anaemia. *Br J Haematol* 1980;45:557–564.
125. Aker M, Varadi G, Slavin S, et al. Fludarabine-based protocol for human umbilical cord blood transplantation in children with Fanconi anemia. *J Pediatr Hematol Oncol* 1999;21:237–239.
126. McCloy M, Almeida A, Daly P, et al. Fludarabine-based stem cell transplantation protocol for Fanconi's anaemia in myelodysplastic transformation. *Br J Haematol* 2001;112: 427–429.
127. Verlinsky Y, Rechitsky S, Schoolcraft W, et al. Preimplantation diagnosis for Fanconi anemia combined with HLA matching. *JAMA* 2001;285:3130–3103.
128. Croop JM, Cooper R, Fernandez C, et al. Mobilization and collection of peripheral blood CD34+ cells from patients with Fanconi anemia. *Blood* 2001;98:2917–2921.
129. Yamada K, Olsen JC, Patel M, et al. Functional correction of Fanconi anemia group C hematopoietic cells by the use of a novel lentiviral vector. *Mol Ther* 2001;3:485–490.

130. Gush KA, Fu KL, Grompe M, et al. Phenotypic correction of Fanconi anemia group C knockout mice. *Blood* 2000;95:700–704.
131. Gregory JJ, Jr., Wagner JE, Verlander PC, et al. Somatic mosaicism in Fanconi anemia: evidence of genotypic reversion in lymphohematopoietic stem cells. *Proc Natl Acad Sci USA* 2001;98:2532–2537.
132. Waisfisz Q, de Winter JP, Kruyt FA, et al. A physical complex of the Fanconi anemia proteins FANCG/XRCC9 and FANCA. *Proc Natl Acad Sci USA* 1999;96:10,320–10,325.
133. Waisfisz Q, Morgan NV, Savino M, et al. Spontaneous functional correction of homozygous Fanconi anaemia alleles reveals novel mechanistic basis for reverse mosaicism. *Nature Genet* 1999;22:379–83.
134. Waisfisz Q, Saar K, Morgan NV, et al. The Fanconi anemia group E gene, FANCE, maps to chromosome 6p. *Am J Hum Genet* 1999;64:1400–1405.
135. Lo Ten Foe JR, Kwee ML, Rooimans MA, et al. Somatic mosaicism in Fanconi anemia: molecular basis and clinical significance. *Eur J Hum Genet* 1997;5:137–148.
136. Gatti R. Ataxia telangiectasia. In: Vogelstein B, Kinzler K, eds. The Genetic Basis of Human Cancer. McGraw-Hill, New York, 1998:275–300.
137. Gilad S, Chessa L, Khosravi R, et al. Genotype-phenotype relationships in ataxia–telangiectasia and variants. *Am J Hum Genet* 1998;62:551–561.
138. Meyn MS. Ataxia–telangiectasia, cancer and the pathobiology of the ATM gene. *Clin Genet* 1999;55:289–304.
139. Chenevix-Trench G, Spurdle AB, Gatei M, et al. Dominant negative ATM mutations in breast cancer families. *J Natl Cancer Inst* 2002;94:205–215.
140. Savitsky K, Sfez S, Tagle DA, et al. The complete sequence of the coding region of the ATM gene reveals similarity to cell cycle regulators in different species. *Hum Mol Genet* 1995;4:2025–2032.
141. Shiloh Y. Ataxia–telangiectasia: closer to unraveling the mystery. *Eur J Hum Genet* 1995;3:116–138.
142. Weyl Ben Arush M, Rosenthal J, Dale J, et al. Ataxia telangiectasia and lymphoma: an indication for individualized chemotherapy dosing—report of treatment in a highly inbred Arab family. *Pediatr Hematol Oncol* 1995;12:163–169.
143. Lavin MF, Shiloh Y. Ataxia–telangiectasia: a multifaceted genetic disorder associated with defective signal transduction. *Curr Opin Immunol* 1996;8:459–464.
144. Johnson RT, Gotoh E, Mullinger AM, et al. Targeting double-strand breaks to replicating DNA identifies a subpathway of DSB repair that is defective in ataxia–telangiectasia cells. *Biochem Biophys Res Commun* 1999;261:317–325.
145. Powell S, Whitaker S, Peacock J, et al. Ataxia telangiectasia: an investigation of the repair defect in the cell line AT5BIVA by plasmid reconstitution. *Mutat Res* 1993;294:9–20.
146. Kastan MB, Zhan Q, el-Deiry WS, et al. A mammalian cell cycle checkpoint pathway utilizing p53 and GADD45 is defective in ataxia–telangiectasia. *Cell* 1992;71:587–597.
147. Khanna KK, Beamish H, Yan J, et al. Nature of G1/S cell cycle checkpoint defect in ataxia–telangiectasia. *Oncogene* 1995;11:609–618.
148. Savitsky K, Bar-Shira A, Gilad S, et al. A single ataxia telangiectasia gene with a product similar to PI-3 kinase. *Science* 1995;268:1749–1753.
149. Pecker I, Avraham KB, Gilbert DJ, et al. Identification and chromosomal localization of Atm, the mouse homolog of the ataxia–telangiectasia gene. *Genomics* 1996;35:39–45.
150. Hartley KO, Gell D, Smith GC, et al. DNA-dependent protein kinase catalytic subunit: a relative of phosphatidylinositol 3-kinase and the ataxia telangiectasia gene product. *Cell* 1995;82:849–856.
151. Hari KL, Santerre A, Sekelsky JJ, et al. The mei-41 gene of *D. melanogaster* is a structural and functional homolog of the human ataxia telangiectasia gene. *Cell* 1995;82:815–821.

152. Jimenez G, Yucel J, Rowley R, et al. The rad3+ gene of *Schizosaccharomyces pombe* is involved in multiple checkpoint functions and in DNA repair. *Proc Natl Acad Sci USA* 1992;89:4952–4956.

153. Greenwell PW, Kronmal SL, Porter SE, et al. TEL1, a gene involved in controlling telomere length in *S. cerevisiae*, is homologous to the human ataxia telangiectasia gene. *Cell* 1995;82:823–829.

154. Ahn JY, Schwarz JK, Piwnica-Worms H, et al. Threonine 68 phosphorylation by ataxia telangiectasia mutated is required for efficient activation of Chk2 in response to ionizing radiation. *Cancer Res* 2000;60:5934–5936.

155. Jung M, Kondratyev A, Lee SA, et al. ATM gene product phosphorylates I kappa B-alpha. *Cancer Res* 1997;57:24–27.

156. Xu Y, Ashley T, Brainerd EE, et al. Targeted disruption of ATM leads to growth retardation, chromosomal fragmentation during meiosis, immune defects, and thymic lymphoma. *Genes Dev* 1996;10:2411–2422.

157. Barlow C, Hirotsune S, Paylor R, et al. Atm-deficient mice: a paradigm of ataxia telangiectasia. *Cell* 1996;86:159–171.

158. Plug AW, Peters AH, Xu Y, et al. ATM and RPA in meiotic chromosome synapsis and recombination. *Nature Genet* 1997;17:457–461.

159. Waterman MJ, Stavridi ES, Waterman JL, et al. ATM-dependent activation of p53 involves dephosphorylation and association with 14-3-3 proteins. *Nature Genet* 1998;19:175–178.

160. Smith GC, Cary RB, Lakin ND, et al. Purification and DNA binding properties of the ataxia–telangiectasia gene product ATM. *Proc Natl Acad Sci USA* 1999;96:11,134–11,139.

161. Sandoval C, Swift M. Treatment of lymphoid malignancies in patients with ataxia–telangiectasia. *Med Pediatr Oncol* 1998;31:491–7.

162. Eyre JA, Gardner-Medwin D, Summerfield GP. Leukoencephalopathy after prophylactic radiation for leukaemia in ataxia telangiectasia. *Arch Dis Child* 1988;63:1079–1080.

163. Cunlift PN, Mann JR, Cameron AH, et al. Radiosensitivity in ataxia–telangiectasia. *Br J Radiol* 1975;48:374–346.

164. Pernin D, Bay JO, Uhrhammer N, et al. ATM heterozygote cells exhibit intermediate levels of apoptosis in response to streptonigrin and etoposide. *Eur J Cancer* 1999;35:1130–1135.

165. Janin N, Andrieu N, Ossian K, et al. Breast cancer risk in ataxia telangiectasia (AT) heterozygotes: haplotype study in French AT families. *Br J Cancer* 1999;80:1042–1045.

166. Inskip HM, Kinlen LJ, Taylor AM, et al. Risk of breast cancer and other cancers in heterozygotes for ataxia–telangiectasia. *Br J Cancer* 1999;79:1304–1307.

167. Yoshida M, Hosoi Y, Miyachi H, et al. Roles of DNA-dependent protein kinase and ATM in cell-cycle-dependent radiation sensitivity in human cells. *Int J Radiat Biol* 2002;78:503–512.

168. Chernikova SB, Wells RL, Elkind MM. Wortmannin sensitizes mammalian cells to radiation by inhibiting the DNA-dependent protein kinase-mediated rejoining of double-strand breaks. *Radiat Res* 1999;151:159–166.

169. Khanna KK, Gatti RA, Concannon P, et al. Cellular responses to DNA damage. In: Nickoloff JA, Hoekstra MF, eds. *DNA Damage and Repair.* Humana, Totowa, NJ, 1998; vol. 2.

170. German J, Ellis NA. Bloom Syndrome. In: Vogelstein B, Kinzler K, eds. *The Genetic Basis of Human Cancer.* McGraw-Hill, New York, 1998:731.

171. Schellenberg GD, Miki T, Yu CE, et al. Werner syndrome. In: Vogelstein B, Kinzler K, eds. *The Genetic Basis of Human Cancer.* McGraw-Hill, New York, 1998.

172. van Brabant AJ, Stan R, Ellis NA. DNA helicases, genomic instability, and human genetic disease. *Annu Rev Genomics Hum Genet* 2000;1:409–459.

173. Ellis NA, Ciocci S, Proytcheva M, et al. The Ashkenazic Jewish Bloom syndrome mutation blmAsh is present in non-Jewish Americans of Spanish ancestry. *Am J Hum Genet* 1998;63:1685–1693.

174. Jungerman M, Sedzimirska M, Lange A, et al. Application of microsatellite markers for monitoring of reconstitution pattern in patients after bone marrow transplantation/peripheral blood progenitor cell transplantation. *Arch Immunol Ther Exp* 1998;46:39–43.

175. Ellis NA, Groden J, Ye TZ, et al. The Bloom's syndrome gene product is homologous to RecQ helicases. *Cell* 1995;83:655–666.

176. Ellis NA, Lennon DJ, Proytcheva M, et al. Somatic intragenic recombination within the mutated locus BLM can correct the high sister-chromatid exchange phenotype of Bloom syndrome cells. *Am J Hum Genet* 1995;57:1019–1027.

177. Straughen J, Ciocci S, Ye TZ, et al. Physical mapping of the Bloom syndrome region by the identification of YAC and P1 clones from human chromosome 15 band q26.1. *Genomics* 1996;35:118–128.

178. Ababou M, Dumaire V, Lecluse Y, et al. Bloom's syndrome protein response to ultraviolet-C radiation and hydroxyurea-mediated DNA synthesis inhibition. *Oncogene* 2002;21:2079–2088.

179. Gaymes TJ, North PS, Brady N, et al. Increased error-prone non homologous DNA end-joining—a proposed mechanism of chromosomal instability in Bloom's syndrome. *Oncogene* 2002;21:2525–2533.

180. Franchitto A, Pichierri P. Bloom's syndrome protein is required for correct relocalization of RAD50/MRE11/NBS1 complex after replication fork arrest. *J Cell Biol* 2002;157:19–30.

181. Sanz MM, Proytcheva M, Ellis NA, et al. BLM, the Bloom's syndrome protein, varies during the cell cycle in its amount, distribution, and co-localization with other nuclear proteins. *Cytogenet Cell Genet* 2000;91:217–223.

182. Brosh RM, Jr., Li JL, Kenny MK, et al. Replication protein A physically interacts with the Bloom's syndrome protein and stimulates its helicase activity. *J Biol Chem* 2000;275: 23,500–23,508.

183. Moens PB, Kolas NK, Tarsounas M, et al. The time course and chromosomal localization of recombination-related proteins at meiosis in the mouse are compatible with models that can resolve the early DNA–DNA interactions without reciprocal recombination. *J Cell Sci* 2002;115:1611–1622.

184. Harmon FG, DiGate RJ, Kowalczykowski SC. RecQ helicase and topoisomerase III comprise a novel DNA strand passage function: a conserved mechanism for control of DNA recombination. *Mol Cell* 1999;3:611 620.

185. Azuma E, Hirayama M, Bonno M, et al. Successful immunization following cord blood transplantation in a child with Diamond–Blackfan anemia. *Pediatr Hematol Oncol* 2001;18:193–197.

186. Vanderschueren-Lodeweyckx M, Fryns JP, Van den Berghe H, et al. Bloom's syndrome. Possible pitfalls in clinical diagnosis. *Am J Dis Child* 1984;138:812–816.

187. Grasemann H, Kremens B, Passarge E. Experience treating a patient with Bloom syndrome and acute myelogenous leukemia. *Med Pediatr Oncol* 1998;30:309–310.

188. Vijayalaxmi, Evans HJ, Ray JH, et al. Bloom's syndrome: evidence for an increased mutation frequency in vivo. *Science* 1983;221:851–853.

189. Karnak I, Ciftci AO, Senocak ME, et al. Colorectal carcinoma in children. *J Pediatr Surg* 1999;34:1499–1504.

190. Epstein CJ, Martin GM, Schultz AL, et al. Werner's syndrome: a review of its symptomatology, natural history, pathologic features, genetics and relationship to the natural aging process. *Medicine (Baltimore)* 1966;45:177–221.

191. Sato K, Goto M, Nishioka K, et al. Werner's syndrome associated with malignancies: five case reports with a survey of case histories in Japan. *Gerontology* 1988;34:212–218.

192. Prince PR, Ogburn CE, Moser MJ, et al. Cell fusion corrects the 4-nitroquinoline 1-oxide sensitivity of Werner syndrome fibroblast cell lines. *Hum Genet* 1999;105:132–138.

193. Ogburn CE, Oshima J, Poot M, et al. An apoptosis-inducing genotoxin differentiates heterozygotic carriers for Werner helicase mutations from wild-type and homozygous mutants. *Hum Genet* 1997;101:121–125.

194. Fukuchi K, Martin GM, Monnat RJ Jr. Mutator phenotype of Werner syndrome is characterized by extensive deletions. *Proc Natl Acad Sci USA* 1989;86:5893–5897.

195. Gray MD, Shen JC, Kamath-Loeb AS, et al. The Werner syndrome protein is a DNA helicase. *Nature Genet* 1997;17:100–103.

196. Shen JC, Gray MD, Oshima J, et al. Werner syndrome protein. I. DNA helicase and dna exonuclease reside on the same polypeptide. *J Biol Chem* 1998;273:34,139–34,144.

197. Yu CE, Oshima J, Fu YH, et al. Positional cloning of the Werner's syndrome gene. *Science* 1996;272:258–262.

198. Goto M, Imamura O, Kuromitsu J, et al. Analysis of helicase gene mutations in Japanese Werner's syndrome patients. *Hum Genet* 1997;99:191–193.

199. Matsumoto T, Imamura O, Yamabe Y, et al. Mutation and haplotype analyses of the Werner's syndrome gene based on its genomic structure: genetic epidemiology in the Japanese population. *Hum Genet* 1997;100:123–130.

200. Moser MJ, Oshima J, Monnat RJ Jr. WRN mutations in Werner syndrome. *Hum Mutat* 1999;13:271–279.

201. Marciniak RA, Lombard DB, Johnson FB, et al. Nucleolar localization of the Werner syndrome protein in human cells. *Proc Natl Acad Sci USA* 1998;95:6887–6892.

202. Shiratori M, Sakamoto S, Suzuki N, et al. Detection by epitope-defined monoclonal antibodies of Werner DNA helicases in the nucleoplasm and their upregulation by cell transformation and immortalization. *J Cell Biol* 1999;144:1–9.

203. Blander G, Kipnis J, Leal JF, et al. Physical and functional interaction between p53 and the Werner's syndrome protein. *J Biol Chem* 1999;274:29,463–29,469.

204. Lebel M, Spillare EA, Harris CC, et al. The Werner syndrome gene product co-purifies with the DNA replication complex and interacts with PCNA and topoisomerase I. *J Biol Chem* 1999;274:37,795–37,799.

205. Spillare EA, Robles AI, Wang XW, et al. p53-mediated apoptosis is attenuated in Werner syndrome cells. *Genes Dev* 1999;13:1355–1360.

206. Pichierri P, Franchitto A, Mosesso P, et al. Werner's syndrome protein is required for correct recovery after replication arrest and DNA damage induced in S-phase of cell cycle. *Mol Biol Cell* 2001;12:2412–2421.

207. Sakamoto S, Nishikawa K, Heo SJ, et al. Werner helicase relocates into nuclear foci in response to DNA damaging agents and co-localizes with RPA and Rad51. *Genes Cells* 2001;6:421–430.

208. Kamath-Loeb AS, Shen JC, Loeb LA, et al. Werner syndrome protein. II. Characterization of the integral 3' $\rightarrow$ 5' DNA exonuclease. *J Biol Chem* 1998;273:34,145–34,150.

209. Fry M, Loeb LA. The three faces of the WS helicase. *Nature Genet* 1998;19:308–309.

210. Li B, Comai L. Functional interaction between Ku and the Werner syndrome protein in DNA end processing. *J Biol Chem* 2000;275:39,800.

211. Kamath-Loeb AS, Loeb LA, Johansson E, et al. Interactions between the Werner syndrome helicase and DNA polymerase delta specifically facilitate copying of tetraplex and hairpin structures of the d(CGG)n trinucleotide repeat sequence. *J Biol Chem* 2001;276:16,439–16,446.

212. Szekely AM, Chen YH, Zhang C, et al. Werner protein recruits DNA polymerase delta to the nucleolus. *Proc Natl Acad Sci USA* 2000;97:11,365–11,370.

213. Kamath-Loeb AS, Johansson E, Burgers PM, et al. Functional interaction between the Werner Syndrome protein and DNA polymerase delta. *Proc Natl Acad Sci USA* 2000;97:4603–4608.

214. Mohaghegh P, Hickson ID. DNA helicase deficiencies associated with cancer predisposition and premature ageing disorders. *Hum Mol Genet* 2001;10:741–746.

215. Kitao S, Shimamoto A, Goto M, et al. Mutations in RECQL4 cause a subset of cases of Rothmund–Thomson syndrome. *Nature Genet* 1999;22:82–84.
216. Weemaes CM, Hustinx TW, Scheres JM, et al. A new chromosomal instability disorder: the Nijmegen breakage syndrome. *Acta Paediatr Scand* 1981;70:557–564.
217. Taalman RD, Hustinx TW, Weemaes CM, et al. Further delineation of the Nijmegen breakage syndrome. *Am J Med Genet* 1989;32:425–431.
218. Conley ME, Spinner NB, Emanuel BS, et al. A chromosomal breakage syndrome with profound immunodeficiency. *Blood* 1986;67:1251–1256.
219. Jaspers NG, Gatti RA, Baan C, et al. Genetic complementation analysis of ataxia telangiectasia and Nijmegen breakage syndrome: a survey of 50 patients. *Cytogenet Cell Genet* 1988;49:259–263.
220. Pitts SA, Kullar HS, Stankovic T, et al. hMRE11: genomic structure and a null mutation identified in a transcript protected from nonsense-mediated mRNA decay. *Hum Mol Genet* 2001;10:1155–1162.
221. D'Amours D, Jackson SP. The mre11 complex: at the crossroads of dna repair and checkpoint signalling. *Nat Rev Mol Cell Biol* 2002;3:317–327.
222. Zhu XD, Kuster B, Mann M, et al. Cell-cycle-regulated association of RAD50/MRE11/NBS1 with TRF2 and human telomeres. *Nature Genet* 2000;25:347–352.
223. Saito S, Goodarzi AA, Higashimoto Y, et al. ATM mediates phosphorylation at multiple p53 sites, including Ser(46), in response to ionizing radiation. *J Biol Chem* 2002;277:12,491–12,494.
224. Gatei M, Scott SP, Filippovitch I, et al. Role for ATM in DNA damage-induced phosphorylation of BRCA1. *Cancer Res* 2000;60:3299–3304.
225. Gatei M, Young D, Cerosaletti KM, et al. ATM-dependent phosphorylation of nibrin in response to radiation exposure. *Nature Genet* 2000;25:115–119.

14 Role of BRCA1 and BRCA2 in DNA Repair and Potential Impact on Therapeutics

Razqallah Hakem, PhD

CONTENTS

1. BACKGROUND

DNA-damage repair pathways are essential for the normal growth and development of all organisms *(1)*. In mammals, defective repair of damaged DNA is associated with various diseases, including cancer *(2,3)*. DNA damage results from exogenous insults such as ionizing radiation (IR) and chemical agents, as well as from endogenous occurrences such as intracellular oxidative stress and chromosomal breaks. Errors during DNA replication and V(D)J recombination of the T- and B-lymphocyte antigen-receptor genes also give rise to DNA damage *(1)*. Mammalian cells have evolved a network of proteins to sense DNA damage and signal the repair of damaged DNA, or the induction of apoptosis if the damage is beyond repair. Defective repair of damaged DNA allows cells to acquire mutations that could confer a growth advantage, an important step toward malignancy. Particularly important in the maintenance of a normal genome are certain tumor suppressor genes (TSGs), including *ATM* (ataxia–telangiectasia mutated) *(4)* and the DNA mismatch repair (MMR) genes *MSH2* and *MSH6 (5)*. The function of some TSGs as guardians of genomic stability has led to their designation as "caretaker" genes *(6)*.

From: *Cancer Drug Discovery and Development: DNA Repair in Cancer Therapy*
Edited by: L. C. Panasci and M. A. Alaoui-Jamali © Humana Press Inc., Totowa, NJ

Our understanding of the mechanisms leading to sporadic and hereditary cancers has progressed significantly over the past decade. Studies of hereditary cancers in large kindreds with multiple affected individuals have facilitated the identification of novel cancer susceptibility genes. Interestingly, these cancer susceptibility genes are frequently also involved in sporadic cancer. For example, the TSG *p53* mutated in the rare autosomal dominantly inherited Li–Fraumeni syndrome is also the most frequently mutated TSG in all human cancers *(7)*.

Breast cancer is one of the commonest malignancies in women. It is estimated that one in eight women in the United States will develop breast cancer *(8)*. The majority of breast cancers are sporadic, and only 5–10% of all female breast cancers are considered to be hereditary *(9)*. Mutations of genes such as breast cancer susceptibility genes 1 and 2 *(BRCA1, BRCA2)*, *p53*, ataxia telangiectasia mutated *(ATM)* and phosphatase and tensin homolog deleted from chromosome 10 *(PTEN)* have been shown to predispose for breast cancer *(10)*.

The majority of the hereditary breast cancers that typically occur at a younger age, and are more frequently bilateral, are associated with *BRCA1/2* mutations *(8)*. Mutations of the *BRCA1* gene are associated with 80% of hereditary breast and ovarian cancers and half of all hereditary breast cancers *(11)*. *BRCA2* mutations are responsible for about 30% of hereditary breast cancers in women and also predispose affected men to male breast cancer *(11)*. In addition, carriers of *BRCA1* germline mutations have an elevated risk of developing prostate or colorectal cancer, and carriers of *BRCA2* germline mutations are prone to developing prostate or pancreatic cancer, or ocular melanoma *(8)*.

Hundreds of mutations of *BRCA1* and *BRCA2* *(BRCA1/2)* have been identified. Mutations associated with hereditary cancers are scattered over the coding regions of these genes (www.nhgri.nih.gov/Intramural_research/Lab_transfer/Bic). These mutations include insertions, deletions, and missense and nonsense mutations, all leading to either lack of protein expression or the synthesis of nonfunctional proteins. Furthermore, tumors isolated from individuals heterozygous for *BRCA1* or *BRCA2* mutations frequently display a loss of heterozygosity (LOH) for the remaining wild-type allele, indicating that *BRCA1* and *BRCA2* are typical TSGs according to Knudson's model (reviewed in ref. *12*). In contrast to hereditary cancers, mutations of *BRCA1* or *BRCA2* are rare in individuals with sporadic cancers; however, defective expression of these genes does occur in some cases. For example, BRCA1 expression was found reduced or totally lost in the majority of sporadic high-grade, ductal breast carcinomas studied *(13)*.

Ever since they were cloned, the *BRCA1/2* genes have been subject to extensive study to determine their function. Roles in both transcriptional regulation and DNA-damage repair have been ascribed to these two TSGs *(14)*. In this review, we describe the evidence supporting roles for BRCA1/2 in DNA damage repair pathways and the maintenance of the genome stability.

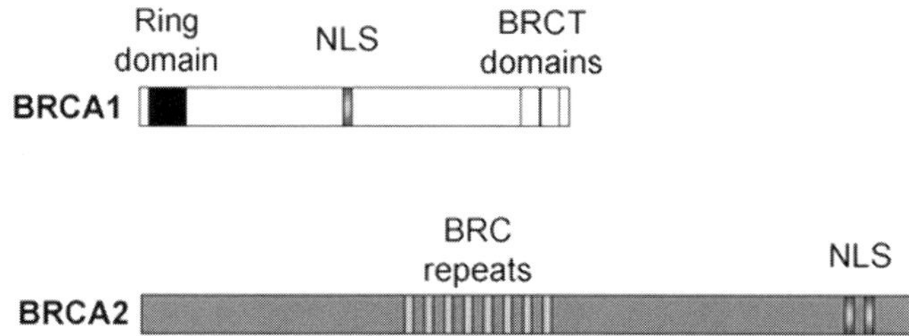

Fig. 1. Schematic representation of the BRCA1 and BRCA2 proteins showing some of their domains. The *BRCA1* and *BRCA2* genes and their proteins are defined in the text.

2. *BRCA1* AND *BRCA2* GENES AND PROTEINS

The human *BRCA1* and *BRCA2* genes encode two large, unrelated proteins of 1863 and 3418 amino acids, respectively *(8)*. The human BRCA1/2 proteins are approx 60% identical to their respective mouse orthologs, Brca1 and Brca2. The *BRCA1* gene, located on chromosome 17q21, is composed of 24 exons and encodes a major transcript of 7.8 kb. This transcript gives rise to a highly charged nuclear protein of 220 kDa which contains an N-terminal ring-finger domain, two nuclear localization signals, and two copies of the C-terminal BRCT (BRCA1 carboxy terminal) domain *(see* Fig. 1*)*. The BRCA1 ring-finger domain mediates its interaction with various proteins, including the BARD1 protein implicated in the control of RNA processing following DNA damage *(15)*. The BRCT domain is a phylogenetically conserved sequence found in many proteins involved in DNA repair and cell cycle regulation. The two BRCT domains of BRCA1 allow it to interact with proteins such as RNA Pol II, RNA helicase A, and CtIP, a protein that associates with the CtBP transcriptional corepressor *(16)*. Various other proteins interact with BRCA1 as described in the following.

The *BRCA2* gene is located on chromosome 13q12.3 and contains 27 exons that encode a 384-kDa protein. The BRCA2 protein contains nuclear localization signals (NLS) at the carboxyl terminus *(see* Fig. 1*)*. Eight internal repeats (BRC motifs), each composed of 30–80 amino acids, are encoded by *BRCA2* exon 11. These BRC repeats are essential for mediating the association of BRCA2 with the DNA repair protein RAD51 *(see* Subheading 3) *(17)*.

The *BRCA1/2* genes are expressed in most human tissues and their expression is increased in the late-G_1/early S phase of the cell cycle *(18)*. Various spliced forms of *BRCA1/2* transcripts have been identified, but their precise physiological functions remain under investigation. Whereas BRCA1 is expressed during human embryogenesis *(19)*, the embryonic expression of BRCA2 has not been reported. In mice, the *Brca1/2* genes are expressed in various adult tissues and throughout embryonic development. Interestingly, expression of *Brca1/2* is higher in proliferating murine tissues and cells than in their resting counterparts.

3. INVOLVEMENT OF BRCA1/2 IN DNA REPAIR

3.1. Interaction of BRCA1/2 With DNA Repair Proteins

Several lines of evidence support the hypothesis that BRCA1/2 are caretakers involved in DNA-damage repair and the maintenance of genomic stability. The first clues suggesting this involvement came from the finding that BRCA1 and BRCA2 both interact with RAD51 *(20)*. Subsequently, BRCA1 was shown to also interact with the DNA repair complex RAD50/MRE11/NBS1 *(21)*. In yeast, *Rad50, Rad51,* and *Mre11* (meiotic recombination 11) belong to the *Rad52* family of genes essential for repairing double strand breaks (DSBs) in DNA via homologous and nonhomologous recombination *(3)*. Interestingly, mutation of the *NBS1* gene in humans is associated with Nijmengen breakage syndrome (NBS), an ataxia–telangiectasia-like disorder characterized by chromosomal instability, increased radiosensitivity, and predisposition to cancer *(22)*. Furthermore, in humans, mutation of *Mre11* is also associated with another ataxia–telangiectasia-like disorder *(23)*.

Components of the DNA-damage repair machinery accumulate in large nuclear domains (foci) after DNA damage or blockage of replication. BRCA1 and BRCA2 colocalize with RAD51 in nuclear foci in the S- and G_2-phases of the somatic cell cycle and on the axial elements of developing synaptonemal complexes during meiosis *(24)*. Furthermore, BRCA1, BRCA2, and RAD51 all relocate to the proliferating cell nuclear antigen (PCNA)-positive replication sites following exposure of S-phase cells to hydroxyurea (HU) or ultraviolet (UV) irradiation *(24,25)*. In addition to its relocalization to sites of abnormal DNA structures generated during defective replication, BRCA1 also relocates to sites of DNA damage such as DSBs resulting from ionizing radiation (IR) *(25)*. These findings have led to the hypothesis that BRCA1/2 play a role in the repair of abnormal DNA structures generated at sites of DNA synthesis and DNA damage.

BRCA1 has also been shown to colocalize with other components of the DNA-damage sensing and repair machinery such as ATM, ATR (ATM and Rad3-related), Chk2/hCds1, BLM (Bloom's syndrome helicase), and RAD50/MRE11/NBS1 *(26)*. Moreover, BRCA1 is one of several proteins that are sequentially assembled at sites of DNA damage *(27)*. Within 1–3 min following DSB, the histone H2AX is phosphorylated (γ-H2AX) and forms early foci at DNA break sites. BRCA1 is then recruited to these foci and colocalizes with the γ-H2AX. The density of the γ-H2AX/BRCA1 foci reaches its maximum by 2 h postirradiation. Either RAD50 or RAD51 is subsequently recruited to the γ-H2AX in the foci, some time after BRCA1 is recruited. Various studies have indicated that the foci in an individual cell's nucleus contain BRCA1 and RAD50, or BRCA1 and RAD51, but not both complexes *(28)*. The precise significance of the selective interaction of BRCA1 with either RAD50 or RAD51 on the DNA-damage repair pathways remain to be investigated.

The BASC (BRCA1-associated genome surveillance complex) is a large complex containing proteins involved in DNA repair *(26)*. In addition to BRCA1, the components of the BASC include ATM, MSH2, MSH6, MLH1, BLM, and the RAD50/MRE11/NBS1 protein complex. BASC also contains three subunits of replication factor C (RFC), an ATPase that functions in DNA replication, DNA repair, and DNA metabolism-related checkpoints. The presence of BRCA1 in the BASC in association with proteins known to be involved in various aspects of DNA repair is convincing evidence that BRCA1 functions in a DNA-damage repair pathway. Nevertheless, it remains to be investigated whether the composition of the BASC varies according to cell type, cell cycle stage or type of DNA insult.

Other BRCA1-interacting proteins that have a putative DNA-damage repair function are BACH1, a putative DNA helicase, and FANCD2, a novel protein associated with Fanconi anemia (FA) *(29,30)*. Fanconi anemia is a hereditary disorder associated with developmental abnormalities, bone marrow failure, and susceptibility to myeloid leukemia and oral cancer *(31,32)*. Cells from FA patients exhibit chromosomal instability and hypersensitivity to agents that induce interstrand DNA crosslinks. The significance of the interaction between FANCD2 and BRCA1 is still unclear.

In vivo evidence directly demonstrating the binding of BRCA1 to DNA is lacking. However, a recent in vitro study showed that BRCA1 can bind to DNA with high affinity in a sequence nonspecific manner but with a preference for branched DNA *(33)*.

The evidence for BRCA2's involvement in DNA repair parallels that described earlier for BRCA1. BRCA2 colocalizes with RAD51 in nuclear foci after DNA damage and at synaptonemal complexes early in meiotic prophase. Recently, BRCA2 was identified as a component of a novel 2-MDa multiprotein complex *(34)*. A key protein in this large complex is BRAF35 (BRCA2-associated factor), a 35-kDa protein that interacts with BRCA2. BRAF35 specifically binds to cruciform DNA (as occurs in four-way junctions), suggesting that this protein might be the DNA-binding component of this multiprotein complex. In addition, BRCA2 and BRAF35 associate with chromatin during early phases of chromosome condensation during mitosis. The precise role of BRCA2 and BRAF35 interaction and the composition, and the function of the large protein complex that contains them, remain to be determined.

3.2. DNA-Damage-Induced Phosphorylation of BRCA1/2 Proteins

BRCA1 is a target for phosphorylation by the kinases ATM *(35)*, Chk2/hCds1 *(36)*, and ATR *(37)*. These kinases are key regulators of DNA-damage responses, and their mutations are associated with increased susceptibility to cancers *(38,39)*.

Mutations of the *ATM* gene are responsible for the rare human genetic disorder ataxia–telangiectasia (AT) *(40)*. AT patients are characterized by

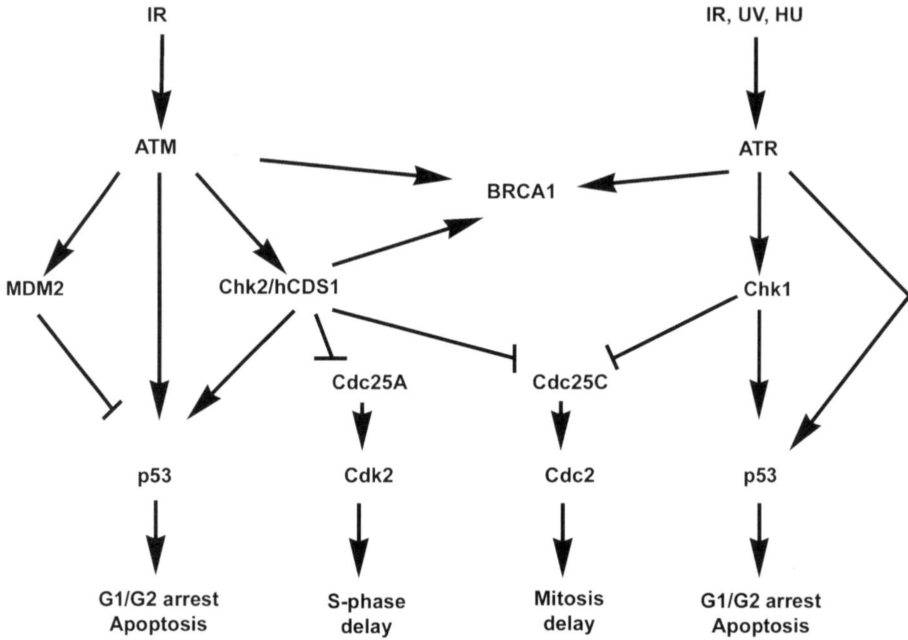

Fig. 2. Overview of the DNA-damage response pathways. Arrows indicate activating reactions, and T- bars denote inhibitory biochemical events. The proteins and the pathways are described in the text.

immunodeficiency, progressive cerebellar ataxia, radiosensitivity, defects in cell cycle checkpoints, and predisposition to leukemia and lymphoma. In response to DSBs, ATM phosphorylates several targets, including p53 (Ser15), Chk2/hCds1, NBS1 and H2AX *(3)*. DSB also trigger ATM to phosphorylate MDM2, preventing p53 nuclear transport and degradation *(see* Fig. 2).

Evidence is accumulating to support a role for *Chk2/hCds1* mutations in breast cancer *(41)*. Phosphorylation of Chk2/hCds1 is triggered by DNA-damage induced by IR or HU *(42)*. Activated Chk2/hCds1 phosphorylates Ser 20 of p53, inhibiting p53/MDM2 interaction and stabilizing p53 *(see* Fig. 2). Furthermore, activated Chk2/hCds1 also mediates phosphorylation of the phosphatase Cdc25A (Ser123), leading to its degradation and to a transient blockade at the S-phase *(43)*. A recent study has indicated that Chk2/hCds1 interacts with Mus81, a novel mammalian protein involved in resolving Holliday junctions that arise during homology-directed recombination and when DNA replication is blocked *(44)*.

ATR is activated by DSB, UV damage, and replication arrest, differentiating it from ATM which responds exclusively to DSBs *(see* Fig. 2). Activated ATR controls the G_1 checkpoint through its phosphorylation of p53 (Ser 15), and the G_2 checkpoint by phosphorylation/activation of Chk1 (Ser 345) *(45)*. Activated Chk1, like Chk2, inactivates Cdc25C by phosphorylation, triggering the G_2/M check-point. ATR also phosphorylates H2AX in response to replicational stress *(46)*.

A significant number of studies have investigated the phosphorylation of BRCA1 during cell cycle progression or in response to DNA insults. BRCA1 is a serine phosphoprotein that undergoes hyperphosphorylation during the late G_1- and S-phase of the cell cycle and transient dephosphorylation soon after the M-phase *(47)*. BRCA1 also becomes hyperphosphorylated in response to IR, UV, or HU and relocalizes rapidly from its native foci to sites of aberrant DNA structures *(24)*. Depending on the type of DNA insult and which DNA damage signaling pathway is activated, BRCA1 can become a target for phosphorylation by ATM, Chk2/hCds1, or ATR (*see* Fig. 2).

In contrast to BRCA1, the phosphorylation status of BRCA2 in the absence or presence of DNA insults remains poorly documented. This very large protein has been difficult to express in vitro, and, until recently, specific antisera against BRCA2 were not available. Nevertheless, in vitro kinase assays have indicated that BRCA2 peptides are phosphorylation targets for ATR, but not for ATM or DNA-PK (DNA-dependent protein kinase) *(48)*. BRCA2 has also been reported as an in vitro phosphorylation substrate for the mitotic checkpoint protein hBUBR1 *(49)*. The in vivo phosphorylation status of BRCA2 and the effect of this phosphorylation on BRCA2 function, if any, remains to be established.

3.3. Roles for BRCA1/2 in Homology-Directed Recombination

Double-stranded breaks are probably the most dangerous DNA insults that target eukaryotic cells *(3)*. DSBs are intentionally generated during normal cel-lular processes such as meiosis and V(D)J recombination of the T-cell receptor (TCR) and immunoglobulin (Ig) genes. DSB also naturally occur when replica-tion forks encounter single-strand DNA breaks. Exogenous insults such as IR and chemotherapeutic agents also generate DSBs. Regardless of their origin, efficient repair of DSBs is crucial for cells because DSBs either induce apoptosis or increase genomic instability that can lead to tumorigenesis *(1)*.

Mammalian cells have evolved highly regulated and complex systems to deal with damaged DNA. The repair of DSBs is mediated by two major pathways: homology-directed recombination (HDR) and nonhomologous end-joining (NHEJ) *(1)*. HDR of DSBs is more accurate than NHEJ because of the use of the undamaged sister chromatid as a template. HDR involves the RAD52 group of proteins, whereas NHEJ is mediated by Ku70, Ku80, DNA-PK, ligase IV, and XRCC4 *(1)*. The role of BRCA1/2 proteins in HDR and NHEJ has been inves-tigated in mouse Brca1/2-deficient cells. Compared to wild-type mouse embry-onic stem (ES) cells, ES cells harboring hypomorphic mutations of either *Brca1* or *Brca2* exhibit reduction in HDR and the repair of DSBs *(50)*. Similar defects have been identified in human cells harboring *BRCA2* mutations *(51)*. Taken together, these observations support a requirement for BRCA1 and BRCA2 in HDR.

The role of BRCA1/2 in the NHEJ pathway of DNA repair has been addressed by in vivo studies of the effect of *Brca1* and *Brca2* mutations on V(D)J recombination. During the initiation of V(D)J recombination in the TCR and Ig loci, DSBs are generated by the endonucleases RAG_1 and RAG_2 *(1)*. The repair of these DSBs is necessary for TCR and Ig expression, which, in turn, are prerequisites for T- and B-cell maturation, respectively. In vivo studies of hypomorph *Brca2* mutant mice *(52)* and mice harboring a T-cell-specific null mutation of *Brca1* *(53)* have indicated that V(D)J recombination of the TCR and Ig loci proceeds normally in the absence of Brca1 or Brca2. Thus, BRCA proteins are required for HDR but are dispensable for NHEJ.

3.4. BRCA1/2 Proteins and Transcription-Coupled Repair

Transcription-coupled repair (TrCR) mediates preferential repair of the transcribed strand of active genes *(54)*. In particular, TrCR is responsible for repairing DNA damage induced by oxidation or UV. Evidence is emerging that links the BRCA1/2 proteins to TrCR of DNA damaged by at least some agents. For example, ES cells and mouse embryonic fibroblasts (MEFs) from mice harboring a hypomorphic mutation of *Brca1* failed to perform TrCR of oxidative DNA damage, but were able to repair UV-induced damage *(55,56)*. Defective TrCR has also been observed in human cells carrying a *BRCA1* or *BRCA2* mutation *(57,58)*.

The effect of BRCA1 on TrCR may be mediated in part by BRCA1's interaction with the MMR protein MSH2, which is known to be involved in TrCR. Msh2–deficient mouse cells, like Brca1-deficient cells, have a defect in TrCR of oxidation-induced DNA damage *(59)*. However, unlike Brca1, Msh2 is also required for the TrCR of UV-induced DNA damage. The exact role of BRCA1/2 in TrCR and how they interact with the transcription machinery remain to be investigated.

3.5. BRCA1/2 Proteins and Maintenance of Genomic Stability

Damaged DNA (and particularly DSB) have been shown to lead to genomic instability *(60)*, a hallmark of cancer. Mammalian cells have developed several checkpoint proteins that, when activated, promote the elimination of cells with unrepaired damaged DNA. Therefore, only cells with incompetent checkpoints survive with damaged DNA and accumulate mutations. Defective HDR or NHEJ pathways are also often associated with increased genomic instability. Because BRCA1/2 are required for HDR, it is not surprising that cells deficient for BRCA1 or BRCA2 function exhibit signs of genomic instability. Cells from mice with mutations of either *Brca1* or *Brca2* accumulate chromosomal aberrations (*see* Fig. 3). These aberrations include polyploidy, chromosomal breaks and nonreciprocal translocations *(53)*. Similarly, tumors deficient for BRCA1 or BRCA2 exhibit increased genomic instability characterized by aneuploidy and chromosomal rearrangements *(61)*. These observations further support a "caretaker" role for BRCA1/2 in maintaining genomic stability.

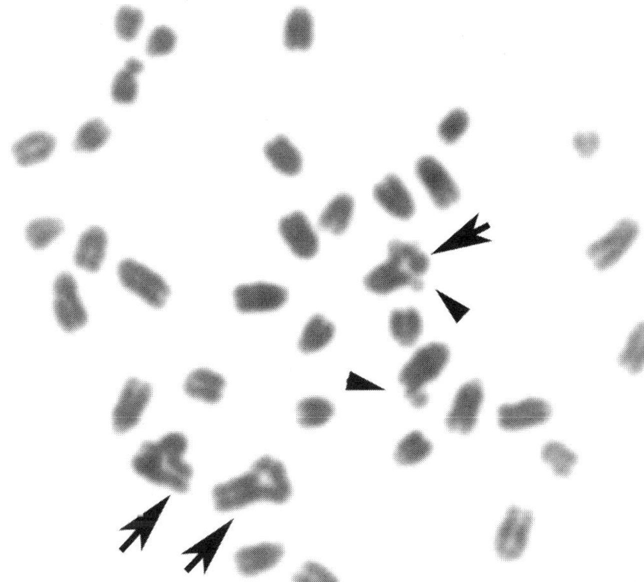

Fig. 3. Accumulation of chromosomal abberations in thymocytes deficient for Brca1. Brca1-deficient thymocytes, which overexpress the antiapoptotic protein Bcl-2, show a high level of chromosomal aberrations and DSBs. Arrows point to quadriradial chromosomes and arrowheads indicate DSBs.

4. IN VIVO MODEL FOR BRCA1/2-ASSOCIATED CANCERS

A standard method for attempting to create in vivo models for human cancers is to use gene-targeting techniques to "knock out" the orthologous TSG in mice *(62)*. Mice with mutation of a specific TSG have been generated this way, including mice deficient for Brca1 or Brca2. However, unlike humans with inherited *BRCA* mutations, mice hemizygous for a *Brca1* or *Brca2* mutation do not show an increased incidence of tumors *(62)*. Homozygous null mutation of *Brca1* or *Brca2* leads to early embryonic lethality at about E7.5–E9.5 *(63–67)*. In contrast to the null mutation, hypomorphic mutation of *Brca1* or *Brca2* results in a milder phenotype and delayed embryonic lethality, with some Brca2 hypomorphs surviving to birth *(52,68,69)*. This relatively mild phenotype is likely the result of incomplete loss of Brca1/2 function.

Brca1 and *Brca2* null mutant embryos exhibit defective cellular proliferation and activation of p53-dependent pathways *(67)*. A delay in the lethality of *Brca1* and *Brca2* null mutant embryos can be induced by breeding the mutation into a *p53* or *p21* null genetic background *(67,70,71)*. Furthermore, a hypomorphic *Brca1* mutant that normally dies at late stages of gestation was able to survive once crossed into a *p53* null background *(71)*.

Mice carrying conditional mutations of the *Brca1* or *Brca2* genes have been generated to study the effect of loss of these genes on adult tissues and tumorigenesis. Mice with a mammary gland hypomorphic *Brca1* mutation show increased apoptosis in mammary glands accompanied by abnormal ductal development *(72)*. Mammary gland tumors develop in these mice at a low frequency and with a long latency *(72)*. Similarly, mice with a T-cell lineage-specific mutation of *Brca1* show increased T-cell apoptosis; however, only a small subset of these animals develops thymic lymphomas *(53;* Hakem, unpublished data). Interestingly, p53 mutation drastically increases the frequency and decreases the latency of mammary tumors and thymic lymphomas in mice with a *Brca1* mutation targeted to the mammary glands or T-cell lineage, respectively *(72;* Hakem, unpublished data). The predisposition of Brca1-deficient mice to a wide range of cancers has been confirmed in a study of a hypomorph *Brca1* mutant strain that survives to adulthood *(73)*. These mutant animals develop lymphomas, sarcomas, and carcinomas, including mammary gland tumors.

Mouse models for *Brca2*-associated cancers have been generated also. A small subset of Brca2 hypomorph mutants survives to adulthood, but all of these animals succumb to thymic lymphomas by 12–14 wk of age *(69)*. Interestingly, mice with *Brca2* mutations (deletion of exons 3 and 4) targeted specifically to mammary gland epithelial cells develop mammary adenocarcinomas with high incidence and after a long latency period (over 1 yr) *(74)*. Another study of a different strain of *Brca2* conditional mutant mice carrying a deletion of exon 11 in epithelial tissues has indicated that these mice are healthy and are not predisposed to cancer. However, in a p53 null background, mice with the conditional *Brca2* exon 11 mutation develop mammary and skin cancers *(75)*. The basis for the different outcome of these two *Brca2* mutations is unclear and could be the result of alternative splicing.

Overall, these studies indicate that in mice, as in humans, mutation of *Brca1* or *Brca2* predisposes for cancer (including breast cancer), and this tumorigenesis is enhanced in p53 mutant backgrounds. These results are consistent with the finding that tumors from BRCA1 or BRCA2 patients frequently exhibit loss of p53 function *(76–79)*. Defective p53 checkpoint allows BRCA1 and BRCA2 mutant cells to survive the accumulation of DNA damage and genomic instability and therefore increase their risk to acquire other oncogenic mutations.

5. POTENTIAL IMPACT ON THERAPEUTICS

DNA-based testing for *BRCA1* and *BRCA2* cancer-predisposing mutations is available for families with high risk of familial breast cancer. Several strategies have been suggested to reduce cancer risk in individuals with *BRCA1* or *BRCA2* cancer-predisposing mutations. These strategies include cancer screening, prophylactic mastectomy and/or oophorectomy, and chemoprevention *(80)*.

Although our knowledge regarding the function of BRCA1/2 has significantly increased over the years, there are still various questions that remain to be answered. BRCA1 and BRCA2 are tumor suppressors that are required for various essential functions of normal cells. Certain BRCA1/2 functions such as their role in transcription and differentiation remain to be better understood. Also, their precise role in DNA repair is yet to be unraveled.

More trials are required to address the efficiency of the available therapeutic strategies for BRCA1/2 cancer patients. It is our hope that further clinical knowledge and better characterization of the various functions of BRCA1/2 will help improve the prevention and the treatment of BRCA1/2-associated tumors. Molecular profiling is promising in helping make more accurate and patient-specific therapeutic decisions. A recent study has shown that molecular profiling could efficiently predict breast cancer prognosis *(81)*. Molecular profiling will help identify genes that are overexpressed in BRCA1 and BRCA2 tumors and these could be potential targets for the rational development of new BRCA1/2 cancer drugs. Finally, although normal cells deficient for BRCA1/2 are hypersensitive to irradiation because of their defective DNA-damage repair, the frequent loss of function of p53 in the BRCA1/2 tumors hamper the use of irradiation or other DNA-damaging agents as potent therapeutic approaches. A better understanding of the cellular pathways involving BRCA1/2 and p53 could lead to the identification of novel downstream targets for therapeutic intervention in familial BRCA1/2 associated cancers.

ACKNOWLEDGMENTS

I thank members of my laboratory for reviewing the manuscript. I apologize to those whose work was not cited directly owing to space limitations. Research in my laboratory is supported by the Canadian Institute of Health Research and National Canadian Institute of Cancer.

REFERENCES

1. van Gent, DC, Hoeijmakers JH, Kanaar R. Chromosomal stability and the DNA double-stranded break connection. *Nature Rev Genet* 2001;2(3):196–206.
2. Hoeijmakers JH. Genome maintenance mechanisms for preventing cancer. *Nature* 2001; 411(6835):366–374.
3. Khanna KK, Jackson SP. DNA double-strand breaks: signaling, repair and the cancer connection. *Nature Genet* 2001;27(3):247–254.
4. Kastan MB, Lim DS. The many substrates and functions of ATM. *Nature Rev Mol Cell Biol* 2000;1(3):179–186.
5. Harfe BD, Jinks-Robertson S. DNA mismatch repair and genetic instability. *Annu Rev Genet* 2000;34:359–399.
6. Kinzler KW, Vogelstein B. Cancer-susceptibility genes. Gatekeepers and caretakers. *Nature* 1997;386(6627):761, 763.
7. Vogelstein B, Lane D, Levine AJ. Surfing the p53 network. *Nature* 2000;408(6810):307–310.

8. Rahman N, Stratton MR. The genetics of breast cancer susceptibility. *Annu Rev Genet* 1998;32:95–121.

9. Easton DF, Bishop DT, Ford D, et al. Genetic linkage analysis in familial breast and ovarian cancer: results from 214 families. The Breast Cancer Linkage Consortium. *Am J Hum Genet* 1993;52(4):678–701.

10. Arver B, Du Q, Chen J, et al. Hereditary breast cancer: a review. *Semin Cancer Biol* 2000;10(4):271–288.

11. Ford D, Easton DF, Stratton M, et al. Genetic heterogeneity and penetrance analysis of the BRCA1 and BRCA2 genes in breast cancer families. The Breast Cancer Linkage Consortium. *Am J Hum Genet* 1998;62(3):676–689.

12. Knudson AG. Chasing the cancer demon. *Annu Rev Genet* 2000;34:1–19.

13. Wilson CA, Ramos L, Villasenor MR, et al. Localization of human BRCA1 and its loss in high-grade, non-inherited breast carcinomas. *Nature Genet* 1999;21(2):236–240.

14. Kerr P and Ashworth A, New complexities for BRCA1 and BRCA2. *Curr Biol* 2001;11(16): R668–R676.

15. Kleiman FE, Manley JL. The BARD1-CstF-50 interaction links mRNA 3' end formation to DNA damage and tumor suppression. *Cell* 2001;104(5):743–753.

16. Welcsh PL, King MC. BRCA1 and BRCA2 and the genetics of breast and ovarian cancer. *Hum Mol Genet* 2001;10(7):705–713.

17. Chen PL, Chen CF, Chen Y, et al. The BRC repeats in BRCA2 are critical for RAD51 binding and resistance to methyl methanesulfonate treatment. *Proc Natl Acad Sci USA* 1998;95(9): 5287–5292.

18. Gudas JM, Li T, Nguyen H, et al. Cell cycle regulation of BRCA1 messenger RNA in human breast epithelial cells. *Cell Growth Differ* 1996;7(6):717–723.

19. Buraczynska MJ, Van Keuren ML, Buraczynska KM, et al. Construction of human embryonic cDNA libraries: HD, PKD1 and BRCA1 are transcribed widely during embryogenesis. *Cytogenet Cell Genet* 1995;71(2):197–202.

20. Scully R, Chen J, Plug A, et al. Association of BRCA1 with Rad51 in mitotic and meiotic cells. *Cell* 1997;88(2):265–275.

21. Zhong Q, Chen CF, Li S, et al. Association of BRCA1 with the hRad50–hMre11–p95 complex and the DNA damage response. *Science* 1999;285(5428):747–750.

22. Matsuura S, Tauchi H, Nakamura A, et al. Positional cloning of the gene for Nijmegen breakage syndrome. *Nature Genet* 1998;19(2):179–181.

23. Stewart GS, Maser RS, Stankovic T, et al. The DNA double-strand break repair gene hMRE11 is mutated in individuals with an ataxia-telangiectasia-like disorder. *Cell* 1999;99(6): 577–587.

24. Chen J, Silver DP, Walpita D, et al. Stable interaction between the products of the BRCA1 and BRCA2 tumor suppressor genes in mitotic and meiotic cells. *Mol Cell* 1998;2(3):317–328.

25. Scully R, Chen J, Ochs RL, et al. Dynamic changes of BRCA1 subnuclear location and phosphorylation state are initiated by DNA damage. *Cell* 1997;90(3):425–435.

26. Wang Y, Cortez D, Yazdi P, et al. BASC, a super complex of BRCA1-associated proteins involved in the recognition and repair of aberrant DNA structures. *Genes Dev* 2000;14(8):927–939.

27. Paull TT, Rogakou EP, Yamazaki V, et al. A critical role for histone H2AX in recruitment of repair factors to nuclear foci after DNA damage. *Curr Biol* 2000;10(15):886–895.

28. Maser RS, Monsen KJ, Nelms BE, et al. hMre11 and hRad50 nuclear foci are induced during the normal cellular response to DNA double-strand breaks. *Mol Cell Biol* 1997;17(10): 6087–6096.

29. Cantor SB, Bell DW, Garesah S, et al. BACH1, a novel helicase-like protein, interacts directly with BRCA1 and contributes to its DNA repair function. *Cell* 2001;105(1):149–160.

30. Garcia-Higuera I, Taniguchi T, Ganesan S, et al. Interaction of the Fanconi anemia proteins and BRCA1 in a common pathway. *Mol Cell* 2001;7(2):249–262.

31. Grompe M, D'Andrea A. Fanconi anemia and DNA repair. *Hum Mol Genet* 2001;10(20): 2253–2259.
32. Giampietro PF, Auerbach AD, Elias ER, et al. New recessive syndrome characterized by increased chromosomal breakage and several findings which overlap with Fanconi anemia. *Am J Med Genet* 1998;78(1):70–75.
33. Paull TT, Cortez D, Bowers B, et al. From the Cover: Direct DNA binding by Brca1. *Proc Natl Acad Sci USA* 2001;98(11):6086–6091.
34. Marmorstein LY, Kinev AV, Chan GK, et al. A human BRCA2 complex containing a structural DNA binding component influences cell cycle progression. *Cell* 2001;104(2): 247–257.
35. Cortez D, Wang Y, Qin J, et al. Requirement of ATM-dependent phosphorylation of brca1 in the DNA damage response to double-strand breaks. *Science* 1999;286(5442):1162–1166.
36. Lee JS, Collins KM, Brown AL, et al. hCds1-mediated phosphorylation of BRCA1 regulates the DNA damage response. *Nature* 2000;404(6774):201–204.
37. Tibbetts RS, Cortez D, Brumbaugh, KM, et al. Functional interactions between BRCA1 and the checkpoint kinase ATR during genotoxic stress. *Genes Dev* 2000;14(23):2989–3002.
38. Sullivan A, Yuille M, Repellin C, et al. Concomitant inactivation of p53 and Chk2 in breast cancer. *Oncogene* 2002;21(9):1316–1324.
39. Lee SB, Kim SH, Bell DW, et al. Destabilization of CHK2 by a missense mutation associated with Li–Fraumeni Syndrome. *Cancer Res* 2001;61(22):8062–807.
40. Moses RE. DNA damage processing defects and disease. *Annu Rev Genomics Hum Genet* 2001;2:41–68.
41. Meijers-Heijboer H, van den Ouweland A, Klijn J, et al. Low-penetrance susceptibility to breast cancer due to CHEK2(*)1100delC in noncarriers of BRCA1 or BRCA2 mutations. *Nature Genet* 2002;31(1):55–59.
42. Bartek J, Falck J, Lukas J. CHK2 kinase—a busy messenger. *Nature Rev Mol Cell Biol* 2001;2(12):877–86.
43. Falck J, Mailand N, Syljuasen RG, et al. The ATM–Chk2–Cdc25A checkpoint pathway guards against radioresistant DNA synthesis. *Nature* 2001;410(6830):842–847.
44. Chen XB, Melchionna R, Denis CM, et al. Human Mus81-associated endonuclease cleaves Holliday junctions in vitro. *Mol Cell* 2001;8(5):1117–1127.
45. Abraham RT. Cell cycle checkpoint signaling through the ATM and ATR kinases. *Genes Dev* 2001;15(17):2177–2196.
46. Ward IM, Chen J. Histone H2AX is phosphorylated in an ATR-dependent manner in response to replicational stress. *J Biol Chem* 2001;276(51):47,759–47,762.
47. Ruffner H. Verma IM. BRCA1 is a cell cycle-regulated nuclear phosphoprotein. *Proc Natl Acad Sci USA* 1997;94(14):7138–7143.
48. Kim ST, Lim DS, Canman CE, et al. Substrate specificities and identification of putative substrates of ATM kinase family members. *J Biol Chem* 1999;274(53):37,538–37,543.
49. Futamura M, Arakawa H, Matsuda K, et al. Potential role of BRCA2 in a mitotic checkpoint after phosphorylation by hBUBR1. *Cancer Res* 2000;60(6):1531–1535.
50. Moynahan ME, Chiu JW, Koller BH, et al. Brca1 controls homology-directed DNA repair. *Mol Cell* 1999;4(4):511–518.
51. Moynahan ME, Pierce AJ, Jasin M. BRCA2 is required for homology-directed repair of chromosomal breaks. *Mol Cell* 2001;7(2):263–272.
52. Friedman LS, Thistlethwaite FC, Patel KJ, et al. Thymic lymphomas in mice with a truncating mutation in Brca2. *Cancer Res* 1998;58(7):1338–1343.
53. Mak TW, Hakem A, McPherson JP, et al. Brca1 required for T cell lineage development but not TCR loci rearrangement. *Nature Immunol* 2000;1(1):77–82.
54. Svejstrup JQ. Mechanisms of transcription-coupled DNA repair. *Nature Rev Mol Cell Biol* 2002;3(1):21–29.

55. Gowen, L.C, Aurutskaya AV, Latour AM, et al. BRCA1 required for transcription-coupled repair of oxidative DNA damage. *Science* 1998;281(5379):1009–1012.
56. Cressman VL, Backlund DC, Aurutskaya AV, et al. Growth retardation, DNA repair defects, and lack of spermatogenesis in BRCA1-deficient mice. *Mol Cell Biol* 1999;19(10): 7061–7075.
57. Abbott DW, Thompson ME, Robinson-Benion C, et al. BRCA1 expression restores radiation resistance in BRCA1-defective cancer cells through enhancement of transcription-coupled DNA repair. *J Biol Chem* 1999;274(26):18,808–18,812.
58. Le Page F, Randrianarison V, Marot D, et al. BRCA1 and BRCA2 are necessary for the transcription-coupled repair of the oxidative 8-oxoguanine lesion in human cells. *Cancer Res* 2000;60(19):5548–5552.
59. Mellon I, Rajpal DK, Koi M, et al. Transcription-coupled repair deficiency and mutations in human mismatch repair genes. *Science* 1996;272(5261):557–560.
60. Richardson C, Jasin M. Frequent chromosomal translocations induced by DNA double-strand breaks. *Nature* 2000;405(6787):697–700.
61. Tirkkonen M, Kainu T, Loman N, et al. Distinct somatic genetic changes associated with tumor progression in carriers of BRCA1 and BRCA2 germ-line mutations. *Cancer Res* 1997;57(7):1222–1227.
62. Hakem R, Mak TW. Animal models of tumor-suppressor genes. *Annu Rev Genet* 2001;35: 209–241.
63. Hakem R, de la Pompa JL, Sirard C, et al. The tumor suppressor gene Brca1 is required for embryonic cellular proliferation in the mouse. *Cell* 1996;85(7):1009–1023.
64. Liu CY, Flesken-Nikitin A, Li S, et al. Inactivation of the mouse Brca1 gene leads to failure in the morphogenesis of the egg cylinder in early postimplantation development. *Genes Dev* 1996;10(14):1835–1843.
65. Suzuki A, de la Pompa JL, Hakem R, et al. Brca2 is required for embryonic cellular proliferation in the mouse. *Genes Dev* 1997;11(10):1242–1252.
66. Sharan SK, Morimatsu M, Albrecht U, et al. Embryonic lethality and radiation hypersensitivity mediated by Rad51 in mice lacking Brca2. *Nature* 1997;386(6627):804–810.
67. Ludwig T, Chapman DL, Papaioannou VE, et al. Targeted mutations of breast cancer susceptibility gene homologs in mice: lethal phenotypes of Brca1, Brca2, Brca1/Brca2, Brca1/p53, and Brca2/p53 nullizygous embryos. *Genes Dev* 1997;11(10):1226–1241.
68. Gowen LC, Johnson BL, Latour AM, et al. Brca1 deficiency results in early embryonic lethality characterized by neuroepithelial abnormalities. *Nature Genet* 1996;12(2):191–194.
69. Connor F, Bertwistle D, Mee PJ, et al. Tumorigenesis and a DNA repair defect in mice with a truncating Brca2 mutation. *Nature Genet* 1997;17(4):423–430.
70. Hakem R, de la Pompa JL, Elia A, et al. Partial rescue of Brca1 (5–6) early embryonic lethality by p53 or p21 null mutation. *Nature Genet* 1997;16(3):298–302.
71. Xu X, Qiao W, Linke SP, et al. Genetic interactions between tumor suppressors Brca1 and p53 in apoptosis, cell cycle and tumorigenesis. *Nature Genet* 2001;28(3):266–71.
72. Xu X, Wagner KV, Larson D, et al. Conditional mutation of Brca1 in mammary epithelial cells results in blunted ductal morphogenesis and tumour formation. *Nature Genet* 1999;22(1):37–43.
73. Ludwig T, Fisher P, Ganesan S, et al. Tumorigenesis in mice carrying a truncating Brca1 mutation. *Genes Dev* 2001;15(10):1188–1193.
74. Ludwig T, Fisher P, Murty V, et al. Development of mammary adenocarcinomas by tissue-specific knockout of Brca2 in mice. *Oncogene* 2001;20(30):3937–3948.
75. Jonkers J, Meuwissen R, van der Gulden H, et al. Synergistic tumor suppressor activity of BRCA2 and p53 in a conditional mouse model for breast cancer. *Nature Genet* 2001;29(4): 418–425.
76. Crook T, Crossland S, Crompton MR, et al. p53 mutations in BRCA1-associated familial breast cancer. *Lancet* 1997;350(9078):638–639.

77. Buller RE, Lallas TA, Shlain MS, et al. The p53 mutational spectrum associated with BRCA1 mutant ovarian cancer. *Clin Cancer Res* 2001;7(4):831–838.
78. Phillips KA, Nichol K, Ozcelik H, et al. Frequency of p53 mutations in breast carcinomas from Ashkenazi Jewish carriers of BRCA1 mutations. *J Natl Cancer Inst* 1999;91(5):469–473.
79. Gretarsdottir S, Thorlacius S, Valgardsdottir R, et al. BRCA2 and p53 mutations in primary breast cancer in relation to genetic instability. *Cancer Res* 1998;58(5):859–862.
80. Chlebowski RT. Breast cancer risk reduction: strategies for Women at Increased Risk. *Annu Rev Med* 2002;53:519–540.
81. van 't Veer LJ, Dai H, van de Vijver MJ, et al. Gene expression profiling predicts clinical outcome of breast cancer. *Nature* 2002;415(6871):530–536

15 Genetic Variations in DNA Repair

Their Implications in Human Cancer Risk, Prevention, and Therapy

Jennifer J. Hu, PhD

1. INTRODUCTION

Mammalian cells are constantly exposed to endogenous and exogenous genotoxic agents, and distinct repair pathways have evolved to handle the different types of DNA damage. The six major repair pathways are base excision repair (BER), direct reversal of damage, mismatch excision repair (MMR), nucleotide excision repair (NER), homologous recombination (HR), and nonhomologous end-joining (NHEJ). Over 130 human DNA repair genes have been reported, and more will be identified. Many single-nucleotide polymorphisms (SNPs) of DNA repair genes have also been reported and some of them may affect DNA repair functions. Although, on an individual basis, SNPs may cause only small changes in repair activity, their presence at polymorphic frequencies in the human population may contribute to a high proportion of cancer cases and the variation of interindividual responses to cancer therapies.

This chapter focuses on the implications of genetic variations in DNA repair for assessing, preventing, and treating human cancer. Using case–case and

From: *Cancer Drug Discovery and Development: DNA Repair in Cancer Therapy*
Edited by: L. C. Panasci and M. A. Alaoui-Jamali © Humana Press Inc., Totowa, NJ

339

case–control studies, researchers have identified genetic variations in DNA repair that may predispose (sub)populations to certain cancers and explain the variability of individual responses to cancer treatment. The ultimate goal is to apply DNA repair SNP information to the development of "personalized medicine" that tailors cancer prevention strategies and treatments to an individual's genetic profile.

2. MAJOR DNA REPAIR PATHWAYS

DNA repair functions are a major defense against environmental damage to cells in all organisms. Efficient DNA repair is critical in processes that minimize cell death, mutations, replication errors, and genomic instability. Three different types of mammalian repair pathway are (1) single-step reactions with a single protein (i.e., direct reversal), (2) single and multistep reactions with a single or multiple proteins (i.e., BER), and (3) multistep reactions with multiple protein components (i.e., NER, MMR, and double-strand breaks [DSB]s repair). More than 130 human DNA repair genes have been reported *(1)*, and the updated table is available at www.cgal.icnet.uk.

2.1. Base Excision Repair

Base excision repair proteins excise and replace damaged DNA bases, mainly those arising from endogenous oxidative and hydrolytic decay of DNA. The human genome is exposed to as many as 10,000 oxidative insults every day, most of which are removed by BER. Therefore, efficient BER is critical in maintaining genomic integrity. Human genes involved in BER are *UNG, SMUG1, TDG, OGG1, MYH, NTH1, MPG, APE1/2, ADPRT, PCNA, POL-β, POLD1*, and *XRCC1 (1)*.

Most recent research focuses on the roles of SNPs in *OGG1, XRCC1*, and *APE1* in cancer susceptibility and treatment response *(2–10)*. Glycosylases initiate BER by releasing the modified base. Cleavage of the sugar–phosphate chain, excision of the abasic residue, and local DNA synthesis and ligation follow. The human 8-oxoguanine DNA glycosylase 1 (*hOGG1*) gene encodes a DNA repair enzyme that excises 7,8-dihydro-8-oxoguanine (8oxoG), which is highly mutagenic, from the DNA, yielding G:C to T:A transversions *(11,12)*. Therefore, decreased OGG1 activity could lead to increased oxidative-stress-related cancer risk.

XRCC1 functions in the repair of single-strand DNA breaks in mammalian cells and forms a repair complex with polymerase β (pol β), DNA ligase III, and poly(ADP-ribose) polymerase (PARP) *(13)*. *XRCC1*-mutant cells have increased sensitivity to ionizing radiation (IR), ultraviolet (UV), hydrogen peroxide, and mitomycin C *(13)*.

APE/Ref-1 is the rate-limiting enzyme in the BER pathway *(14)*. It cleaves 5'- of DNA abasic sugar residues, generated from exogenous factors, such as IR, carcinogens, and endogenous agents from normal cellular metabolism.

2.2. Direct Reversal of Damage

O^6-Methylguanine DNA methyltransferase (MGMT) acts as a single protein that reverses specific types of alkylation damage simply by removing the alkyl group *(15)*. A number of DNA-damaging chemotherapeutic agents attack the *O(6)* position on guanine, forming cytotoxic DNA adducts, such as carmustine, temozolomide, streptozotocin, and dacarbazine *(16)*. The types of damage that MGMT repair are generated by both carcinogenic and chemotherapeutic alkylating agents. Therefore, MGMT plays a critical role in both carcinogenesis and cancer chemotherapy.

2.3. Mismatch Excision Repair

Mismatch excision repair is responsible for repairing DNA damage resulting from replication errors, spontaneous or induced base modifications, and recombination *(17)*. A subset of cancers with mutations of MMR genes exhibit a particular phenotype called microsatellite instability (MSI) *(18–20)*. Human MMR genes include *MSH2/3/4/5/6*, *MLH1/3*, and *PMS2/L3/L4 (1)*. Mutations in *hMSH2* and *hMLH1* genes contribute to 80% of the hereditary nonpolyposis colorectal cancer (HNPCC) syndrome *(21)*.

2.4. Nucleotide Excision Repair

Nucleotide excision repair acts on a wide variety of DNA-damaging agents, such as the pyrimidine dimers caused by UV light and chemical adducts induced by 4-nitroquinoline 1-oxide and cisplatin *(22)*. However, NER can also process nonbulky lesions, such as oxidized and alkylated bases *(23)*. There are two NER subpathways: (1) transcription-coupled repair, which removes DNA damage preferentially from actively transcribed genes, and (2) global genome repair, which removes damage throughout the genome *(24)*.

The most important function of NER in humans is to remove UV-induced damage from DNA. This critical role is apparent in the inherited disorder xeroderma pigmentosum (XP), where NER-defective individuals have a 1000 times greater risk of skin cancer than normal individuals *(25)*. Human NER genes include *XPC*, *RAD23B, CETN2, RAD23A, XPA, RPA1/2/3*, TFIIH *(XPB, XPD, CTF2H1/2/3/4, CDK7, CCNH, MNAT1), XPG, ERCC1, XPF*, and *LIG1 (1)*. The SNPs in XPD have been evaluated extensively in several human studies *(26–28)*. *XPD* is the 5' helicase in the two human transcription factor IIH (TFIIH) complex, which participates in NER *(29)*. Mammalian TFIIH carrying an active-site mutant *XPD* subunit does not repair effectively, despite the induction of substantial damage-dependent DNA synthesis, which is consistent with the decreased cell survival in XPD patients *(30)*.

2.5. Homologous Recombination and Nonhomologous End-Joining

DNA DSBs are a deleterious form of DNA damage that can result in loss and rearrangement of genomic material, which, in turn, lead to cell death or carcinogenesis. DSBs are induced by IR, but they also arise as intermediates in normal endogenous processes, including DNA replication and meiotic and V(D)J recombination. The induction of DSBs elicits a cascade of events, including break sensing, signal transduction, and effector functions. The cellular mechanisms involved include DNA repair, cell cycle checkpoint arrest, and apoptosis, which combine to limit the generation of mutations and the proliferation of damaged cells.

Two distinct mechanisms function to repair DSBs in mammalian cells. First, HR repairs a DSB in one DNA molecule by using a second, homologous molecule as a template for DNA synthesis and repair, thereby enabling retrieval of coding information lost at the site of the break. Although HR has the potential to be a highly accurate and efficient pathway for repairing DSBs, in higher eukaryotes it does not seem to play a major role in the repair of radiation-induced DSBs. Mammalian mutants defective in HR display only mild radiosensitivity. However, such mutants are hypersensitive to DNA crosslinking agents *(31)*. Human HR genes include *RAD51, DMC1, XRCC2, XRCC3, RAD52, RAD54L/B, BRCA1/2, RAD50, MRE11A*, and *NBS1 (1)*.

The second and potentially more significant pathway for repair of radiation-induced DSBs in mammalian cells is NHEJ. During NHEJ, the ends of a break are often modified by the addition and deletion of nucleotides and then ligated to restore covalent continuity to a broken chromosome that requires little or no sequence homology at its ends *(32)*. In contrast to HR, mammalian cells with defects in NHEJ components display marked radiosensitivity *(31)*. Five genes function in human NHEJ: *DNA-PK* complex, *Ku70, Ku80, XRCC4,* and *LIG4 (1)*.

3. DNA REPAIR SNPS IN CANCER RISK ASSESSMENT AND PREVENTION

Interindividual variability in responses to carcinogens or treatments has been described. Much attention has been devoted to genetic polymorphisms in drug metabolism and detoxification. Another important source of variability is in DNA repair, including the genetic instability syndromes. These are rare, recessive traits that include ataxia–telangiectasia (AT), Fanconia anemia (FA), Bloom's syndrome, XP, and HNPCC. More common DNA repair polymorphisms occur at a population frequency of at least 1–5% *(29,33–36)*. Polymorphisms are defined as sequence variations resulting from nucleotide substitutions, deletions/insertions, and gene duplications/deletions. They may or may not result in changes in protein function or phenotype. If a polymorphism is in an exon within a gene's coding region, amino acid substitution may result in changes in protein activity. When a polymorphism is located in the promoter region, it may alter gene expression. Finally, if a polymorphism is located at an intron/exon boundary, it may produce incomplete or inactive proteins as a result of incorrect mRNA splicing.

Inherited defective DNA repair may be associated with human cancer risk. In the general population, there is probably a mixture of high-risk, rare cancer predisposition genes, such as *XP*, *ATM*, *BRCA1/2*, and *p53*, and low-risk but more common genetic polymorphisms, such as those involved in DNA repair. Although it is important to evaluate genetic variants of DNA repair in human cancer risk, their functional significance must be elucidated first. At the present time, the most rapidly growing research areas are focusing on the association between DNA repair genotype and phenotype in human cancer susceptibility and response to cancer therapy.

3.1. DNA Repair SNPs and Human Cancer Risk

3.1.1. *XRCC1* SNPs and Cancer Risk

XRCC1 plays an important role in BER and participates as a scaffolding intermediate by interacting with Lig III, DNA pol β, and PARP in the C-terminal, N-terminal, and central regions, respectively *(37,38)*. *XRCC1* mutant cells have increased sensitivity to IR, UV, hydrogen peroxide, and mitomycin C *(13)*. The *XRCC1*-variant allele, Arg194Trp, results in a nonconservative substitution in a hydrophobic region of *XRCC1*, and the SNP of Arg399Gln occurs within the BRCA1 C-terminal domain known to interact with PARP *(29)*. The *XRCC1* 194Arg allele was associated with increased risk for oral cavity and pharyngeal cancers as well as bladder cancer *(2,3)*. The *XRCC1* 399Gln/Gln genotype was associated with adenocarcinoma of the lung and breast cancer *(4,5)*. Interestingly, when gene–exposure interactions were considered, the *XRCC1* 399Arg/Arg genotype was associated with breast cancer risk in subjects with occupational IR exposure *(4)* and esophageal cancer risk among alcohol drinkers, respectively *(6)*. The conflicting results may stem from the complexity of cancer etiology, encompassing potential relationships between carcinogenic exposures and other genetic factors. Because XRCC1 interacts with many other DNA repair proteins, studies to elucidate the functions of *XRCC1* variants, and their effects on protein/protein complex formation are needed.

3.1.2. *XRCC3* SNPs and Cancer Risk

XRCC3 is a member of the *Rad51* DNA repair gene family. It functions in the homologous recombinational repair (HRR) pathway by repairing DSBs and is required for maintaining chromosome stability in mammalian cells *(39,40)*. *XRCC3* mutant cells show moderate hypersensitivity to IR, UV, and monofunctional alkylating agents but extreme sensitivity to DNA crosslinking drugs, such as mitomycin C or cisplatin *(41–43)*. The *XRCC3* Thr241Met-variant may contribute to melanoma skin cancer and bladder cancer risk *(26,44)*. In addition, two polymorphisms in XRCC3, T241M and IVS5 A>G at nt 17893, were associated with breast cancer risk *(45)*.

3.1.3. *OGG1* SNPs and Cancer Risk

OGG1 is a DNA repair enzyme that excises the mutagenic (8oxoG) from DNA produced by reactive oxygen species *(11,12)*. The *hOGG1* Ser326Cys polymorphism was associated with risk of esophageal and lung cancers *(8,9)*. In addition, the genotype frequency of two *OGG1* sequence variants (11657A/G and Ser326Cys) significantly differed between prostate cancer cases and controls *(10)*.

3.2. The DNA Repair Genotype/Phenotype Relationship

In contrast to rare, highly penetrant alleles, low-penetrance susceptibility alleles may contribute to a substantial proportion of cancer cases, because some are very common in the general population. It is clearly important to evaluate genetic variants of DNA repair in human cancer risk, but, first, their functional significance must be elucidated. Current research on DNA repair is also working to determine the association between sequence variations with heritable phenotypes and cancer susceptibility.

Two approaches have been applied to study the genotype/phenotype relationship of DNA repair SNPs. First, most human studies have evaluated the association between DNA repair SNPs and damage/repair end points. Several examples are described here. *XRCC1* codon 280 polymorphism is linked to hypersensitivity to bleomycin *(46)*. Amino acid substitution variants of *XRCC1* and *APE1* may contribute to hypersensitivity to IR in breast cancer risk *(47)*. SNPs of *XPC* and *XPD* may contribute to deficient repair of UV-induced damage *(28)*. *XRCC1*, *XRCC3*, and *XPD* gene polymorphisms are associated with DNA adducts *(48)*. The *XRCC3* Thr241Met-variant allele was also associated with elevated micronuclei *(49)*. However, association studies cannot provide direct evidence for a genotype/phenotype relationship.

The second approach is to clone and to express the protein from either the wild type or the variant allele; these proteins are then used to study the alleles' functions. Results showed that four (L104R, E126D, R237A, and D283G) of seven substitutions in the repair domain of APE1 have reduced endonuclease activities *(50)*. The *APE1* 148 Glu-variant allele has a small but nonsignificant effect on endonuclease activity (94% of wild type) and DNA-binding activity (K_d: 20.3 ± 3.4 vs 25.8 ± 12.2 nM in wild type) *(50)*. Although the small differences in binding and lower endonuclease activity may be within experimental error, the lower K_d of the variant may imply a higher affinity between APE1 protein and damaged DNA after catalysis and, therefore, turns over less effectively. Isolated protein was studied but not the effects of APE1 interaction with other BER components. These other proteins may also affect APE1 turnover. Future studies are needed to discover whether the *APE1* 148 Glu allele alters the APE1 protein's ability to communicate with other BER proteins, thereby influencing BER efficiency.

The data from a recent study showed that cells expressing the *XRCC3* Thr241Met-variant allele were active for homology-directed repair of DSBs and not more sensitive to the interstrand crosslinking agent mitomycin C *(51)*. The authors suggest that the increased cancer risk associated with *XRCC3* Thr241Met may not be the result of an intrinsic homology-directed repair defect. However, their data showed that the *XRCC3* Thr241Met-variant allele has a small (13% decreased) effect on homology-directed repair activity (0.15 ± 0.02% for wild type vs 0.13 ± 0.04% for variant) and mitomycin C hypersensitivity (survival fraction of 0.18 ± 0.005 for wild type vs 0.17–0.001 for variant, 6% difference) *(51)*. Although the small differences may be within experimental error, their data suggest a potential functional role of the *XRCC3* Thr241Met-variant. In contrast to rare *XP* and *ATM* mutations, this variant is very common in the general population and, therefore, unlikely to affect DNA repair function in a significant way.

Finally, two *hOGG1* SNPs were evaluated for their functional implications *(52–54)*. In a bacterial complementation assay system, the amino substitution variant *hOGG1* Ser326Cys may affect enzyme activity *(52)*. However, the 326Cys allele was not associated with altered *hOGG1* activity in a human study *(53)*. Therefore, it is still not clear whether this *hOGG1* SNP has any functional significance. Furthermore, another study showed that the *hOGG1* 1245C/G polymorphism was related to leucocyte 8-OHdG contents *(54)*.

3.3. DNA Damage/Repair As Target for Cancer Prevention

The long-term goals in targeting DNA damage/repair in cancer prevention are (1) to evaluate the roles of genetic polymorphisms of DNA repair that may contribute to human cancer risk, (2) to characterize gene–gene and gene–environment interactions in cancer risk, (3) to identify high-risk populations by using combined genetic and environmental risk factors, and (4) to reduce human cancer risk by targeting modifiable risk factors, such as diet and other environmental exposures. This approach will allow us to devise etiology-based preventive strategies that are more closely tailored to the specific genetic defects conferring individual risk. To date, considerable attention has been given to gene–environment interactions as they affect the efficacy of DNA repair and DNA metabolism. Environmental exposures may be equally important in determining genomic stability and its impact on cancer risk.

Because inherited genotypes cannot be modified, cancer prevention focuses on reducing the damage they cause and enhancing DNA repair. Some naturally occurring agents, such as vitamins and minerals (i.e., folate, vitamin E, vitamin D, calcium, selenium), natural phytochemicals (i.e., curcumin, genistein, indole-3-carbinol, and L-perillyl alcohol), and synthetic compounds (i.e., retinoids, selective estrogen-receptor modulators, and cyclooxygenase-2 inhibitors) show promise for preventing, arresting, and reversing cancer development *(55)*. Many agents have multiple chemopreventive activities. For example, some of the functions of *N*-acetylcysteine protect against DNA damage, carcinogenesis, and inflammation, enhance DNA repair, and inhibit genotoxicity and cell transformation *(56)*. In humans, supplementation of antioxidants, such as α-tocopherol, may protect cells against oxidative DNA damage, upregulating the network of an antioxidant-defense system *(57)*. Oxidative DNA damage can also be modulated by dietary supplementation of antioxidants *(58,59)*. Two recent studies also show that selenomethionine and cytokine interleukin (IL)-12 may induce DNA repair and protect cells from DNA damage *(60,61)*.

4. DNA REPAIR AND RESPONSE TO CANCER THERAPY

It is well established that somatic mutations in tumors may correlate with survival of cancer patients, but little is known about the prognostic significance of germ-line polymorphisms. DNA repair mechanisms are important to correct the DNA damage induced by anticancer treatments. The responsiveness of cancer cells to treatments can be influenced by altered repair efficiency *(62,63)*. BER, MMR, and NER were shown to contribute to the drug-resistance phenotypes *(64,65)*.

4.1. Expression of DNA Repair Genes and Response to Cancer Therapies

Much effort has been directed into finding molecular markers in human tumors that would help to identify patients most likely to benefit from a particular therapy. Several studies have demonstrated that expression of DNA repair genes in tumors may affect responses to cancer therapy. For example, MGMT can inhibit the killing of tumor cells by alkylating agents. Silencing MGMT expression by promoter hypermethylation is a favorable prognostic marker in patients with brain tumors who are treated with alkylating agents and in patients with diffuse large B-cell lymphoma who are treated with multidrug regimens, including cyclophosphamide *(66,67)*. In patients with gallbladder carcinomas and extrahepatic bile duct carcinomas, lack of expression of MGMT and MMR protein (hMLH1 and hMSH2) was associated with poor prognosis *(68)*. The subcellular localization of Ape1/ref-1 served as a prognostic marker in patients with non-small-cell lung carcinomas *(69,70)*. It is also associated with the aggressiveness of breast carcinomas *(71)*.

4.2. DNA Repair Variants As Prognostic Markers

Depending on the types of damage introduced by anticancer treatments and the specific repair pathway(s) involved, DNA repair genetic variants may also influence clinical outcomes and survival. In patients with metastatic colorectal cancer treated with 5-fluorouracil (5-FU)/oxaliplatin, the median survival rate was significantly associated with the *XPD* Lys751Gln variant *(27)*. In patients with advanced colorectal cancer, the *XRCC1* 399Gln variant allele was associated with a 5.2-fold (95% confidence interal = 1.21–22.07) increased risk to fail the 5-FU/oxaliplatin chemotherapy *(72)*. These results suggest that *XPD* Lys751Gln and *XRCC1* Arg399Gln variants may influence an individual's response to platinum-based chemotherapy. Finally, in a large British population-based study ($n = 2430$), the effects of germline polymorphisms on survival among women with breast cancer were assessed *(73)*. The most significant finding associated a silent polymorphism, D501D (t → c) in *LIG4*, a gene involved in DNA DSB repair, with survival. However, because the D501D polymorphism in *LIG4* does not alter the amino acid sequence, it is unlikely to have functional significance. Therefore, this SNP in *LIG4* may be in linkage disequilibrium with other functional polymorphisms that may influence survival. Although several studies demonstrated the potential application of germline polymorphisms in predicting responses to treatment and prognosis, we have to interpret their results with caution because of their limited sample size. In summary, DNA repair variants are unlikely to replace conventional prognostic markers. However, understanding their molecular mechanisms in DNA repair will provide insight into their effects on treatment and prognosis and open up the possibility of a mechanism-based, targeted strategy in cancer treatment based on a combination of DNA repair genotypes and tumor characteristics.

5. CONCLUDING REMARKS AND FUTURE DIRECTIONS

Over the past several years, substantial evidence has suggested roles for DNA repair SNPs in identifying populations at increased risk for certain cancers. In addition, DNA repair SNPs can help to determine how patients will respond to cancer therapies and to reveal specific genetic profiles that affect drug efficacy and toxicity. Considering multiple pathways for repairing the diverse DNA damages induced by carcinogens or cancer therapies, genetic variants in multiple repair pathways should be evaluated simultaneously, because they may have a joint or additive effect. However, even with the rapid development of high-throughput technology, it is questionable whether it is feasible to screen every one of the 1.42–4 million known SNPs or the estimated 10 million common human SNPs *(74,75)*. The solution may rely on the construction of a *haplotype* map of the human genome *(76)*. Haplotypes are segments of chromosomes that contain many genetic variations inherited together as a set or a block and they can be used to determine genetic susceptibility differences to carcinogens or treatments. The major attraction of haplotype methods is based on the concept that common haplotypes capture most of the genetic variation across sizable regions and that they can be tested *(76)*. Moreover, a large majority of untested markers showed strong correlation to the haplotype framework within blocks. Finally, they imply that it may not be necessary to discover and test millions of SNPs

individually; a haplotype map (HapMap) would make finding cancer susceptibility and therapy-response genes a manageable task *(76)*. At the present time,
the National Human Genome Research Institute focuses on the development of
a HapMap of the genome. which is a map of haplotype blocks, including the
specific SNPs that identify the haplotypes. The HapMap will enable researchers
to identify high-risk populations for cancer and to determine drug efficacy associated with specific genetic profiles.

ACKNOWLEDGMENTS

This work was supported by grants from the National Cancer Institute
(CA73629, CA91221, and CA90898) and the American Cancer Society (RPG-
97-115-01 and CNE-101119).

REFERENCES

1. Wood RD, Mitchell M, Sgouros J, et al. Human DNA repair genes. *Science* 2001;291:
1284–1289.
2. Sturgis EM, Castillo EJ, Li L, et al. Polymorphisms of DNA repair gene XRCC1 in squamous
cell carcinoma of the head and neck. *Carcinogenesis* 1999;20:2125–2129.
3. Stern MC, Umbach DM, van Gils CH, et al. DNA repair gene XRCC1 polymorphisms,
smoking, and bladder cancer risk. *Cancer Epidemiol Biomarkers Prev* 2001;10:125–131.
4. Duell EJ, Millikan RC, Pittman GS, et al. Polymorphisms in the DNA repair gene XRCC1 and
breast cancer. *Cancer Epidemiol Biomarkers Prev* 2001;10:217–222.
5. Divine KK, Gilliland FD, Crowell RE, et al. The XRCC1 399 glutamine allele is a risk factor
for adenocarcinoma of the lung. *Mutat Res* 2001;461:271–278.
6. Lee JM, Lee YC, Yang SY, et al. Genetic polymorphisms of XRCC1 and risk of the esophageal
cancer. *Int J Cancer* 2001;95:240–246.
7. Hu JJ, Smith TR, Miller MS, et al. Amino acid substitution variants of APE1 and XRCC1
genes associated with ionizing radiation sensitivity. *Carcinogenesis* 2001;22:917–922.
8. Xing DY, Tan W, Song N, et al. Ser326Cys polymorphism in *hOGG1* gene and risk of
esophageal cancer in a Chinese population. *Int J Cancer* 2001;95:140–143.
9. Le Marchand L, Donlon T, Lum-Jones A, et al. Association of the hOGG1 Ser326Cys polymorphism with lung cancer risk. *Cancer Epidemiol Biomarkers Prev* 2002;11:409–412.
10. Xu J, Zheng SL, Turner A, et al. Associations between hOGG1 sequence variants and prostate
cancer susceptibility. *Cancer Res* 2002;62:2253–2257.
11. Shibutani S, Takeshita M, Grollman AP. Insertion of specific bases during DNA synthesis past
the oxidation-damaged base 8-oxoG. *Nature* 1991;349:431–434.
12. Cheng KC, Cahill DS, Kasai H, et al. 8-Hydroxyguanine, an abundant form of oxidative DNA
damage, causes G—T and A—C substitutions. *J Biol Chem* 1992;267:166–172.
13. Thompson LH, West MG. XRCC1 keeps DNA from getting stranded. *Mutat Res* 2000;459:1–18.
14. Ramana CV, Boldogh I, Izumi T, et al. Activation of apurinic/apyrimidinic endonuclease in
human cells by reactive oxygen species and its correlation with their adaptive response to
genotoxicity of free radicals. *Proc Natl Acad Sci USA* 1998;95:5061–5066.
15. Margison GP, Santibanez-Koref MF. O^6-alkylguanine-DNA alkyltransferase: role in carcinogenesis and chemotherapy. *BioEssays* 2002;24:255–266.
16. Gerson SL. Clinical relevance of MGMT in the treatment of cancer. *J Clin Oncol*
2002;20:2388–2399.

17. Marti TM, Kunz C, Fleck O. DNA mismatch repair and mutation avoidance pathways. *J Cell Physiol* 2002;191:28–41.
18. Aaltonen LA, Peltomäki P, Leach FS, et al. Clues to the pathogenesis of familial colorectal cancer. *Science* 1993;260:812–816.
19. Thibodeau SN, Bren G, Schaid D. Microsatellite instability in cancer of the proximal colon. *Science* 1993;260: 816–819.
20. Ionov Y, Peinado M, Malkhosyan S, et al. Ubiquitous somatic mutations in simple repeated sequences reveal a new mechanism for colonic carcinogenesis. *Nature* 1993;363: 558–561.
21. Muller A, Fishel R. Mismatch repair and the hereditary non-polyposis colorectal cancer syndrome (HNPCC). *Cancer Invest* 2002;20:102–109.
22. Balajee AS, Bohr VA. Genomic heterogeneity of nucleotide excision repair. *Gene* 2000;250:15–30.
23. Petit C, Sancar A. Nucleotide excision repair: from *E. coli* to man. *Biochimie* 1999;81:15–25.
24. Mullenders LH, Berneburg M. Photoimmunology and nucleotide excision repair: impact of transcription coupled and global genome excision repair. *J Photochem Photobiol B* 2001;65:97–100.
25. Kraemer KH, Lee MM, Scotto J. DNA repair protects against cutaneous and internal neoplasia: evidence from xeroderma pigmentosum. *Carcinogenesis* 1984;5:511–514.
26. Matullo G, Guarrera S, Carturan S, et al. DNA repair gene polymorphisms, bulky DNA adducts in white blood cells and bladder cancer risk in a case-control study. *Int J Cancer* 2001;92:562–567.
27. Park DJ, Stoehlmacher J, Zhang W, et al. A xeroderma pigmentosum group D gene polymorphism predicts clinical outcome to platinum-based chemotherapy in patients with advanced colorectal cancer. *Cancer Res* 2001;61:8654–8658.
28. Qiao Y, Spitz MR, Shen H, et al. Modulation of repair of ultraviolet damage in the host-cell reactivation assay by polymorphic XPC and XPD/ERCC2 genotypes. *Carcinogenesis* 2002;23:295–299.
29. Shen MR, Jones IM, Mohrenweiser H. Nonconservative amino acid substitution variants exist at polymorphic frequency in DNA repair genes in healthy humans. *Cancer Res* 1998;58:604–608.
30. Winkler GS, Araujo SJ, Fiedler U, et al. TFIIH with inactive XPD helicase functions in transcription initiation but is defective in DNA repair. *J Biol Chem* 2000;275:4258–4266.
31. Jeggo PA, Concannon P. Immune diversity and genomic stability: opposite goals but similar paths. *J Photochem Photobiol B* 2001;65:88 96.
32. van Gent DC, Hoeijmakers JH, Kanaar R. Chromosomal stability and the DNA double-stranded break connection. *Nature Rev Genet* 2001; 2:196–206.
33. Broughton BC, Steingrimsdottir H, Lehmann AR. Five polymorphisms in the coding sequence of the xeroderma pigmentosum group D gene. *Mutat Res* 1996;362:209–211.
34. Bell DW, Wahrer DC, Kang DH, et al. Common nonsense mutations in RAD52. *Cancer Res* 1999;59:3883–3888.
35. Fan F, Liu C, Tavare S, Arnheim N. Polymorphisms in the human DNA repair gene XPF. *Mutat Res* 1999;406,115–120.
36. Ford BN, Ruttan CC, Kyle VL, et al. Identification of single nucleotide polymorphisms in human DNA repair genes. *Carcinogenesis* 2000;21:1977–1981.
37. Nash RA, Caldecott KW, Barnes DE, et al. XRCC1 protein interacts with one of two distinct forms of DNA ligase III. *Biochemistry* 1997;36:5207–5211.
38. Masson M, Niedergang C, Schreiber V, et al. XRCC1 is specifically associated with poly(ADP-ribose) polymerase and negatively regulates its activity following DNA damage. *Mol Cell Biol* 1998;18:3563–3571.
39. Liu N, Lamerdin JE, Tebbs RS, et al. XRCC2 and XRCC3, new human Rad51-family members, promote chromosome stability and protect against DNA cross-links and other damages. *Mol Cell* 1998;1:783–793.

40. Brenneman MA, Weiss AE, Nickoloff JA, et al. XRCC3 is required for efficient repair of chromosome breaks by homologous recombination. *Mutat Res* 2000;459:89–97.
41. Caldecott K, Jeggo P. Cross-sensitivity of gamma-ray-sensitive hamster mutants to cross-linking agents. *Mutat Res* 1991;255:111–121.
42. Fuller LF, Painter RB. A Chinese hamster ovary cell line hypersensitive to ionizing radiation and deficient in repair replication. *Mutat Res* 1988;193:109–21.
43. Cui X, Brenneman M, Meyne J, et al. The XRCC2 and XRCC3 repair genes are required for chromosome stability in mammalian cells. *Mutat Res* 1999;434:75–88.
44. Winsey SL, Haldar NA, Marsh HP, et al. A variant within the DNA repair gene XRCC3 is associated with the development of melanoma skin cancer. *Cancer Res* 2000;60:5612–5616.
45. Kuschel B, Auranen A, McBride S, et al. Variants in DNA double-strand break repair genes and breast cancer susceptibility. *Hum Mol Genet* 2002;11:1399–1407.
46. Tuimala J, Szekely G, Gundy S, et al. Genetic polymorphisms of DNA repair and xenobiotic-metabolizing enzymes: role in mutagen sensitivity. *Carcinogenesis* 2002;23:1003–1008.
47. Hu JJ, Smith TR, Miller MS, et al. Genetic regulation of ionizing radiation sensitivity and breast cancer risk. *Environ Mol Mutagen* 2002;39:208–215.
48. Matullo G, Palli D, Peluso M, et al. XRCC1, XRCC3, XPD gene polymorphisms, smoking and (32)P-DNA adducts in a sample of healthy subjects. *Carcinogenesis* 2001;22:1437–1445.
49. Lunn R, Langlois R, Hsieh L, et al. XRCC1 polymorphisms: effects on aflatoxin B_1-DNA adducts and glycophorin A variant frequency. *Cancer Res* 1999;59:2557–2561.
50. Hadi MZ, Coleman MA, Fidelis K, et al. Functional characterization of Ape1 variants identified in the human population. *Nucl Acids Res* 2000;28:3871–3879.
51. Araujo FD, Pierce AJ, Stark JM, et al. Variant XRCC3 implicated in cancer is functional in homology-directed repair of double-strand breaks. *Oncogene* 2002;21:4176–4180.
52. Kohno T, Shinmura K, Tosaka M, et al. Genetic polymorphisms and alternative splicing of the hOGG1 gene, that is involved in the repair of 8-hydroxyguanine in damaged DNA. *Oncogene* 1998;16:3219–3225.
53. Janssen K, Schlink K, Gotte W, et al. DNA repair activity of 8-oxoguanine DNA glycosylase 1 (OGG1) in human lymphocytes is not dependent on genetic polymorphism Ser326/Cys326. *Mutat Res* 2001;486:207–216.
54. Tarng DC, Tsai TJ, Chen WT, et al. Effect of human OGG1 1245C→G gene polymorphism on 8-hydroxy-2'-deoxyguanosine levels of leukocyte DNA among patients undergoing chronic hemodialysis. *J Am Soc Nephrol* 2001;12:2338–2347.
55. Greenwald P. Cancer chemoprevention. *BMJ* 2002;324:714–718.
56. De Flora S, Izzotti A, D'Agostini F, et al. Mechanisms of N-acetylcysteine in the prevention of DNA damage and cancer, with special reference to smoking-related end-points. *Carcinogenesis* 2001;22:999–1013.
57. Hu JJ, Berwick M, Dubin N, et al. Dietary supplementation of α-tocopherol increases plasma glutathione and DNA repair. *Cancer Epidemiol Biomarkers Prev* 1996;5:263–270.
58. Hu JJ, Chi CX, Frenkel K, et al. α-Tocopherol dietary supplement decreases titers of antibody against 5-hydroxymethyl-2'-deoxyuridine (HMdU). *Cancer Epidemiol Biomarkers Prev* 1999;8:693–698.
59. Porrini M, Riso P. Lymphocyte lycopene concentration and DNA protection from oxidative damage is increased in women after a short period of tomato consumption. *J Nutr* 2000;130:189–192.
60. Seo YR, Sweeney C, Smith ML. Selenomethionine induction of DNA repair response in human fibroblasts. *Oncogene* 2002;21:3663–3669.
61. Schwarz A, Stander S, Berneburg M, et al. Interleukin-12 suppresses ultraviolet radiation-induced apoptosis by inducing DNA repair. *Nature Cell Biol* 2002;4:26–31.
62. Chaney SG, Sancar A. DNA repair: enzymatic mechanisms and relevance to drug response. *J Natl Cancer Inst* 1996;88:1346–1360.

63. Aebi S, Fink D, Gordon R, et al. Resistance to cytotoxic drugs in DNA mismatch repair-deficient cells. *Clin Cancer Res* 1997;3:1763–1767.
64. Rajewsky MF, Engelbergs J, Thomale J, et al. DNA repair: counteragent in mutagenesis and carcinogenesis—accomplice in cancer therapy resistance. Mutat. *Res* 2000;462:101–105.
65. Runger TM, Emmert S, Schadendorf D, et al. Alterations of DNA repair in melanoma cell lines resistant to cisplatin, fotemustine, or etoposide. *J Investig Dermatol* 2000;114:34–39.
66. Esteller M, Jesus Garcia-Foncillas J, Andion E, et al. Inactivation of the DNA-repair gene MGMT and the clinical response of gliomas to alkylating agents. *N Engl J Med* 2000;343: 1350–1354.
67. Esteller M, Gaidano G, Goodman SN, et al. Hypermethylation of the DNA repair gene O(6)-methylguanine DNA methyltransferase and survival of patients with diffuse large B-cell lymphoma. *J Natl Cancer Inst* 2002;94:26–32.
68. Kohya N, Miyazaki K, Matsukura S, et al. Deficient expression of O(6)-methylguanine-DNA methyltransferase combined with mismatch-repair proteins hMLH1 and hMSH2 is related to poor prognosis in human biliary tract carcinoma. *Ann Surg Oncol* 2002;9:371–379.
69. Kakolyris S, Giatromanolaki A, Koukourakis M, et al. Nuclear localization of human AP endonuclease 1 (HAP1/Ref-1) associates with prognosis in early operable non-small cell lung cancer (NSCLC). *J Pathol* 1999;189:351–357.
70. Puglisi F, Aprile G, Minisini AM, et al. Prognostic significance of Ape1/ref-1 subcellular localization in non-small cell lung carcinomas. *Anticancer Res* 2001;21:4041–4049.
71. Puglisi F, Barbone F, Tell G, et al. Prognostic role of Ape/Ref-1 subcellular expression in stage I-III breast carcinomas. *Oncol Rep* 2002;9:11–17.
72. Stoehlmacher J, Ghaderi V, Iobal S, et al. A polymorphism of the XRCC1 gene predicts for response to platinum based treatment in advanced colorectal cancer. *Anticancer Res* 2001;21:3075–3079.
73. Goode EL, Dunning AM, Kuschel B, et al. Effect of germ-line genetic variation on breast cancer survival in a population-based study. *Cancer Res* 2002;62:3052–3057.
74. Sachidanandam R, Weissman D, Schmidt SC, et al. A map of human genome sequence variation containing 1.42 million single nucleotide polymorphisms. *Nature* 2001;409:928–933.
75. Kruglyak L, Nickerson DA. Variation is the spice of life. *Nat Genet* 2001;27:234–236.
76. Gabriel SB, Schaffner SF, Nguyen H, et al. The structure of haplotype blocks in the human genome. *Science* 2002;296:2225–2229.

Index

About the Editors

Lawrence C. Panasci is Associate Director of the Oncology Centre for The Jewish General Hospital and Director of Medical Oncology at McGill University.

Dr. Panasci received his medical degree at Georgetown University and trained in medical oncology at Georgetown University and for a short time at the National Cancer Institute in Milan, Italy. He was an Assistant Professor at Upstate Medical Center in Syracuse University before joining the Faculty at McGill in 1980.

He is now a full Professor of Medicine. His research endeavors over the past two decades have focused on various topics in anti-cancer drug development. Specifically he has developed a new anti-cancer agent which is in clinical trials. In addition, his laboratory has extensively examined the mechanism of DNA crosslinking agent drug resistance in both cell lines and clinical samples. The results of these investigations suggest that DNA repair plays a central role in the mechanism of DNA cross linking agent drug resistance.

Dr. Moulay Alaoui-Jamali is Professor of Experimental Oncology/Pharmacology at the Departments of Oncology and Medicine, Faculty of Medicine, McGill University. He received his PhD in Oncology in 1986 from the University of René Descartes and Sorbonne in Paris, followed by a postdoctoral training at the Cancer Center of the Roswell Park Memorial Institute in Buffalo, USA. His research career has included interdisciplinary studies spanning the fields of drug resistance in cancer, DNA repair vis-à-vis chemotherapy response, discovery of molecular switches for early cancer invasion, and novel therapeutics. He received several National

and International awards, and served on many grant review panels, research forums, and boards. Dr. Alaoui-Jamali has authored more than 125 publications.